Bioprocessing and Biotechnology

Bioprocessing and Biotechnology

Edited by **Edgardo Turner**

Syrawood Publishing House
New York

Published by Syrawood Publishing House,
750 Third Avenue, 9th Floor,
New York, NY 10017, USA
www.syrawoodpublishinghouse.com

Bioprocessing and Biotechnology
Edited by Edgardo Turner

International Standard Book Number: 978-1-68286-207-0 (Hardback)

Printed in China.

Contents

Preface

Bioprocessing and biotechnology are interdisciplinary fields that have experienced consistent growth in the recent times. The chapters covered in this book encompass some important topics such as bioprocess systems, enzyme manufacturing, renewable fuels and chemicals, environmental engineering, genetic engineering, etc. It compiles state-of-the-art researches and developments in the field of bioprocessing and biotechnology. Coherent flow of topics, student-friendly language and extensive use of examples make this book an invaluable source of knowledge.

After months of intensive research and writing, this book is the end result of all who devoted their time and efforts in the initiation and progress of this book. It will surely be a source of reference in enhancing the required knowledge of the new developments in the area. During the course of developing this book, certain measures such as accuracy, authenticity and research focused analytical studies were given preference in order to produce a comprehensive book in the area of study.

This book would not have been possible without the efforts of the authors and the publisher. I extend my sincere thanks to them. Secondly, I express my gratitude to my family and well-wishers. And most importantly, I thank my students for constantly expressing their willingness and curiosity in enhancing their knowledge in the field, which encourages me to take up further research projects for the advancement of the area.

Editor

Fungal Lipase Production by Solid-State Fermentation

Luis Beltrán Ramos-Sánchez[1]*, Mario César Cujilema-Quitio[1], Maria Caridad Julian-Ricardo[1], Jesus Cordova[2] and Patrick Fickers[3,4]

[1]*Department of Chemical Engineering, University of Camagüey "Ignacio Agramonte Loynaz" Circunvalación norte, Km 5.5, 74650 Camagüey, Cuba*
[2]*Department of Chemistry, University of Guadalajara, Blvd. M. García Barragán 1421, CP 44430, Guadalajara, Jalisco, Mexico*
[3]*Biotechnology and Bioprocess Unit, Université libre de Bruxelles, Av. F.-D. Roosevelt, 50, CP165 / 61, 1050 Brussels, Belgium*
[4]*Microbial Processes and Interactions, Gembloux Agrobiotech, Univeristy of Liège, Passage des déportés, 2, 5030 Gembloux, Belgium*

Abstract

Lipases are one of the most promising enzymes in the chemical and biopharmaceutical industries. Numerous applications have been reported including fine chemistry, detergents formulation and biodiesel synthesis. Lipases are commonly produced by a wide variety of yeasts and filamentous fungi in submerged fermentation or solid-state fermentation. Filamentous fungi and yeasts usually behave more efficiently in solid-state fermentation and show greater productivities when compared to submerged fermentation. Although filamentous fungi adequately growth in solid-state fermentation, there are some limitations for cultivating them by this process. Here, a review is made about lipase production in solid-state fermentation. This includes the analysis of solid-state fermentation as a promising technology, characterization of growth media and ambient factors such as moisture, pH and temperature.

Keywords: Lipase production; Solid-state fermentation; Yeast; Filamentous fungi; Bioreactor

Introduction

Lipases are serine hydrolases that catalyze in nature the hydrolysis of ester bonds of tri-, di- and mono- glycerides into fatty acids and glycerol. They are defined as triacylglycerol acyl hydrolases (E. C. 3.1.1.3) and differ from the closely related esterase's (E. C. 3.1.1.1) due to their ability to hydrolyze triglycerides at a lipid-water interface [1]. This phenomenon, the so-called "interfacial activation", was first proposed to distinguish lipases from esterases [2]. Unfortunately, this criterion was not sufficient to characterize the behavior of several lipases [3]. They were subsequently redefined as long chain fatty acid triglyceride hydrolases in apposition of esterases that are able to hydrolyze only short chain (i.e. less than 10 carbon atoms) fatty acid triglycerides.

The rising interest in lipase production mainly lies on the wide industrial applications of the enzyme [4-6]. They recognize a wide variety of substrates and may catalyze many different reactions, such as hydrolysis or synthesis of esters bonds, alcoholysis, aminolysis, peroxidations, epoxidations and interesterifications. The great number of recently published papers concerning different aspects of lipase production and applications also demonstrates the increasing importance of these enzymes [7].

Fungi and bacteria produce most lipases used in biotechnological applications. Filamentous fungi are an interesting source of lipases because they produce extracellular enzymes [8]. Lipases can be produced by Submerged Fermentation (SmF) or by Solid-State Fermentation (SSF) processes, although fungi are better adapted than yeast and bacteria to grow on SSF [8]. Despite SmF is widely used in the enzyme industry and has advantages in process control and production yields, SSF represents an interesting alternative to produce industrial enzymes at lower costs [8,9]. SSF has many preferences to SmF for microbial enzyme production including: superior production yields and productivities, lower operating costs, less demands for asepsis control, cheaper fermentation media, higher oxygen distribution, fewer operational troubles, simpler equipments and control systems, lower energy consumption [6,10].

The aim of present paper is to review literature concerning fungal lipase production by SSF in order to characterize different aspects that could be used to improve the process.

Solid-state Fermentation as a Proper System for Fungal Culture

SSF could be defined as a fermentation process in which microorganisms grow on a solid material, natural or synthetic, in absence or in a very limited amount of free liquid phase [11,12]. Bioreactors for this process could be operated in batch or in continuous mode (CSSF). In this later case, there is a flow of the solid phase within the reactor [13,14]. There are specific challenges to operate CSSF bioreactors that are not faced in classical continuous SmF. Consequently, CSSF processes are currently scarce in industry [14]. However, improved design procedures and sensors development promise a better future for CSSF at industrial scale.

SSF offers a quite different environment than SmF for cell growth and metabolism. In that regard, SSF is well adapted for yeasts and filamentous fungi cultivation since they can grow at low moisture content by contrast to bacteria [15]. Technologically, SSF presents several advantages compared to SmF [16]. Oxygen transfer limitations often encountered in SmF are eluded by a much more enlarged surface (external and/or internal) of the solid support for mass transfer [12,17]. In addition, cells and especially aerial hyphae are exposed directly to the gas phase, increasing significantly the overall oxygen transfer. Since SSF is performed at low water activity, it permits to prevent bacterial contaminations that require a high water activity to grow [18]. Moreover, the control of the optimal water activity in regards to a specific fungal species together with the utilization of an adequate cell inoculum, also permit to avoid or at least to reduce contamination by other fungal species. Beside these advantages, SSF also present several drawbacks such as heat generation by the cellular metabolic activity.

***Corresponding author:** Luis Beltrán Ramos-Sanchez, Department of Chemical Engineering, University of Camagüey "Ignacio Agramonte Loynaz" Circunvalación norte, Km 5.5, 74650 Camagüey, Cuba

One way to dissipate this heat consists in the mechanical agitation/mixing of the solid medium [19]. However, this action could be harmful for filamentous fungi. It requires, thus a great compromise between cell survival and heat dissipation for process viability. This problem is not encountered for yeasts, thus offering a good opportunity for producing enzymes such as lipases with enhanced possibilities of competing in enzyme market.

SSF versus SmF for Lipase Production

SSF has many advantages over SmF for microbial enzyme production namely high yield and productivities, lower operating costs, low-cost fermentation media (raw biomass, agro industrial residues), higher oxygen distribution, fewer operational troubles, simpler equipment and control system [10,20].

A comparative study on alkaline lipase production by a strain of *Aspergillus fumigatus* under SSF and SmF was made in which the maximum concentration of the enzyme was comparable in both cases [21]. However, lipase produced by SSF was stable over a period of 15 days whereas lipase production in SmF decreased from day 5. Other authors reported lipase produced by SSF to be 24% more active and 64% more thermally stable than lipase produced by SmF using a strain of *Rhizopus homothallicus* [22].

Although most of the industrial enzymes are currently produced by submerged fermentations, SSF appears a promising technology due to the advantages it offers, such as lower cost of production [12,23-25]. Compared to the cost of culture medium in SmF, SSF of agro-industrial solid wastes is inexpensive, and therefore this process is industrially and economically attractive [16,21]. Coradi et al. [20] reported a cost of lipase production by *Trichoderma viride* in SSF ten times lower than in SmF. For instance, total capital investment needed for a production scale of 100 m^3 of lipase concentrate per year is 78% higher for SmF compared to SSF [17]. These results confirm the interesting potential of SSF for producing lipase at low cost.

Industrial Applications of Lipases

Lipases are essential components in the modern industrial process due to their ability to catalyze, depending on the thermodynamic conditions, hydrolysis reactions as well as synthesis reactions such as esterification and transesterification (Figure 1) [26]. Remarkably, in the last decade, yeast lipases were also applied in the manufacture of pharmaceuticals, pesticides, biosensors and in waste management [27]. Furthermore, the genes encoding lipases in *Candida* sp., *Geotrichum* sp., *Trichosporon* sp. and *Yarrowia lipolytica* have been cloned and over-expressed [7,28]. The most important lipase producers among yeasts used in industrial processes are *Candida* species namely *C. antarctica*, *C. cylindracea*, *C. lipolytica* and *C. rugosa* [7,29,30]. *C. rugosa* lipase is one of the enzymes most frequently used in bio-transformation and has the great advantage of being considered as safe for food applications. The crude enzyme extract contains at least five isoenzymes, denominated Lip1 to Lip5. Despite they share high sequence homology, they have different catalytic characteristics, substrate specificities and thermal stabilities [31]. CAL B from *C. antartica* is the second most studied yeast lipase, having several applications, such as synthesis of flavors and fragrance esters, surfactants, biodiesel, waxes, acylated flavonoids and modified glycerides as well as kinetic resolution of racemic esters and amines [32]. Its stability has also been improved using ionic liquids or no aqueous solvents for enzymatic synthesis [31]. *Y. lipolytica* lipase produces high value compounds, being used for bio-transformation of steroids and synthesis of pharmaceutical intermediates and fine chemicals [28]. Lipases from yeast are used in many different industrial processes as illustrated in the following sections.

C. antarctica lipases were used for esterification of free fatty acids in the absence of organic solvent and for transesterification of fatty acid methyl esters in hexane with isopropylidene glycerol [33]. Immobilized lipases from *C. antartica*, *C. cylindracea* and *G. candidum* were used for the esterification of functionalized phenols for synthesis of lipophilic antioxidants [34]. *C. rugosa* lipase is used by Nippon Oils & Fats (Tokyo, Japan) for the preparation of highly pure unsaturated fatty acids (oleic, linoleic, linoleic, etc.) [35].

Large amounts of lipases, nearly 1000 tons per year, are used in the detergent industry for removal of oil-based stains [29]. For this purpose, lipases from *C. cylindracea*, *Y. lipolytica* and *C. antarctica* are preferred because they can operate at low temperatures and alkaline pH, compared with bacterial and other fungal lipases [36,37]. Miyoshi Yushi, a company in Nagoya (Japan), produces sizable amounts of soap through lipase-based hydrolysis of oils and fats with *C. rugosa* lipase [35]. Enzymatic synthesis of surfactants has been carried out at moderate temperatures (60-80°C) with excellent regioselectivity. Unichem International (USA) has manufactured isopropyl myristate, isopropyl palmitate and 2-ethyl palmitate, as emollient for personal care products using *C. cylindracea* lipase [38].

Lipases are used for flavor development and improving quality of foods. *C. parapsilosis* and *C. antarctica* lipases were used to synthesize hydroxamic acids, a food additive, and short chain flavor thioester, respectively. Additionally, *C. rugosa* lipase was used in the cheese manufacture [27]. The flavor development in concentrated milk and creams by microbial lipases was investigated, finding that each lipase developed a characteristic flavor. *C. rugosa* lipases were found as the most suitable for this purpose [39]. Several structural lipids were synthesized by yeast lipases and included in infant formula and nutraceuticals [29,30]. *C. antarctica* lipase is employed as a robust biocatalyst for esterification reactions, obtaining high conversions of flavoring esters of short-chain [40]. *C. cylindracea* lipase is used in bakery products manufacturing [41].

Biocatalytic processes are used to prepare chiral intermediates for pharmaceuticals. (S)[1-(acetoxyl)-4-(3-phenyl) butyl] phosphonic acid diethyl ester, a key chiral intermediate required for chemical synthesis of BMS-188494 (an anticholesterol drug) was prepared by stereoselective acetylation of racemic [1- (hydroxy)-4-(3-phenyl) butyl] phosphonic acid diethyl ester using *Geotrichum candidum* lipase [42]. CAL-B lipase was used for the enzymatic resolution of a racemic mixture of 2-pentanol, a key chiral intermediate required for synthesis of anti-Alzheimer's drugs [42]. *C. cylindracea* lipase was applied to the resolution of 2-bromopropionic acids and 2-cloropropionic acids which are starting materials for the synthesis of phenoxy propionic herbicides [43]. *C. cylindracea* and *C. antartica* lipases have been used to resolve the enantiomers of flurbiprofen, naproxen, ibuprofen, suprofen and baclofen [43]. Both enzymes were successfully used to synthesize lobucavir, hepatitis B antiviral and ribavarin antiviral [44]. These lipases are also suitable for transesterification reactions, displaying both wide substrate specificity and regiospecificity [45].

Croda Universal Ltd. (Sweden) uses *C. cylindracea* lipase to synthesize wax esters (esters of fatty acids and fatty alcohols) as additives for personal care products. According to the manufacturer, the overall production cost is slightly higher than that of the conventional method, however the cost is justified by the improved quality of the final product [42]. *C. rugosa* lipase was selected to hydrolyze triacylglycerol for clinical

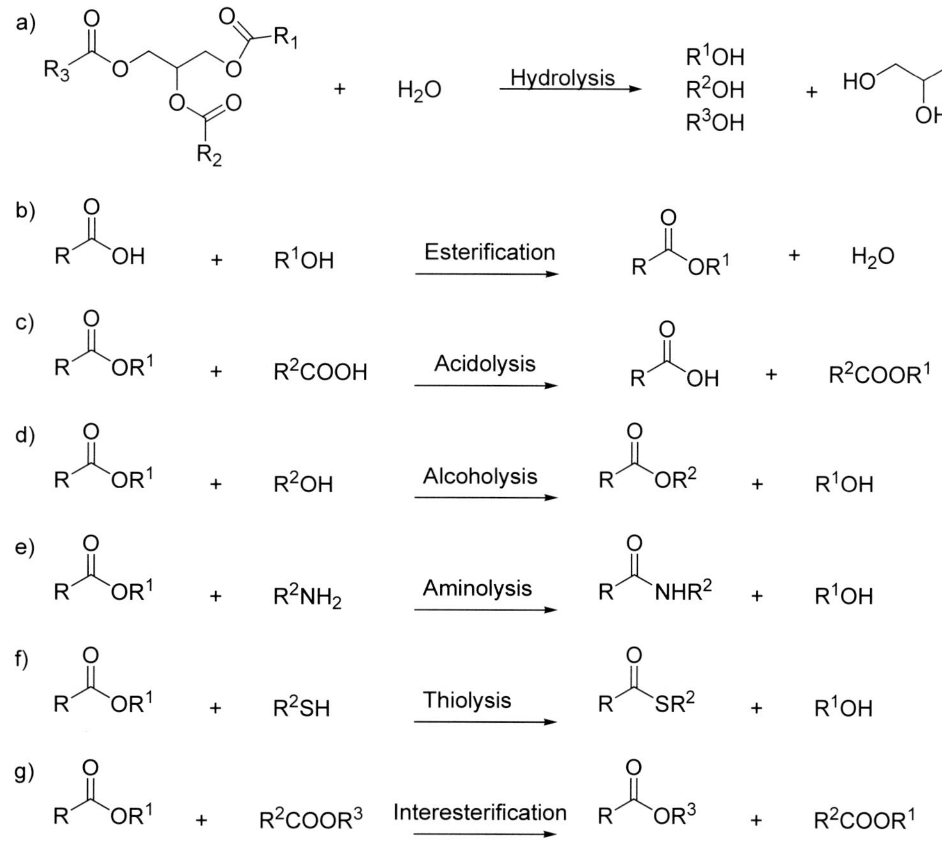

Figure 1: Reactions catalyzed by lipases. a) hydrolysis, b) esterification, c) acidolysis, d) alcoholysis, e) aminolysis, f) thiolysis, g) interesterifications

lipid analysis and developed as a biosensor [38,43]. *C. rugosa* lipase was also used to synthesize lovastatin, a drug lowing serum cholesterol level. The asymmetric hydrolysis of 3-phenylglycidic acid ester, which is a key intermediate in the synthesis of diltiazem hydrochloride, a coronary vasodilator, was carried out with *S. marcescens* lipase [46].

Lipases from different sources are able to catalyze biodiesel synthesis but those from *C. antarctica* and *C. rugosa* are the most used [45]. The Novozym 435 (CAL-B) lipase is able to convert at high yield used olive oil which contained high levels of free fatty acids and water content into biodiesel [45]. Simple alkyl ester derivatives of restaurant grease waste were prepared using immobilized lipases from *Thermomyces lanuginosus* and *C. antarctica* [47].

Lipases are utilized in activated sludge and other aerobic waste processes, where thin layers of fats must be continuously removed to allow oxygenation. Effective breakdown of solids, clearing and prevention of fat blockage or filming in waste systems are important in many industrial operations such as degradation of organic debris using a commercial mixture of lipase, cellulase, protease and amylase or sewage treatment, cleaning of holding tanks, septic tanks, grease traps. *C. rugosa* lipase is a good option to hydrolyze residual fats [37]. *C. rugosa* lipase was also used by Nippon Paper Industry to hydrolyze 90% of wood triglycerides as a pitch control method [43,48].

Regulation of Lipase Production Concerning the Medium Composition

Yields in lipase synthesis in SSF vary according the strain, the growth medium (solid support), culture medium (carbon and nitrogen sources) and cultivation conditions (pH, temperature and moisture). The main important lipases from filamentous fungi produced in SSF belong to the genera *Aspergillus*, *Rhizopus* and *Penicillium* while those from yeast origin belong to *Candida* and *Yarrowia* (Table 1). Most of the lipases produced by yeast and filamentous fungi are extracellular and their production is modulated by the culture conditions. Expression of lipase encoding genes is regulated by the nutrients from the culture medium and/or by the solid substrate used. Lipids such as triglycerides, fatty acid or long chain fatty acid esters are requested for high lipase production yields. Oleic acid (cis-9-Octadecenoic acid) was reported as the most suitable inducer for the production of the main extracellular Lip2p lipase in *Y. lipolytica* [51,52]. However, pure oleic acid is not convenient for the development of industrial processes owing to its excessive cost. With the aim of developing a cost-friendly process, different oleic acid esters and triglycerides have been tested for their ability to support lipase synthesis. Olive oil, mainly constituted by triolein as the main triglyceride, and methyloleate led to lipase activities equal to 1130 and 1190 U ml^{-1}, respectively [53]. These results are in the range of that obtained with pure oleic acid (i.e. 1045 U ml^{-1}). By contrast, ethyloleate led to a significant decrease of lipase production (195 U ml^{-1}). Expression of the *ylLIP2* lipase-encoding gene, and thus Lip2p synthesis appeared also to be markedly influenced by oleic acid concentration [51]. By contrast, carbon sources such as glucose and glycerol were found to repress lipase production [54].

Nitrogen sources used to supplement the culture medium also greatly influence lipases production. Various mineral and organic nitrogen sources were tested for their capacity to support cell growth and lipase production [51]. Mineral nitrogen such as ammonium chloride

Strain	Solid support / substrate	Lipase activity	References
Yeast			
Candida rugosa	Coconut oil cake	88 U g^{-1}	[65]
	Sesame oil cake	32 U ml^{-1}	[85]
	Groundnut oil cake	57 U ml^{-1}	[85]
	Rice Bran	37 U g^{-1}	[80]
Yarrowia lipolytica	Sugarcane bagasse/wheat bran	9 U g^{-1}	[90]
	Palm kernal cake	19 U g^{-1}	[84]
Mold			
Aspergillus niger	Wheat bran	650 U g^{-1}	[50]
	Wheat bran/oleic acid	49 U g^{-1}	[6]
Aspergillus fumigatus	Defatted rice bran	9 U ml^{-1}	[21]
Aspergillus carneu	Sunflower	13 U g^{-1}	[69]
Rhizopus homothalicus	Sugarcane bagasse/olive cake	43 U ml^{-1}	[81]
Rhizopus oryzae	Sugarcane bagasse	215 U g^{-1}	[16]
Penicillium brevicompactum	Babassu oil cake	49 U g^{-1}	[75]
	Castor meal	88 U g^{-1}	[75]
Penicillium restrictum	Babassu oil cake	30 U g^{-1}	[74]
Pennicillium sp	Soybean meal	140 U g^{-1}	[79]
Rhizomucor pusillus	Sugarcane bagasse/olive cake	21 U ml^{-1}	[81]
	Sugarcane bagasse	5 U g^{-1}	[81]
Rhizomucor rhizopodiformis	Sugarcane bagasse	3 U g^{-1}	[81]

Table 1: Strain used in solid-state fermentation for lipase production. Lipase activity, unless stated otherwise, is expressed in unit per gram of dry solid support

or ammonium sulfate had no influence on either cell growth or lipase production. By contrast, a significant increase in lipase productivity was observed upon addition of organic nitrogen sources such as urea or casein hydrolysates. For instance, tryptone N1 (Organotechnie, La Courneuve, France) allowed a 166-fold increased lipase production compared to the non-supplemented medium. However, cultures in the presence of casamino acid, which result from a complete hydrolysis of casein, led to a low lipolytic activity. These results suggest that specific peptides constituting tryptone N1 could regulate lipase production [51]. This hypothesis was confirmed by testing separately different tryptone constituting peptides [55].

Recently, lipase production by *Y. lipolytica* in SSF using different agro-industrial residues was reported [56]. When soybean cake and its sludge were used, lipase activity reached 139 Ug^{-1} after 14 h of fermentation whereas a lipase activity of 102 Ug^{-1} was obtained after 28 h of fermentation when cottonseed cake was used as substrate without any supplement [56].

C. rugosa is able to produce three lipase isoenzymes, namely Lip1, Lip2 and Lip3. Their production is modulated in yield (enzymatic activity) and in quality (isoenzyme synthesis) by the culture conditions and medium composition. Glucose and glycerol repress lipase synthesis [57,58] whereas oleic acid, ester of oleic acid and olive oil enhance its synthesis [59]. At low concentrations, oleic acid seems to favor the synthesis and secretion of Lip2 and Lip3 lipase isoenzymes while Lip1 is only produced at high oleic acid concentration [60]. Lipase production was also reported in the presence of carbon sources such as sterols [61], hexadecane [61], carboxylic acids [59], vegetable oils [62] and dodecanol [63]. By contrast, no lipase production could be detected in the presence of alkane, despite this carbon source could support cell growth [58]. Corn steep liquor [64] was reported has a suitable nitrogen source for lipase production in *C. rugosa* together with urea [65] and peptone [66].

In SSF, lipase production is also modulated by the solid support/ medium used. Twelve solid substrates were tested in regard to lipase production in *Aspergillus niger* [67]. The highest lipase production was obtained with wheat bran (198 Ug^{-1}) and gingelly oil cake (169 Ug^{-1}). Addition of various nitrogen sources, carbohydrates and inducers to the substrate did not yielded to any significant increase of lipase synthesis. This demonstrates that in some cases, the solid substrate used is sufficient to support both cell growth and lipase production at high yield. Other authors reported that lipase synthesis by *A. niger* is enhanced in the presence of Tween 80, soybean meal, urea and ammonium sulfate [68]. Lipase production in *Aspergillus carneus* was improved by a response surface approach using a one-at-a-time method [3,69]. The highest lipase productions were obtained in a medium composed of sunflower oil (1%), glucose (0.8%) and peptone (0.8%). Similarly, the highest lipases production by *Aspergillus terrus* was obtained in a medium containing corn oil (2%) and casein hydrolysate (0.1%) [70]. The presence of calcium and magnesium ions enhanced lipase secretion [70]. Beside this, the highest lipase production by *Aspergillus oryzae* was obtained in a medium containing sunflower oil (3%), glucose (2%), yeast extract (1%) and polypeptone (2%) [71]. Similarly to *Aspergillus sp*, lipases synthesis by *Rhizopus arrhizus* is inhibited by glucose whereas a 2.5 fold increase in lipase production was obtained in the presence of corn oil used as inducer [72]. For *Rhizopus oryzae* growing on rice straw, olive oil (1.38%) and soybean meal (1.27%) are the optimal inducer and nitrogen source, respectively [73]. In these conditions, lipase production reached 59 Ug^{-1}. Lipase production by *Penicillium restrictum* in SSF was investigated on solid waste from the babassu oil industry. The medium was supplemented with peptone, olive oil or starch at different C/N ratios. The highest lipase production was obtained with 1% of peptone enrichment (C/N ratio of 11.7, 28 Ug^{-1}) and with 2% olive oil enrichment (C/N ratio of 14.1, 30 Ug^{-1}) [74]. Molasses, urea and soybean oil concentrations as well as moisture content were tested for lipase production by *Penicillium brevicompactum* in SSF using babassu cake and castor meal as solid support [75]. For babassu cake, the highest lipase activity was obtained for a moisture content of 70% and 2% of soybean oil. This result was in accordance with those obtained for *Burkholderia* sp. using olive oil [76]. Beside this, lipase production by *Penicillium citrinum* was investigated on medium containing 1% groundnut-oil refinery residues as carbon source and various nitrogen contents. Ammonium chloride (0.75%) was found to be the best nitrogen source compared to ammonium sulfate and urea. This latter yielded to a decrease of lipase synthesis by 85% [77]. Compared to oleic acid, a two-fold increased lipase activity was obtained in the presence of this oil refinery residue due to its high carbohydrates content (35%). Production of lipase by *Penicillium simplicissium* in SSF, using babassu cake as the basal medium, was investigated in the presence of olive oil, sugarcane molasses, corn steep liquor and yeast extract. Olive oil and corn steep liquor were found the most suited carbon and nitrogen sources for lipase synthesis [78]. Rigo et al. [79] reported that the medium supplementations with urea and soybean oil significantly increased lipase production by *Penicillium* sp. grown in SSF with soybean meal as the solid substrate [79]. All these data highlight clearly that production of fungal lipase depends of several factors such as the solid substrate used and the nature of the carbon and nitrogen sources used as supplement. Additionally, the moisture content, pH and temperature should also be taken into account for process optimization.

Solid Media, Solid Support and Ambient Factors used for Lipase Production

As stated above, one of the main characteristics of SSF is the utilization of solid substrates as a support for cell growth. For some applications, these solid supports are also the only sources of carbon and nitrogen available to sustain microbial growth. These supports could be from artificial nature or of biological origin. In the recent years, considerable researches have been carried out using agricultural wastes, which are renewable and abundantly available, to produce value-added products by SSF. Utilization of agro-industrial wastes provides alternative substrates and may help to solve pollution problems. The nature of the substrate is the most important factor affecting fermentative processes.

The choice of a substrate depends on several factors, mainly related to their cost and availability but also to the nature of the compound to produce (enzyme, metabolite, biomass) [7]. Many substrates such as rice bran [80], coconut oil cake [65], olive cake and sugar cane bagasse [81], gingelly oil cake [67], babassu oil cake [74], wheat bran [82], *Jatropha curcas* seed cake [83], niger seed oil cake [84], palm kernal cake [84], groundnut oil cake [85] and mustard oil cake [86] have been used successfully for lipase production. The most important process parameters that have been investigated and optimized up to now are the incubation time, inoculum age, inoculum size, initial moisture content, nature and concentration of carbon and nitrogen sources as well as the C/N ratio of the medium [87].

The importance of water content in SSF has long been recognized [88,89]. This could be viewed in terms of moisture or water activity. Moisture represents the fraction of water present in the medium while water activity represents the fraction of free water in available to microorganisms [90]. The moisture content of the solid substrate used can vary due to liquid evaporation, microbial metabolism and by the water-binding characteristics of the substrate [12]. This moisture content has a significant impact on microbial growth and metabolic activities, including lipase production [90]. In general, the moisture content used in SSF processes vary between 30 and 85% [91]. The influence of initial moisture content on lipase production by *Y. lipolytica* from niger seed oil cake was studied in the range of 50 to 100% (v/w) [84]. An initial moisture content of 60% (v/w) lead to the maximal value of lipase production (18.5 Ug^{-1}). At higher moisture contents, the lower lipase production observed could be due to a decrease of the substrate porosity and hence in a decrease of the gaseous exchange that lead to suboptimal conditions for cell growth and enzyme production. By contrast, low moisture content could lead to a reduction of the nutrients solubility and availability, a reduced degree of swelling and a higher water tension. For the production of lipase in SSF by *Y. lipolytica* from mustard oil cake, the maximum lipase activity was obtained at 50% v/w initial moisture content [86]. The effect of initial moisture content from 40 to 90% (v/w) on lipase production of *C. rugosa* using groundnut oil cake was investigated. A moisture content of 60% v/w was reported as the optimal value [85,86]. Although moisture content is a commonly used factor to optimize growth medium, it is known that microorganisms are more sensitive and specific to water activity [88].

Another important parameter to consider in SSF is the pH variation during the process since this latter is difficult to control due to the heterogeneity of the system. During cell growth, pH variation is mainly caused by the secretion of organic acids or by the assimilation of nitrogen sources such as ammonium, nitrate or urea. Each microorganism has an optimal pH range for growth and metabolic activity, including lipase production and secretion. However, they are able to grow in a wide pH range from 2.5 to 8.5. This versatility of fungi in regards to pH can be beneficially exploited to prevent or minimize bacterial contaminations [91]. Lipase production has been investigated for different pH values ranging from 3.0 to 9.0. The maximum lipase production has been observed between pH 6.0 and 7.0 [5].

Temperature control is also an important factor to consider. During SSF, it is difficult to remove the excess of heat generated by the cellular metabolism, mainly due to the low thermal conductivity of the solid medium [15]. In some cases, this heat acccumulates within the solid medium, causing a decrease in microbial growth and even, in some cases cell death. The temperature at the center of the fermenting solid medium could be higher up to twenty degrees than the incubation temperature [92]. An airflow through the system is often applied to dissipate metabolic heat from the solid medium and also to supply the oxygen requirements for cultures [92]. Remarkably, optimal temperatures for lipase production have been reported in the range of 28 to 45°C [12,86] .

Lipase Production using different SSF Bioreactors

Lipase production by SSF has been mainly performed in tray bioreactors (TB) and packed-bed bioreactor (PBB). TB are the simplest design for SSF [93,94]. They can be considered as a modification of a column bioreactor in which the height of the bed is short enough to prevent temperature profiles to form in the axial direction. Tray bioreactors consist of a chamber where controlled air (flow, temperature and relative humidity) is circulated around a number of trays (Figure 2). Each tray contains a layer of solid substrate, typically between 5 and

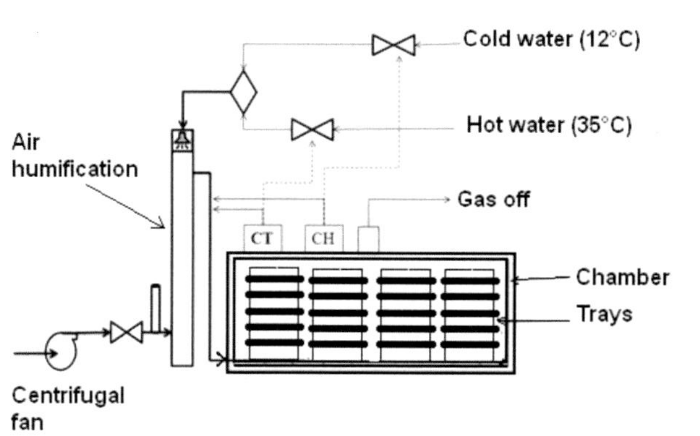

Figure 2: Schematic representation of a tray chamber with air temperature and moisture control

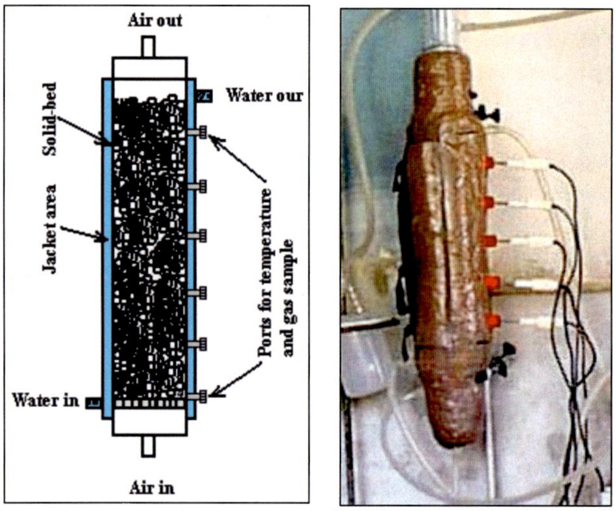

(a) (b)

Figure 3: Packed-bed bioreactors used at different stages of process development: (a) schematic representation; (b) Real lab-scale bioreactor [7]

Strain	Bioreactor type	Scale (g)	Lipase activity (Ug^{-1})	Time (h)	References
Yeast					
Yarrowia lipolytica	TB	25	26	96	[84]
Candida rugosa	TB	12	11	120	[85]
	TB	10	88	96	[65]
Mold					
Penicillium Sp.	TB	10	140	48	[79]
Penicillium simplicissimum	TB	10	90	72	[78]
Penicillium chrysogenum	TB	17	65	72	[98]
Aspergillus niger	TB	33	1934	96	[99]
	TB	20	63	48	[6]
	TB	10	623	72	[9]
	TB	100	598	72	[9]
	TB	1000	523	72	[9]
	TB	10	142	96	[100]
Rhizopus chinensis	TB	30	350	36	[101]
Rhizopus homothallicus	PBB	20	826	12	[102]
	PBB	50	1500	13	[103]
Rhizopus microsporus var. tuberosus	PBB	20	74	20	[104]
Rhizopus microsporus var. chinensis	PBB	20	72	20	[104]
Rhizopus oligosporous	TB	10	48	48	[105]
	TB	8	77	120	[106]
Rhizopus oryzae	TB	5	236	72	[16]

Table 2: Lipase activity achieved in tray (TB) and packed-bed bioreactors (PBB)

15 cm deep. Trays usually have an open top and perforated bottom to favor exchange with the gaseous phase. Intermittent mixing of the solid substrate by hand could be carried out but in general this will occur only once per day. The main drawback of TB is their low volumetric efficiency compared to column bioreactor [95]. Indeed, the thickness of the bed must remain low while the space between the trays must remain large enough to ensure an optimal gas and heat transfer. There have been no significant advances in tray design over the last decade.

PBB typically involves a static bed on top of a perforated plate through which conditioned air is blown (Figure 3). In some PBB design, air is blown through a perforated rod inserted into the center of the bed. Over the last 25 years, PBB has received much experimental and modeling attention. The main attractive characteristic is that it has no mechanical (i.e. moving) parts, thus reducing the cost of construction, operation and maintenance. In PBB processes, axial dynamical temperature profiles could be observed [96]. Since the water-carrying capacity of the air increases with temperature, these axial temperature gradients promote evaporation within the bed even if water-saturated air is used to aerate the column. Despite, this evaporation phenomenon can remove up to 65% of the heat generated by the cellular metabolism, it lowers the moisture of the solid substrate and thus limit the cellular growth [97]. In PBB, moisture regulation is almost impossible since the bed is unmixed. Therefore, it is desirable to increase heat removal by other means such as water jacket as shown in Figure 3.

In the literature, reports on large-scale production are very scarce and most of them are in the scale of few grams of solid support [11]. At small scale, comparison of the yields of lipase productions obtained for different bioreactors design is not significant. Indeed, the process at that scale is not influenced by axial temperature profiles that form within the solid substrate. Therefore, the differences in lipase productions, presented in Table 2, seem to be more dependent on strains and growth conditions than other factors. The data shown in Table 2 indicate that lipase productions are below level of 500 Ug^{-1} in 71% of the cases while less that 10% are over 1500 Ug^{-1}. This indicates that very high productions and productivities could be reached under controlled SSF process. However at higher scale, a decrease in process productivity should be expected. Nevertheless, Edwinoliver and coll. reported only a slight decrease in lipase production when scaling up the process from 10 g to 1000 g [9].

Acknowledgments

This work was supported by WBI (Wallonie-Bruxelles International) grant SUB/2012/86836 to P. Fickers and L. B. Ramos-Sánchez.

References

1. Sarda L, Desnuelle P (1958) [Actions of pancreatic lipase on esters in emulsions]. Biochim Biophys Acta 30: 513-521.
2. Verger R, De Haas GH (1973) Enzyme reactions in a membrane model. 1. A new technique to study enzyme reactions in monolayers. Chem Phys Lipids 10: 127-136.
3. Verger R (1997) 'Interfacial activation' of lipases: facts and artifacts. Trends Biotechnol 15: 32-38.
4. Rodrigues RC, Fernandez-Lafuente R (2010) Lipase from *Rhizomucor miehei* as an industrial biocatalyst in chemical process. J Mol Catal B: Enzymatic 64: 1-22.
5. Liu CH, Chen CY, Wang YW, Chang JS (2011) Fermentation strategies for the production of lipase by an indigenous isolate *Burkholderia* sp. C20. Biochem Eng J 58-59: 96-102.
6. Damaso MC, Passianoto MA, de Freitas SC, Freire DM, Lago RC, et al. (2008) Utilization of agroindustrial residues for lipase production by solid-state fermentation. Braz J Microbiol 39: 676-681.
7. Thakur S (2012) Lipases, its sources, properties and applications: A Review. Int J Sci Energ Res 3: 1-29.
8. Gutarra ML, Godoy MG, Maugeri F, Rodrigues MI, Freire DM, et al. (2009) Production of an acidic and thermostable lipase of the mesophilic fungus *Penicillium simplicissimum* by solid-state fermentation. Bioresour Technol 100: 5249-5254.
9. Edwinoliver NG, Thirunavukarasu K, Naidu RB, Gowthaman MK, Kambe TN, et al. (2010) Scale up of a novel tri-substrate fermentation for enhanced production of *Aspergillus niger* lipase for tallow hydrolysis. Bioresour Technol 101: 6791-6796.
10. Hosseinpour MN, Najafpour GD, Younesi H, Khorrami M, Vaseghi Z (2012) Lipase production in solid state fermentation using *Aspergillus niger*: Response surface methodology. IJE Transactions B: Applications 25: 151-159.

11. Bhargav S, Panda BP, Ali M, Javed S (2008) Solid-state fermentation: An overview. Chem Biochem Eng Q 22: 49-70.

12. Kumar A, Kanwar S (2012) Lipase production in solid-sate fermentation (SSF): Recent developments and biotechnological applications. Dynam Biochem Proc Biotecnol Mol Biol 6: 13-27.

13. Khanahmadi M, Mitchell DA, Beheshti M, Roostaazad R, Sánchez LR (2006) Continuous solid-state fermentation as affected by substrate flowpattern. Chem Eng Sci 61: 2675-2687.

14. Ramos-Sanchez LB, Khanahmadi M, Mitchell D (2006) Continuous solid-state fermentation bioreactors. Solid-state fermentation bioreactors- fundamentals of design and operation. D. A. Mitchell, N. Krieger & M. Berovic (Editors) Springer edition, Germany 142-158.

15. Pandey A, Soccol CR, Mitchell DA (2000) New developments in solid state fermentation: I-bioprocesses and products. Proc Bioch 35: 1153-1169.

16. Vaseghi Z, Najafpour GD, Mohseni S, Mahjoub S (2013) Production of active lipase by *Rhizopus oryzae* from sugarcane bagasse: solid state fermentation in a tray bioreactor. Int J Food Sci Technol 48: 283-289.

17. Couto SR, Sanroman MA (2006) Application of solid-state fermentation to food industry-A review. J Food Eng 76: 291-302.

18. Mitchell D, Berovi M, Krieger N (2006) Introduction to solid-state fermentation bioreactors. Solid-state fermentation bioreactors-fundamentals of design and operation. Mitchell DA, Krieger N, Berovic M (Editors) Springer edition, Germany, 33-44.

19. Mitchell D, Berovi M, Nopharatana M (2006) The Bioreactor Step of SSF: A Complex interaction of phenomena. Solid-state fermentation bioreactors-fundamentals of design and operation. Mitchell DA, Krieger N, Berovic M (Editors) Springer edition, Germany, 13-32.

20. Coradi GV, Visitacao VL, de Lima EA, Saito LYT, Palmieri DA, et al. (2013) Comparing submerged and solid-state fermentation of agro-industrial residues for the production and characterization of lipase by *Trichoderma harzianum*. Ann Microbiol 63: 533-540.

21. Rajan A, Nair AJ (2011) A comparative study on alkaline lipase production by a newly isolated *Aspergillus fumigatus* MTCC 9657 in submerged and solid-state fermentation using economically and industrially feasible substrate. Turkish J Biol 35: 569-574.

22. Mateos Diaz JC, Rodríguez JA, Roussos S, Cordova J, Abousalham A, et al. (2006) Lipase from the thermotolerant fungus *Rhizopus homothallicus* is more thermostable when produced using solid state fermentation than liquid fermentation procedures. Enz Microb Technol 39: 1042-1050.

23. Nwuche CO, Aoyagi H, Ogbonna JC (2013) Lipase production from palm oil mill effluent by *Aspergillus terreus* immobilized on Luffa sponge. J Appl Sci 24: 5661-5671.

24. Khoramnia A, Ebrahimpour A, Beh BK, Lai OM (2011) Production of a solvent, detergent, and thermotolerant lipase by a newly isolated *Acinetobacter* sp. in submerged and solid-state fermentations. J Biomed Biotechnol 2011: 702179.

25. Godoy MG, Gutarra ML, Castro AM, Machado OL, Freire DM (2011) Adding value to a toxic residue from the biodiesel industry: production of two distinct pool of lipases from *Penicillium simplicissimum* in castor bean waste. J Ind Microbiol Biotechnol 38: 945-953.

26. Casas-Godoy L, Duquesne S, Bordes F, Sandoval G, Marty A (2012) Lipases: an overview. Methods Mol Biol 861: 3-30.

27. Aravindan R, Anbumathi P, Viruthagiri T (2007) Lipase applications in food industry. Indian J Biotechnol 6: 141-158.

28. Fickers P, Marty A, Nicaud JM (2011) The lipases from *Yarrowia lipolytica*: genetics, production, regulation, biochemical characterization and biotechnological applications. Biotechnol Adv 29: 632-644.

29. Erik B, Steinborn G, Florsch K (2009) Yeast Biotechnology: Diversity and Applications. Yeast Biotechnol Div Appl: 323-355.

30. Ray A (2012) Application of Lipase in Industry. Asian J Pharm Tech 2: 33-37.

31. Lock LL, Corbellini VA, Valente P (2007) Lipases produced by yeasts: Powerful biocatalysts for industrial purposes. Tecno-Logica 11: 18-25.

32. Tanino T, Aoki T, Chung WY, Watanabe Y, Ogino C, et al. (2009) Improvement of a *Candida antarctica* lipase B-displaying yeast whole-cell biocatalyst and its application to the polyester synthesis reaction. Appl Microbiol Biotechnol 82: 59-66.

33. Akoh CC (1993) Lipase-catalyzed synthesis of partial glyceride. Biotechnol Lett 15: 949-954.

34. Bakirci H, Li M, He B (1995) Immobilization of *Candida cylindracea* lipase on methyl acrylate-divinyl benzene copolymer and its derivatives. Enz Microb Technol 17: 194-199.

35. Schmid R, Verger R (1998) Lipasen: Grenzflächen-Enzyme mit attraktiven Anwendungen. Ang Chemie 110: 1694-1720.

36. Pierce G, Wick C, Palmer D (1990) Unique microbial lipases with activity at temperatures and pHs suitable for use in detergents. EP 0385401 A1.

37. Hasan F, Shah AA, Javed S, Hameed A (2010) Enzymes used in detergents: lipases. J Biotechnol 9: 4836-4844.

38. Vakhlu J, Kour A (2006) Yeast lipases: enzyme purification, biochemical properties and gene cloning. Elect J Biotechnol 9: 69-85.

39. Pandey A, Benjamin S, Soccol CR, Nigam P, Krieger N, et al. (1999) The realm of microbial lipases in biotechnology. Biotechnol Appl Biochem 29 : 119-131.

40. Larios A, García HS, Oliart RM, Valerio-Alfaro G (2004) Synthesis of flavor and fragrance esters using *Candida antarctica* lipase. Appl Microbiol Biotechnol 65: 373-376.

41. Shay LK, Fisher TJ (1990) Enhancing the flavor of protein products derived from microorganisms. EP Patent US 5071762 A.

42. Hasan F, Shah AA, Hameed A (2006) Industrial applications of microbial lipases. Enz Microbial Technol 39: 235-251.

43. Sharma D, Sharma B, Shukla AK (2011) Biotechnological approach of microbial lipase: A Review. Biotechnology 10: 23-40.

44. Johnson EA1 (2013) Biotechnology of non-*Saccharomyces* yeasts-the basidiomycetes. Appl Microbiol Biotechnol 97: 7563-7577.

45. Yücel S, Terzioglu P, Özçimen D (2012) Lipase applications in biodiesel production. Biodiesel-feedstocks, production and application. InTech Publishers, Croatia.

46. Matsumae H, Furui M, Shibatani T (1993) Lipase-catalyzed asymmetric hydrolysis of 3-phenylglycidic acid ester, the key intermediate in the synthesis of diltiazem hydrochloride. J Ferm Bioeng 75: 93-98.

47. Hsu AF, Jones K, Foglia TA, Marmer WN (2002) Immobilized lipase-catalysed production of alkyl esters of restaurant grease as biodiesel. Biotechnol Appl Biochem 36: 181-186.

48. Verma N, Thakur S, Bhatt AK (2012) Microbial Lipases: Industrial Applications and Properties (A Review). Int Res J Biol Sci 1: 88-92.

49. Abada EA (2008) Production and characterization of a mesophilic lipase isolated from *Bacillus stearothermophilus* AB-1. Pak J Biol Sci 11: 1100-1106.

50. Dutra JC, da C Terzi S, Bevilaqua JV, Damaso MC, Couri S, et al. (2008) Lipase production in solid state fermentation monitoring biomass growth of *Aspergillus niger* using digital image processing. Appl Biochem Biotechnol 147: 63-75.

51. Fickers P, Nicaud JM, Gaillardin C, Destain J, Thonart P (2004) Carbon and nitrogen sources modulate lipase production in the yeast *Yarrowia lipolytica*. J Appl Microbiol 96: 742-749.

52. Fickers P, Fudalej F, Nicaud JM, Destain J, Thonart P (2005) Selection of new over-producing derivatives for the improvement of extracellular lipase production by the non-conventional yeast *Yarrowia lipolytica*. J Biotechnol 115: 379-386.

53. Destain J, Fickers P, Weekers F, Moreau B, Thonart P (2005) Utilization of methyloleate in production of microbial lipase. Appl Biochem Biotechnol 121-124: 269-77.

54. Fickers P, Nicaud JM, Destain J, Thonart P (2003) Overproduction of lipase by *Yarrowia lipolytica* mutants. Appl Microbiol Biotechnol 63: 136-142.

55. Turki S, Kraeim IB, Weeckers F, Thonart P, Kallel H (2009) Isolation of bioactive peptides from tryptone that modulate lipase production in *Yarrowia lipolytica*. Bioresour Technol 100: 2724-2731.

56. Farias MA, Valoni EA, Castro AM, Coelho MAZ (2014) Lipase production by *Yarrowia lipolytica* in solid state fermentation using different agro industrial residues. Chem Eng Trans 38: 301-306.

57. del Rio JL, Serra P, Valero F et al. (1990) Reaction scheme of lipase production by *Candida rugosa* growing on olive oil. Biotechnol Lett 12: 835-838.

58. de María PD, Sánchez-Montero JM, Alcántara AR, Valero F, Sinisterra JV (2005) Rational strategy for the production of new crude lipases from *Candida rugosa*. Biotechnol Lett 27: 499-503.

59. Wei D, Zhang LY, Song Q (2004) Studies on a novel carbon source and cosolvent for lipase production by *Candida rugosa*. J Ind Microbiol Biotechnol 31: 133-136.

60. Ferrer P, Montesinos JL, Valero F, Solà C (2001) Production of native and recombinant lipases by *Candida rugosa*: a review. Appl Biochem Biotechnol 95: 221-255.

61. Ota Y, Miyairi S, Yamada K (1968) Sterol requirements for the lipase production by *Candida cylindracea*. Agric Biol Chem 32: 1476-1478.

62. Lakshmi BS, Kangueane P, Abraham B, Pennathur A (1999) Effect of vegetable oils in the secretion of lipase from *Candida rugosa* (DSM2031). Lett Appl Microbiol 29: 66-70.

63. Alcantara AR, de Maria PD, Fernandez M, Hernaiz MJ, Montero S, et al. (2004) Resolution of racemic acids, esters and amines by *Candida rugosa* lipase in slightly hydrated organic medium. Food Technol Biotechnol 42: 343-354.

64. Korbekandi H, Abedi D, Pourhossein M, Motovali-Bashi M (2008) Optimisation of *Candida rugosa* lipase esterase activity. Biotechnology 7: 112-117.

65. Benjamin S, Pandey A (1997) Coconut cake-a potent substrate for the production of lipase by *Candida rugosa* in solid-state fermentation. Acta Biotechnologica 17: 241-251.

66. Rajendran A, Palanisamy A, Thangavelu V (2008) Evaluation of medium components by Plackett-Burman statistical design for lipase production by *Candida rugosa* and kinetic modeling. Sheng Wu Gong Cheng Xue Bao 24: 436-444.

67. Kamini NR, Mala JGS, Puvanakrishnan R (1998) Lipase production from *Aspergillus niger* by solid-state fermentation using gingelly oil cake. Process Biochem 33: 505-511.

68. Sarkar D, Laha S (2013) Optimization of extracellular lipase enzyme production from Aspergillus niger by submerged and solid-state fermentation process. Int J Pharm Bio Sci 4: 978-985.

69. Kaushik R, Saran S, Isar J, Saxena RK (2006) Statistical optimization of medium components and growth conditions by response surface methodology to enhance lipase production by *Aspergillus carneus*. J Mol Catal B: Enz 40: 121-126.

70. Yadav RP, Saxena RK, Gupta R, Davidson S (1998) Lipase production by *Aspergillus* and *Penicillium* species. Folia Microbiol (Praha) 43: 373-378.

71. Ohnishi K, Yoshida Y, Sekiguchi J (1994) Lipase production of *Aspergillus oryzae*. J Ferment Bioing 77: 490-495.

72. Elibol M, Ozer D (2000) Influence of oxygen transfer on lipase production by *Rhizopus arrhizus*. Process Biochem 36: 325-329.

73. Khayati G, Kiyani F (2012) A statistical approach for optimization of lipase production by using rice straw: analysis of different inducers and nitrogen sources effect. Minerva Biotechnol 24: 83-89.

74. Gombert AK, Pinto AL, Castilho LR, Freire DMG (1999) Lipase production by *Penicillium restrictum* in solid-state fermentation using babassu oil cake as substrate. Proc Bioch 35: 85-90.

75. Silva MF, Freire DM, de Castro AM, Di Luccio M, Mazutti MA, et al. (2011) Production of multifunctional lipases by *Penicillium verrucosum* and *Penicillium brevicompactum* under solid state fermentation of babassu cake and castor meal. Bioprocess Biosyst Eng 34: 145-152.

76. Liu CH, Lu WB, Chang JS (2006) Optimizing lipase production of *Burkholderia* sp. by response surface methodology. Process Biochem 41: 1940-1944.

77. Miranda OA, Salgueiro AA, Pimentel MCB, Filho JLL, Melo EHM, et al. (1999) Lipase production by a Brazilian strain of *Penicillium citrinum* using an industrial residue. Biores Technol 69: 145-147.

78. Gutarra ML, Cavalcanti ED, Castilho LR, Freire DM, Sant'Anna GL (2005) Lipase production by solid-state fermentation: cultivation conditions and operation of tray and packed-bed bioreactors. Appl Biochem Biotechnol 121-124: 105-16.

79. Rigo E, Ninow JL, Di Luccio M, Vladimir Oliveira J, Polloni AE, et al. (2010) Lipase production by solid fermentation of soybean meal with different supplements. LWT- Food Sci Technol 43: 1132-1137.

80. Rao PV, Jayaraman K, Lakshmanan CM (1993) Production of lipase by *Candida rugosa* in solid-state fermentation. I: determination of significant process variables. Proc Biochem 28: 385-389.

81. Cordova J, Nemmaoui M, Ismaili-Alaoui M, Morin A, Roussos S, et al. (1998) Lipase production by solid state fermentation of olive cake and sugar cane bagasse. J Mol Catal B: Enz 5: 75-78.

82. Mahadik ND, Puntambekar US, Bastawde KB et al. (2002) Production of acidic lipase by *Aspergillus niger* in solid state fermentation. Proc Biochem 38: 715-721.

83. Mahanta N, Gupta A, Khare SK (2008) Production of protease and lipase by solvent tolerant *Pseudomonas aeruginosa* PseA in solid-state fermentation using *Jatropha curcas* seed cake as substrate. Bioresour Technol 99: 1729-1735.

84. Imandi SB, Karanam SK, Garapati HR (2014) Use of Plackett-Burman design for rapid screening of nitrogen and carbon sources for the production of lipase in solid state fermentation by *Yarrowia lipolytica* from mustard oil cake (*Brassica napus*). Braz J Microbiol 44: 915-921.

85. Rekha K, Lakshmi M, Sridevi V (2012) An overview of microbial lipases. J Chem Biol Phys Sci 2: 1379-1389.

86. Imandi SB, Karanam SK, Garapati HR (2014) Use of Plackett-Burman design for rapid screening of nitrogen and carbon sources for the production of lipase in solid state fermentation by *Yarrowia lipolytica* from mustard oil cake (*Brassica napus*). Braz J Microbiol 44: 915-921.

87. Imandi SB, Karanam SK, Garapati HR (2010) Optimization of media constituents for the production of lipase in solid state fermentation by *Yarrowia lipolytica* from palm Kernal cake (*Elaeis guineensis*). Adv Biosci Biotechnol 1: 115-121.

88. Gervais P, Molin P (2003) The role of water in solid-state fermentation. Biochem Eng J 13: 83-110.

89. Rocha NRAF, Barros MA, Fisher J, Filho UC, Cardoso VL (2013) Ethanol production from agroindustrial biomass using a crude enzyme complex produced by *Aspergillus niger*. Renew Energ 57: 432-435.

90. Babu I, Rao G (2007) Lipase production by *Yarrowia lipofyäca* NCIM 3589 in solid state fermentation using mixed substrate. Res J Microbiol 2: 469-474.

91. Krishna C1 (2005) Solid-state fermentation systems-an overview. Crit Rev Biotechnol 25: 1-30.

92. Pandey A (2003) Solid-state fermentation. Biochem Eng J 13: 81-84.

93. Ali H, Zulkali MMD (2011) Utilization of agro-residual ligno-cellulosic substances by using solid state fermentation: A review. Croatian J Food Technol Biotechnol Nut 6: 5-12.

94. Vaseghi Z, Najafpour GD (2014) An Investigation on lipase production from soybean meal and sugarcane bagasse in solid state fermentation using *Rhizopus oryzae*. IJE Trans B: Applications 27: 171-176.

95. Mirón J, Vázquez J, González P, Murado MA (2010) Enhancement glucose oxidase production by solid-state fermentation of *Aspergillus niger* on polyurethane foams using mussel processing wastewaters. Enz Microb Technol 46: 21-27.

96. Ashley VM, Mitchell DA, Howes T (1999) Evaluating strategies for overcoming overheating problems during solid state fermentation in packed bed bioreactors. Biochem Eng J 3: 141-150.

97. Gutierrez M, Amar S, Auria R, Revah S, Torres EF (1996) Heat transfer in citric acid production by solid state fermentation. Proc Biochem 31: 363-369.

98. Iftikhar T, Niaz M, Nisa Z, Tariq A, Khalid MN, et al. (2011) Optimization of cultural conditions for the biosynthesis of lipases by *Penicillium chrysogenum* (MBL 22) through solid state fermentation. Pak J Bot 43: 2201-2206.

99. Adinarayana K, Raju KVVSN, Zargar MI, Bhavani DR, Jhansi LP, et al. (2004) Optimization of process parameters for production of lipase in solid-state fermentation by newly isolated *Aspergillus* species. Ind J Biotechnol 3: 65-69.

100. Mohseni S, Najafpour G, Vaseghi Z, Mahjoub S (2012) Solid state fermentation of agricultural residues for lipase production in a tray-bioreactor. World Appl Sci J 16: 1034-1039.

101. Sun SY, Xu Y (2008) Solid-state fermentation for "whole-cell synthetic lipase"production from *Rhizopus chinensis* and identification of the functional enzyme. Process Biochem 43: 219-224.

102. Rodriguez JA, Mateos JC, Nungaray J, Gonzalez V, Bhagnagar T, et al. (2006) Improving lipase production by nutrient source modification using *Rhizopus homothallicus* cultured in solid state fermentation. Process Biochem 41: 2264-2269.

103. Diaz JC, Rodríguez JA, Roussosb S, Cordovac J, Abousalhamd A, et al. (2006) Lipase from thermotolerant fungus *Rhizopus homothalicus* is more stable when produced using solidstate fermentation than in liquid fermentation procedures. Enz Microb Technol 39: 1042-1050.

104. Hernández B, Córdova J, Bárzana E, Favela E (2009) Effects of organic solvents on activity and stability of lipases produced by thermotolerant fungi in solid-state fermentation. J Mol Catal B Enz 61: 136-142.

105. Ul-Haq I, Idrees S, Rajoka M (2002) Production of lipase by *Rhizopus oligosporous* by solid state fermentation. Process Biochem 37: 637-641.

106. Mahapatra P, Kumari A, Garlapati VK, Banerjee R, Nag A (2010) Optimization of Process Variables for Lipase Biosynthesis from *Rhizopus oligosporus* NRRL 5905 Using Evolutionary Operation Factorial Design Technique. Indian J Microbiol 50: 396-403.

2

Immobilization of *Gluconobacter oxydans* by Entrapment in Porous Chitosan Sponge

Cunxun Wang, Kefeng Ni, Xu Zhou, Dongzhi Wei and Yuhong Ren*

State Key Laboratory of Bioreactor Engineering, New World Institute of Biotechnology, East China University of Science and Technology, Shanghai 200237, China

Abstract

The porous chitosan sponge was prepared using $NaHCO_3$ as the porogen and used to immobilize *Gluconobacter oxydans*. Under the optimum conditions, the activity recovery of the immobilized cells reached 92%. The morphology characterization of the immobilized cells revealed that the cells were attached to the surface of the pores (100-400μm) which were well distributed in the chitosan sponge. The valuation of cell activity showed that the immobilized cells displayed enhanced pH and thermal stability compared to free cells. Furthermore, the immobilized cells retained 74% of its origin activity after 12 repeated reaction cycles separated by filtration.

Keywords: Chitosan sponge; Gluconobacter oxydans; Immobilization; Porogen

Introduction

Gluconobacter oxydans have been widely used to incompletely oxidize sugar, alcohol and aldehyde to produce aldehyde, ketone and acid by its dehydrogenases connected to the respiratory chain [1]. However, *G. oxydans* is a very small size bacterial leading to difficulties in the reuse or recycling of the cells for large-scale application.

Whole cell immobilization, which provides cells with easy separation, enhanced stability and reusability, has been proved to be an efficient solution [2]. As the immobilization matrices, natural polymers such as agar, alginate, carrageenan and chitosan have received considerable attention due to their high biocompatibility [3], and a variant of the whole cell immobilization technique have been developed. Chitosan, the second most abundant natural polymer found in the exoskeleton of marine crustaceans, has gained great interest in immobilization technology [4]. It is also supposed to be very interest substance for diverse applications in biomaterial such as in preparing films [5], beads [6], scaffolds [7], hydrogels [8], nanofibers [9] and nanoparticles [10] in the pharmaceutical field due to its biocompatibility, biodegradability and bioactivity.

In this study, we proposed a novel method to immobilize *G. oxydans* in the chitosan spongy material which was synthesized using $NaHCO_3$ as the porogen to react with acetic acid producing CO_2 and forming pores in chitosan sponge. Compared with other porogen such as polyethylene glycol [11] and silica [12,13], $NaHCO_3$ can be removed in mild condition leading to a higher cell activity recovery. After immobilized, the cells were easily separated from the solution by filtering, and their thermal and pH stability and reusability were increased.

Materials and Methods

Microorganism and cultivation conditions

The strain *G.oxydans DSM 2003* was used in this study. The cells were cultured in 500-mL flasks containing 50 mL sterile fermentation medium, which contained 80 g sorbitol/L, 20 g yeast extract/L, 1 g KH_2PO_4/L, 0.5 g $MgSO_4$/L, 0.1 g Glutamine/L. The cells were incubated at 30°C with shaking at 200 rpm for 24 h and collected by centrifugation.

Immobilization of *G. oxydans*

Chitosan (1.5 g) was dissolved in 2% (v/v) acetic acid (100 mL), and then the pH of the solution was adjusted to 5.0 using 2 M NaOH. 5 mL *G. oxydans* suspension (10 g cells/L, dry weight) was added to the above solution followed by the addition of 10 mL the mixture of $NaHCO_3$ (10% w/v) and glutaraldehyde (5% w/v). The reaction mixture was further stirred at room temperature until formatting the porous chitosan sponge with cells embeded. The immobilized cells were filtered and washed several times with phosphate buffer (pH 6.0). The density of free cells in the solution was determined by measuring the optical density of the cell suspension at 600 nm (OD600) with a spectrophotometer (U-2001; Hitachi, Tokyo). The difference in cell density was used to calculate immobilization efficiency. The morphology of the immobilized cells was observed on a Scanning electron microscopy (SEM) (JEOL Japan). Samples were lyophilized and sputter-coated with gold prior to scanning.

Activity assay

The activity of cells was determined by measuring the production of dihydroxy acetone (DHA) from glycerol. The reaction was carried out at 30°C in phosphate buffer (pH 6.0,10 mM) containing 10 g glycerol/L. After reacting for 1 hour, the free cells were separated by centrifugation and the immobilized cells were separated by filtration. The reaction products were analyzed by HPLC using a COREGEL 87H3 column (Transgenomic, USA) with isocratic elution of 4 mM H_2SO_4.

Stability and reusability

Effect of temperature and pH on the activity of immobilized cells were determined from 20-50°C and pH 5-8.5 and compared with free cells. The reusability of the immobilized cells was assessed under the same conditions as described in activity assay section.

***Corresponding author:** Yuhong Ren, State Key Laboratory of Bioreactor Engineering, New World Institute of Biotechnology, East China University of Science and Technology, Shanghai 200237, China

Results and Discussion

Morphology of the biocarrier

Figure 1 showed the SEM images of porous chitosan sponge with and without immobilization of *G. oxydans*. The biocarrier was highly porous and the pores (ranging from 100 to 400μm) were well distributed with an interconnected pore wall structure. The porous structure of the chitosan sponge was beneficial for cell adhesion and substrates diffusion. As shown in Figure 1d, the chitosan sponge surface without cells entrapment is relatively smoother than the one with cells in Figure 1c which revealed the cells attached to the chitosan sponge and covered a majority of the inner surface of the pores. The internal cells distribution of chitosan sponge was confirmed by Laser Scanning Confocal Microscopy (LSCM) (Supplementary Figure S1) and the results displayed that the cells were well distributed inside the chitosan sponge.

Immobilization of *G. oxydans*

The effect of glutaraldehyde concentrations showed the cells' activity decreased with the increase of glutaraldehyde (Supplementary Figure S2). However, the chitosan couldn't form a cross-linked sponge when the concentration of glutaraldehyde was less than 0.5% (w/v), leading to cells leakage. Hence, the optimum glutaraldehyde concentration was 0.5% and the cell activity retained more than 90% at this concentration.

As the porogen, $NaHCO_3$ reacted with acetic acid producing CO_2, which formed pores in chitosan sponge. The porosity of chitosan sponge increased with the increase of the concentration of $NaHCO_3$ (Supplementary Table S1). Figure 2a showed that the activity recovery of immobilized cells increased from 21% to 92% when the concentration of $NaHCO_3$ increased from 0 to 10%, but decreased to 62% as the concentration of $NaHCO_3$ continued to rise to 15%. The improvement of the activity recovery of immobilized cells was attributed to the increase of the porosity of chitosan sponge which was favorable for the diffusion of substrates. However, when the concentration of $NaHCO_3$ was up to 15% the activity recovery of immobilized cells was decreased quickly due to the alkaline internal condition of chitosan sponge which exceeded the optimum catalytic pH conditions.

The effect of the concentration of chitosan on the activity recovery of immobilized cells was investigated and the result was showed in Figure 2b. Increasing the concentration of chitosan in the range 0 - 1.5% resulted in an increase in the activity recovery from 41% to 87.5%. However, the activity recovery began to decrease above chitosan concentration of 1.5%, which supposed to be due to the decrease of porosity of the chitosan sponge (Supplementary Table S2).

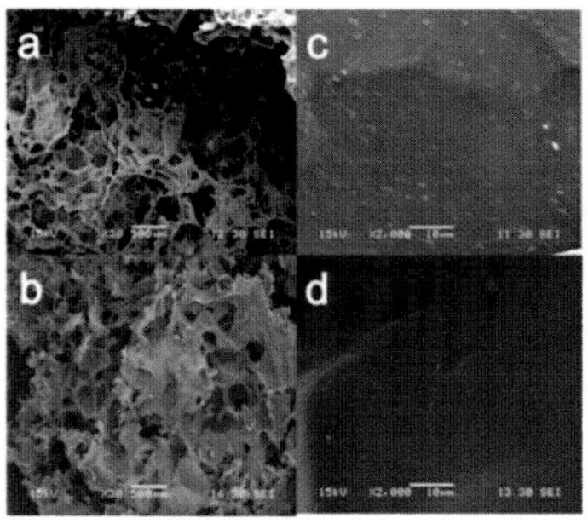

Figure 1: SEM images of porous chitosan sponge. (a) magnification~30x, (b) magnification~30x, (c) magnification~2,000x, (d) magnification~2,000x.

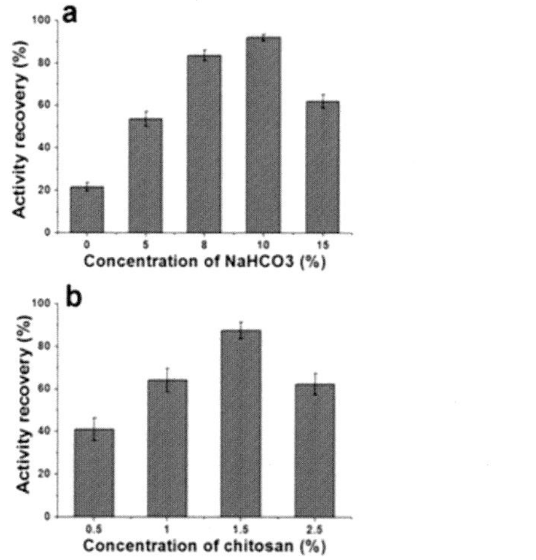

Figure 2: The effect of the concentration of $NaHCO_3$ (a) and chitosan (b) on the activity recovery of immobilized cells. The activity of free cells was defined as 100%.
a: 1.5% (w/v) glutaraldehyde,10% (w/v) $NaHCO_3$, 5 mg (dry weight) cells; b: 1.5% (w/v) glutaraldehyde,1.5% (w/v) chitosan, 5 mg (dry weight) cells.

Effect of pH and temperature

Figures 3a and 3b showed that the optimal temperature and pH for both free and immobilized cells to achieve the highest activity were 30°C and pH 6.0. Compared to the free cells, the immobilized cells retained higher thermal and pH stability after incubated at various temperatures (Figure 3c) and pH (Figure 3d). Because the enzyme responsible for DHA synthesis was a membrane bound dehydrogenase, the enhanced stability of immobilized cells may be attributed to the decrease damaged of cell's membrane after the cells were covalently linked to the chitosan sponge by glutaraldehyde.

Reusability

The reusability of immobilized cells is a very important property in their application. Compared to the free cells separated by centrifugation, the chitosan sponge entrapped cells can be easily separated by filtration and applied for another cycle. As shown in Figure 4, the immobilized cells retained 74% of its initial activity after 12 cycles, whereas the free cells only retained 50% of its initial activity. The decrease of activity was caused by the cell inactivation, but not the cell leakage (no cell loss was observed in this process). Furthermore, the cells were hardly leaked from the chitosan sponge even after shaking at 200 rpm for 3 days (data not shown). It was suggested that the cells were covalently linked with chitosan sponge by glutaraldehyde which prevented cells detaching from the carrier.

Conclusion

Gluconobacter oxydans was immobilized on the porous chitosan sponge with a 92% activity recovery after optimization. The immobilized cells displayed higher thermal and pH stability than the free cells. More importantly, the immobilized cells can be easily separated and reused, and retained 74% of the initial activity even after 12 cycles. These results make the porous chitosan sponge a promising material in cells immobilization applications.

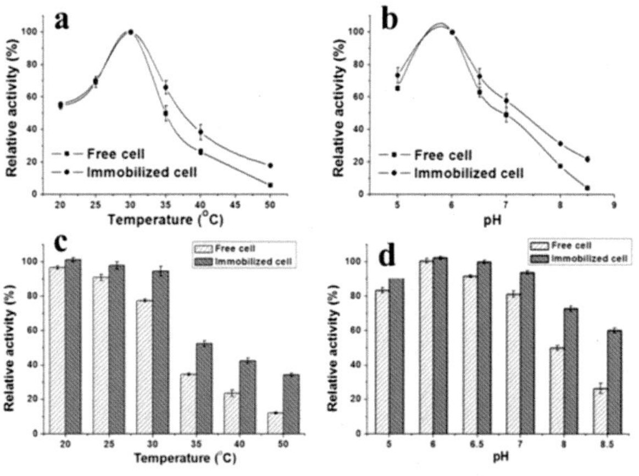

Figure 3: Effects of temperature (a) and pH (b) on the activity of the free and immobilized cells, Thermal (c) and pH (d) stability of the free and immobilized cells.
a: 5 mg cells, 10 g glycerol/l, sodium phosphate buffer (pH 6), temperature ranging from 20-50°C; b: 5 mg cells, 10 g glycerol/l, 100 mM sodium phosphate buffer (pH from 5to 8.5), at 30°C; c: The samples were pre-incubated at temperature ranging from 25 to 50°C for 12 hours and the residual activity was measured at pH 6.0 at 30°C; d: The samples were pre-incubated at pH ranging from 5 to 8.5 for 12 hours and the residual activity was measured at pH 6.0 at 30°C.

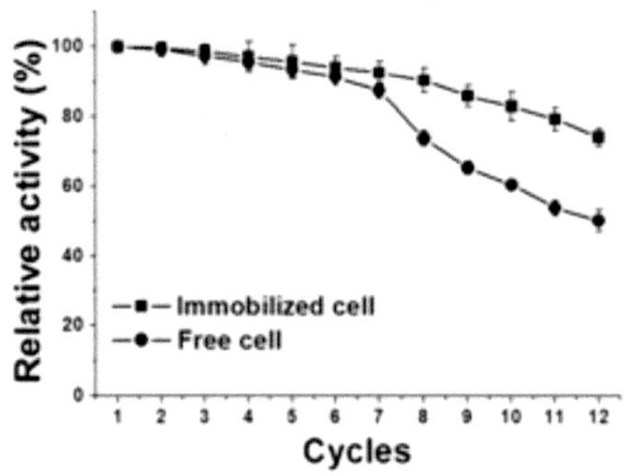

Figure 4: Reusability of the immobilized cells. Reaction conditions: 5 mg cells, 100 mM sodium phosphate buffer (pH 6), at 30°C. After each run, the immobilized cells were separated by filtration and the free cells were separated by centrifugation (10,000 rpm), washed with phosphate buffer (pH 6.0) for 2 times. The initial activities of the free and immobilized cells were defined as 100%.

Acknowledgements

This work was funded by The National Natural Foundation of China (NO.21076079), Open Funding Project of the State Key Laboratory of Bioreactor Engineering and the Fundamental Research Funds for the Central Universities.

References

1. Mogi T, Matsushita K, Murase Y, Kawahara K, Miyoshi H, et al. (2009) Identification of new inhibitors for alternative NADH dehydrogenase (NDH-II). FEMS Microbiol Lett 291: 157-161.
2. Behera S, Kar S, Mohanty RC, Ray RC (2010) Comparative study of bio-ethanol production from mahula (Madhuca latifolia L.) flowers by Saccharomyces cerevisiae cells immobilized in agar agar and Ca-alginate matrices. Appl Energ 87: 96-100.
3. Wang JY, Chao YP (2006) Immobilization of Cells with Surface-Displayed Chitin-Binding Domain. Appl Environ Microb 72: 927-931.
4. Orrego CE, Salgado N, Valencia JS, Giraldo GI, Giraldo OH, et al. (2010) Novel chitosan membranes as support for lipases immobilization: characterization aspects. Carbohyd Polym 79: 9-16.
5. Jayakumar R, Prabaharan M, Reis RL, Mano JF (2005) Graft copolymerized chitosan-present status and applications. Carbohyd Polym 62: 142-158.
6. Kumar V, Jahan F, Raghuwanshi S, Mahajan RV, Saxena RK (2013) Immobilization of Rhizopus oryzae lipase on magnetic Fe_3O_4-chitosan beads and its potential in phenolic acids ester synthesis. Biotechnol Bioproc Eng 18: 787-795.
7. Seol YJ, Lee JY, Park YJ, Lee YM, Young-Ku, et al. (2004) Chitosan sponges as tissue engineering scaffolds for bone formation. Biotechnol Lett 26: 1037-1041.
8. Ramos R, Carvalho V, Gama M (2006) Novel hydrogel obtained by chitosan and dextrin-VA copolymerization. Biotechnol Lett 28:1279-1284.
9. Chu XH, Shi XL, Feng ZQ, Gu ZZ, Ding YT (2009) Chitosan nanofiber scaffold enhances hepatocyte adhesion and function. Biotechnol Lett 31: 347-352.
10. Anitha A, Divya Rani VV, Krishna R, Sreeja V, Selvamurugan N, et al. (2009) Synthesis, characterization, cytotoxicity and antibacterial studies of chitosan, O-carboxymethyl, N, O-carboxymethyl chitosan nanoparticles. Carbohydr Polymers 78: 672-677.
11. Tang CY, Zhang Q, Wang K., Fu Q, Zhang CL (2009) Water transport behavior of chitosan porous membranes containing multi-walled carbon nanotubes (MWNTs). J Membrane Sci 337: 240-247.
12. Santos DES, Neto CGT, Fonseca JLC, Pereira MR (2008) Chitosan macroporous asymmetric membranes: preparation, characterization and transport of drugs. J Membrane Sci 325: 362-370.
13. Zeng XF, Ruckenstein E (1996) Control of Pore Sizes in Macroporous Chitosan and Chitin Membranes. Ind Eng Chem Res 35: 4169-4175.

3

Enhance Vitamin B_{12} Production by Online CO_2 Concentration Control Optimization in 120 m^3 Fermentation

Ze-Jian Wang[1], Hui-Yuan Wang[1], Ping Wang[1], Yi-ming Zhang[2], Ju Chu[1], Ying-Ping Zhuang[1*] and Si-Liang Zhang[1]

[1]State Key Laboratory of Bioreactor Engineering, East China University of Science and Technology, Shanghai 200237, China
[2]Department of Chemical and Biological Engineering, Chalmers University of Technology, Kemivägen 10, SE-412 96 Göteborg, Sweden

Abstract

Great amounts of carbon dioxide generated by *Pseudomonas denitrificans* during high aerobic vitamin B_{12} fermentation, while the influence of CO_2 concentration on vitamin B_{12} production remains unclear. In this paper, we present parallel experiments to investigate various levels of inlet CO_2 fractions on the physiological metabolism of *P. denitrificans* in laboratory scale fermentation. The results demonstrated that the oxygen transfer rate, cell growth and glucose consumption were inhibited with CO_2 fraction elevated from 0.03% to 8.86 ± 0.24%, while the most exciting results showed that the specific vitamin B_{12} production rate and the yield to glucose were greatly stimulated when dissolved CO_2 increased to 8.86 ± 0.24%. Therefore, the optimal exhausted CO_2 fraction control strategy in 120 m^3 fermenter was established. With the exhaust CO_2 concentration was well controlled at 7.5 ± 0.25% on-line, vitamin B_{12} production greatly improved to 223.7 ± 3.7 mg/L, which was 11.2% higher than that of control. This strategy was proved to be significant necessary and effective for successfully scale up optimization in industrial vitamin B_{12} fermentation.

Keywords: Dissolved carbon dioxide; *Pseudomonas denitrificans;* Vitamin B_{12}; Fermentation; Scale up; Optimization

Introduction

Vitamin B_{12} is an important growth factor and has many applications in medicine and nutrition. Because of its increasing need in the world, vitamin B_{12} has received much more attention recently. Most of the vitamin B_{12} was produced by the culture of aerobic bacterium *P. denitrificans* in industrial production [1]. Due to the actively respiring characters of *P. denitrificans*, the poorly ventilated systems, high back pressure and hydrostatic pressure in large-scale fermenters, high dissolved CO_2 concentration were always encountered in industrial process. As for the significant influence of dissolved CO_2 on the growth and metabolism of many microorganisms [1], it must be taken into consideration in large-scale industrial fermentation.

The effects of CO_2 on growth and product formation in submerged cultures have been investigated in various microorganisms, and the controversy results were always encountered [2,3]. Researches of Gill and Lacoursiere had revealed that cell growth and metabolism of *Pseudomonas fluorescens* and *Escherichia coli* could be stimulated at low dissolved CO_2 level not more than 100 mm Hg pressure or 5% inlet gas phase respectively, while it could be greatly inhibited by the increased CO_2 concentration [4,5]. In yeasts cultivation, the growth rates were seriously decreased under CO_2-enriched condition [6,7]. For filamentous fungi, in which the secondary metabolites formation was always accompanied with morphology changes [1], it was proved that both the morphology and product formation were negatively affected by the elevated CO_2 conditions [1,8,9]. However, many other reports demonstrated that CO_2 could also have positive effect on the metabolism of some microorganisms [10], in which existed high CO_2-fix pathway activity, the anaplerotic reactions were very important for replenishing the metabolites generated from tricarboxylic acid cycle's pool, Researches had shown that enhanced CO_2 could stimulate the activity of key enzymes for the anaplerotic reaction in *Corynebacterium glutamicum* [11,12] and succinate-producing strain [13] by accelerating the accumulation of glutamate and succinate. The former studies on industrial vitamin B12 fermentation have shown that *P. denitrificans* has high affinity to oxygen, and accompanied with higher CO_2 evolution rate during fermentation [14]. However, at present, there was not clearly understood of the effect of CO_2 on metabolism of vitamin B_{12} biosynthesis.

Therefore, the aim of present study was to evaluate the effect of CO_2 on production and metabolism of vitamin B_{12} produced by *P. denitrificans*. Furthermore, an optimal industrial fermentation strategy based on online and CO_2 controlling strategy was successfully applied to enhance industrial vitamin B_{12} production.

Materials and Methods

Microorganisms and medium

The industrial strain *P. denitrificans* used in this study was donated by Huarong pharmacy corporation (Shijiazhuang, China). The cultivations were carried out at a temperature of 32°C and agitation speed of 260 rpm. The seed medium was composed of (g/L): sucrose 40, corn steep liquor 20, betaine 5, $(NH_4)_2SO_4$ 1, $(NH_4)_2HPO_4$ 2, $MnSO_4{\cdot}H_2O$ 0.8, $COCl_2{\cdot}6H_2O$ 0.02, MgO 0.3, 6-dimethyl-benzimidazole (DMBI) 0.01, $ZnSO_4{\cdot}7H_2O$ 0.01, $CaCO_3$ 1.5, the pH was adjusted to 7.2-7.4 by NaOH.

The fermentation medium *lab scale and industrial scale* was composed of (g/L): glucose 80, corn steep liquor 45, betaine 14, $(NH_4)_2SO_4$ 1, KH_2PO_4 0.75, $COCl_2{\cdot}6H_2O$ 0.075, MgO 0.5, DMBI 0.05, $ZnSO_4{\cdot}7H_2O$ 0.08, CaCO3 1, pH 7.2-7.4. Feed media (FM) for the fed-batch fermentation were as follows (g/L), FM-1: Glucose 200, DMBI 0.15, $COCl_2{\cdot}6H_2O$ 0.15 and FM-2: betaine 30, $COCl_2{\cdot}6H_2O$ 0.3, DMBI 0.3.

***Corresponding author:** Ying-Ping Zhuang, State Key Laboratory of Bioreactor, Engineering, East China University of Science & Technology, P. O. Box 329, 130 Meilong Road, Shanghai 200237, People's Republic of China

Determination of dry cell weight (DCW) and residual sugar concentration

For determination of biomass concentration, 10 ml ferment broth was centrifuged at 4,000 g for 10 min. After washing twice with distilled water, the cell precipitate was dried to a constant weight at 105°C. The residual sugar concentration was measured by anthrone method [15].

Determination of Pyruvate Carboxylase and intracellular δ-aminolevulinic acid (δ-ALA)

Vitamin B_{12} is a highly complicated molecular structure [16,17] that is biosynthesized from eight molecules of δ-aminolevulinic acid (δ-ALA). This δ- ALA precursor can be generated by either of two pathways (Figure 1): In the C4 pathway, δ-ALA is made from glycine and succinyl CoA by the action of ALA synthase [1]. The δ-ALA was determined according to the literature [18]. For determination of Pyruvate Carboxylase, 10 ml of samples were taken quickly and centrifuged at 4°C, 12,000 g for 5 min, the obtained cells were frozen in -80°C nitrogen. After several times of repeated freezing and thawing, cell pellets were washed with 20 mM Tris–HCl (pH 7.6) for ultrasonication. The sonicated cell suspension was centrifuged and the cell extract was used for the enzyme assay. Pyruvate carboxylase activity was determined by the method of Uy [19].

Measurement of organic acids and amino acids

For the analysis of the extracellular organic acids and amino acids

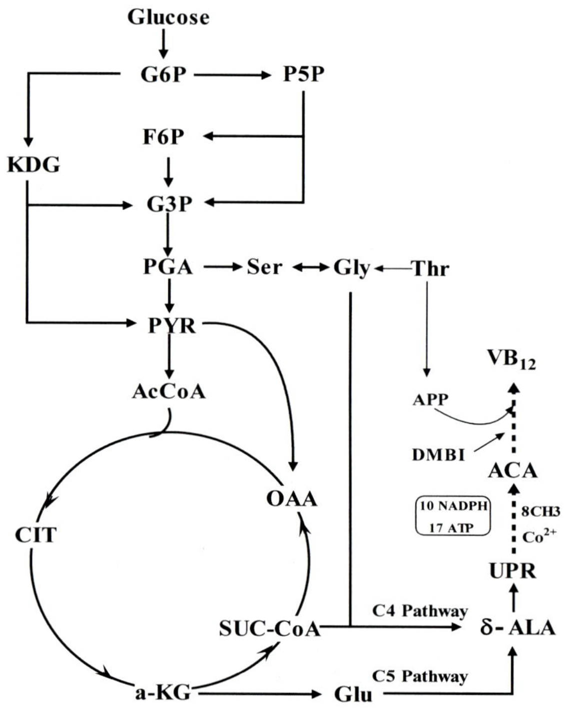

Figure 1: Proposed overview of the pathway of vitamin B12 biosynthesis by *P. denitrificans* connects with central metabolism. Abbreviations: ACA, adenosyl cobyric acid; AcCoA, acetyl-CoA; APP, aminopropanol-o-2-phosphate; CIT, citrate; DMBI, dimethyl benzimidazole; F6P, fructose-6-phosphate; Gly, glycine; Glu, glutamate; G3P, glyceraldehyde-3-phosphate; G6P, glucose-6-phosphate; KDG, 2-Keto-3-deoxy-6-phosphogluconate; OAA, oxaloacetate; P5P, pentose-5-phosphate; PGA, 3-phosphate-glycerate; PYR, pyruvate; Ser, serine; SUC-CoA, succinyl-CoA; Thr, threonine; UPR, uroporphyrinogen III; α-KG, α-ketoglutarate; δ-ALA, δ-aminolevulinic acid.

Parameters	
Diameter (T, mm)	3,800
Bottom type	Dished
Filled volume (VL, L)	105,000
Capacity (V, L)	120,000
Baffle length (L, mm)	7600
Baffle width (w, mm)	15
Gas pipe inlet vent diameter (d, mm)	125
Impellers:	
6-Blade concave disc turbine impellers (D, mm)	1,150
3-Blade Rushton disc turbine impellers (D, mm)	1,219
3-Curved-blade disc turbine impellers (D, mm)	1,219
Standard pitched blade disc turbine impellers (D, mm)	1,219

Table 1: Geometric parameters of the 120 m^3 fermenter.

in metabolism, the HPLC systems (Agilent 1100, USA) were equipped with an Aqua Sep C18 (250 × 4.6 mm, 5 µm, ES, USA) and Zorbax Eclipse AAA column (150 × 4.6 mm, 5 µm, Agilent, USA) respectively, the determination conditions were the same to that of previously reported [14].

Quantification of vitamin B_{12} in the broth

The vitamin B_{12} concentration was determined by HPLC. Broth samples (25 ml), into which 2.5 ml of 8% $NaNO_2$ and 2.5 ml of glacial acetic acid were added, were boiled for 30 min. Then the upper aqueous phase was measured using HPLC system equipped with C_{18} column (4.6 mm, 25 cm, 5 µm) and UV detector (361 nm). 5% of acetonitrile was used as the mobile phase with flow rate of 1.0 ml min^{-1} at 25°C combined with 95% of 0.25 M sodium acetate anhydrous, the pH was adjusted to 3.6 by acetic acid.

Fermentation process design

Firstly, fed-batch fermentation was carried out in a 50 L turbine-agitated bioreactor (Shanghai Guoqiang Inc., China) with 30 L working volume. The cultivation conditions were same to that of previous report. When the cultivation turned into higher vitamin B_{12} biosynthesis phase, 2.5 L broth was incubated into four parallel 5 L bioreactors respectively (Shanghai Guoqiang Inc., China) for further investigation. Fermentations were performed at 32°C, stirrer speed was 450 rpm, and the pressure in the tar was all kept at 0.1 MP. The inlet gas stream was consisted of a defined mixture of air and pure carbon dioxide with the aeration rate at 1 vvm. The mixture was prepared by thermal mass flow controllers (Mass-Trak, SIERRA, Netherlands). In addition to the control process where air was the input gas, processes were also performed with 3.32 ± 0.12%, 8.86 ± 0.24%, and 13.84 ± 0.27% CO_2 addition (v/v air) in the inlet gas mixture. Meanwhile, the dissolved CO_2 concentration in fermentation was determined with using dissolved carbon dioxide concentration electrode (Mettler).

Fermentation in 120 m^3 fermenter

Three-stage fermentation was performed in 150 L, 9 m^3, and 120 m^3 fermenters gradually, involving two stages of seed growth and one stage of fed-batch vitamin B_{12} fermentation. The primary seed, secondary seed and large-scale fermentation were cultivated with the methods as previous reported [14]. The geometry and dimensions of 120 m^3 fermenter tanks and agitators were given in Table 1.

The inlet and exhaust gas ingredients were analyzed by a mass spectrometer (MAX300-LG, Extrel). Oxygen Uptake Rate (OUR) and Carbon dioxide Evolution Rate (CER) was calculated and collected online Oxygen Uptake Rate (OUR), and were controlled by adjusting agitation and aeration conditions.

Carbon dioxide control strategy model in 120 m^3 fermenter

The culture medium always enriched with CO_2 due to the high evolution rate of growing cells during fermentation process [20], especially in industrial scale fermenters with high hydrostatic pressures, high cell density or high pressure [21,22]. Previous researches revealed that *P. denitrificans* has high affinity to oxygen, with the dissolved oxygen concentration maintained nearly zero, therefore the oxygen transfer rate (OTR) was equal to OUR during the steady fermentation process. The OTR was determined by the agitation, aeration and the pressure, so the equation was used to create their relationship. Meanwhile, the active respiration make the CO_2 levels rise rapidly in large scale fermenters. In order to implement online control the oxygen supply levels and exhausted carbon dioxide concentration, we establish a response model about the OTR and exhausted CO_2 concentration related to aeration and agitation

$$OUR \approx OTR = m \times F^{\alpha} \times R^{\beta} \times P^{r} \text{ (Eq.1)}$$

Where m=correction coefficient; α, β, γ=response coefficient of flow, agitation and pressure relate to oxygen transfer rate; OUR= oxygen uptake rate (mmol L/h); OTR=oxygen transfer rate (mmol L/h); F=air flow (L/min); P=pressure (MPa); R=agitator speed (rpm).

Moreover, the relation between oxygen uptake rate and carbon dioxide evolution rate has presented by RQ which shown in Equation 2. In addition, equation 3 shows online determination formula of carbon dioxide evolution rate.

$$RQ = \frac{CER}{OUR} \quad \text{(Eq.2)}$$

$$CER = \frac{F}{V}\frac{(E_{CO_2} - 0.03)}{22.4} \quad \text{(Eq.3)}$$

Results and Discussion

Effect of CO_2 on the dynamic of vitamin B_{12} fermentation

For investigating the effects of dissolved CO_2 levels on vitamin B_{12} biosynthesis, the stationary cell growth broth in 50 L fermenter was transferred to the four 5 L reactors with 2.5 L working volume simultaneously. The fermentations were performed with operating conditions held constant to ensure that dissolved carbon dioxide was the only major variable. Samples were taken individually for analysis of biomass, glucose, inter-metabolites, and vitamin B_{12} concentrations; and the data points represent mean values from duplicate experiments. A gas-mixing unit was used to give input CO_2 concentration in the influent gas of 0.03%, 3.32 ± 0.12%, 8.86 ± 0.24%, and 13.84 ± 0.27% respectively.

The time course of cell growth and pH under the four different CO_2

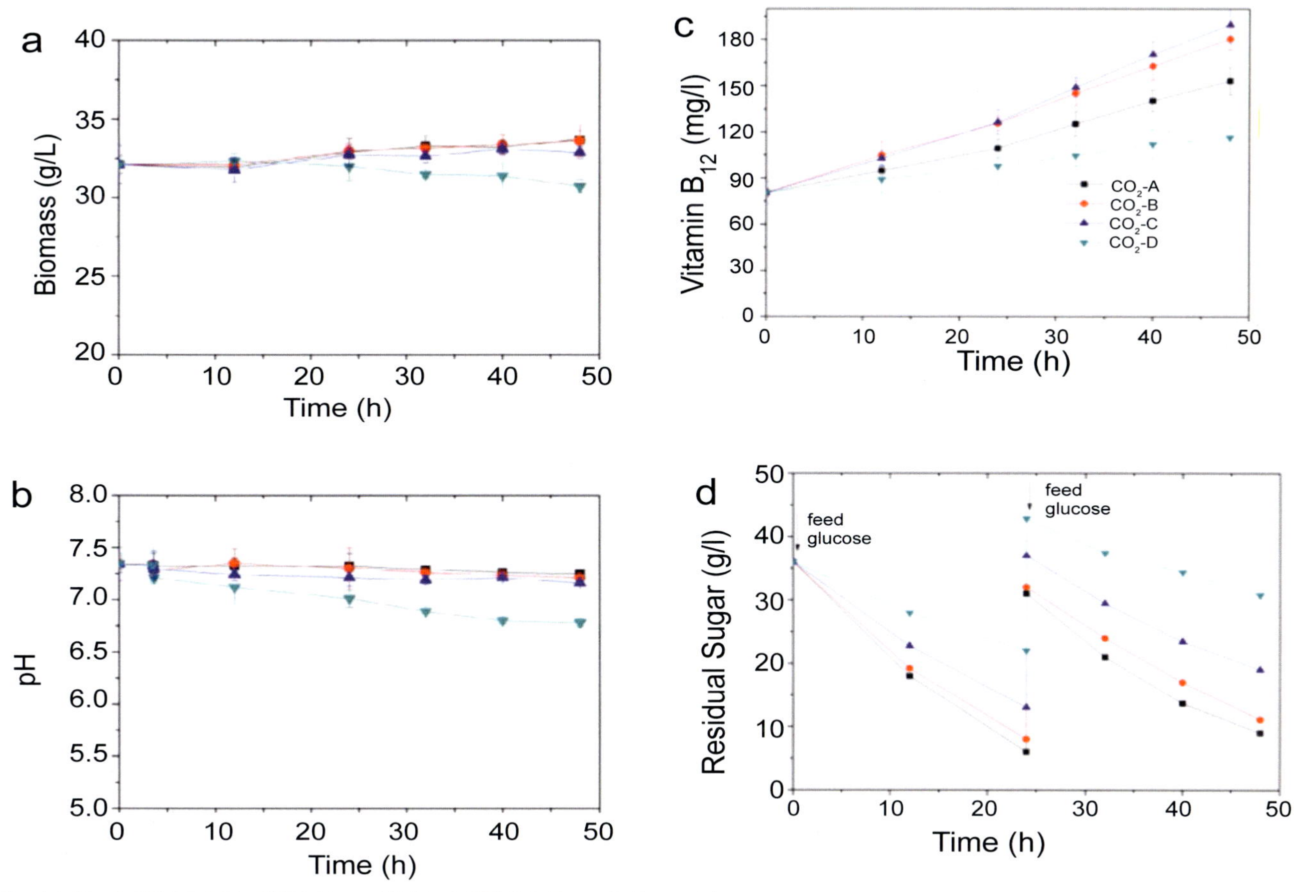

(Air, ■), 3.32 ± 0.12% (●), 8.86 ± 0.24% (▲), and13.84 ± 0.27% ( ) in 5L parallel fermentaters

Figure 2: The effect of different carbon dioxide content in the inlet gas stream on biomass and PH. Symbols: the inlet CO_2 concentrations were controlled at 0.03%.

addition were illustrated in Figure 2. During those four conditions, the dissolved oxygen concentration were all maintained at nearly zero levels, while the cell growth haven't any effect when the CO_2 addition elevated from 0.01% to 8.86 ± 0.24%; However, with influent CO_2 concentration increased to 13.84 ± 0.27%, the cell growth rate was greatly inhibited and biomass concentration decreased to 30.7 g/L after 48h cultivation, which was 8.9% lower than that of control (33.68 g/L) (Figure 2a). These might be caused by the respiration inhibition of CO_2 on biological membranes and cytoplasmic enzymes [23], and which may also interact with lipids of the cell membrane, the higher CO_2 concentration would greatly inhibit cell growth [1,9]. It could be clear seen from Figure 2b that pH had hardly any different to the control when CO_2 addition increased to 8.86 ± 0.24%, while the pH was remarkably decreased when the influent CO_2 elevated to 13.84 ± 0.27%.

The effect of influent CO_2 on vitamin B_{12} production rate, glucose consumption rate and specific carbon dioxide evolution rate were summarized in Table 2. Results revealed the inhibitory effects of CO_2 concentration on the specific glucose consumption rate (SGCR). Little decrease of SGCR was appeared with CO_2 addition increased to 3.32 ± 0.12%; However, the SGCR were dramatically decreased to 0.14 and 0.10 mmol/gDCW/h when the influent CO_2 increased to 8.86 ± 0.24% and 13.84 ± 0.27% respectively, which were 19.2% and 46.7% lower than that of control (0.18 mmol/gDCW/h).

Just different from cell growth and SGCR, vitamin B_{12} productions were greatly stimulated when influent CO_2 increased from 0.03% to 8.86 ± 0.24% (Figure 2c). The highest specific vitamin B_{12} production rate reached to 69.4 μg/gDCW/h under 8.86 ± 0.24% CO_2 addition, 53.5% higher than that of control (45.2 μg/gDCW/h). While vitamin B_{12} productions and specific production rate were all significantly inhibited when the CO_2 concentration raised to 13.84 ± 0.27%. Analysis of vitamin B_{12} yields to the glucose consumption ($Y_{p/s}$) (Figure 2d) showed that the maximum yield of 480.5 μg/mmol (vitamin B_{12}/ glucose consumption) were obtained when the influent CO_2 reached 8.86 ± 0.24%, while 13.84 ± 0.27% CO_2 addition led the yields dramatically decreased to lowest levels of 248.3 μg/mmol. Exhaust gas analysis showed that specific

CO_2 fraction	%	0.03	3.32 ± 0.12	8.86 ± 0.24	13.84 ± 0.27
CO_2 *conc.*	mmol/L	0.001 ± 0.0006	1.16± 0.07	3.05 ± 0.09	4.81 ± 0.12
P_{ra}	ug/L/h	1521.9	2164.2	2279.1	750.0
***SPR*[b]**	ug/ g(DCW)/h	45.2 ± 0.9	64.4 ± 2.1	69.4 ± 1.2	23.7 ± 1.6
***SGCR*[c]**	mmol/ g(DCW)/h	0.18	0.17	0.14	0.10
$Y_{p/s}$	ug/mmol	252.9	382.4	480.5	248.3
***SCER*[d]**	mmol/ g(DCW)/h	1.03	0.90	0.73	0.54
CO_2/Glucose	C(mol)/ C(mol) %	94.6 ± 2.1	89.2 ± 1.9	85.6 ± 1.5	93.3 ± 1.3

[a], P_r: Vitamin B_{12} production rate
[b], *SPR:* specific vitamin B_{12} production rate
[c], *SGCR*: specific glucose consumption rate
[d], *SCER:* specific carbon dioxide evolution rate

Table 2: Yield factors of vitamin B12 fermentation under various influent CO_2 concentration for 48h with the cultivations of stationary phase biomass transfered from a well cultured 50L fermentor.

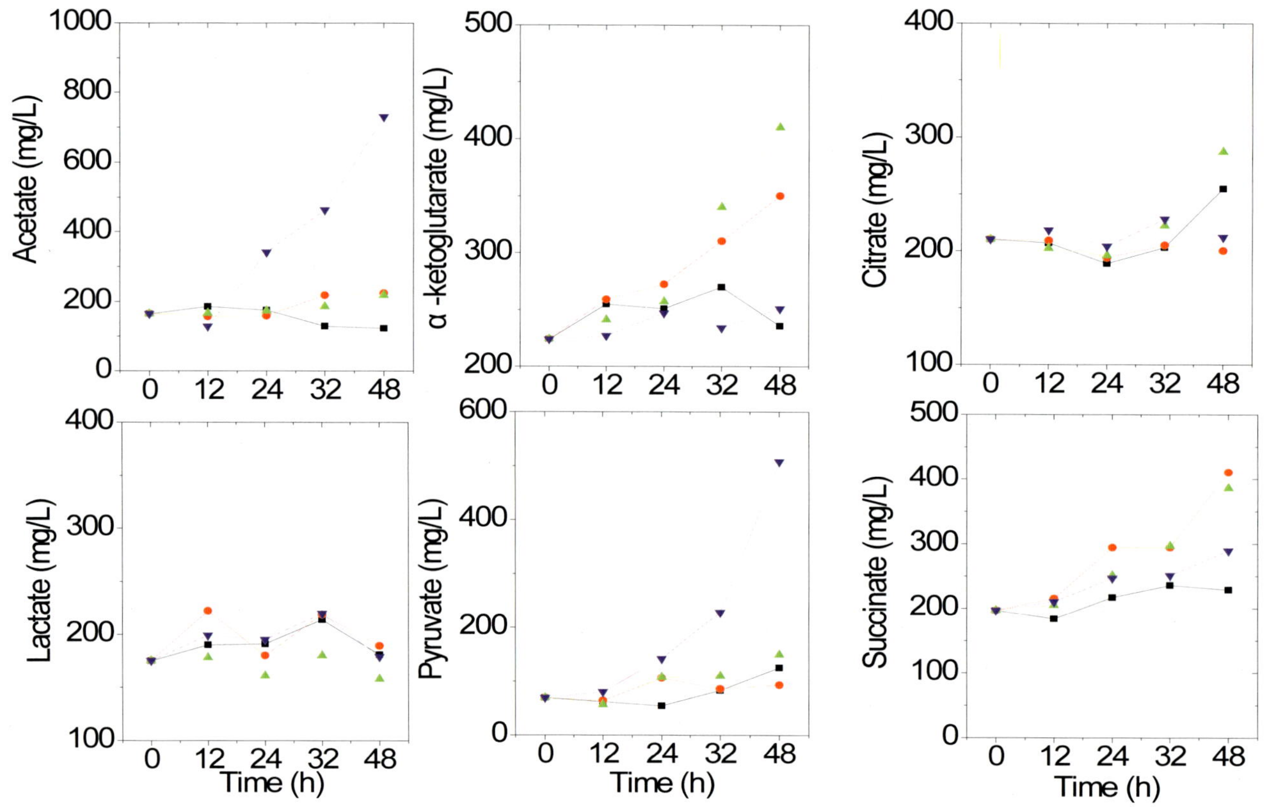

(Air, ■), 3.32 ± 0.12%(●), 8.86 ± 0.24%(▲), and 13.84 ± 0.27%( ) in 5L parallel fermenters

Figure 3: Influence of CO_2 content in the inlet gas stream on extracellular organic acids. Symbols: the inlet CO_2 concentrations were controlled at 0.03%.

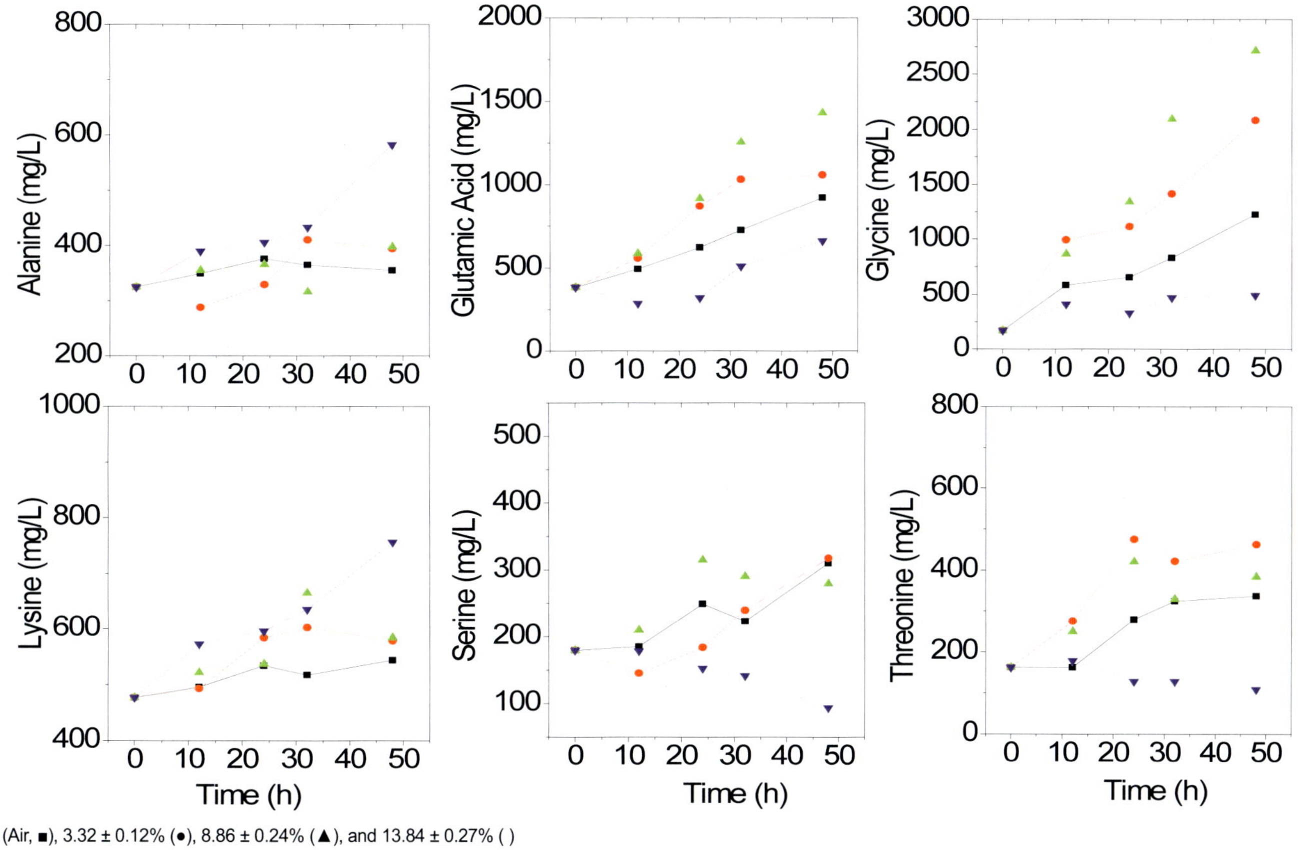

Figure 4: Effect of dissolved carbon dioxide concentration on metabolism of extracellular amino acids in 5L parallel fermenters. Symbols: the inlet CO_2 concentrations were controlled at 0.03%.

carbon dioxide evolution rate (SCER) of *P. denitrificans* decreased with elevated CO_2 concentration in input gas. The statistics of the percent of glucose used for CO_2 generation were decreased from 94 ± 2.1% to 85 ± 1.5% with influent CO_2 increased from 0.03% to 8.86 ± 0.24%; while under the highest CO_2 concentration of 13.84 ± 0.27%, nearly 93.3 ± 1.3% of the glucose consumption was seriously transformed to CO_2. In conclusion, maintained an appropriate input CO_2 concentration around 8.86 ± 0.24% was positive for the economical vitamin B_{12} production.

Previous researches on the effects of elevated CO_2 on the industrial producing strains of bacteria, yeasts, and fungi [1,9] revealed that the inhibitory action of product biosynthesis was probably caused by the decreased substrate consumption with elevated CO_2 concentration. In our research, however, glucose consumption decreased with the increased influent CO_2 concentration, while the yields of vitamin B_{12} based on glucose consumption increased greatly with the elevated influent CO_2 concentration. These results indicated that there must exist abnormal metabolism characters of *P. denitrificans* under the various influent CO_2 concentrations. The reasons were further investigated by organic acid and amino acid analysis of the central metabolism.

Effect of CO_2 on the inter-metabolites in fermentation

The researches on vitamin B_{12} biosynthesis had demonstrated that the concentrations of the precursor of δ-aminolevulinic acid (δ-ALA) played an important role on vitamin B_{12} biosynthesis [16]. δ-ALA could be synthesized from C4-pathway (from glycine and succinyl CoA by δ-ALA synthesis) or C5-pathway (from glutamate by glutamyl-tRNA reductase) associated with central metabolism [24]. The dynamic processes of extracellular organic acids concentration under different CO_2 levels were shown in Figures 3 and 4.

When influent CO_2 increased from 0.03% to 8.86 ± 0.24%, α-ketoglutarate and succinate concentrations greatly increased, which were actually accelerated the biosynthesis of glutamate, and the other organic acids such as pyruvate, acetate, lactate and citrate hadn't any different. However, the acetate and pyruvate were accumulated much higher when influent CO_2 reached 13.84 ± 0.27%. These were probably caused by the inhibition of respiration under high CO_2 concentration, and the resulted high NADH accumulation in the cell really repressed the activity of pyruvate dehydrogenase and TCA cycle, and then attributed to the accumulation of acetate and pyruvate.

The dynamic analysis on the concentrations of extracellular amino acids (Figure 3) showed that glutamate and glycine synthesis rate was significantly enhanced with elevated CO_2 concentration. Under 8.86 ± 0.24% CO_2 addition, glutamate and glycine concentration reached 1430.2 mgL^{-1} and 2716.4 mg/L, respectively, which were 55% and 122% higher than that of control. Threonine reached highest synthesis rate when inlet CO_2 increased to 3.32 ± 0.12%. These metabolites accumulation could be accelerate the precursor's biosynthesis for vitamin B_{12} synthesis.

δ-ALA is the critical precursors of vitamin B_{12} biosynthesis, the time course of δ-ALA concentration under different levels of inlet CO_2

were shown in Figure 5a, experiment results demonstrated the positive effects of moderate CO_2 concentration exerted to the δ-ALA synthesis. In bioprocess sparged with 8.86 ± 0.24% CO_2 addition, the δ-ALA concentration reached the top point at 26.4 ± 1.2 mg/gDCW, which was 40.4% higher compared to that of control (18.6 ± 0.9 mg/gDCW). However, the δ-ALA biosynthesis was seriously inhibited with13.84 ± 0.27% CO_2 addition, the final δ-ALA concentration was lowest as 13.02 ± 0.5 mg/gDCW, 30% lower than that of control.

Experiment results showed that the biosynthesis of glycine, succinic acid, and glutamate were elevated with the increased influent CO_2 concentration from 0.03% to 8.86 ± 0.24%, which actually augmented δ-ALA generation for higher vitamin B_{12} biosynthesis.

Effect of CO_2 on pyruvate carboxylase activity

Previous researches have illustrated that anaplerotic reaction catalysed by pyruvate carboxylase was mostly existed for replenishing the TCA cycle in *Pseudomonas* spp [25]. In this work, the dynamic analyses of specific pyruvate carboxylase activity in different input CO_2 addition were shown in Figure 5b. In the process sparged with 8.86 ± 0.24% CO_2, the maximal specific activity reached 37.2 U/gDCW, which was 35% higher than that of control (27.5 U/gDCW).

Tricarboxylic acid (TCA) cycle is one of the main central pathways in aerobic bacteria. It is responsible for the complete oxidation of acetyl-CoA and provision of precursors for many TCA-related metabolites biosynthesis. As precursors of the metabolites succinyl-CoA and glutamate for δ-ALA biosynthesis were all generated from

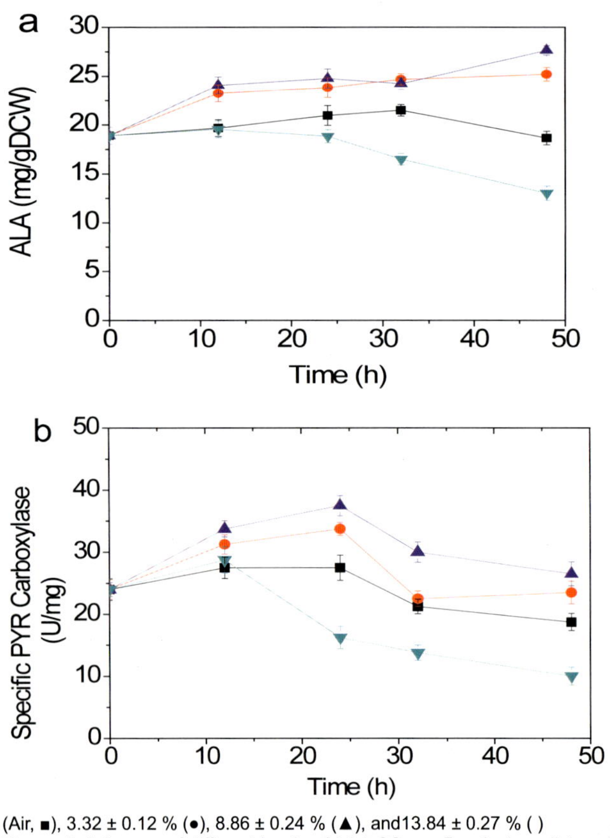

(Air, ■), 3.32 ± 0.12 % (●), 8.86 ± 0.24 % (▲), and13.84 ± 0.27 % ( )

Figure 5: Influence of different levels of inlet CO_2 on δ-aminolaevulinic acid and specific pyruvate carboxylase activity in 5L parallel fermentaters. Symbols: the inlet CO2 concentrations were controlled at 0.03%.

F(L/min)	R(rpm)	P(MPa)	OUR (mmol/L/h) Determined	Predicteda
1263.1	78	0.031	47.38	45.71
929.9	78	0.034	40.66	41.77
406.4	50	0.031	24.73	24.20
600.5	90	0.041	39.10	39.72
850.4	78	0.032	37.51	39.80
1523.2	110	0.040	63.2	61.68
1224.5	81	0.032	46.07	46.41
746.3	79	0.029	36.65	37.23
960.0	55	0.035	38.42	36.21
959.4	78	0.033	40.77	41.95
709.2	77	0.028	36.74	35.77
625.5	81	0.033	42.58	36.39
1121.8	78	0.027	43.82	42.32

a calculated from the obtained equation 2 with the corresponding process parameters

Table 3: The determined and predicted OUR values at various aeration, agitation and pressure levels in the 120 m³ bioreactor.

TCA cycle. In the rapid VB_{12} synthesis phase, the relative high demand of succinyl-CoA and glutamate lead to an "overflow" of corresponding compounds of TCA cycle. Therefore, TCA cycle has to be replenished continuously through anaplerotic reaction for oxalacetate generation to maintain the cycle running.

The experiment results demonstrated that elevated levels of CO_2 concentration had significant effect on vitamin B_{12} fermentation. When the influent CO_2 concentration increased from 0.03% to 8.86 ± 0.24%, the specific vitamin B_{12} production rate and the yields of vitamin B_{12} to glucose were stimulated. However, cell growth, glucose utilization, specific vitamin B_{12} production rate, and the yields of vitamin B_{12} to glucose were all greatly inhibited when CO_2 addition up to 13.84 ± 0.27%. Therefore, the influences of CO_2 on industrial vitamin B_{12} production have to take into consideration.

Industrial fermentation strategy based on online exhaust CO_2 controlling

Exhaust CO_2 concentration controlling strategy: With the process parameters determined under various agitation speed, air flow, and pressure obtained from 120 m³ fermentation tank are shows in Table 3, we obtained relevant respond coefficient through regression analysis of these parameters between OTR and Process parameters determined (Eq.1). Then, oxygen transfer rate and oxygen uptake rate could be calculated according to these parameters (Eq.4).

$$OUR \approx OTR = 1.0074 \times F^{0.3729} \times R^{0.4635} \times P^{0.2472} \qquad \text{(Eq.4)}$$

The carbon dioxide fraction in exhausted gas was calculated from a mutual relation between RQ, CER, and OUR (Eq. 2,3,5)

$$Eco_2 = RQ \times F^{-0.36271} \times R^{0.4635} \times P^{0.2472} \times V \times 22.57 + 0.03 \qquad \text{(Eq.5)}$$

During the vitamin B_{12} fermentation process, glucose was used as the main carbon substrate, the respiration quotient (RQ) of this strain was always maintained at 1.03 ± 0.02 in microorganism growth period and RQ was calculated as 0.95 ± 0.04 in vitamin B_{12} synthesize period. Then OUR and exhausted CO_2 concentration from 120 m³ fermentation tank could be obtained by adjusting agitator speed and air flow according to the created model above (Figure 6).

Scale-up fermentation based on exhaust CO_2 concentration control: Based on the results obtained from 5 L vitamin B_{12} fermentation, the optimal controlling strategies based on exhausted

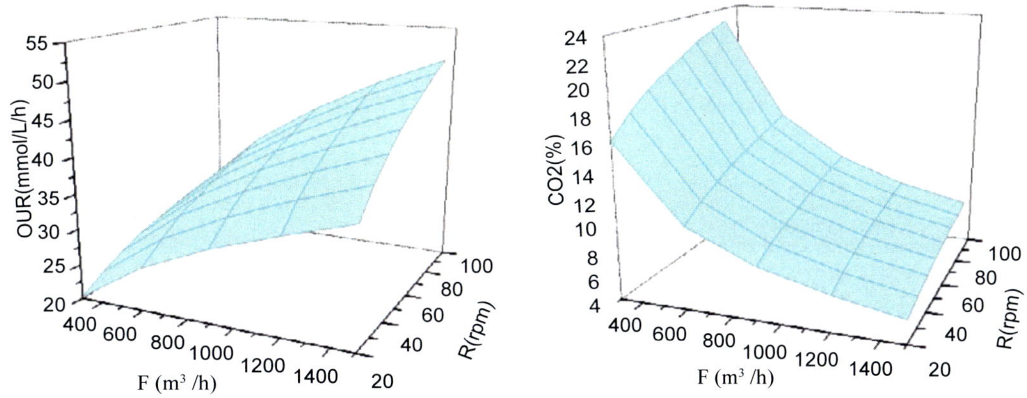

Figure 6: The response of oxygen uptake rates and exhaust CO_2 concentrations as a function of aeration and agitation in120 m^3 large scale fermenter (based on the obtained equation4 and equation 5.

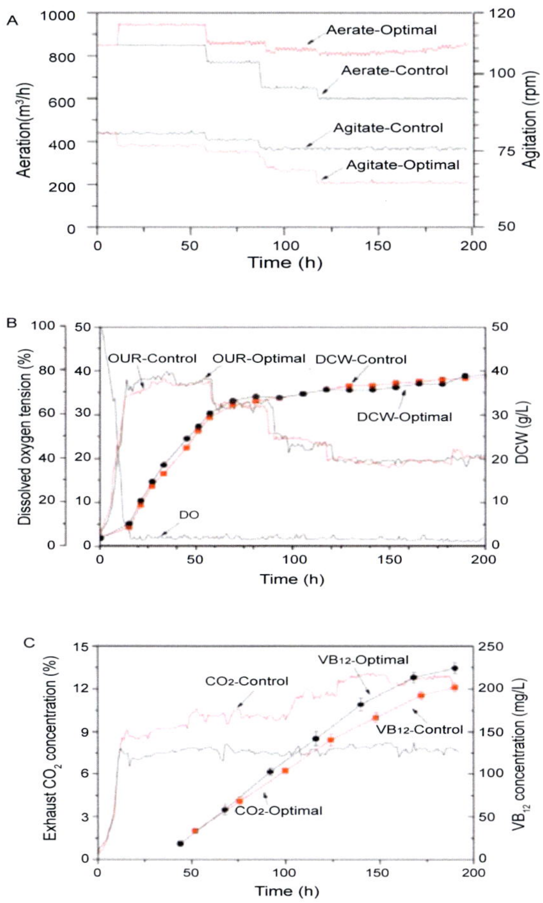

Figure 7: Time course of industrial VB12 fermentation with the exhaust CO_2 concentration controlled at throughout at 7.5 ± 0.25 %, (optimal) and the control without exhaust CO_2 controlled (control) in 120 m^3 fermenters and the effect of CO_2 concentration on DCW (b) and VB_{12} concentration (c) by correlation adjustment of oxygen supply conditions (a).

CO_2 concentration were implemented for vitamin B_{12} fermentation in 120 m^3 fermenters. The exhausted CO_2 concentration was controlled at 7.5 ± 0.25% by adjusting aeration and agitator speed, also OUR was maintained at needed level. This strategy greatly prevented the inhibition of high dissolved CO_2 concentration on vitamin B_{12} synthesis.

Time course of OUR, dry cell weight, vitamin B_{12} production, and exhausted CO_2 concentration were shown in Figure 7. By control the aeration and agitation throughout the fermentation process, the changes of OUR were maintained at similar levels in fermentation processes (Figure 7b), the optimal strategy by stepwise reduction of oxygen uptake rate under dissolved oxygen limiting level during fermentation process was used for oxygen supply levels control [11]. Meanwhile, the exhausted CO_2 concentration changed greatly and maintained at the suitable levels with the adjustment of agitation and aeration conditions. In the contrast conditions, the exhaust CO_2 concentration increased greatly to 9.1 ± 0.15%, cell growth rate was lower than that of optimal strategy, especially when it elevated to 12.0 ± 0.24% after 75 h cultivation, the vitamin B_{12} production rate was gravely inhibited (Figure 7c).

The results presented demonstrated that cell growth and vitamin B_{12} production were greatly stimulated by controlling a lower CO_2 concentration of 6.0 ± 0.24% during the early fermentation phase. The highest vitamin B_{12} production rate was achieved when exhausted CO_2 concentration were controlled at 7.5 ± 0.25%, The maximum vitamin B_{12} production reached 223.7 ± 3.7 mg/L at 190 h, which was 11.2% higher than that of without CO_2 concentration controlled (201.4 ± 2.9 mg/L). It can be observed that higher production would be realized through simultaneously regulate the appropriate OUR and CO_2 concentration levels are more beneficial to vitamin B_{12} synthesis.

Conclusion

CO_2 was produced in aerobic and anaerobic fermentations through decarboxylation reactions in a number of metabolic pathways, and hence has an influence on the performance of microbial cultivations [1,9]. Especially in industrial large bioreactors with high total pressure, poorly ventilated system, and inefficiently aerated fermenters. In this work, we investigated the effects of dissolved CO_2 concentration on vitamin B_{12} fermentation of *P. denitrificans,* established the correlation of exhausted CO_2 concentration and the activity of the microbial metabolism by inter-scale observation and data association. The novel and optimal fermentation controlling strategy based on OUR and exhausted CO_2 concentration was successfully applied in industrial scale up of vitamin B_{12} fermentation.

Acknowledgement

This work was financially supported by a Grant from the National Natural Science Foundation of China (Grant No. 31200024), Hebei Key Technology R&D Program (No. 13272803D), the Major State Basic Research Development Program of China (973 Program), No. 2012CB721000, and Doctoral fund No. 20110074110015. We also thank Huarong Pharmacy Corporation (Shijiazhuang, China) for donating the industrial strain *P. denitrificans*.

References

1. El-Sabbagh N, McNeil B, Harvey LM (2006) Dissolved carbon dioxide effects on growth, nutrient consumption, penicillin synthesis and morphology in batch cultures of Penicillium chrysogenum. Enzyme and Microbial Technology 39: 185-190.
2. Onken U, Liefke E (1989) Effect of total and partial pressure (oxygen and carbon dioxide) on aerobic microbial processes. Adv Biochem Eng Biotechnol 40: 137-169.
3. Kuriyama H, Mahakarnchanakul W, Matsui S, Kobayashi H (1993) The effects of pCO2 on yeast growth and metabolism under continuous fermentation. Biotechnol Lett 15: 189-194.
4. Gill CO, Tan KH (1979) Effect of carbon dioxide on growth of Pseudomonas fluorescens. Appl Environ Microbiol 38: 237-240.
5. Lacoursiere A, Thompson B, Kole MM, Ward D (1986) Effects of carbon dioxide concentration on anaerobic fermentations of Escherichia coli. Applied Microbiology and Biotechnology 23: 404-406.
6. Chen SL, Gutmanis F (1976) Carbon dioxide inhibition of yeast growth in biomass production. Biotechnology and Bioengineering 18: 1455-1462.
7. Jones RP, Greenfield PF (1982) Effect of carbon dioxide on yeast growth and fermentation. Enzyme Microb Technol 4: 210-223.
8. McIntyre M, McNeil B (1997) Dissolved carbon dioxide effects on morphology, growth, and citrate production in Aspergillus niger A60. Enzyme and Microbial Technology 20: 135-142.
9. El-Sabbagh N, Harvey LM, McNeil B (2008) Effects of dissolved carbon dioxide on growth, nutrient consumption, cephalosporin C synthesis and morphology of Acremonium chrysogenum in batch cultures. Enzyme and Microbial Technology 42: 315-324.
10. Ho SH, Kondo A, Hasunuma T, Chang JS (2013) Engineering strategies for improving the CO_2 fixation and carbohydrate productivity of Scenedesmus obliquus CNW-N used for bioethanol fermentation. Bioresource Technology 143: 163-171.
11. Rados D, Turner DL, Fonseca LL, Carvalho AL, Blombach B, et al. (2014) Carbon Flux Analysis by 13C Nuclear Magnetic Resonance To Determine the Effect of CO_2 on Anaerobic Succinate Production by Corynebacterium glutamicum. Appl Environ Microbiol 80: 3015-3024.
12. Baumchen C, Knoll A, Husemann B, Seletzky J, Maier B, et al. (2007) Effect of elevated dissolved carbon dioxide concentrations on growth of Corynebacterium glutamicum on D-glucose and L-lactate. J Biotechnol 128: 868-874.
13. Lu S, Eiteman MA, Altman E (2009) Effect of CO_2 on succinate production in dual-phase Escherichia coli fermentations. J Biotechnol 143: 213-223.
14. Wang ZJ, Wang HY, Li YL, Chu J, Huang MZ, et al. (2010) Improved vitamin B_{12} production by step-wise reduction of oxygen uptake rate under dissolved oxygen limiting level during fermentation process. Bioresource Technology 101: 2845-2852.
15. Jermyn MA (1975) Increasing the sensitivity of the anthrone method for carbohydrate. Anal Biochem 68: 332-335.
16. Warren MJ, Raux E, Schubert HL, Escalante-Semerena JC (2002) The biosynthesis of adenosylcobalamin (vitamin B_{12}). Nat Prod Rep 19: 390-412.
17. Raux E, Lanois A, Warren MJ, Rambach A, Thermes C (1998) Cobalamin (vitamin B_{12}) biosynthesis: identification and characterization of a Bacillus megaterium cobI operon. Biochem J 335: 159-166.
18. Kim JN, Yun JS, Ryu HW (2001) Simultaneous determination of delta-aminolevulinic acid, porphobilinogen, levulinic acid and glycine in culture broth by capillary electrophoresis. J Chromatogr A 938: 137-143.
19. Uy D, Delaunay S, Engasser JM, Goergen JL (1999) A method for the determination of pyruvate carboxylase activity during the glutamic acid fermentation with Corynebacterium glutamicum. Journal of Microbiological Methods 39: 91-96.
20. Gros JB, Dussap CG, Catte M (1999) Estimation of O2 and CO_2 Solubility in Microbial Culture Media. Biotechnol Prog 15: 923 927.
21. Bylinkina ES, Nikitina TS, Biryukov VV, Cherkasova ON (1973) Effect of dissolved carbon dioxide on life activity of antibiotic-producing microorganisms. Biotechnol Bioeng Symp 2: 197-207.
22. Jones RP, Greenfield PF (1982) Effect of carbon dioxide on yeast growth and fermentation. Enzyme Microb Technol 4: 210-223.
23. Dixon NM, Kell DB (1989) The inhibition by CO_2 of the growth and metabolism of micro-organisms. J Appl Bacteriol67: 109-136.
24. Piao Y, Yamashita M, Kawaraichi N, Asegawa R, Ono H, et al. (2004) Production of vitamin B_{12} in genetically engineered Propionibacterium freudenreichii. J Biosci Bioeng 98: 167-173.
25. Fuhrer T, Fischer E, Sauer U (2005) Experimental Identification and Quantification of Glucose Metabolism in Seven Bacterial Species. J Bacteriol 187: 1581-1590.

Heavy Metals Biosorption in Liquid Solid Fluidized Bed by Immobilized Consortia in Alginate Beads

R Ilamathi[1]*, GS Nirmala[2] and L Muruganandam[3]

[1]*Department of Biotechnology, SNIST, Yamnampet, Hyderabad, Andhra Pradesh, India*
[2,3]*Chemical Engineering Division, School of Mechanical and Building Sciences, Vellore Institute of Technology, Vellore, India*

Abstract

The adsorptive removal chromium, nickel, copper and cadmium by alginate beads containing a mixed consortium of Yeast, *Pseudomonas aeruginosa, Bacillus subtilis* and *Escherichia coli* in batch and fluidized bed column reactor was investigated. Under optimized conditions (pH 4.5; contact time 3hrs; initial metal concentration of 150mg/L) batch experiments showed that the immobilized mixed culture was successfully used for the removal of these metal ions in waste water. Fluidized bed studies were carried out in with an adsorbent dosage of 1g/L, a flow rate of 132 LPH, a bed height of length of the reactor. Efficiency of biosorption for copper, cadmium, chromium and nickel was found to be 84.62%, 67.17%, 49.25% and 61.02%. Desorption of the exhausted beads was found to be successful, however with a reduced biosorption capacity.

Keywords: Liquid solid fluidized bed; Immobilization; Heavy metals; Biosorption

Introduction

One of the most challenging environmental problems is the removal of heavy metals and other toxic contaminants from industrial wastewater of the important metals, Mercury, lead, cadmium, Arsenic and Chromium (VI) are regarded as toxic; whereas, others such as copper, nickel, cobalt and zinc are not as toxic, but their extensive usage and increasing levels in the environment are of serious concerns [1,2,3]. Several methods are being used for the removal of heavy metals ions from aqueous wastes (Chemical Precipitation, Ion Exchange, Electrochemical Treatment, Membrane Technologies, adsorption on activated Carbon. etc. [4]. Each of these methods has its own merits and demerits. But the search for new eco-friendly and cost-effective technology for the removal of heavy metals from wastewaters has been directed towards biosorption.

Biosorption using potential metal biosorbents like algae, bacteria, fungi, and yeast can be an effective technique to decrease the concentration of heavy metal ions in solution [5]. Reduction of hexavalent chromium Cr(VI) to Cr(III) by bacteria such as Pseudomonas aeruginosa [6], Bacillus sp. [7] and Escherichia Coli [8] is already reported. However, application of free bacterial cells at industrial scale is disadvantageous due to the difficulty of biomass/effluent separation [9] etc., which may be overcome by using immobilized bacterial cells with the advantages of stability, regeneration, solid–liquid separation and minimal clogging in continuous systems [10]. Immobilization of microorganisms in a suitable matrix like polyvinyl alcohol, agar media and sol–gel materials has been proven to be an efficient solution to this problem [11,12].

Adsorption processes are traditionally carried out in fixed beds [13] due to the high concentration of solids and the obtainable uniform residence time. However since the wastewater to be treated often contains solid impurities leading to a plugging of the fixed bed, the liquid must be clear to avoid column blocking. Recently, many experimental studies have been conducted in fluidized beds, which allow treatment of turbid liquids while avoiding the channeling problems [14]. Fluidized beds are common and important reactors in process engineering because of the good mass and heat transfer rate between the fluid and the particles, and between the particles and the side wall of the column. The term fluidization is used to describe the condition of fully suspended particles. Liquids or gases are passed at certain velocity up through a bed of solid particles, at this velocity the pressure drop across the bed counter balances the force of gravity on the particles and further increase in velocity achieve fluidization at a minimum fluidization velocity. Fluidization quality is closely related to the intrinsic properties of particles, e.g. particle density, particle size and size distribution, and also their surface characteristics [15]. From the previous literature related to the biosorption of heavy metal using bacteria, it can be concluded that there is a lack in literature of using immobilized bacterial consortia in liquid solid fluidized bed to study the behavior of biosorbents and its efficiency in heavy metal adsorption. Hence, this work aims to study the adsorption of heavy metals like chromium, nickel, copper and cadmium in liquid solid fluidized bed using immobilized sol-gels as a solid catalyst containing mixed cultures of Yeast, *Pseudomonas aeruginosa, Bacillus subtilis and Escherichia coli.*

Materials and Methods

Materials

Microorganisms: *Pseudomonas aeruginosa, Bacillus subtilis, E.coli,* Yeast obtained from the laboratory culture collection was maintained in the specific medium and appropriate proportions used for the experiment. Standard sterile techniques were used for inoculation of cultures. Medium used for the microorganism and all the glassware were properly sterilized autoclaved at 15 lb/in^2 pressure and 121^{0}C for 30 minutes.

***Corresponding author:** R Ilamathi, Department of Biotechnology, SNIST, Yamnampet, Hyderabad, Andhra Pradesh, India

Methods

Preparation of metal solutions: Different metal concentrations were prepared by dissolving of $CuCl_2$, $CdCl_2$, $NiSO_4$ and $K_2Cr_2O_7$ salts in double distilled water in equal ratio to have metal concentrations of 50, 100, 150, 200,250 and 300 mg/L. A stock solution of 1000mg/L was prepared all other concentrations are obtained from it. All glassware washed with 0.1 M HCl before and after each experiment to avoid binding of the metal to it.

Preparation of biosorbent: The culture was transferred and grown on specific media (Bromifield medium-*Bacillus subtilis;* Cetrimide medium-*Pseudomonas aeruginosa;* YPD-Yeast; LB medium-*E.coli*) for subculture. 100 ml of sterilized culture media was transferred to 250 ml Erlenmeyer flask. The media was allowed to cool and then the 100µl microbial solution was inoculated into the medium in laminar air flow chamber. The inoculated flasks were incubated in an orbital shaker at 250 rpm at 32°C for 2 days to obtain the biomass. Mixed cultures were prepared by adding equal amounts of individual cultures. Biomass was harvested from the medium by centrifugation at 9000 rpm for 10 min. The supernatant was discarded and the cells were re-suspended in double distilled water (MilliQ) for washing and again centrifuged as above to make sure that no media remain on the cell surface. This biomass was used for sorption studies.

Immobilization of biosorbent: A 4% (w/v) solution of sodium alginate was prepared by continuous stirring in hot (60°C) distilled water. After cooling, 5% (w/v) of bacterial consortia was added and stirred. The mixture was extruded using a burette into 0.5 M $CaCl_2.{}_2H_2O$. The resultant beads were kept in the polymerizing medium for 4 h. The beads were washed in saline and distilled deionized water. The beads were frozen at 20°C for 24 h, kept at 4°C for 12 h, and thawed at room temperature for further use.

Batch biosorption studies

Optimization of parameters: Batch studies were done using biomass as a function of various parameters such as pH, time, temperature, initial metal concentration and bed height. Biosorption experiments were conducted at an initial metal concentration of 100 mg/L and 100 mg sorbent in 100 ml of metal solution at 30°C for 3 hours at pH varying from 1.0 to 7.0 by adding 0.01 N HCl. Effect of contact time was studied at an initial metal concentration of 100 mg/L and 100 mg sorbent in 100 ml solution at 30°C and optimized pH. Samples were analyzed for the concentration of metal at regular interval of one hour for 24 hours. The effect of temperature on sorption was determined through batch experiments carried out at 10, 20, 30, 40 and 50°C. Sorption studies at optimized conditions were also carried out with initial metal concentrations in the range of 50–300 mg/L of metal solutions prepared as stated in section 2.2.1. Biosorbent dosage was optimized by using biosorbent amounts of 25, 50, 75, 100, 125 and 150 mg in100 ml of 100 mg/L of metals solutions. The optimization studies were carried out for mixed culture immobilized in alginate beads.

Experimental Protocol for LSFB

Experimental setup: Figure 1 shows the fabricated LSFB consists of a homogenizing section 14 cm long for the uniform mixing of the inlet wastewater before it enters the reactor. Above the homogenizing section is the distributor plate of diameter 5 cm, with pores of diameter 1.5mm arranged in triangular pitch. Above the distributor plate is an acrylic riser of 88 cm, which functions as the LSFB. There is a solid disengaging section above the riser, from where the liquid effluent is withdrawn. Pressure tapings were made at different heights.

Ball valves are used for the inlet and the bypass. The LSFB requires a ½ HP pump for the inlet. A rotameter of range 0-300 LPH was used to vary the flow rate. Immobilized Sodium Alginate beads were selected as the fluidizing particle. The main advantage of using Sodium Alginate is that it doesn't react with water. Also, it has small diameter and low density, and hence, it is easier for fluidization and entrainment. Another advantage is that it fluidizes at low liquid flow rate. Hydrodynamics studies were carried out on the LSFB using Alginate beads as the fluidizing particle to study its behavior and to check its functioning.

Experimental Procedure: The fluidized bed is initially filled with beads up to a 1/4th of the total riser volume. Tap water is pumped from the reservoir into the reactor column using a ½ HP pump. The flow rate of the liquid is measured using a rotameter with a range of 0 to 300 LPH. At each flow rate, the bed height is measured and tabulated. The pressure drop across the column is also measured using a digital manometer and tabulated. The void age is calculated at minimum fluidization velocity and at different flow rates. The pressure drop across the bed was found to be the same for different flow rates, thus indicating the proper construction of the LSFB.

Biosorption studies in LSFB: Immobilized biosorbent prepared as per section 2.2.3 was filled in the fluidized bed till 1/4th of the riser volume. 30 liters of synthetic heavy metal solution was prepared as described in section 2.2.1. The prepared heavy metal solution (100mg/L) was pumped through the column at desired flow rate of 132 LPH which was determined by the hydrodynamic studies in the column. Optimized parameters were used, and the heavy metal concentration was determined using atomic absorption spectrophotometer at 5 min interval in the start of the experiment and then at 15 min intervals subsequently from the sampling port. The fluidized bed studies were carried out at pH 4.5 at room temperature. The effect of pressure drop and bed height on different flow rate on heavy metal adsorption was studied. Samples were collected at pre-defined time intervals, centrifuged as above and the amount of metal in the supernatant was determined.

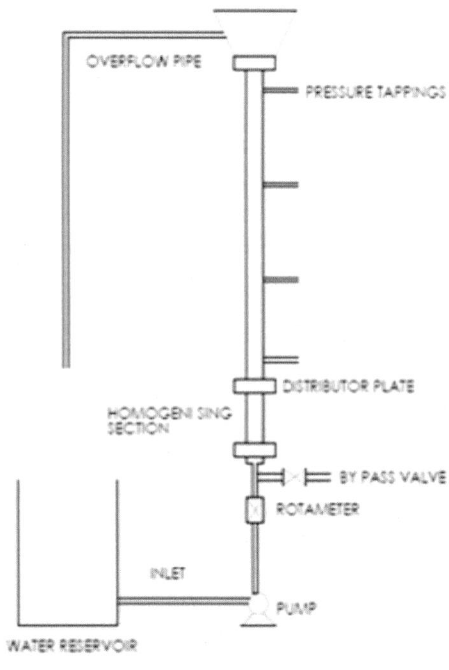

Figure 1: Schematic diagram of the experimental setup.

Determination of metal concentration in the supernatant: The heavy metal concentration was determined by the use of atomic absorption spectrophotometer, Perkin Elmer Analyst 300. Determination of copper, chromium, cadmium and nickel was done by using its specific lamp for each metal and at a specific wavelength.

Data evaluation: The amount of metal bound by the biosorbents was calculated as follows:

$$Q = v(Ci\text{-}Cf)/m$$

Where Q is the metal uptake (mg metal per g biosorbent), v the liquid sample volume (ml), C_i the initial concentration of the metal in the solution (mg/L), C_f the final (equilibrium) concentration of the metal in the solution (mg/L) and m the amount of the added immobilized biosorbent on the dry basis (mg).

Desorption studies: The exhausted Alginate beads containing immobilized microorganisms after heavy metal biosorption were removed from the LSFB. The beads were then treated with Nitric Acid, and allowed to stay for an hour and loaded back into the LSFB.10 ml of samples were withdrawn every half hour. The samples were then analyzed in the Atomic Absorption Spectrophotometer to determine the heavy metal concentration.

Sample analysis: Metal concentration at initial and at regular intervals was examined using atomic absorption spectrophotometer. 10ml of the samples were taken out for every 5 minutes in the first hour and continued by every 15 min in the subsequent hour. The metal concentrations were determined and the results are tabulated.

Results

Batch sorption studies

Effect of batch sorption parameters

Effect of pH: The experimental results of chromium, cadmium, nickel and copper using mixed cultures of Yeast, *Pseudomonas aeruginosa, Bacillus subtilis* and *Escherichia coli* at varying pH was shown in the Figure 2(A). Effect of pH on biosorption has been studied over a range of 1 to 7. The highest removal of cadmium and copper was found at pH 4, while at pH 3 and 5 the highest removal of chromium and nickel obtained respectively. At pH<3, lowest biosorption was recorded and the experiments beyond pH 5 were hindered due to precipitation of metals during the biosorption process.

Effect of contact time: Before the starting of the batch experiments, it was very important to know the required time to reach equilibrium biosorption. In other words, for all batch experiments, the contact time should be sufficient to ensure reaching an equilibrium concentration. Figure 2(B) shows the results of biosorption removal of heavy metals concentration versus contact time for the four metal solutions. It can be seen that a 3 hour contact time is sufficient to reach equilibrium for all four heavy metals.

Effect of temperature: Biosorption studies of Metals solutions using immobilized mixed cultures were carried out at different temperatures ranging from 10°C to 50°C. The effect of temperature on metal sorption was presented in Figure 2(C). The percentage of metal sorption was increased from 10°C to30°C and then showed decrease in sorption percentage with increase in temperature. Figure 3.4 shows a maximum percent of sorption around 90% achieved at 30°C at fixed pH 4.5 and biomass concentrations 2.0mg/ml. This decrease in biosorption efficiency at temperatures 40-50°C may be attributed to: an increase in the relative desorption of the heavy metals from the solid phase to the liquid phase, deactivation of the biosorbent surface, destruction of active sites on the biosorbent surface due to bond disruption, or due to the weakness of the sorbent active site binding forces and the sorbate species and also between the adjacent molecules of the sorbed phase. From Figure 3 it was concluded that experiments can be carried out at room temperature.

Effect of initial metal concentration: Figure 2(D) shows the variation of percentage sorption of immobilized microbial culture at different initial metal concentrations. The results indicated that the percentage biosorption of metals was not altered greatly from concentrations between 50-150mg/L whereas still increasing in the initial metal concentration beyond 150mg/L shows a decrease in the sorption capacity, this may be due to the sorbent dosage depicted in Figure 2(E) which was found to be 1g/L of immobilized biosorbent containing enough biosorption surfaces and sites for this concentration.

LSFB studies for biosorption

Effect of pressure drop and flow rate: From Figure 3(A), it is evident that the pressure drop across the LSFB increases continuously till the flow rate of the liquid attains that of the minimum fluidization velocity, which is the velocity required to fluidize the particles, in this case, the Alginate beads. On further increase of the flow rate of the liquid beyond minimum fluidization velocity (43 LPH), the pressure drop across the LSFB is found to remain nearly constant, hence obtaining a straight line in the latter part of the graph. This signifies that once the fluidization starts, the pressure drop across the LSFB remains the same for different flow rate.

Effect of bed height and flow rate: Figure 3(B) shows that the bed height of the constructed Liquid Solid Fluidized Bed (LSFB) remains constant till the flow rate reaches the minimum fluidization velocity, which is 43 LPH for the above mentioned bed. Since the beads are supposed to fill 25% of the entire volume of the reactor, the static bed height is 22cm. Once the flow rate of the liquid exceeds the minimum fluidization velocity, the fluidization of the beads starts and the bed expands. The bed height was found to increase more or less linearly with the liquid flow rate. This indicates the proper construction of the LSFB.

The optimum flow rate was determined as 132 LPH since it was found that at this flow rate, the bed height is the length of the reactor. In other words, at a liquid flow rate of 132 LPH, the beads filled the entire length of the column, without any overflow.

LSFB biosorption studies under optimized conditions: Biosorption studies in LSFB were conducted, which was packed with the known quantity of immobilized alginate beads. The synthetic metals solution was pumped at desired flow rate. Optimized parameters were used and the metal concentrations were determined by atomic absorption spectrophotometer at an interval of 15 min from the sampling port. Figure 3(D) shows the plot of metals concentration in the LSFB against the residence time in the fluidized bed. The maximum amount of copper was adsorbed in the first 170 minutes. Nickel, Chromium and cadmium were adsorbed at 70 min. The maximum efficiency of biosorption of Cadmium in the LSFB was found to be 67.17%. Biosorption of Chromium using in the LSFB was also found to increase up to 90 minutes, after which the biosorbed metal leached back into the solution. The maximum biosorption efficiency for uptake of Chromium (VI) metal was calculated to be 49.25%. Biosorption of copper in the LSFB were found to be high than other metals this may be due to the high suitability of optimized parameters and enough surfaces

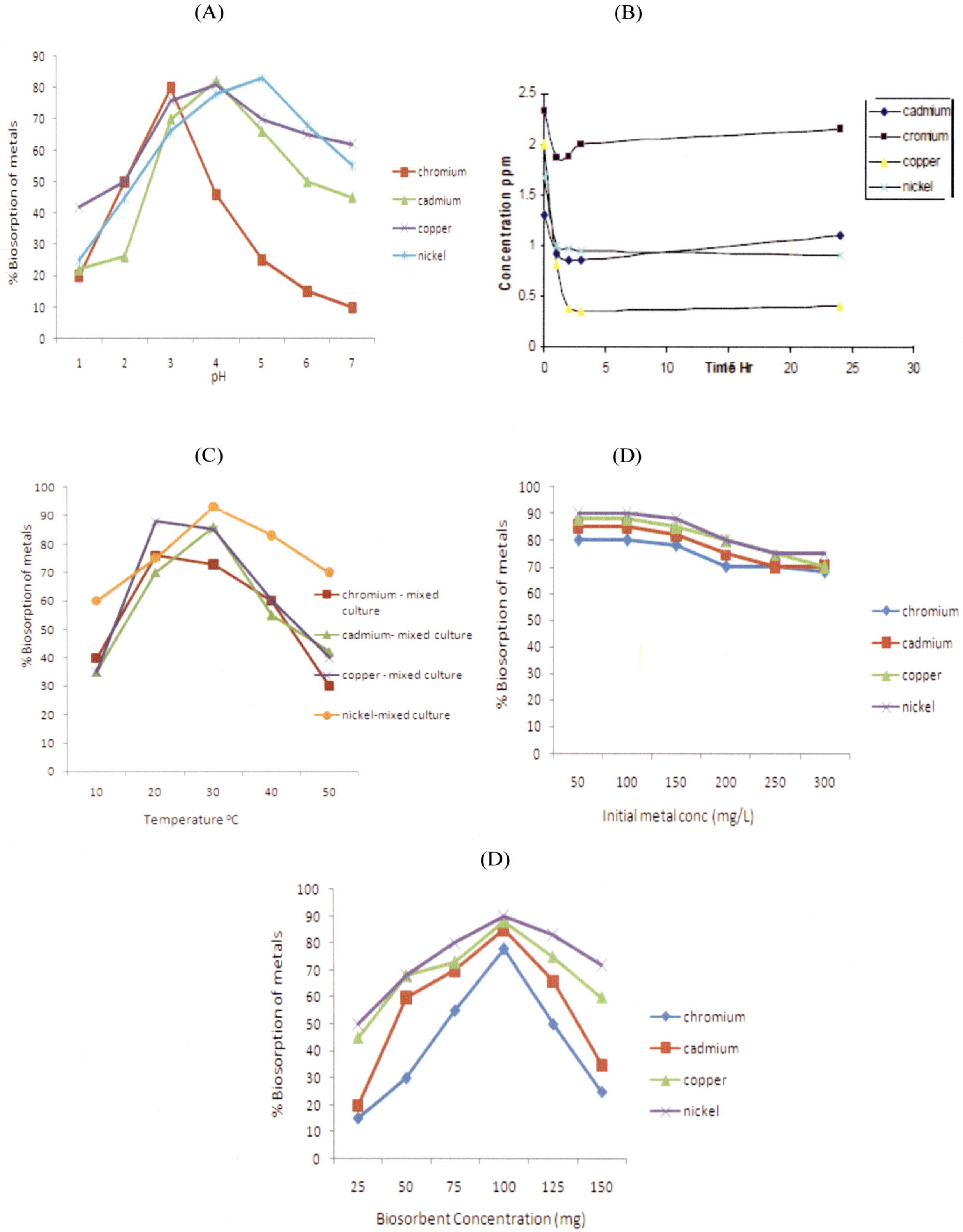

Figure 2: (A) Effect of pH. (B) Contact time (C) Temperature (D) Initial Metal Concentration (E) Biosorbent Concentration of the bacterial consortium immobilized in alginate beads.

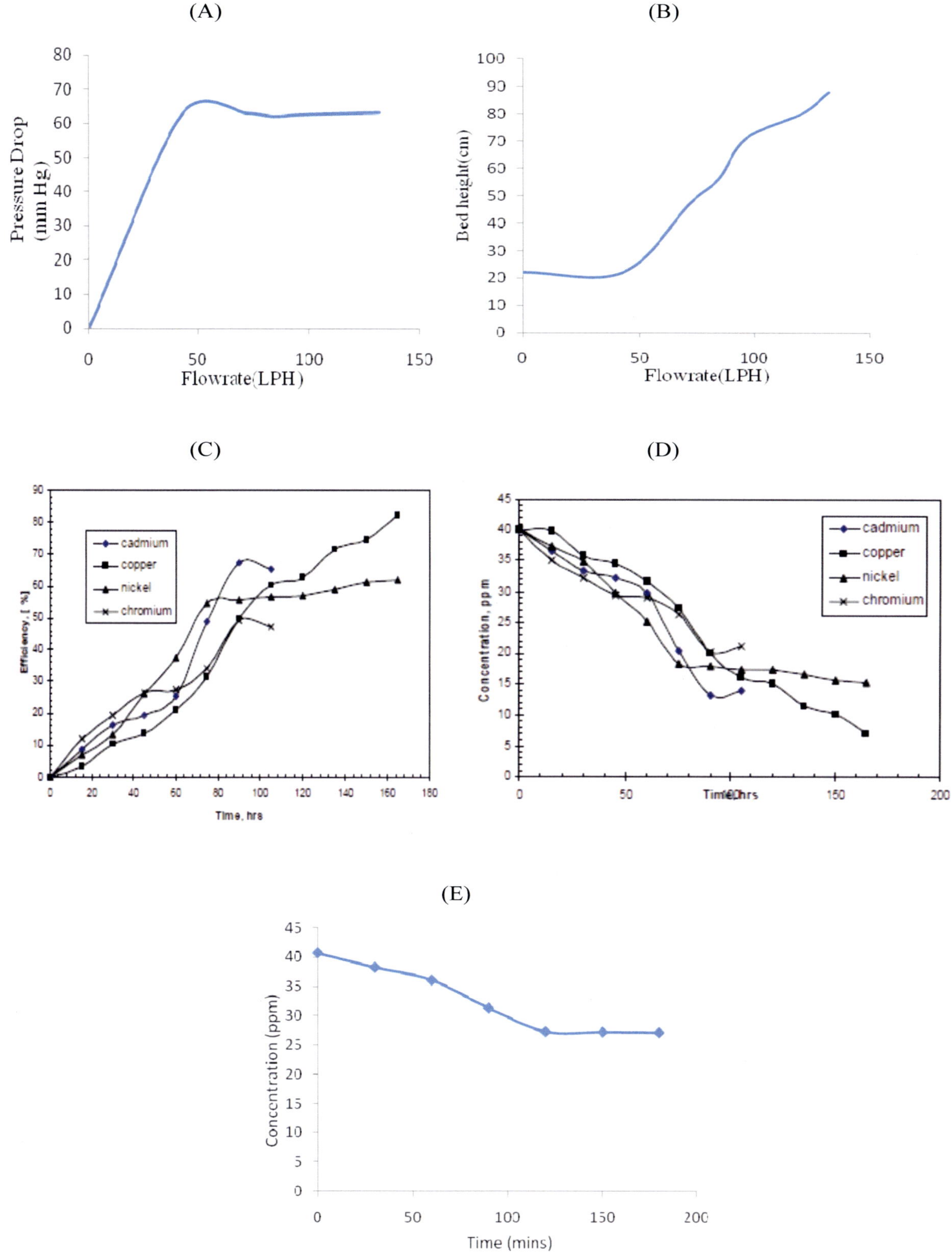

Figure 3: (A) Effect of Pressure drop and flow rate (B) Bed height on flow rate (C) Efficiency of Fluidized bed sorption studies (D) Removal of metal concentration in LSFB on time (E) Desorption studies on removal of copper.

and sites for this concentration. The optimum time for the biosorption was determined to be around nearly 3 hours, after which the biosorption efficiency remains more or less constant which is same as that of the batch studies. Similar to the batch studies performed, the concentration of Copper in the solution was found to continuously decrease in the first 3 hours. Figure 3(C) gives the maximum biosorption efficiency of metals in which the copper biosorption efficiency in the LSFB was determined to be high as 84.62%. Biosorption efficiency of the other metals may increase by optimizing the biomass concentration of each microorganism with respect to the metals solution before immobilizing in alginate beads.

Desorption and column regeneration: The desorption of exhausted Alginate beads containing mixed culture using 0.1 N Nitric Acid. Figure 3(E) shows the concentration of was found to decrease from 40 ppm to 27.12 ppm for copper. The efficiency was found to decrease from 82.4% to 33.46%. Reusability of biosorbent was studied by three cycles of alternating sorption/desorption experiments with the supplement of 100 mg/L of metals solution at the beginning of each cycle. The sorption capacities for each cycle were shown to be decreasing by 5%.

Conclusion

The present study evaluated the removal of copper, chromium, nickel and cadmium from synthetic metals solutions using immobilized consortium of Yeast, *Pseudomonas aeruginosa, Bacillus subtilis* and *Escherichia coli* in alginate beads in fluidized bed reactor. Batch experiments showed that the mixed culture immobilized in alginate beads can be successfully used for the removal of these metal ions. LSFB studies showed that the removal of copper and nickel was high compared to chromium and cadmium. This concluded that biosorption was found to depend significantly on pH of the solution. Optimum flow rate 132 LPH obtained from batch studies results in good contact time between the adsorbent and adsorbate in LSFB also. It can thus be concluded from this project work that biosorption of heavy metals in an LSFB is not only a cost effective method, but also a highly efficient alternative to chemical and physical treatments employed for the removal of heavy metal pollutants from contaminated water. Further work will be carried out in developing a model with higher fluidization velocity with different particle size and with different biomass concentration.

References

1. Brown B, Absanullah M (1971) Effects of Heavy Metals on Mortality and Growth. Mar Pollut Bull 2: 182-187.
2. Moore JW (1991) Inorganic Contaminants of Surface Water. Research and Monitoring Priorities.
3. Volesky B, Holan ZR (1995) Biosorption of heavy metals. Biotechnol Prog 11: 235-250.
4. Matheickal JT, Yu Q (1999) Biosorption of lead (II) and Copper (II) from aqueous solution by pretreated biomass of Australian marine algae. Bioresour Technol 69: 223-229.
5. Volesky B (1986) Biosorbent materials. Biotechnol Bioeng Symp 16: 12-126.
6. Song HX, Liu YG, Xu WH, Zeng GM, Aibibu N, et. al, (2009) Simultaneous Cr(VI) reduction and phenol degradation in pure cultures of Pseudomonas aeruginosa CCTCC AB91095. Bioresour Technol 100: 5079-5084.
7. Wang YT, Xiao C (1995) Factors Affecting Hexavalent Chromium Reduction in Pure Cultures of Bacteria. Water Res 24: 2467-247.
8. Liu GF, Yang H, Wang J, Jin RF, Zhou JT, et. al., (2010) Enhanced Chromate Reduction by Resting Escherichia Coli Cells in the Presence of Quinone Redox Mediators. Bioresour Technol 101: 8127–8131.
9. White C, Wilkinson SC, Gadd GM (1995) The Role of Microorganisms in Biosorption of Toxic Metals and Radionuclides. Int Biodeterior Biodegradation 35: 17-40.
10. Viera M, Curutchet G, Donati E (2003) A Combined Bacterial Process For The Reduction And Immobilization of Chromium. Int Biodeterior Biodegrad 52: 31-34.
11. Saraj MA, Abdel-Latif MS, El-Nahal I, Baraka R (1999) Bioaccumulation of Some Hazardous Metals by Sol-Gel Entrapped Microorganisms. J Non-Cryst Solids 248: 137-140.
12. Xu J, Song XC, Zhang Q, Pana H, Liang Y, et. al., (2011) Characterization of Metal Removal of Immobilized Bacillus Strain CR-7 Biomass from Aqueous Solutions. J Hazard Mater 187: 450-458.
13. McDougall GJ, Fleming CA (1987) Extraction of precious metals on activated carbon, in Ion Exchange and Sorption Processes in Hydrometallurgy, Ed. by Streat, M. and Naden, D., Critical Reports on Applied Chemistry. 19: 56-126.
14. Fu Y, Liu D (2007) Novel Experimental Phenomena of Fine-Particle Fluidized Bed. Exp Therm Fluid Sci 32: 341-344
15. Richardson JF, Harker JH, Bachurst JR (2002) Coulson and Richardson's Chemical Engineering, 5th Edn. Particle Technology and Separation Processes, Butterworth-Heinemann.

Chemical Synthesis, Characterization and Bioactivity Evaluation of Hydroxyapatite Prepared from Garden snail (*Helix aspersa*)

Anjuvan Singh[1]*and K. M. Purohit[2]

[1]*Department of Biotechnology, Lovely School of Sciences, Lovely Professional University, Phagwara, India.*
[2]*Department of Chemistry, National Institute of Technology, Rourkela –769008, Orissa, India*

Abstract

A novel attempt is made to convert the calcium carbonate skeleton of widely available garden snail shell (*Helix aspersa*) to hydroxyapatite based bioceramics. The snail shell was found to decompose within 850°C to all the carbonate phases. The calcined snail shells were then treated with acids followed by different chemicals in ammoniacal media maintaining proper stoichemetry to produce fine Hydroxyapatite (HAP) as filter cake with Ca/P molar ratio of 1.67. The dried HAP powder was extremely pure with specific surface area of 15m²/g. The different characterization techniques were adopted both for calcined snail shell and HAP synthesized like X-ray Diffraction (XRD), Thermal Analysis (DTA/TGA), Fourier Transform Infra red Spectroscopy (FTIR) and Scanning Electron Microscopy (SEM). The surface area and the particle size, of the HAP powder prepared by chemical precipitation route, were also determined by BET and Malvern particle size analyzer respectively. The synthesized powder was soaked in stimulated body fluid (SBF) medium for various periods of time in order to evaluate its bioactivity. The changes of the pH of SBF medium were measured. High bioactivity of prepared HAP powder due to the formation of apatite on its surface was observed.

Keywords: Snail shell; Hydroxyapatite; Calcination; β-tricalcium phosphate; Bioactivity

Introduction

Hydroxyapatite [$Ca_{10}(PO_4)_6(OH)_2$] based bioceramics are successfully used as implants as it is chemically similar with the inorganic constituent of biological hard tissue [1]. It is present in bone, teeth and tendons to give these organs stability, hardness and function. On account of its chemical similarity with the biological calcified tissue it is remarkably biocompatible [2]. Due to the formation of strong bond with the hard tissue, it is widely used in orthopedics or in dental implants.

HAP is also a potential implant material due to its excellent osteoconductive properties [3,4]. HAP has been shown to stimulate osteoconduction and is a material that can be integrated into bone without provoking an immune reaction. The biological response to HAP implants is influenced by its properties. The application of HAP as useful biocompatible materials largely depends on the purity and morphology of the powder. HAP can be prepared by different routes like chemical precipitation, sol-gel route, combustion synthesis, plasma etc [5-10]. The purity in the final HAP powder and stoichemetry (molar ratio of Ca/P = 1.67) can be well controlled in chemical precipitation route. The different chemical processes use precursors like $Ca(NO_3)_2$, $Ca(OH)_2$ etc. as the source of Calcium[Ca] and $(NH_4)_2HPO_4$, H_3PO_4 etc. as the source of Phosphorus [P] during synthesis of HAP. The extremely pure HAP powder is very costly and needs high quality precursors. The most of the sources of Ca^{2+} contains different types and level of impurities mainly silica. Snail shell consists of $CaCO_3$ with minor amount of $MgCO_3$ and other matters, can be potential precursors for the production of HAP.

The composition of human bone is an inorganic/organic hybrid consisting of 70% (wt) apatitic calcium phosphates and 30% (wt) organic (largely collagen) [11]. The apatitic calcium phosphate of bone mineral consists of carbonate, small amount of sodium, magnesium and other trace elements. The submicroscopic crystals of calcium phosphates in bone resemble the crystal structure of synthetic HAP[12].

The objective of the present paper is to synthesize pure and biocompatible HAP using snail shell as precursor following chemical precipitation method [13]. The powders are characterized using XRD, DTA/TGA, FTIR, SEM and BET surface area. Hydroxyapatite prepared by precipitation route also has the feature of small size, low crystallinity and high surfacial activation which can meet different demands [14].

Keeping the above points in mind, the present study was aimed to produce and to enhance the bioactivity of stoichemetric HAP prepared from garden snail shell (*Helix aspersa*) and to evaluate its bioactivity in simulated body fluid.

Materials and Methods

Garden snail shells (SS) were collected and their shell covering was removed carefully. Shells were washed with tap water followed by distilled water to remove the mud, sand and other impurities. The cleaned shells were dried in the direct sunlight for 2 days. Dry and cleaned SS were calcined at 1000°C for 2 hours so that all organic matters and proteins escape out. The calcined SS was treated with concentrated nitric acid to convert it to $Ca(NO_3)_2$. 130 ml. of 1.63(N) ammonical $Ca(NO_3)_2$ solution was added drop wise to a mixture of ammonical $(NH_4)_2HPO_4$ solution with constant stirring with the help of magnetic stirrer. The pH of the solution was maintained at 10. Hydroxyapatite was formed as per the following reaction:

$$10Ca(NO_3)_2 + 6(NH_4)_2HPO_4 + 2H_2O \longrightarrow Ca10(PO_4)_6(OH)_2 + 12NH_4NO_3 + 8HNO_3$$

The resulting suspension was boiled for 10 minutes and cooled in an ice bath overnight to obtain a white gelatinous precipitate. The pre-

***Corresponding authors:** A. Singh, Lecturer, Department of Biotechnology, Lovely School of Sciences Lovely Professional University, Phagwara, India

cipitate was filtered and filtered cake (residue) was dried in the oven at 800C. The dried sample of hydroxyapatite was ground to powder.

The thermal analysis of the Snail shell was performed by (NETZSCH-Geratebau GmbH Thermal Analyser) at a heating rate of 10^0C/minutes from ambient to 1200^0C to study the weight loss and thermal behavior. The powder samples of SS, calcined SS and synthesized HAP samples were examined with high resolution X-ray Diffractometer (PW-1830, Philips, Netherlands) using Cu-Kα radiation. The X-ray diffraction (XRD) patterns were recorded in the steps of 0.01^0 interval with 1s counting time at each step. Fourier transform infra red (FTIR) spectrum of synthesized HAP powder was obtained over the wave numbers 400 – 4500cm^{-1}. The powder was dispersed into pellets of KBr (mixed in 1:4 ratio) and the spectra was recorded with a Perkin-Elmer (S2000) IR spectrometer. The particle size analysis of HAP was done by Malvern Particle Size Analyzer (Model – Micro-P, UK). The surface morphology of SS and the prepared HAP powder were studied by Scanning Electron Microscope (SEM) while the surface area of the prepared HAP samples was determined with BET surface area analyzer (QUANTACHROME Model:Autosorb1).

In vitro bioactivity evaluation

The *in vitro* bioactivity evaluation of synthesized HAP powder from garden snail shell was performed in a stimulated body fluid (SBF) media of pH 7.4 at a ratio of 1 mg/ml in a water bath at 37°C. The changes in the pH of SBF medium were measured at pre-determined time intervals using a pH meter. Scanning electron microscopy (SEM) was used to identify the apatite formation on surface of the samples and to evaluate the surface morphology of the samples after immersion in SBF medium for 2, 8 and 15 days respectively.

Preparation of (Synthetic Body Fluid) SBF

SBF is known to be a metastable buffer solutions [15,16] and even a small, undesired variance in both of the preparation steps and the storage temperatures, may drastically affect the phase purity and high-temperature stability of the produced HA powders, as well as the kinetics of the precipitation processes.

Merck-grade NaCl (99.5%), $NaHCO_3$ (99.5%), KCl (99.0%), $Na_2HPO_4).2H_2O$ (99.5%), $MgCl_2).6H_2O$(99.0%), Na_2SO_4, $(CH_2OH)_3CNH_2$ (99.5%), $CaCl_2.H_2O$(99.0%) and HCl (37 vol%, Carlo-Erba, Rome, Italy) were used in the preparation of the SBF of this study.

SBF solutions [17-21] were prepared by dissolving appropriate quantities of the above chemicals in deionized water. Reagents were added, one by one after each reagent was completely dissolved in 700 ml of water, in the order given in Table 1. A total of 40 ml of 1 M HCl solution was consumed for pH adjustments during the preparation of 1 l of SBF solutions. A 15 ml aliquot of this acid solution was added just before the addition of the sixth reagent, viz., $(CaCl_2)$ $2H_2O$. Otherwise, the solution would display slight turbidity. The remaining part of the HCl solution was used during subsequent titration. Following the addition of the eight reagent (*tris*(hydroxymethyl) aminomethane), the solution temperature was raised from ambient to 37^0C. This solution was then titrated with 1 M HCl to a pH of 7.4 at 37^0C. During the titration process, the solution was also continuously diluted with consecutive additions of de-ionized water to make the final volume equal to 1 l. It was observed in this study that the prepared SBF solutions can be stored at 5^0C for a month without degradation.

Biodegradation Test

Biodegradation test of calcined HAP prepared from garden snail shell (*Helix aspersa*) was done by taking Tris-HCl buffer solution. 0.05MTris- HCl solution was prepared using distilled water. The pH of solution was maintained 7.4 at 37^0c by adding 1MHCl. Calcined HAP in the form of pallets were soaked in Tris-HCl buffer solution for one week then the samples were dried at 100^0c and final weight loss of sample was determined by the formulae as given below:.

$$\% \text{ Weight Loss} = \frac{W_1 - W_2 \text{ x } 100}{W_2}$$

where, W_1 = initial weight of sample

W_2 = final weight of sample after soaking in Tris-HCl solution.

Results and Discussion

Dta/tga of snail shell (ss)

Figure 1 Showed the DTA/TGA analysis of Snail Shell showed the weight loss at temperature between 90^0C – 120^0C that is due to the physically adsorbed water. Over a wide range of temperature from 250^0C – 400^0C the weight loss is due to the decomposition of $MgCO_3$ combined with the combustion of hydrocarbons. The weight loss along with endothermic peak at 750^0C - 850^0C indicates the decomposition of $CaCo_3$ following the reaction.

$$CaCO_3 \longrightarrow CaO + CO_2\uparrow$$

So it is confirmed from the thermal analysis that Snail Shell mainly contains $CaCO_3$ along with small amount of $MgCO_3$ and other organic matters.

XRD analysis

A typical XRD profile of SS and calcined SS HAP has been shown in Figure 2. The raw SS showed the presence of $CaCO_3$ phase, where as CaO was detected in the calcined Snail shell. The appearance of calcined SS was soft, porous and white in colour. However, due to delay in recording some amount of CaO was converted to $Ca(OH)_2$ by adsorbing moisture from the atmosphere which is depicted in Figure 2b.

Order	Reagent	Amount (gpl)
1	NaCl	6.547
2	NaHCO	2.268
3	KCl	0.373
4	$Na_2HPO_4.2H_2O$	0.178
5	$MgCl_2.6H_2O$	0.305
6	$CaCl_2.\ 2H_2O$	0.368
7	Na_2SO_4	0.071
8	$(CH_2OH)_3CNH_2$	6.057

Patent pending. "Turkish Patent Institute," Turkey, *Appl.* No. 99-0037, 11 January (1999)

Table 1: Chemical composition of SBF solutions.

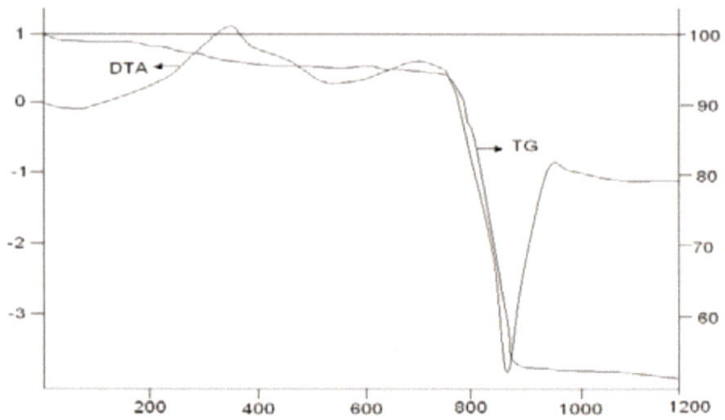

Figure1: Shows the TGA/DTA analysis of as dried Snail shell.

The XRD phase analysis of HAP powder has been shown in Figure 3. Three-high-intensity peaks located at 2θ = 31.7°, 32.2° and 32.9° with Cu-Kα radiation are difficult to be exactly recognized from their diffraction patterns. XRD patterns reveals the formation of HAP and is well resembled with the standard JCPDS file. The unindexed peak at 30.75° Figure 3b may be due to β-tricalcium phosphate; which indicates the initiation of conversion of HAP to β-tricalcium phosphate on heating HAP above 800°C. The calcined HAP exhibits well crystallized sharp peaks of characteristics HAP. The HAP powders, thus synthesized from Snail Shell precursor, are very pure and chemical analysis of powders confirms the same observation.

FTIR analysis

Infrared characterization was carried out for the sample to study the spectral characteristics indicative of the chemical bonding in the synthesized HAP powder. The spectrum Figure 4 can be divided into four regions with peaks having wave numbers around 3500, 1420, 1100 and 600cm^{-1}. The peak observed around 3431.8cm^{-1} is due to the presence of -OH bond [22]. This peak is mainly due to O-H stretching vibration in HAP [23]. The peak at 1036.2 cm^{-1} is associated with the stretching modes of the P-O bonds of HAP [23,24]. The double peak at 603.1 cm^{-1} and 567.4cm^{-1} are due to bending modes of P-O bonds in phosphate groups [24]. Thus, the presence of PO_4^{3-} - group in HAP is almost confirm from IR studies. The pH of the medium during synthesis of HAP was maintained using ammonium solution and it was removed from the suspension with repeated washing with distilled water. In spite of all efforts to remove ammonia from the solution, there is a possibility of small amount of it in the HAP powder. The IR analysis shows a small broad peak at 1422.6cm^{-1}; which is characteristics peak of NH_4^+- group [25-27].

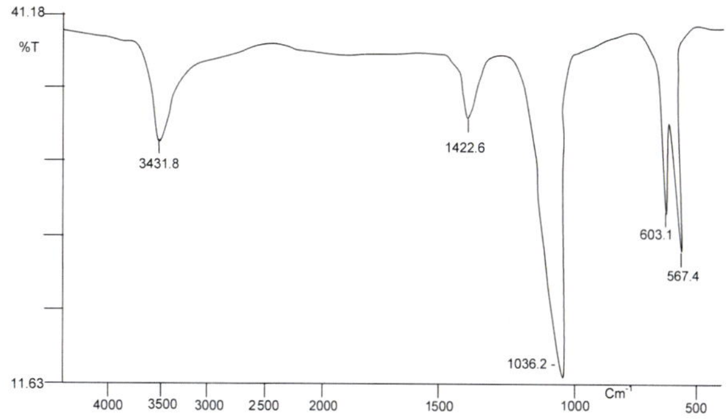

Figure 4: FTIR Spectrum of HAP powder.

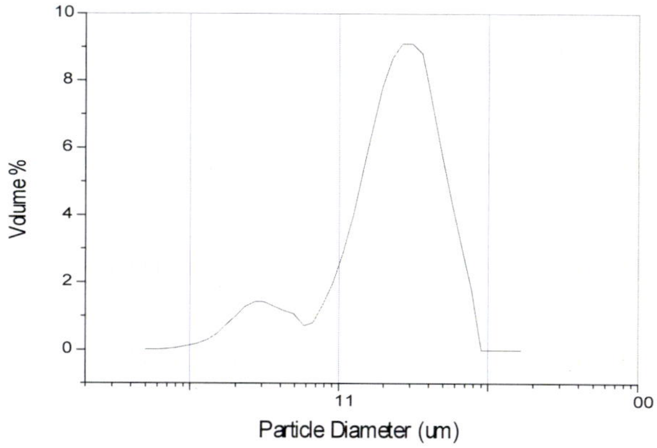

Figure 5: Particle Size Analysis of HAP powder.

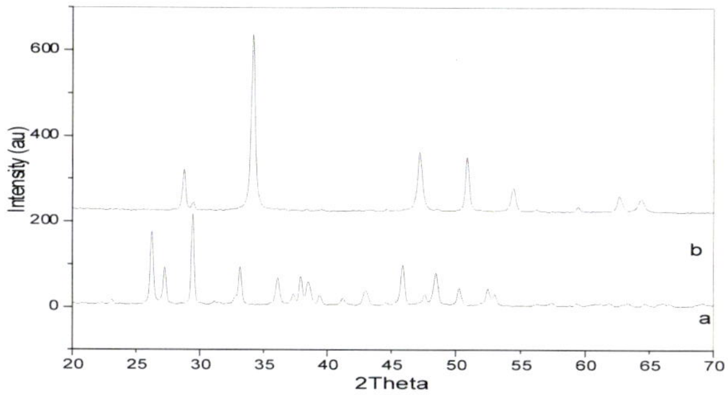

Figure 2: XRD patterns of (a), Snail shell; (b), Calcined snail shell at 1000°C/2Hr.

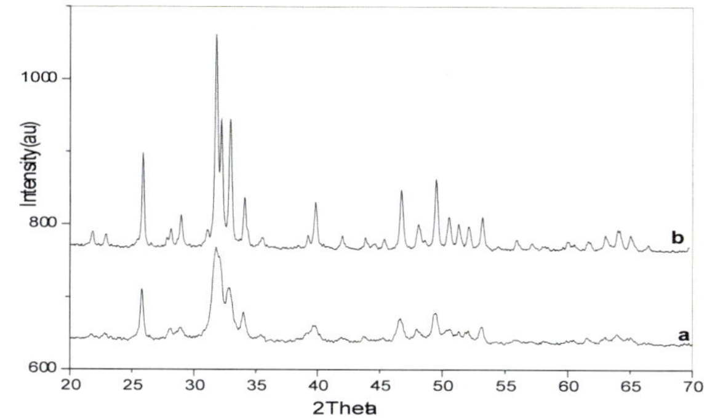

Figure 3: XRD Pattern of: (a) HAP(synthesized), (b) Calcined HAP (synthesized) 800°C/2Hr.

Figure 6: SEM Micrograph of (a) Synthesized HAP powder (b) Calcined (850°C/2h) HAP powder.

Particle size analysis

Particle size analysis of HAP powder was carried out following Laser technique and pattern of particle size distribution is plotted in Figure 5. Average particle size was found to be 2.63 μm. Small amount of fine particles (0.2-0.3 μm) are also present in the synthesized powder.

Surface area measurement

The surface areas of the hydroxyapatite powder and calcined HAP are determined are 83 and 15m^2/gm respectively. Powders are agglomerated during calcinations; but HAP powders have to be calcined to remove volatile impurities like ammonia.

Scanning electron microscope (SEM)

The morphologies of as synthesized and calcined HAP powders are shown in Figure 6. Uncalcined HAP powders are almost regular and round in shape; where as calcined HAP powders are agglomerated. The microstructure as reveals from SEM is in well- agreement with the particle size analysis and BET surface area analyzer results.

Figure 7: SEM micrographs of (a) the surfaces of synthesized HAP powders after soaking in SBF for 2 Days (b) the surfaces of synthesized HAP powders after soaking in SBF for 8 Days (c) the surfaces of synthesized HAP powders after soaking in SBF for 15 Days.

Bioactivity evaluation

The SEM micrographs of the surfaces of the immersed HAP powder after soaking in SBF for various periods of time are shown in Figure 7. Tiny agglomerated bone-like apatite particles could be formed on the surface of the HAP powders.

Figure 7 shows that tiny agglomerated bone-like apatite particles could be formed on the surface of the HA powders soaked for 2 days, 8 days and 15 days respectively. The number and the size of these agglomerated particles increased with increasing soaking times. Identification and evaluation of apatite formation on the surface of a material in SBF is useful for predicting the *in vivo* bone bioactivity of the material, not only qualitatively but also quantitatively [28-31]. The results indicated that the synthesized HAP powder from Garden Snail shell (*Helix aspersa*) showed the high bioactivity in SBF solution.

In-vitro biodegradation

Biodegradation of calcined HAP samples in the form of pallets were carried out in Tris-HCl solution. HAP samples were soaked in Tris-buffer solution at pH 7.4 and temperature 37°C for 7 days [32]. When porous HA was soaked in Tris-buffer solution, the loss of calcium ion took place which resulted in the increase in pH of the buffer from 7.4 to 8.2 which confirms the biodegradation of HAP. The calcined HAP prepared from garden snail shell by chemical precipitation method showed weight loss of 4.5%. Thus it appears that the ageing time in Tris-HCl solution may also affect the weight loss behavior.

Conclusions

A stoichiometric, pure and thermally stable hydroxyapatite powder was synthesized from Snail shell (*Helix aspersa*) by chemical precipitation method. XRD analysis indicated the phase purity and crystallinity of hydroxyapatite powder. TG/DTA result showed that Snail shell is mainly composed of calcium carbonate ($CaCO_3$). Fine particle size of hydroxyapatite was produced. The present work is based on the utilization of biological waste (Snail shell) to produce hydroxyapatite for Bio-medical applications. The prepared HAP powder showed high bioactivity similar to that in biological apatite and higher bioactivity in comparison with conventional HAP. Thus, prepared HAP from garden snail shell (*Helix aspersa*) might be more useful for treatment of oral bone defects in comparison with conventional HAP, and might be more effective as a bone replacement material to promote bone formation.

An attempt will be made in future to synthesize porous HAP and study its bio-compatibility. Mass production of biocompatible HAP for biological application may be possible at simple and low cost through this route.

Acknowledgement

The authors like to acknowledge H.O.D, Department of Chemistry, National Institute of Technology, Rourkela, for providing laboratory facilities for carrying out research work.

Competing Interests

The authors have no competing interests with the work presented in this manuscript.

References

1. Layrotte P, Ito A and Tateishi T (1998) Sol-gel Synthesis of Amorphous Calcium Phosphate and Sintering into Microporous Hydrozyapatite Bioceramics. J. Am. Ceram. Soc 81: 1421-1428.
2. Li P, Ohtsuki C, Kokubo T, Nakanishi K, Soga N, et al. (1992) Apatite Formation Induced by Silica Gel in a Simulated Body Fluid. J. Am. Ceram. Soc 75: 2094-2097.
3. de Groot K (1900) Bioceramics Consisting of Calcium Phosphate Salts . Biomaterials 1: 47-50.
4. Jarcho M (1981) Calcium Phosphate Ceramics as Hard Tissue Prastetics. Clin Orthop Rel Res157: 259-278.
5. Brown PW, Fulmer M (1991) Kinetics of Hydroxyapatite Formation at Low Temperature. J. Am. Ceram. Soc 74: 934-940.
6. Young RA, Holcomb DW(1982) Variability of Hydroxyapatite Preparations. Calcified Tissue Int 34: 17-32.
7. Arita IH, Wilkinson DS, Mondragon MA, Castano VM (1995) Chemistry and Sintering Behaviour of Thin Hydroxyapatite Ceramics with Controlled Porosity Biomaterials 16: 403-408.
8. Brendel T, Engel A, Russel C (1992) Hydroxyapatite Coatings by Polymeric Route. J. Mater Sci: Mater. Med 3: 175-179.
9. Partenfelder UME, Engel A, Russel CA (1991) Pyrolutic Route for the Formation of Hydroxyapatite/Fluoroapatite Solid Solutions. J Mater. Sci: Mater. Med 4: 292-295.
10. Roy DM, Linnehan SK (1974) Hydroxyapatite Formed From Coral Skeletal Carbonate by Hydrothermal Exchange Nature 247: 220-222.
11. Cowin SC, Vanburskirk WC, Ashaman RB (1987) Handbook of Bioengineering (eds) R Skalak and Schien C Newyork (McGRawHill)
12. Sinha A, Ingle A, Munim KR, Vaidya SN, Sharma BP, et al. (2001) Development of Calcium Phosphate based Bioceramics. Bull. Mater. Sci 24: 653-657.
13. Jarcho M (1978) Hydroxyapatite Ceramic U.S Patent, 4097935.
14. Liu C, Huang Y, Shen W, Cui J (2001) Kinetics of Hydroxyapatite Precipitation at pH 10 to 11. Biomaterials 22: 301-306.
15. Ohtsuki C, Kokubo T, Yamamuro T (1992) Mechanism of HA formation of CaO{SiO_2}P_2O_5 glasses in simulated body Fluid .J Non- Cryst Solids 143: 84-92.
16. W Neuman, M Neuman (1958) The chemical dynamics of bone mineral Chicago University of Chicago Press p.34.
17. Li P, Kangasniemi I, de Groot K, Kokubo T (1994) Bone-like hydroxyapatite induction by a gel-derived titania on a titanium substrate. J Am Ceram So 77: 1307-12.
18. Li P, Nakanishi K, Kokubo T, de Groot K (1993) Induction and morphology of HA precipitated from metastable simulated body Fluids on sol gel prepared silica. Biomaterials 8: 963-8.
19. Li P,Kangasniemi I, de Groot K, Kokubo T, Yli-Urpo AU(1994) Apatite crystallization from metastable calcium phosphate solution on sol gel prepared silica. J Non-Cryst Solids 6: 281-286.
20. SO Cho, Nakanishi K, Kokubo T, Soga N, Ohtsuki C, et al. (1995) Dependence of apatite formation on silica gel on its structure: e!ect of heat treatment. J Am Ceram Soc 74: 1769-1774.
21. Kokubo T, Miyaji F, Kim HM, Nakamura T (1996) Spontaneous formation of bonelike apatite layer on chemically treated titanium metals. J Am Ceram Soc 9: 1127-1129.
22. Russell SW, Luptak KA, Suchicital CTA, Alford TLA, Pizziconi VB(1996) Chemical and Structural Evolution of Sol-Gel-Derived Hydroxyapatite Thin Films under Rapid Thermal Processing. J Am Ceram Soc 79: 837- 843.
23. Sargin Y, Kizilyalli M, Telli C, Guler H (1997) A New Method for the Solid-State Synthesis of Tetracalcium Phosphate: A Dental Cement: X-ray Powder Diffraction and IR Studies. J Eur Ceram Soc 17: 963-970.
24. Ramanan SR, Ramannan V (2004) A Study of Hydroxyapatite Fibers Prepared via Sol–Gel Route Material Letters 58: 3320-3323.
25. Tas Cuneyt (2000) Synthesis of Biomimetic Ca-Hydroxyapatite Powders at 37°C in Synthetic Body Fluids. Biomaterials 21: 1429-1438.
26. Caroline Victoria E, Gnanam FD (2002) Synthesis and Characterisation of Biphasic Calcium Phosphate Trends. Biomater Artif Organs 16: 12-14.
27. Gomez-Morales J, Torrent-Burgues J, Boix T, Fraile J, Rodriguez-Clement R (2001) Precipitation of Stoichiometric Hydroxyapatite by a Continuous Method. Cryst Res Technol 36: 15-16.
28. Williamson GK, Hall WH (1953) Acta Metall Sinica 1: 22.
29. Hench LL, Wilson J (1993) An Introduction to Bioceramics 2nd ed. London: World Scientific Publishing Co Ltd
30. Murugan R, Ramakrishna S (2005) Aqueous mediated synthesis of bioresorbable nanocrystalline hydroxyapatite. J Cryst Growth 274: 209-213.
31. Kokubo T, Takadama H (2006) How useful is SBF in predicting in vivo bone bioactivity? Biomaterials 27: 2907-15.
32. Soon-Ho Kwon, Youn-Ki Jun, Seong-Hyeon Hong, In-Seop Lee, Hyoun-Ee Kim, et al. (2002) Calcium phosphate bioceramics with various porosities and dissolution rate. J Am Ceram Soc 85: 3129-31.

6

Effectiveness of a Biological Insecticide Derived from Scorpion Fish Offal against the Chickpea Weevil

Laribi-Habchi H[1], Biche M[2], Drouiche N[1,3]*, Khalfi W[2], Abdi N[1] and Mameri N[1]

[1]*National Polytechnic school of Algiers, B.P. 182-16200, El Harrach, Algiers, Algeria*
[2]*Institut National Agronomique d'Alger, El Harrach, Algeria*
[3]*Centre de Recherche en Technologie des Semi-conducteurs pour l'énergétique (CRTSE). 2, Bd Frantz Fanon BP140, Alger – 7 Merveilles, 16000, Algeria*

Abstract

The aim of this study is to evaluate the biological activity of a crude chitinase extracted from fish offal (scorpion) on the chickpea weevil *Callosobruchus maculatus* (Coleoptera: Bruchidae) at various concentrations: 3%, 6%, 9%, 12%, 15%, 18%, 21% and 24% (V/V). The biological tests were performed under the typical experimental conditions: temperature and relative humidity at 28°C and 75%, respectively. The results have shown that the concentration of 24% of crude chitinase extracted has 100% of insecticide activity after a contact time of 1 h. The lethal dose (LD50) is estimated for every sample and for a range time from 1 to 72 h. The lower percentage of mortality (10%) is obtained for low values of concentrations and contact time (dose of 3% of crude chitinase and 1 h of contact time). The maximum mortality rate is observed for the high values of two parameters (dose of 24% of crude chitinase and 72 h contact time). However, it is observed that for a higher dose (24%) and for a lower contact time (1 h), the mortality rate increased rapidly. The novelty of this work consists in the use of a new bioinsecticide obtained from offal scorpion fish.

Keywords: Crude chitinase; Scorpion fish; Extraction; Bio-insecticide; Callosobruchus maculates

Introduction

Around the world, stored products are attacked by several species of pests. Chemical control against the various pests, using essentially conventional pesticides, continues to be the major means of protecting foodstuffs. Secondary effects of insecticides and environmental imperatives have encouraged the search for new approaches to protect these shifts. To fight against the food insecurity and to protect the environment, we should increase agricultural production through the judicious use of scientific and biotechnological tools. Leguminous seeds represent the main source of protein in many developing countries [1]. Unfortunately, they suffered considerable losses during storage [2]. The stored products can be destroyed by insects, fungi and rodents. Losses caused by insects are considerable in countries where modern techniques of storage are not yet introduced. Larvae of cleopter Bruchidae consume and grow only in the seeds of leguminous [3,4]. Also, the chickpea weevil *Callosobruchus maculatus* may infest the plant *Cicer arietinum* in fields and stocks, in addition it infects other leguminous non-host plants which have an economic importance in developing countries as cowpea (*Vigna unguiculata)* and sap (*Vicia faba*) [5].

Chemical control using synthetic insecticides has dominated control tactics against insect pests of stored products [6-8]. In this study a new bio-insecticide obtained from offal scorpion fish was utilized. For this purpose, chitinase was extracted from scorpion's stomach and tested as a protective film of chickpea weevil.

Experimental design and statistical analysis were utilized to study the effects of two independent variables, namely concentrations of crude chitinases (X_1) and contact time (X_2) on mortality rate of *Callosobruchus maculatus.* The planning of the experiments was carried out in order to evaluate the effects of the parameters and their interactions using the two levels factorial planning of the two studied variables 2^2. A total of 8 experiments were formulated including 2^2 full factorial design (±1) with four replicates at central point (coded level 0). The factors and the experimental domain of variation of factors (coded and uncoded levels) are given in Table1.

The functional relationship between the two independents variables (X_1, X_2) and the response (Y) is expressed by a first order polynomial function (Eq.1):

$$Y= a_0 + a_1X_1 + a_2X_2 + a_{1.2}X_1X_2 \quad (1)$$

Y represents the mortality rate of *Callosobruchus maculates*, X_1, X_2 are independent variables representing the coded values of the factors according to Table 1, a_0 (center point of system) is the average of the results of the replicated central point, a_1, a_2 are coefficients of linear effects and $a_{1.2}$ is coefficient of interactive effect.

In the present work, the influence of the concentration of the crude chitinase and the contact time on the mortality rate of insects has been estimated.

Material and Methods

The insect cuticle is composed of chitin. Chitin is a polymer of *N*-acetyl glucosamine and the second abundant component on the

Parameter	Lower level (-1)	Higher level (+1)	Centre Co-ordinates (0)
Doses of crude chitinases (%) X_1	3	24	14
Contact time (h) X_2	1	72	37

Table 1: Domain of variation of the parameters for the factorial planning 2^2.

***Corresponding author:** Dr. Nadjib Drouiche, Technology of Semi-conductor for the Energetic Research Center, Department of Environmental Engineering, 2, Bd Frantz Fanon BP140 Alger-7-merveilles, Algiers, Algeria

surface of the earth after cellulose [9]. It is present in bacterial, yeasts and fungi cell walls [10]. In animal reign, chitin is essentially present in exoskeleton of arthropods, and cuticle, trachea, diaphragm, peritrophic membrane and exochorion of insect [11]. However, chitinases may be used as antagonist to pests. Chitinases (EC3.2.1.14) are enzymes that are widely distributed in nature and play an important part in degradation of chitin. They occur in many fishes where they are produced by the digestive glands for the digestion of chitinous foods [12].

Choice of biological material and insects breeding

The choice of chitinases from offal scorpion fish is due to their high enzymatic activity (40%) compared to others fishes as Red mullet (24%), burbot (14%), bougue (13%) and silt (5%). Among the insects, *Callosobruchus maculatus* (chickpea weevil) has been chosen for several reasons: the species is cosmopolitan and infests stored products of economic importance in Algeria, and may be easily cultivated in laboratory for testing the efficiency of crude chitinase on a large number of individuals.

The insects were mass reared on chickpea in jars (18 cm in height and 11 cm in diameter). The jars are kept at a controlled temperature T=28 ± 2°C, in obscurity and at a relative humidity of about 75 ± 2%.

Extraction of crude extracts chitinase

The samples were prepared at 1/6 ratio [13] in weight of offal per volume of buffer solution prepared from citric acid (0.15 M) and disodium phosphate (0.3 M) at pH 5. The reaction mixture was incubated at 37°C during 3 h at 120 rpm.

At the end of the incubation, the macerated sample was filtered through a gauze fabric. The obtained filtrate was centrifuged at 6000 g during 30 min. The recovered supernatant is the crude enzyme extract.

Dosage of N-Acetyl-Glucosamine (NAG)

The dosage of NAG was performed using Reissig method [14] modified by Gustowska et al. [15]. This technique requires a reagent A, which is composed of potassium tetraborate solution (0.8 M) adjusted to pH 9.1 with 1 N KOH, and a reagent B involving 1.5 g of dimethylaminbenzaldehyde (DMAB) dissolved in 100 ml of pure glacial acetic acid containing 1.25% of HCl (12 N). The concentration of NAG was determined by adding 0.5 ml of supernatant and a blank (0.5 ml of distilled water) to 0.1 ml of reagent A.

The mixture was heated in a water bath at 100°C for about 3 min and then cooled in a cold water bath, followed by addition of 3 ml of reagent B and homogenization of the solution. The samples were immediately incubated in a thermostatic bath at 37°C during 20 min. The NAG concentration was determined using spectrophotometer at an amount of NAG liberated when chitin was incubated in presence of chitinases solutions. This concentration of NAG exposes the relative importance of process analysis of chitin on simple amines sugars [16].

One unit of chitinase activity (U) is defined as the amount of enzyme that releases 1 µg of NAG per hour [17].

Bioassays

Effect of contact

Different concentrations of crude chitinase 3% (10.5 U/ml), 6% (12.4 U/ml), 9% (29 U/ml), 12% (31.6 U/ml), 15% (49 U/ml), 18% (52 U/ml), 21% (69 U/ml) and 24% (72 U/ml) were deposited on Petri boxes with four replicates per dose. The sample is spread out, dried, and then ten couples of insects aged between 0 to 24 h were placed in each Petri box. The mortality rate was calculated after 1 h, 24 h, 48 h and 72 h.

Contact test on the treated seeds

For each essay, 2 ml of each sample of chitinase crude extract at various concentrations (3%, 6%, 9%, 12%, 15%, 18%, 21% and 24%) were homogeneously dispersed in 100 g of seeds. The experiments were repeated three times. All batches of treated seeds were infested with 10 pairs of insects aged between 0 and 24 h. The counting of dead insects is carried out after 1, 24, 48 and 72 h. The death recorded in batches of treated seeds was calculated and *corrected for mortality* in accordance with Abbott's formula [18] taking into account the natural mortality in samples.

Bio essay of crude chitinase on larval feeding

The method described by Xie [19] to measure chitinases effect on the insect feeding was utilized in our present work. A mix of 50 g of chickpea flour and 14 ml of the lethal dose (24%) was used for making meatballs the size of a chickpea. The seeds used for the control were made with chickpea flour and citrate buffer. The reconstituted seeds were dried for 24 h at a temperature of 30°C (to avoid distortion of chitinases and protein fractions) and at humidity of 75%. Chickpeas well prepared were placed in Petri dishes of 14 cm diameter with 10 insects couples to obtain the eggs. The eggs laid were counted daily until the death of the females and placed in an oven to assess fertility. The experiment was repeated four times for each dose and for the control.

The number of died larvae found in treated population with those concentrations previously cited were counted. The percentage of mortality was calculated and corrected with the control mortality [18],

(Eq.2): $MC(\%) = (M-Mt) / (100-Mt) \times 100$ (2)

MC is the percentage of the mortality corrected, M is the percentage of the mortality observed in the treated population and Mt is the percentage of the mortality in the control population. One way to test the effectiveness of a product is the calculation of Lethal Dose 50 (LD50) which corresponds to the amount killing 50% of individuals of the same batch. It is deduced by drawing the regression line. The percentages of mortality were transformed into corrected probist were plotted as a function of natural logarithm of the LD50 which was determined from the Eq.2 of a straight line obtained theoretically [20].

The experimental design and the results evaluation were performed using Statistica Software (Stat Soft France Version 9.0).

Results and Discussion

Evaluation of the effect of crude chitinase contact

The obtained results show that crude chitinases manifests insecticidal activity by contact. The percentages of mortality obtained ranged from 10 to 100% after only 1 h of contact. The control gave a percentage of mortality close to 0% (Figure 1). Other studies [21] using essential oils from aromatic plants were carried out on the same pest. *Melaleuca quinquenervia*, *Citrus aurantifolia* and *rosemarinus officinalis* revealed mortality rate of 60%, 48% (After a 24 h treatment) and 25% (after 96 h of treatment) respectively. The biopesticide produced in this study appeared more efficient than the essential oils reported in literature.

The results were confirmed by observation of scanning electron microscope SEM (JEOL 6360). Indeed, the deterioration of the insect's

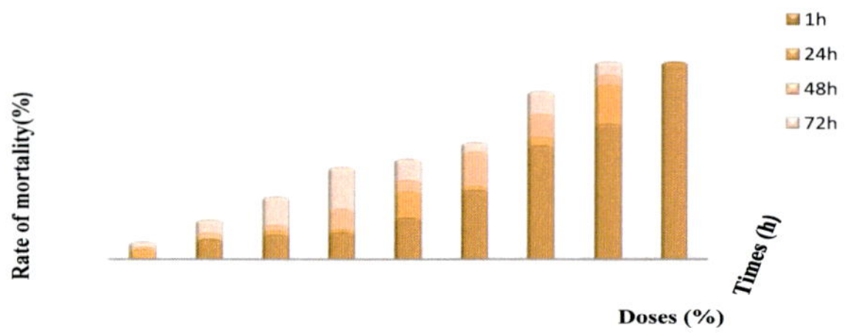

Figure 1: Evolution of percentage mortality (%) of *Callosobruchus maculatus* against doses of crude extract chitinase (%) and contact time (hours).

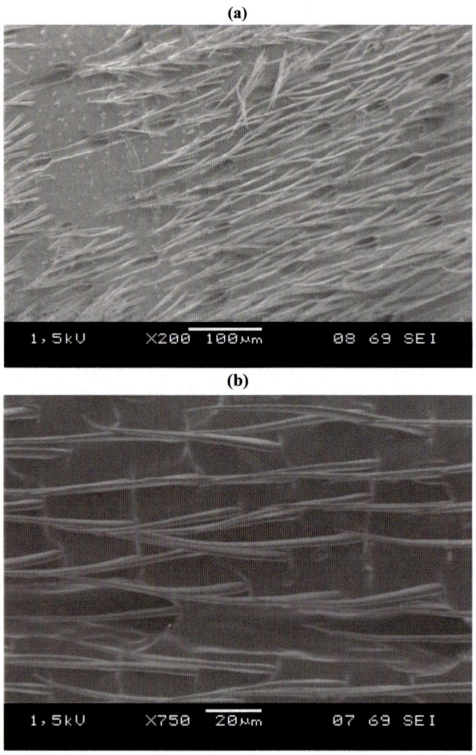

Figure 2: SEM image of a *Callosobruchus maculates* elytron (a):treated by contact with the lethal dose (24%) of crude chitinas, (b): control.

cuticule by the crude chitinases compared to control (Figure 2) was observed.

The LD50 of crude chitinases obtained against the contact time ranging from 1 to 72 h were close to 14%. Indeed, LD50 of 5% (1 h of contact), 15% (24 h of contact), 14% (48 h of contact) and 13% (72 h of contact) were obtained (Figure 3). However, for *Callosobrochus maculatus*, in all studies undertaken the essential oils have been used.

Contact test on the treated seeds and larval feeding

The results obtained after contact tests on the treated seeds showed similar mortality percentages compared to the previous experiments. Nevertheless, chitinases extracted raw from fish offal was very toxic against *Callosobruchus maculatus* with 100% of mortality rate at a dose of 24% after 24 h (Figure 4). Low doses (3%, 6%, 9% and 12%) showed a less active insecticidal effect.

To evaluate the effect of chitinase on larval feeding, five days treatment, all eggs laid entered reconstructed seeds. The results obtained show that after 10 days, all the larvae died after leaving the seeds. In the control, the rate of emergence is about 95% was obtained.

Experiments planning study

The matrix of the experimental design and the experimental results are shown in Table 2. By comparing the different obtained mortality rates, the mortality is mostly affected by the dose's variation of crude chitinases (%) and the contact time (h).

The lower percentage of mortality (10%) is obtained for low values of concentrations and contact time (dose of 3% of crude chitinase and 1 h of contact time). The maximum mortality rate is observed for the high values of two parameters (dose of 24% of crude chitinase and 72 h contact time). However, it is observed that for a higher dose (24%) and for a lower contact time (1 h), the mortality rate increased rapidly.

Table 3 presents the statistical significance of each effect of model's effect. According to the regression analysis, effects lower than 0.05 in the P-value column (95% confidence level) is significant ($P < 0.05$). In this case, it can be seen that the dose of crude chitinases was the most important influence on the mortality rate of *Callosobruchus maculatus*.

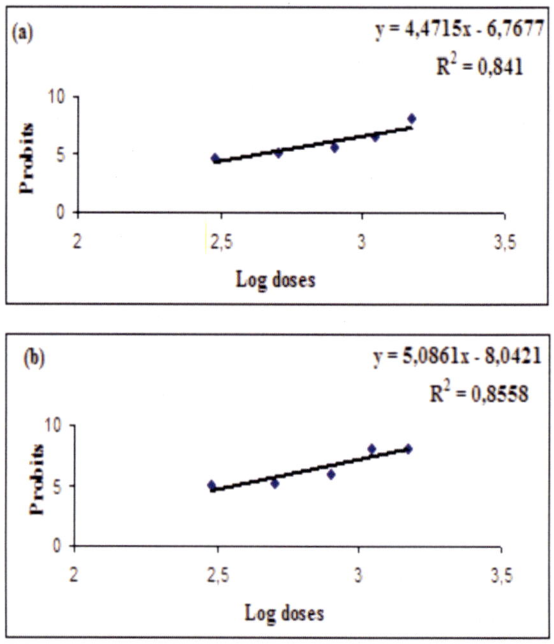

Figure 3: Regression curves: (a): after 24h of contact, (b): after 48h of contact.

Figure 4: Evolution of percentage mortality (%) of *Callosobruchus maculatus at* different doses of crude extract chitinase (%) and contact time (hours): contact test on the treated seeds.

Run n°	Doses of crude chitinases (%)	Contact time (h)	Mortality rate of *Callosobruchusmaculatus* (%)
1	-1	-1	10
2	+1	-1	100
3	-1	+1	12.33
4	+1	+1	100
5	0	0	33
6	0	0	40.8
7	0	0	37
8	0	0	39.5

Table 2: Experimental design and observed responses.

	Coefficient	Statistic (t)	Statistic (t)	Limit confidante (-95%)	Limit confidante (+95%)
Constant	46.57875	10.0754	10.0754	33.7432	59.41425
X_1	44.41750	6.79384	6.79384	26.2654	62.56964
X_2	0.58250	0.08910	0.08910	-17.5696	18.73464
X_1X_2	-0.58250	-0.08910	-0.08910	-18.7346	17.56964

Table 3: Effects of the coefficient of the model.

	Sum square	Freedom degree	Mean square	F	*P*
Regression	7894.372	3	2631.457	15.390072	0.011596
Résidu	683.9076	4	170.9769		
Total	8578.2796	7			

R^2= 0. 920274, Adjusted R^2 = 0.860480

Table 4: Analysis of variance.

The contact time and the interaction between the crude chitinase dose and the contact time were not statistically significant on the response.

The analysis of variance was used for testing the significance of the model, thus various statistical data such as sum of square. F-ratio and p-value are given in Table 4.

The P-value for regression in the ANOVA table is less than 0.05, which means that the model appears to be adequate for the observed data at the 95% confidence level.

The coefficient of determination (R^2) of the model was 0.920274, indicating that the model adequately represented the real relationship between the chosen parameters.

The value of adjusted R^2 (0.860480) suggests that the total mortality rate variation of 86% is attributed to the independent variables, and only about 14% of the total variation cannot be explained by the model.

The influence of the doses and the time on the mortality percentage obtained from the model is represented by the following Eq.3:

$$Y= 46, 57+ 44, 41\ X_1 + 0, 58\ X_2 - 0, 58\ X_1X_2\ (3)$$

The response surface obtained from this equation is presented in Figure 5 in order to show the behaviour of the mortality rate during experiments.

The response surface plot which gives the mortality rate (%) as a function of dose of crude chitinase and the contact time, shows that the mortality rate increases with augmentation of crude chitinase dose and contact time.

Conclusion

The insects were sensitive to biological tests. The crude chitinase extract was used as a biological insecticide and its effect is related to the dose and to the contact time. Thus, the approaches were designed to determine the effectiveness of crude chitinases contact. Interesting results are obtained for 100% of insecticide activity after 1 h contact at a dose of 24%.

The difference noted in the mortality rate of the insects at different doses attracts us to investigate many other aspects in this field. It would be appropriate to extend these trials to all other insects attacking leguminous seeds stocks and other grains such as *Sitophilus oryzae*, to effectively assess their actual action in stored-product protection.

References

1. Boutin JP, Dronne Y, Ducournau S, Gueguen J, Leguen, J, et al. (2006) Les protéagineux. INRA.
2. Erler F, Ceylan F, Erdemir T, Toker C (2009) Preliminary results on evaluation of chickpea, Cecerarietinum, genotypes for resistances to the pulse beetle, Callosobruchusmaculatus. J Insect Sci 9:1-7.
3. Kellouche A, Soltani N (2004) Activité biologique des poudres de cinq plantes et de l'huile essentielle d'une d'entre elles sur Callosobruchusmaculatus (F). Inter J Trop Insect Sci 24: 184-191.
4. Zhou J, Zeng X, Zheng K, Zhu X, Ma L, et al. (2012) Musks and organochlorine pesticides in breast milk from Shanghai, China: Levels, temporal trends and exposure assessment. Ecotoxicol Environ Saf 84:325-333.
5. Regnault R, Hamraoui A(1997) Lutte contre les insectes phytophages par les plantesaromatiques et leurs moleculesallélo chimiques. Edacta Bot Gallica: 401-412.
6. Yang D, Qi S, Zhang J, Wu C, Xing C (2012) Organochlorine pesticides in soil, water and sediment along the Jinjiang River mainstream to Quanzhou Bay, southeast China. Ecotoxicol Environ Saf 89: 59-65.
7. Karabelas AJ, Plakas KV, Solomou ES, Drossou V, Sarigiannis DA (2009) Impact of European legislation on marketed pesticides. A view from the stand point of health impact assessment studies. Environ Inter 35: 1096-1107.
8. Plakas KV, Karabelas AJ (2012) Removal of pesticides from water by NF and RO membranes — A review. Desalination 287: 255-265.
9. Anil K, Seshagirirao K, Podile AR (2007) A simple, rapid and yet less expensive method to detect chitinase in agarose plates. J BiochemBiophys Met 70: 683-684.
10. Baldrian P, Viriskova J, Dobiasova P, Merhautova V, Ludmila L, et al. (2011) Production of extracellular enzymes and degradation of biopolymers by saprotrophic microfungi from the upper layers of forest soil. Plant Soil 338: 111-125.
11. Jasrapuria S, Arakane Y, Osman G, Kramer KJ, Beeman RW, et al.(2010) Gene encoding protein with peritrophin a type chitin binding domains in Triboliumcastaneumare grouped in to three distinct families based on phylogeny, expression and function. Insect Biochem Molecul Biol 40: 214-227.
12. Neetu D, Tewari R, Hoondal GS (2006) Biotechnological aspects of chitinolytic enzymes a review. Appl Microbiol Biotechnol 71: 773-782.
13. Ikeda M, Miyauchi K, Mochizuki A, Matsumiya M (2009) Purification and characterization of chitinase from the stomach of silver croaker Pennahiaargentatus. Protein Expression Purif 65: 214-222.
14. Reissig JL, Strominger JL, Leloir LF (1955) A modified calorimetric method for the estimation of N-acetylamino-sugars. J BiologChem 217: 959-966.
15. Gustowska MA, Drazen JC, Robison BH (2004) Digestive chitinolytic activity in marine fishes of Monterey Bay,California.CompBiochemPhysiol A Mol IntegrPhysiol 208: 351-358.
16. Jeuniaux C (1963) Chitine et chitinolyse. Masson (Eds.), Paris.
17. Fines J, Holt GJ (2010) Chitinase and apparent digestibility of chitin in the digestive tract of Juvenile cobia, Rachycentroncanadum. Aquaculture 303:34-39.
18. Abbot SW (1925) A method of computing the effectiveness of an insecticide. J Economic Entomol 18: 265-267.
19. Xie YS, Bodnaryk RP (1996)A rapid and simple flour disk bioassay for testing natural substance active against stored product insect. CanadEntomolog128: 865-875.
20. Finney DJ (1971)Probit analysis, (3rd Ed.), Cambridge University Press.
21. Nondenot A, Louis R, Boudama P, Seri-Kouassi K, Kouakou H (2010) L'activité insecticide des huiles essentielles de trois plantes aromatiques sur Callosobruchus maculatus F en Cote d'Ivoire. J Europ Recher Sci 39: 243-250.

Evaluation of Survival Rate and Physicochemical Properties of Encapsulated Bacteria in Alginate and Resistant Starch in Mayonnaise Sauce

Bigdelian E[1] and Razavi SH[2*]

[1]*Department of Food Science and Technology, Shahre kord Branch, Islamic Azad University, Shahre kord, Iran*

[2]*Department of Food Science and Technology, Faculty of Biosystem Engineering, Campus of Agriculture and Natural Resources, The University of Tehran, Tehran, Iran*

Abstract

Three types of symbiotic mayonnaise sauces, with free, encapsulated bacteria with calcium alginate (in 4% concentration) and encapsulated with calcium alginate and resistant starch by two strains of *L. acidophilus* and *L. casei* were manufactured in triplicate under the same conditions. The numbers of viable cells, pH, acidity and rheological properties' of symbiotic mayonnaise samples during 91 days of storage in refrigerated (4°C) conditions was evaluated. It was observed that the number of viable cells of Lactobacillus acidophilus and *L. casei* was reduced significantly ($p<0.05$) from day 1 to day 91 of storage period in Free State comparing encapsulated bacteria in both type of them. Reduction of viable cells encapsulated with Hi maize-alginate had a slower rate than probiotic cells encapsulated with calcium alginate (4%) in both *L. acidophilus* and *L. case*i mayonnaise samples container types. About changing in pH and increasing acidity of symbiotic mayonnaise sauces, the samples containing free probiotic cells had the highest changing and samples inoculated encapsulated cells with Hi maize-alginate mixture had lowest changing in these factors between all sample types. Also there weren't any significant changes in rheological properties of encapsulated symbiotic mayonnaise sample compared with control samples but in samples containing free cells a significant differences compared with control samples were seen. Finally, proved microencapsulation could help survival probiotic cells enough in therapeutic effects on consumers at the end of storage and modification encapsulation with adding Hi maize starch to calcium alginate improved preserving chemical qualities of symbiotic mayonnaise sauce.cells enough in therapeutic effects on consumers at the end of storage and modification encapsulation with adding Hi maize starch to calcium alginate improved preserving chemical qualities of symbiotic mayonnaise sauce.

Keywords: Calcium alginate; Resistant starch; Microencapsulation; *Lactobacilli*; Survival; Mayonnaise sauce

Introduction

The expansion of new food products turns out to be increasingly challenging, as it has to comply the consumers' expectations for products that are at once relish and healthy [1] , in this regard, functional food which have health benefits further than their nutritional contents and especially foods containing probiotics are products that are growing in popularity [2].

Probiotics are defined as alive microorganisms which, when employed in sufficient amounts confer a health benefit to the consumers. Prebiotics are non-digestible substances which provide a beneficial physiological effect on the host by selectively stimulating the favorable growth or activity of a limited number of indigenous bacteria [3].

On the protective level, probiotics can act as a 'barricade' against pathogens, by decreasing the luminal pH, producing bacteriocins, competing for limiting nutrients, activate mucosal immune responses and by sticking to the intestinal mucosa, thereby occupying an ideal position at the expense of potentially harmful colonizers [4]. It has been suggested that they must be presented at a minimum level of 106 CFU/g of Probiotic products or 107 CFU/g at point of delivery or be consumed in adequate amounts to comply a daily intake of 108 CFU [5].

Probiotics must be metabolically stable and active in the product, surviving while passing the upper digestive tract in large numbers, and showing the ability to adhere and colonize the intestine system [6]. The physical protection of probiotics by microencapsulation is a new method to improve the probiotic survival. Encapsulation helps to separate the bacterial cells from the effects of the harsh environment and gastrointestinal tract, thus potentially preventing cell from dead [7]. The benefits of applying alginate as an encapsulating agent include: non-toxicity, formation of suitable matrices with calcium chloride to entrap sensitive materials such as living microbial cells, easiness in trapping living microbial cells and inexpensively [8].

Mandal, Puniya and Singh proved an increasing concentration of calcium alginate from 2% to 4% which had a positive effect on the survival of *L. casei* in simulated harsh conditions of GIT and heat processing, and calcium alginate in 4% concentration had the best effects [9].

The prebiotics, indigestible food components, affects the host by selectively instigating either the growth or activity, or both or a limited number of bacterial species already living in the colon [10]. A food product containing both probiotics and prebiotics is named as symbiotic or functional food. There is a synergy between probiotics and prebiotics in symbiotic products [11]. Combining alginate with starch as a prebiotic leads to better effectiveness of different bacterial cells specifically LAB. Also combination of calcium alginate with starch

***Corresponding author:** Razavi SH, Department of Food Science and Technology, Faculty of Biosystem Engineering, Campus of Agriculture and Natural Resources, The University of Tehran, Tehran, Iran

produces beads with good unified structures and prebiotic effect of the bead shell [12].

Food, especially dairy products are Preferred as an ideal mean for delivering probiotic bacteria to the human gastrointestinal tract [13] but restriction of dairy products such as the presence of allergens and the need for cold storage facilities, as well as an increasing request for new foods and tastes have begun a trend in non-dairy probiotic product development. Further, it is important to expand probiotic products with food and drinks that are part of day to day normal diet to maintain minimum beneficial level easily [14]. An alternative strategy to increase the efficacy of a probiotic treatment would be to use a food matrix which naturally contains a higher content of ingredients with protective properties [4]. The development of novel formulations is a challenging task but, regardless of whether the product is newer improved, the product must be stable during storage and transportation, easy to handle and apply, enhance the activity of the organism in the field, be cost-effective and practical.

Mayonnaise is of course not only one of the most favored sauces but also apparently one of the oldest and most greatly used sauce in the today's world [15]. Mayonnaise is a kind of semi-solid, oil-in-water emulsion. It is customarily prepared by carefully mixing egg yolk, vinegar, oil, and spices (especially mustard). Mayonnaise which made in this way typically contains 70–80% fat [16]. Physico-chemical properties of food carriers used for probiotic delivery, such as buffering capacity and pH, are important factors that affect survival of the probiotic and therefore potential probiotic effects during gastric passage. Food formulations with acceptable pH ranges and high buffering capacity would increase the pH of the gastric tract and thereby improve the stability of probiotics [17], so mayonnaise due to having some advantages like its high buffering capacity, similar pH to some dairy products and its good matrix to protect of probiotics (because of having a large amount of protective ingredients like oil & egg) could be a new choice to produce a new functional food.

Material and Method

Preparation and enumeration of free and encapsulated probiotics

Pure freeze-dried probiotic culture of *L. casei (Lc-01)* and *L. acidophilus* (La5) were obtained from CHR-Hansen (Harsholm, Denmark) and were activated by inoculating in the MRS-broth (de Man-Rogasa-Sharpe) at 37°C for 24 h. The probiotic biomass in late-log phase was collected by centrifugation at 600g for 10 min at 4°C (Microcentrifuges SIGMA; model 1-14K; Nr. 10021), then washed twice in sterile 0.9% saline under the same centrifugation conditions, and used in the microencapsulation process.

Bacterial counts were determined immediately after manufacturing of mayonnaise (at time 0) and during 91 days storage at 5°C. The samples of mayonnaise (10 g) were diluted in 100 ml sterile peptone water (0.1%) and 1 ml aliquot dilution was poured into plates of the MRS agar [18-20].

Counting of probiotic bacteria was achieved as described by Haynes and Playne [21]. All container of mayonnaise was sampled on each week of enumeration. All enumerating plates of *L. casei* and *L. acidophilus* were incubated at 37°C for 72 h under aerobic conditions .The averages of all results were expressed as colony-forming units per gram of sample (CFU/g). To count the encapsulated bacteria in mayonnaise, the entrapped bacteria were released from the beads according to the method of Sheu and Marshall [22]. Ten grams of mayonnaise were re-suspended in 100 mL of phosphate buffer (0.1 M, pH 7.0), followed by shaking in a stomacher for 10 min. The mayonnaise sample containing free bacteria was treated in a similar way, in order to maintain the same treatment conditions. The counts (CFU/g) were determined by plating on MRS-agar as discussed above. The mayonnaise sample containing free bacteria was treated in a similar way so as to maintain the same treatment conditions. All experiments were done in triplicate.

Microencapsulation procedure

All glassware's and solutions used in the protocols were sterilized at 121°C for 15 min. Alginate beads were produced using a modified encapsulation method originally reported by Sheu and Marshall [22]. All treatments including a Sodium alginate solution 4% (Merck, Darmstadt, Germany) and a 2% alginate mixture in distilled water containing 2% Hi-maize resistant starch (Sigma, Aldrich, Germany) were prepared, sterilized by autoclaving (120°C for 15 min) and cooled to 38-40°C. Twenty milliliters of this solution and 4mL of cell suspension were transferred into a centrifuge tube (40 mL) and the content was vortexed to homogeneity. Soybean oil (100 mL) containing 0.2% Tween 80(emulsifier) was taken in a beaker (500 mL) and to this the alginate–cell mixture was added drop wise while stirring magnetically. After 5 min, a uniformly turbid emulsion was obtained to which 0.1M calcium chloride (100mL) was quickly added for hardening of microcapsules and breaking the emulsion. The capsules were harvested by centrifuging at 350 g for 10 min at 41°C and washed twice with distilled water. The long-term activity of BL21_MlrA in comparison with wild Sphingomonas strain confirmed much higher potential of the modified bacteria. Immobilization in alginate allowed forming beads with high activity toward MC. A column packed with alginate entrapped cells eliminated MC efficiently from contaminated freshwater. These promising results will broaden the perspective of practical application of microorganisms in bioremediation of freshwater.

Production of mayonnaise sauce

All ingredients used to prepare the mayonnaise, such as soybean oil, eggs, vinegar, mustard, sugar, and salt were purchased from the local supermarket. mayonnaise recipe was modified from Chen [23] with the following ingredients in percentage (w/w): pure whole egg (yolk + albumen)14%, vinegar (5% (w/v) acetic acid) 9%, soybean oil 74%, salt 1.0%, vanillin 0.1%, mustard 0.54%, sugar 1.0% and ground white pepper 0.36% [23].

The preparation was as follows: firstly egg and vinegar were mixed together and then all other ingredients except oil were added and stirred homogeneously by a mechanical overhead stirrer. Finally oil was added very slowly, while stirring at 1600 rpm for 1 min, followed by 2000 rpm for another 4 min. The mixture was cooled to 5-10°C and stored [16]. The mayonnaise was divided to six parts of 250 g (A, B, C, D, E and F). 1% (w/v) free *L. casei* and *L. acidophilus* were added to batches A and B, respectively. The initial counts of these bacteria in mayonnaise for A and B were about 6×10^9 and 5×10^8 (CFU/g), respectively and D portions of mayonnaise sauce mixed with freshly prepared encapsulated *L. casei* and *L. acidophilus* into 4% calcium alginate beads. The initial counts of these bacteria for C and D were about 3×10^8 and 3×10^9 (CFU/g), respectively. Also, microencapsulated bacteria *L. casei* and *L. acidophilus* into beads containing 2% calcium alginate and 2% Hi maize starch incorporated into E and F samples. The counts of probiotic cells in these types of mayonnaise were approximately 9.8×10^7 and 4×10^9 (CFU/g), respectively. All experiment was conducted in triplicate.

Chemical and physical analysis

The pH of the mayonnaise was measured using a digital pH-meter (Microprocessor pH-meter,). Titratable acidity was determined using the following formula at milli moles per 100 grams [24].

Equation 1: $W = V\times0.9 / M$

W = acidity at mill moles per 100 grams

V = milliliters of consumed soda

M = sample weight (preferably100 g)

All chemical measurements were done in triplicate.

Reaction of alginate without borate

One ml of concentrated sulphuric acid and 0.1 ml of 0.5 g/l of alginate solution sample were taken in the test tube. The test tube was cooled in the ice bath and the mixture was shaken and kept in the water bath at 55°C. After 30 min, the sample was cooled in ice bath and 30 µl of carbazole reagent (0.1% in ethanol) was added. The tube was allowed to stand for about 3h and then the absorbance was taken at 546 nm. The color was stable for 2 h.

Rheology analysis

The rheological measurements were evaluated using Brookfield rotational viscometer (DVIII) equipped by ULA. Obtained data analyzed by rheocalc 3.2 software, to determine its rheological properties. The temperature of the test (25 ± 0.1°C) was kept circulating-refrigerating water bath TC502 made by Brookfield Company.

Statistical analysis

The collected data was analyzed by SPSS statistics software, Version 18 edition .The mean values and the standard error were calculated from the data obtained with triplicate trials. These data were then compared by the Duncan's multiple range method.

Result and Discussion

PH changes during mayonnaise sauces storage

The pH changes during mayonnaise sauce storage in experimental mayonnaise sauces during storage at 5°C (per 7-day intervals has been presented) for a period of 91days are shown in Figures 1 and 2.

The pH changes (at end of 91 days storage) of mayonnaise sauces with encapsulated probiotic bacteria was lesser than the mayonnaise sauces inoculated with free probiotic bacteria, also the mayonnaise sauces containing encapsulated cells with Hi maize-alginate in both of *L. casei* and *L. acidophilus* samples had the minimum of pH changes during storage times.

Therefore pH decreasing in symbiotic mayonnaise samples containing *L. acidophilus* in Free states was about 0.3 and in samples containing encapsulated bacterial cells in Hi maize and alginate mixture and encapsulated in alginate 4% were about 0.1 and 0.244 unit respectively. Also *L. casei* inoculated in symbiotic mayonnaise samples in un-encapsulated forms, decreased pH about 0.3579 and bacterial cells in encapsulated with Hi maize and alginate mixture and encapsulated in alginate 4% states were decreased the pH of their environments about 0.1 and 0.17 unit respectively.

Generally Probiotic bacteria are slow acid producers [25] hence changes of pH at storage period time in symbiotic mayonnaise sauces wasn't great.

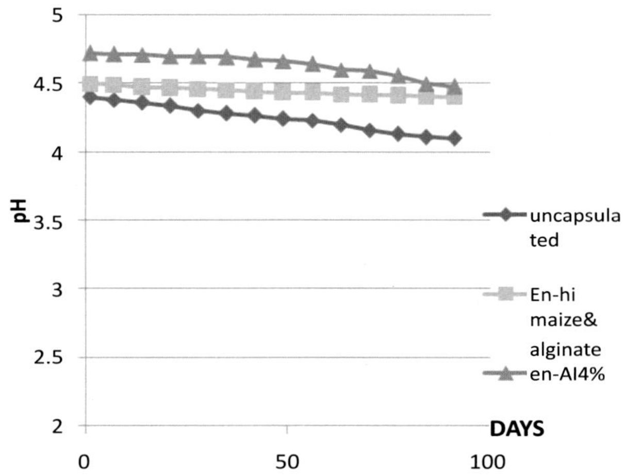

Figure 1: Mayonnaise sauces pH changes containing *L.acidophilus* in Free State, encapsulated by Hi maize-alginate and alginate 4% during storage.

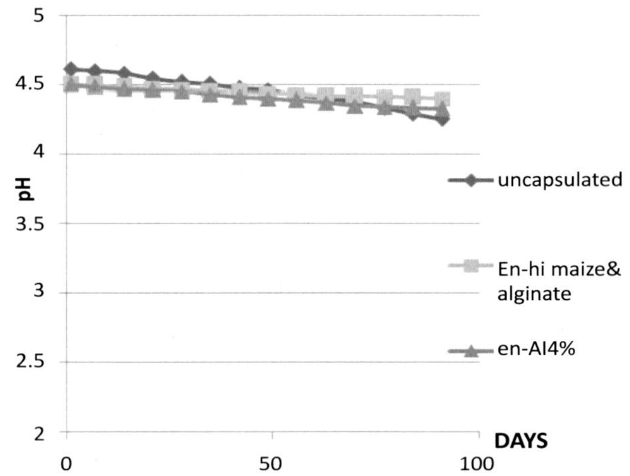

Figure 2: Mayonnaise sauces pH changes containing *L.casei* in Free State, encapsulated by Hi maize-alginate and alginate 4% during storage.

Producing acidity during mayonnaise storage

Measuring the acidity per 7- day intervals (Figures 3 and 4) indicated parallel results that were reached about pH changes during 91 refrigerated in 4°C. According these data's, increasing acidity in mayonnaise samples containing free *L. acidophilus* was approximately 0.25 and increasing acidity in samples containing encapsulated cells (similar strain) in Hi m aize and alginate mixture and sauces samples those blended with encapsulated cells in calcium alginate 4% were 0.142 and 0.21 respectively.

Also increasing acidity in mayonnaise samples containing *L. casei* in free, encapsulated with Hi maize-alginate and calcium alginate 4% were 0.14, 0.1 and 0.11 respectively.

As a result increasing of acidity from starting storage until 91 days in mayonnaise sauces those were containing encapsulated probiotic cells were less than free states, also the probiotic cells in Hi maize-alginate coating is produced the minimum level of acidity in treatment containers in two kind of *L. acidophilus* and *L. casei* probiotic bacteria's.

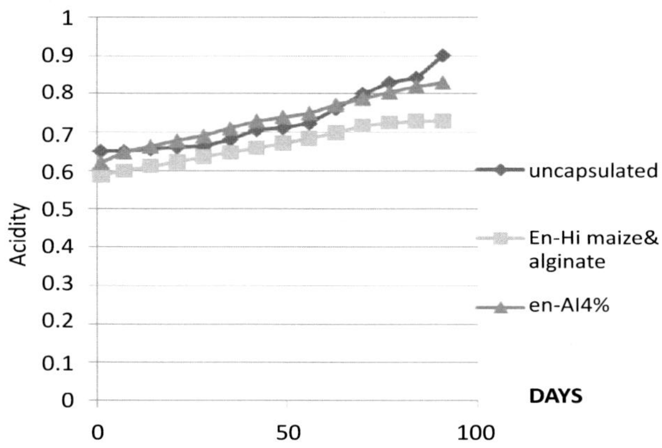

Figure 3: Mayonnaise sauces increasing acidity containing *L. acidophilus* in free and encapsulated states during storage.

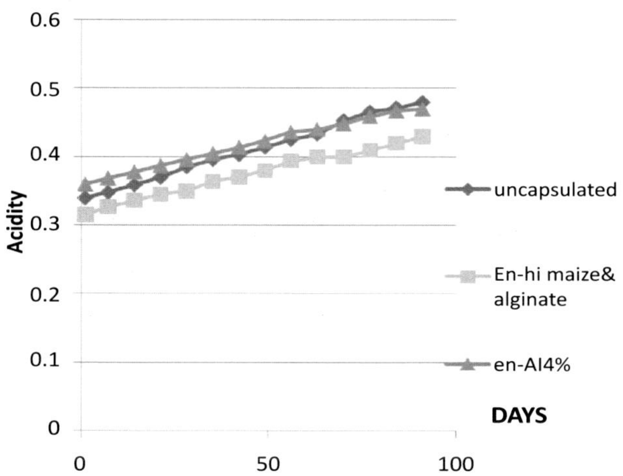

Figure 4: Mayonnaise sauces increasing acidity containing *L. casei* in free and encapsulated states during storage.

Therefore, microencapsulation of free cells greatly, but not completely, restricted their metabolic activity [26]. Even though symbiotic mayonnaise samples inoculated with L. acidophilus had a little more acid producing compared with samples included L. casei bacterial cells.

Survival of free and microencapsulated *L. acidophilus & L. casei* in mayonnaise sauce during storage time

The effect of encapsulation with Hi-maize starch and alginate mixtures (2% concentration) and also by alginate (4% concentration) on the viability of probiotic microorganisms comparing with free cells in the mixture of mayonnaise sauces during 91 days of refrigerated storage period (4°C) was investigated. Logarithmic numbers of surviving bacteria (log cfu/mL) were measured at seven-day intervals are shown in the Table 1.

According to Table 1, viable cells of *L. acidophilus* mixture with mayonnaise sauce showed 2.659 log decreases for the Free State after 91 days, while the encapsulated state with alginate (4% concentration) of the same strains showed a decrease of 1.48 logs. In the case of encapsulated bacteria (*L. acidophilus*) with Hi-maize starch and alginate mixtures, the cell numbers dropped substantially about 1.1497 log numbers by 91 days of storage at 4°C. The loss of *L. acidophilus* showed significant differences (P<0.05) between the free and both type of encapsulated states in symbiotic mayonnaise sauce and there wasn't a significant difference between two type of encapsulated types, but as was observed, using the Hi maize starch along with alginate in micro beads cause improving 22.3 percentage the survival of *L. acidophilus* bacteria during storage period than usage alginate (4%) lonely.

Sultana et al., [20], Homayouni et al., [8] and Mirzaei et al., [13], showed that using the Hi maize in encapsulation probiotics can significantly improving protection of viable bacterial counts in symbiotic ice cream and Iranian white brined cheese respectively [7,12,19]. The mayonnai se sauce samples that were contain of *L. casei* also showed the same results. In the case of free *L. casei*, the cell numbers dropped substantially about 2.52 log numbers by 91days of storage at 4°C, however the decrease of the viable counts of encapsulated bacteria in Hi maze-alginate and encapsulated bacteria in alginate (4%) were 1.1log and 1.377log respectively that indicates 20.11 percentage improving on survival of cells in the case of encapsulated with alginate-Hi maize comparing with cells coated with alginate 4%.

There was a significant difference (P<0.05) between means of the counts of viable cells per 7-days intervals during storage time (91 days) in Free states of *L. casei* and both of encapsulated samples, however there wasn't observed any significant difference in comparison means of viable cells during storage time between Hi maize- alginate and alginate4% encapsulated forms. These results indicate remarkable of increasing viability and resistance of probiotic cells by encapsulation against harsh environmental conditions at storage times.

Increasing cell stability of encapsulated probiotic bacteria with calcium alginate over the refrigerated storage time has been previously reported by Krasaekoopt, Kailasapathy and Mortazavian, that which are in agreement with the results obtained from present research [27-29]. Also these data's again confirmed advantages of encapsulation of probiotic cells with Hi maize in combination with alginate that more than twice protection of viable counts of bacteria compared Free State caused. In addition, it was demonstrated that, encapsulated cells required longer time to decrease one log cycle in viable counts and encapsulated bacterial cells with alginate and resistant starch had the most slow rate of loss viable cell rate.

The supporting impact of microcapsules on probiotic bacteria arises from their barrier properties against the detrimental conditions of mayonnaise environment such as low pH, organic acids and antibacterial effects of mustard [30] and the existence of some antibacterial enzymes' from albumen such as lysozyme, conalbumin and avidin.

Overall the problem of sensitivity to acidity of the probiotic culture is compounded by the fact that acidity may increase during storage, a phenomenon known as 'over acidification' [31]. Increasing rate of viability decline of free probiotic bacteria can be attributed to the adverse effects of accumulated organic acid molecules and hydrogen ions after they grew inside the mayonnaise environment at days 49 and 56 in *L. acidophilus* and *L. casei* respectively.

About encapsulated bacteria accumulating of organic acid molecules and hydrogen ions had a lower rate because of limitation accessibility to nutrients'. This fact is indicated by slower rates of pH drop and acidity increase during refrigeration of both encapsulated types of probiotic bacteria which caused a slower rate of viability decline comparing free probiotic bacteria cells. But the greater loss of survival of encapsulated *L. acidophilus* and *L. casei* cells at the end of storage period time may be due to the gradually increased amount of organic acid molecules inside the capsules after slow diffusion through the capsules pores in mayonnaise environment.

This suggests that the starch grains were presented in the cavities in the alginate matrix [19] may could fulfill the shell surface of beads and due to more restriction of crossing nutrient and metabolites across shell matrix and also reduction of entrance organic acids and other adverse components (such as some antibacterial components of mustard or antibacterial enzymes' of albumen) exist in mayonnaise environment. In this regard, reduction of metabolic activity of probiotic cells in the case of coated cells with resistant starch and alginate caused to a very gentle increasing of acidity and so slowing loss of viable cells during refrigeration period.

With comparing two strains of probiotic cells, it was found that loss of all types of free and coated bacterial cells of L. casei from days 1 to 91th was lower than the same type of *L. acidophilus* cells. In this way the survival of bacteria against unfavorable conditions could be species dependent. This finding is in agreement with those of Haynes, Kailasapathy and Homayouni [23,28].

Comparing rheological properties of mayonnaise sauces before and after storage period

The viscosity of Non-Newtonian fluids cannot be calculated using only the base properties, the main calculation should be in the flow model. In or-der to correlate flow model and medium model with the smallest change possible. Since the flow behavior index is only a function of the states of the fluid such as the temperature, it can be defined and calculated in the medium model. By adding this partial function for the definition of the flow behavior index in the base medium model, implementing the governing equations or tables for all the fluid models is made possible.

In case the fluid is a Newtonian fluid such as water, the flow behavior index should be set to the constant number of "1" and with decline of this index from "1" demonstrate fluid is nearing shear-thinning properties [32,33].

In this work comparing consistency indexes and flow indexes of symbiotic mayonnaise sauces before and after period of storage time were studied.

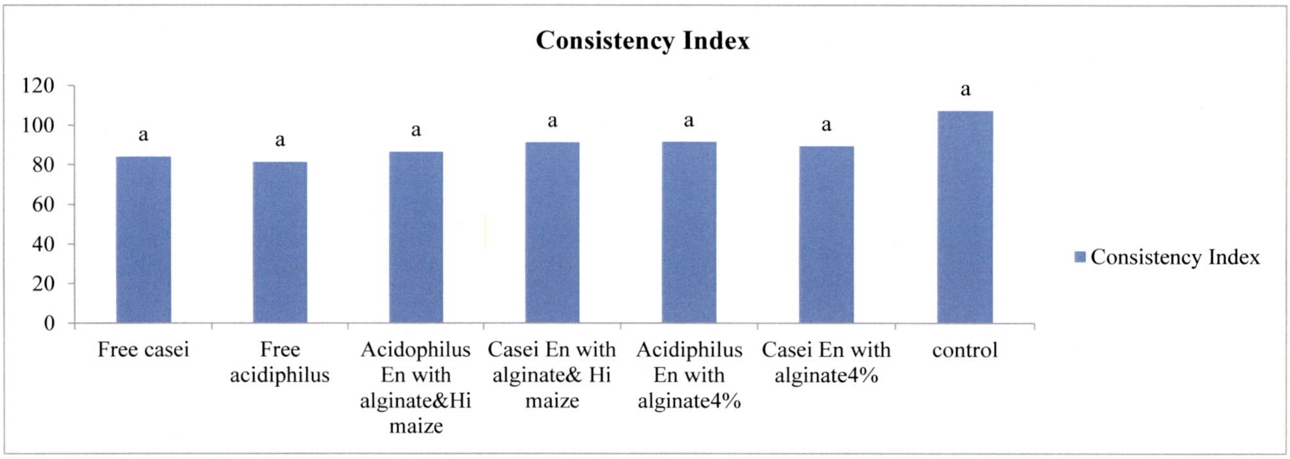

Figure 5: Consistency indexes were compared by the Duncan's multiple range method at the start of storage time (p<0.05).

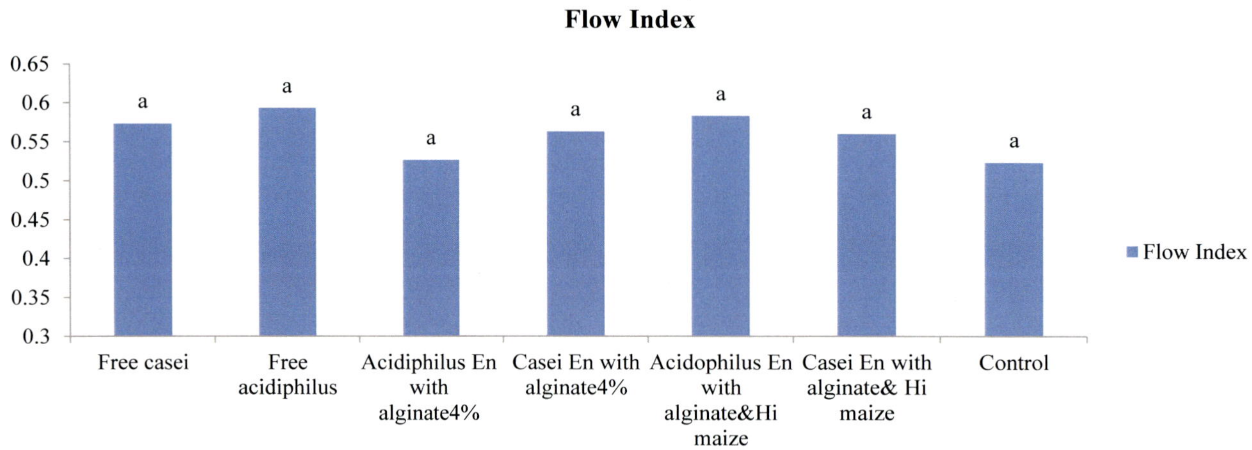

Figure 6: Flow indexes were compared by the Duncan's multiple range method at the start of storage time (p<0.05).

According Figure 5 consistency indexes of symbiotic mayonnaise sauces in neither samples at the start of storage period hadn't significant difference ($p<0.05$) with control treatments (control samples hadn't any probiotic cells) and this fact indicates that adding free probiotic cells or encapsulation operation hadn't significant changes in rheological properties of mayonnaise sauce.

This fact proved while it wasn't any significant differences ($p<0.05$) between flow indexes of control and symbiotic mayonnaise sauces in both free and encapsulated probiotic cell samples (both of *L. casei* and *L. acidophilus*) at the start of storage period (Figure 6).

So there weren't significant changes in rheological properties' of our symbiotic product by inoculation of probiotic cells in free or encapsulated form in mayonnaise sauces samples (Figure 7).

After 91 days storage period rheological properties of symbiotic mayonnaise sauces samples were measured and compared with control samples to determine any changes in consistency indexes and flow indexes.

The results (Figures 8 and 9) indicated that the consistency indexes in control samples and symbiotic mayonnaise sauces containing encapsulated probiotic bacteria's (both *L. casei* and *L. acidophilus* in two kinds of beads) were not any statistical significant difference

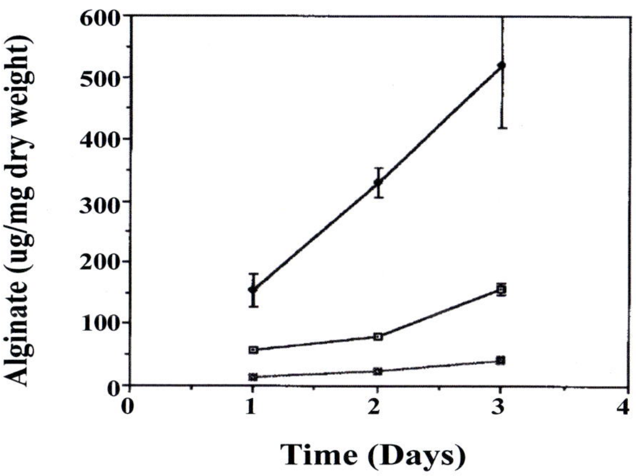

Figure 7: Alginate production.

($p<0.05$). But all symbiotic mayonnaise samples containing free cells (of both species) had a significant differences ($p<0.05$) in consistency indexes compared with control mayonnaise samples, however there were not any significant differences between mayonnaise samples containing encapsulated bacteria and samples containing free cells (both species in two kinds of beads) in consistency indexes .

About flow index, after period of storage time there weren't significant differences ($p<0.05$) between control samples and samples containing encapsulated probiotics (both strains, both coating) but we had significant differences ($p<0.05$) in flow indexes of control samples and mayonnaise containing free cells (both strains). In other wise we saw significant differences ($p<0.05$) in flow indexes neither of samples containing free *L. acidophilus* and samples containing encapsulated cells though nor about samples containing free *L. casei* cells.

Thus after 91 days storage period we didn't see significant changes in rheological properties of encapsulated probiotic mayonnaise samples (Figure 7).

Conclusion

In this study we proved that mayonnaise sauce could be serving as carriers for delivering the probiotic bacteria into the human gut. Also this work demonstrated that microencapsulation of probiotic bacteria significantly and noticeably improves their viability in mayonnaise sauce throughout the 91days refrigerated storage at 4°C.

The high total solids level in mayonnaise sauce including the oil and egg solids may provide protection for the probiotic bacteria. A modified method of encapsulation has been reported in this study. There appears to be much potential for using the prebiotic resistant starch with alginate during encapsulation since it does enhance the survival of the probiotic bacteria.

According to the results of this study, microencapsulation of *L. acidophilus* and *L. casei* cells with calcium alginate gel and resistant starch can successfully keep the count of this probiotic bacterium higher than encapsulated cells of them with calcium alginate 4%, but enough in both methods for the therapeutic minimum in the mayonnaise sauce [34].

So at the end of 91 days of storage, the number of viable probiotic

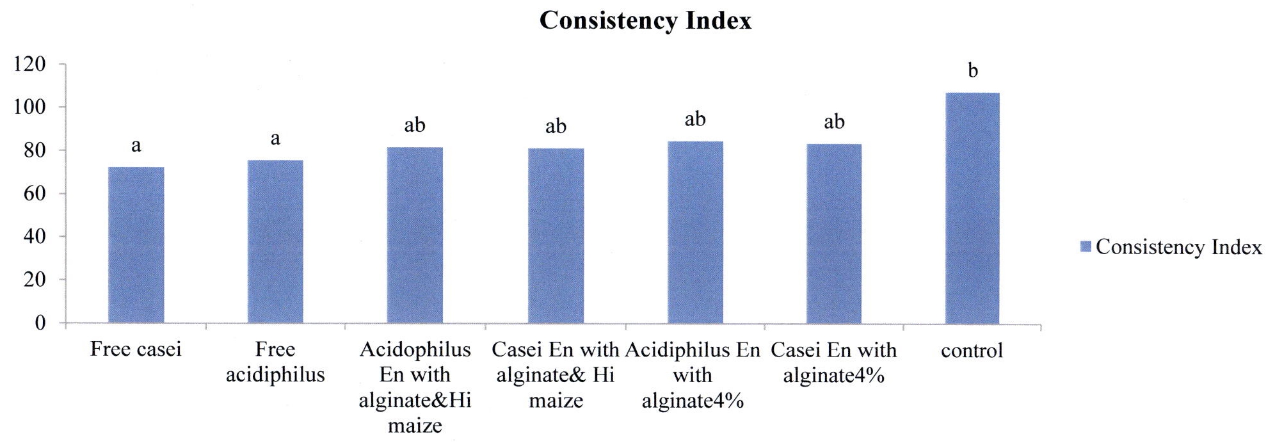

Figure 8: Consistency indexes were compared by the Duncan's multiple range method at the end of storage time ($p<0.05$).

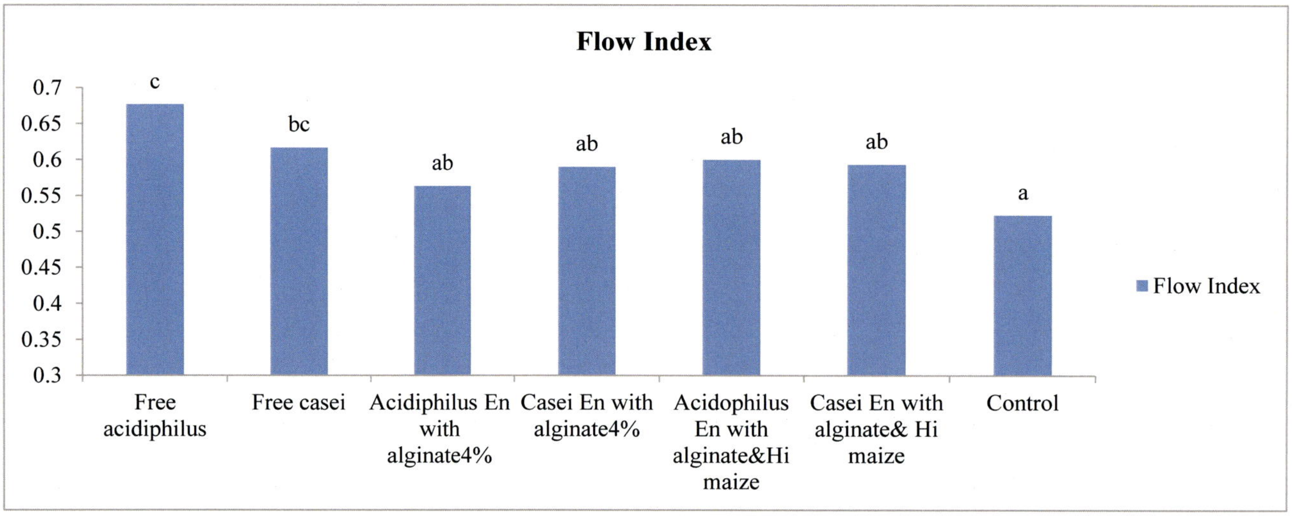

Figure 9: Flow indexes were compared by the Duncan's multiple range method at the end of storage time (p<0.05).

Days		*L.acidophilus*			*L.casei*	
	Free	En-Hi&AL[a]	En-AL4%[b]	Free	En-Hi&AL[a]	En-Al4%[b]
0	8.698	9.601	9.42	8.748	8.301	8.477
7	8.6	9.3	9.397	8	8.278	8.304
14	8.505	9.1	9.389	7.778	8.146	8.291
21	8.468	8.9	9.318	7.568	8.1	8.198
28	8.451	8.85	9.24	7.477	8.079	8.105
35	8.301	8.82	9.113	7.486	7.994	8
42	8	8.77	8.845	7.409	7.929	7.919
49	7.9	8.7	8.654	7.358	7.845	7.786
56	7.568	8.66	8.397	7.278	7.69	7.733
63	7.185	8.602	8.38	7.146	7.582	7.6
70	6.948	8.54	8.25	7.095	7.462	7.337
77	6.711	8.486	8.11	6.845	7.304	7.284
84	6.474	8.47	8	6.544	7.227	7.22
91	6.039	8.45	7.94	6.23	7.204	7.1

a: Encapsulated by Hi maize starch and Calcium alginate (2% concentrations)
b: Encapsulated by Calcium alginate (4%concentrations)

Table 1: Viability of probiotic bacteria (log cfu/mL) in encapsulated and un-encapsulated treatments over a 91-dayrefrigerated storage period (4ºC), per 7-day intervals.

bacteria in mayonnaise types containing *L. acidophilus* and *L. casei* that encapsulated (with Hi maize-alginate and alginate 4%) were higher than what was recommended by the International Dairy Federation (107 CFU/g) (Table 1).

The addition of probiotic cultures in encapsulated forms tend to slow down the post acidification during storage of mayonnaise sauces. Thus according our views encapsulation led to limitation accessibility of probiotic bacterial cells to consuming nutrients and caused to restriction to increasing acidity and pH changing comparing the Free State probiotic cells in mayonnaise sauces samples during storage period [35].

The pH changing and increasing of acidity in symbiotic mayonnaise samples containing *L. acidophilus* and *L. casei* encapsulated with Hi maize and alginate were less than mayonnaise samples that contained encapsulated cells with calcium alginate 4% but we 've seen the most changing in acidity and pH in samples contains free cells (in both kind of probiotic bacteria's). This may be due to the slower uptake of nutrients and the slow release of metabolites across the encapsulated alginate shell and particularly in beads those filled with resistant starch particles [19].

Also about rheological changes in symbiotic mayonnaise sauces it's defined that inclusion of probiotic cells in free or coated states couldn't have a significant effect on consistency and flow indexes of mayonnaise sauces but after the period of storage the high ratio of increasing acidity and producing more of other metabolites in mayonnaise containing free probiotic cells (*L. casei & L. acidophilus*) comparing with control samples we had a significant changes in flow and consistency indexes and these samples hadn't initial textures.

Even though by the microencapsulation of probiotic cells additional of biological and chemical properties, we could save rheological (flow and consistency indexes) qualities' of symbiotic mayonnaise sauces and so there weren't significant changes in these factors comparing control treatment's after 91 days storage period.

According these results, microencapsulation with alginate gel and resistant starch had the best efficiency to survival the probiotic cells enough for therapeutic effects and as well as increasing storage time with less chemical changes in symbiotic mayonnaise sauce during refrigerate in 4°C.

Alginates have a clear antibacterial effect as evidenced by the reduction in growth following the addition of alginates. This inhibitory effect of alginates on bacterial growth was found to be bacteriostatic as the cultures grew well when plated onto agar. The bacteriostatic effect is likely in part to be due to the iron Alginates have a clear antibacterial effect as evidenced by the reduction in growth following the addition of alginates. This inhibitory effect of alginates on bacterial growth was found to be bacteriostatic as the cultures grew well when plated onto agar. The bacteriostatic effect is likely in part to be due to the iron chelating capabilities of alginates as the antibacterial effect was lost when the MIC assay was repeated using alginates loaded with iron. Thus in summary we provide evidence that it is likely that alginates mediate these alterations in the gut micro flora through their potential to sequester iron.

Further studies are needed to evaluate the protection effect of microencapsulation on the probiotic survival in the gastrointestinal tract.

References

1. Cruz AG, Antunes AEC, Sousa ALOP, Faria JAF, Saad SMI (2009) Review Ice-cream as a probiotic food carrier. Food Research International 42: 1233-1239.
2. Muthukumarasamy P, Holley RA (2006) Microbiological and sensory quality of dry fermented sausages containing alginate-microencapsulated Lactobacillus reuteri. International Journal of Food Microbiology 111: 164-169.
3. www.fao.org/es/ESN/Probio/Probio.htm
4. Possemiers S, Marzorati M, Verstraete W, Van de Wiele T (2010) Bacteria and chocolate: A successful combination for probiotic delivery. International Journal of Food Microbiology 141: 97-103.
5. Chan ES, Zhang Z (2002) Encapsulation of probiotic bacteria *Lactobacillus acidophilus* by direct Compression. Institution of Chemical Engineers 80: 78-82.
6. Brinques GB, Ayub M (2011) Effect of microencapsulation on survival of *Lactobacillus plantarum* in simulated gastrointestinal conditions, refrigeration, and yogurt. Journal of Food Engineering 103: 123-128.
7. Mohammadi N, Ahari H, Fahimdanesh M, Khosravi Zanjani MA, Anvar A, et al., (2013) Survival of alginate-prebiotic microencapsulated Lactobacillus acidophilus in mayonnaise sauce. Iranian Journal of Veterinary Medicine 6: 259-264.
8. Homayouni A, Azizi A, Ehsani MR, Yarmand MS, Razavi SH (2008) Effect of microencapsulation and resistant starch on the probiotic survival and sensory properties of synbiotic ice cream. Food Chemistry 111: 50-55.
9. Mandal S, Puniya AK, Singh K (2006) Effect of alginate concentrations on survival of microencapsulated Lactobacillus casei NCDC-298. International Dairy Journal 16: 1190-1195.
10. Chávarri M, Marañón I, Ares R, Ibáñez FC, Marzo F, et al., (2010) Microencapsulation of a probiotic and prebiotic in alginate-chitosan capsules improves survival in simulated gastro-intestinal conditions. International Journal of Food Microbiology 142: 185-189.
11. Gibson GR, Probert HM, Van Loo J, Rastall RA, Roberfroid MB (2004) Dietary modulation of the human colonic microbiota: updating the concept of prebiotics. Nutr Res Rev 17: 259-275.
12. Homayouni A, Ehsani MR, Azizi A, Razavi SH, Yarmand MS (2007) An introduction to functional dairy foods. In Proceedings of the First National Functional Food Congress: 60.
13. Mirzaei H, Pourjafar H, Homayouni A (2012) Effect of calcium alginate and resistant starch microencapsulation on the survival rate of *Lactobacillus acidophilus* La5 and sensory properties in Iranian white brined cheese. Food Chemistry 132: 1966-1970.
14. Ross RP, Fitzgerald GF, Collins JK, Stanton C (2002) Cheese delivering biocultures - probiotic cheese. The Australian Journal of Dairy Technology 57: 71-78.
15. Ranadheera RDCS, Baines SK, Adams SK (2010) Importance of food in probiotic efficacy. Food Research International 43: 1-7.
16. Liua H, Xu XM (2007) Rheological, texture and sensory properties of low-fat mayonnaise with different fat mimetics. LWT 40: 946-954.
17. Shen R, Luo S, Dong J (2011) Application of oat dextrine for fat substitute in mayonnaise. Food Chemistry 126: 65-71.
18. Kailasapathy K, Chin J (2000) Survival and therapeutic potential of probiotic organisms with reference to Lactobacillus acidophilus and Bifidobacterium spp. Immunology & Cell Biology 78: 80-88.
19. Shah NP (2000) Probiotic bacteria: Selective enumeration and survival in dairy foods. Journal of Dairy Science 83: 894-907.
20. Sultana K, Godward G, Reynolds N, Arumugaswamy R, Peiris P, et al. (2000) Encapsulation of probiotic bacteria with alginate-starch and evaluation of survival in simulated gastrointestinal conditions and in yoghurt. International Journal of Food Microbiology 62: 47-55.
21. Fahimdanesh M, Mohammadi N, Ahari H, Khosravi MA, Zanjani FZH, et al. (2012) Effect of microencapsulation plus resistant starch on survival of Lactobacillus casei and Bifidobacterium bifidum in mayonnaise sauce. African Journal of Microbiology Research 6: 6853-6858.
22. Vinderola CG, Prosello W, Ghiberto D, Reinheimer JA (2000) Viability of probiotic (Bifidobacterium, Lactobacillus acidophilus and Lactobacillus casei) and nonprobiotic microflora in Argentinean Fresco cheese. Journal of Dairy Science 83: 1905-1911.
23. Haynes IN, Playne MJ (2002) Survival of probiotic cultures in low fat ice cream. Australian Journal of Dairy Technology 57: 10-14.
24. Sheu TY, Marshall RT (1993) Microencapsulation of Lactobacilli in calcium alginate gels. Journal of Food Science 54: 557-561.
25. Chen MB (2005) Study on processing technology of functional mayonnaise with plasma cholesterol lowering activity. Academic Periodical of Farm Products Processing 32: 22-24.
26. ISIRI number 5222 (2000) Institute of Standards and Industrial Research of Iran. Yoghurt – determination of total Titratable Acidity – potentiometric method, Tehran, Iran.
27. Krasaekoopt W, Bhandari B, Deeth H (2004) The influence of coating materials on some properties of alginate beads and survivability of microencapsulated probiotic bacteria. International Dairy Journal 14: 737-743.
28. Kailasapathy K, Masondole L (2005) Survival of free and microencapsulated Lactobacillus acidophilus and Bifidobacterium lactis and thei effect on texture of feta cheese. Australian Journal of Dairy Technology 60: 252-258.
29. Mortazavian AM, Ehsani MR, Azizi A, Razavi SH, Mousavi SM, et al. (2008) Effect of microencapsulation of probiotic bacteria with calcium alginate on cell stability during the refrigerated storage period in the Iranian yogurt drink (Doogh). MILCAD 63: 233-348.
30. Marshall VM, Tamime AY (1997) Starter cultures employed in the manufacture of biofermented milks. International Dairy Journal 50: 35-39.
31. Adeli milani M, Mizani M, Gavami M (2010) Effects of yellow mustard powder on microbial population, pH and organoleptic properties of mayonnaise. Iranian Journal of Nutrition Sciences & Food Technology 5: 35-44.
32. Kailasapathy K (2006) Survival of free and encapsulated probiotic bacteria and their effect on the sensory properties of yoghurt. LWT 39: 1221-1227.
33. Kailasapathy K, Sultana k (2003) Survival and b-D galactosidase activity of encapsulated and free Lactobacillus acidophilus and Bifidobacterium lactis in ice cream. Australian Journal of Dairy Technology 58: 223-227.
34. Jahangiri P, Streblow R, Müller D (2012) Simulation of Non-Newtonian Fluids using Modelica. Proceedings of the 9th International Modelica Conference: 57-60.
35. Tang HS, Kalyon DM (2004) Estimation of the parameters of Herschel–Bulkley fluid under wall slip using a combination of capillary and squeeze flow viscometers. Rheologica Acta 43: 80-88.

8

The Effect of Novel Rhenium Compounds on Lymphosarcoma, PC-3 Prostate and Myeloid Leukemia Cancer Cell Lines and an Investigation on the DNA Binding Properties of One of these Compounds through Electronic Spectroscopy

Carl Parson[1], Valerie Smith[1], Christopher Krauss[1], Hirendra N Banerjee[1*], Christopher Reilly[2], Jeanette A Krause[3], James M Wachira[4], Dipak Giri[5], Angela Winstead[5] and Santosh K Mandal[5*]

[1]Department of Biology and Pharmaceutical Sciences, Elizabeth City State University-University of North Carolina, Elizabeth City, NC 27909, USA
[2]Department of Biomedical Sciences and Pathobiology, Virginia-Maryland Regional College of Veterinary Medicine, Virginia Polytechnic Institute and State University, Blacksburg, Virginia 24061, USA
[3]Department of Chemistry, University of Cincinnati, Cincinnati, OH 45221, USA
[4]Department of Biology, Morgan State University, Baltimore, MD 21251, USA
[5]Department of Chemistry, Morgan State University, Baltimore, MD 21251, USA

Abstract

Despite the tremendous success of cisplatin and other platinum-based anticancer drugs, severe toxicity and resistance to tumors limit their applications. It is believed that the coordination of the metal to DNA bases causes the ruptures of the cancer as well as normal cells. A search for anticancer drugs with different modes of action resulted in the synthesis of variety of novel compounds. Recently we synthesized a series of novel rhenium pentylcarbonato compounds (PC1-PC6). The rhenium atom in each compound is coordinated to a planar heterocyclic aromatic ligand, thereby forcing each compound to intercalate between the DNA bases. The UV absorption titrations of PC6 with DNA shows hypochromic effect with concomitant bathochromic shift of the charge transfer band at 290 nm. These results suggest that the compound PC6 binds to DNA through intercalation. It is likely that many of the other PC-series of compounds will behave in a similar manner. We have observed that the PC-series of compounds are strong cytotoxic agents against lymphosarcoma (average GI50 ≈ 2.3 ± .6 μM), PC-3 prostate (average GI50 ≈ 2.8 ± 0.6 μM) and myeloid leukemia (average GI50 ≈ 3.0 ± .6 μM) cancer cell lines. The average GI50 values of the PC-series of compounds are much less than the corresponding GI50 values of cisplatin. Also each of the PC-series of compounds exhibits less toxicity than cisplatin in the glomerular mesangial cells.

Keywords: Rhenium Compounds; Lymphosarcoma; PC-3 Prostate; Myeloid Leukemia; Cancer Cells; Mesangial Cells; DNA Binding; Electronic Spectroscopy; Hypochromic Effect; Bathochromic Shift

Abbreviations

GI50: Concentration required to have 50% cell-growth inhibition

Introduction

The serendipitous discovery of the anticancer property of cisplatin [1] contributed to the development of cancer chemotherapy. Due to its severe side effects and resistance to tumor, clinical effectiveness of cisplatin is limited. Second and third generation platinum drugs are not free from side effects and spontaneous or acquired resistance. As a result, it has been proposed that we should move away from the 'platinum paradigm' by developing drugs based on other metals which have different mechanisms of action [2,3]. Non-platinum complexes such as those of ruthenium [4], titanium [5], gold [6], tin [7], palladium [8], rhodium [9], bismuth [10], and gallium [11] exhibit very strong cytotoxicity and possess distinct chemical advantages over cisplatin and related drugs. In fact, some of them entered clinical trials. This opens the door for the exploration of non-platinum anticancer drugs.

Rhenium organometallics are another very new class of promising antiproliferative compounds. For example, several tricarbonylrhenium compounds have been found to be cytotoxic against numerous cancer cell lines [12-15]. In the present study, we synthesized six novel rhenium pentylcarbanato compounds PC1-PC6 [16]. The structure of each compound is depicted in Figure 1. Unlike the rhenium compounds noted above [12-15], the PC-series of compounds exhibit fluorescence [16]. Here we report the cytotoxic effects of the compounds PC1-PC6 on lymphosarcoma, PC-3 prostate, and myeloid leukemia cancer cells and mesangial cells and the DNA-binding study of the compound PC6 using electronic spectroscopy (UV-Visible spectrophotometric titrations).

Materials and Methods:

$Re_2(CO)_{10}$ and the polypyridyls were obtained from commercial sources. Tris buffer was prepared according to standard protocol: 5 mM Tris–HCl, 50 mM NaCl, pH 7.2. Calf thymus DNA (CT-DNA) was purchased from Sigma-Aldrich. Solution of CT-DNA in Tris buffer gave a ratio of UV absorbance (A_{260}/A_{280}) of ≈1.9 at 260 and 280

***Corresponding author:** Hirendra N Banerjee, Department of Biology and Pharmaceutical Sciences, Elizabeth City State University-University of North Carolina, 1704 Weeksville Road, Elizabeth City, NC 27909, USA

Santosh K Mandal, Department of Chemistry, Morgan State University, 1700 East Cold Spring Lane, Baltimore, MD 21251, USA

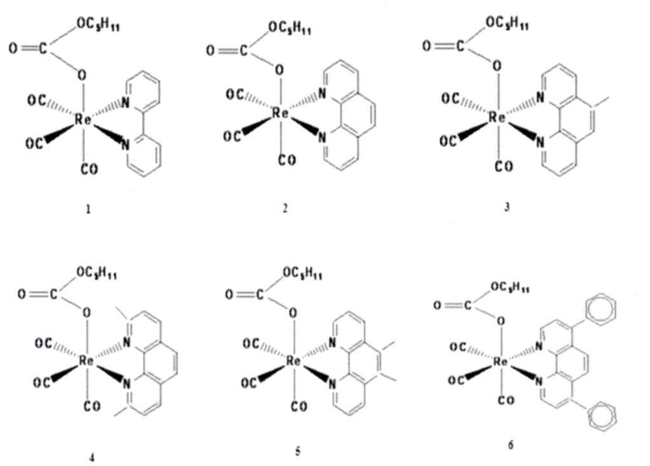

Figure 1: The structural formulae of the PC-series of compounds PC1-PC6.

nm, indicating that the DNA was sufficiently free of protein [17]. The concentration of CT- DNA in nucleotide phosphate was determined spectrophotometrically using a molar absorptivity of 6600 M^{-1} cm^{-1} (260 nm) [18]. The UV-visible spectra were recorded at room temperature using a Varian Cary 50 Scan UV-Visible Spectrophotometer.

Cell Culture

Lymphosarcoma and myeloid leukemia cancer cell lines were obtained from Coriel Institute, MD(USA) and cultured in RPMI 1640 medium supplemented with 10% Fetal Bovine Serum, penicillin and streptomycin antibiotics, grown at 37°C in a CO_2 incubator. PC-3 prostate cancer cell lines were obtained from ATCC, VA (USA) and cultured in the same growth conditions stated above. Mouse glomerular mesangial cells were donated by Dr. C. Reilly at Virginia Tech University, VA (USA). The culture conditions were same as above.

Synthesis of rhenium pentylcarbanato compounds (PC1-PC6)

The rhenium compounds PC1-PC6 were synthesized according to the published procedure [16]. In brief, a mixture of $Re_2(CO)_{10}$ and the corresponding polypyridyls in 1:2 mole ratio was refluxed in 1-pentanol while CO_2 was bubbled through the solutions. The polypyridyls in compounds PC1–PC6 are 2,2′-bipyridyl, 1,10-phenanthroline (phen), 5-methyl-1,10-phenanthroline, 2,9-dimethyl-1,10-phenanthroline (neocuproin), 5,6-dimethyl-1,10-phenanthroline, and 4,7-diphenyl-1,10-phenanthroline (bathophenanthroline), respectively. The reaction was discontinued after 36 hours. After cooling to room temperature, the solids PC1–PC6 were isolated by filtration.

UV Absorption Titrations

We selected only compound PC6 for UV absorption titrations. The titrations of PC6 in Tris buffer were carried out using a fixed concentration (20 μM) of PC6 to which DNA solutions were added. Initially, 3000 μL of the blank buffer and 3000 μL solution of PC6 (20 μM) were placed in the reference and sample cuvettes, respectively. The first spectrum was recorded in the range of 235–700 nm. During the titrations, a measured amount of DNA was added to each cuvette from the DNA stock solution so that the final concentration of the DNA becomes 10, 20, 30, 40, 50, and 60 μM. The solutions were mixed thoroughly by repeated inversion and were allowed to incubate for 10 minutes before the absorption spectra were recorded. The change in concentration of the compound PC6 due to each titration was negligible.

Cytotoxicity Assay

The cancer cell lines were grown in 96-well plates to confluency, the cells were then kept in serum-free medium overnight and then treated with different doses of drugs PC1-PC6 (dissolved in DMSO) along with untreated cells and DMSO vehicular control for 48 hours. DMSO was used as vehicular control, non-drug treated as negative control and cisplatin as positive control. A Trypan blue assay was done to measure cell viability using a standard hemocytometer and light microscope.

Statistical Analysis

Each experiment was repeated three times and statistical t test was used to analyze the results with $P < 0.05$.

Results and Discussion

As shown in Figures 1-4, the rhenium atom in each of the PC-series of compounds is octahedrally coordinated to a chelated (bidentate) polypyridyl ligand, three adjacent CO molecules, and a pentylcarbonate group [$-OC(O)OC_5H_{11}$] in a facial (fac) arrangement. For example, the rhenium atom in PC1 is octahedrally coordinated to chelated 2,2'-bipyridyl ligand, three adjacent CO's, and a pentylcarbonate group. Likewise, the rhenium atom in PC4 is octahedrally coordinated to chelated neocuproin, three adjacent CO's and a pentylcarbonate group and the rhenium atom in PC6 is octahedrally coordinated to chelated bathophenanthroline, three adjacent CO's and a pentylcarbonate group. The PC-series of compounds were characterized spectroscopically. Additionally, PC1, PC4 and PC6 were characterized through X-ray crystallography. The X-ray structures of PC1, PC4, and PC6 are shown in Figures 2-4, respectively. The phenanthroline ring in PC4 is not totally planar (see Figure 5 and Table 1) with a 0.1166 Å deviation from planarity (deviation of 6.4° and 7.6° between the 3 rings). The crystal structures of PC1 and PC6 were published earlier [16]. A complete crystallographic data for PC4 will be published elsewhere.

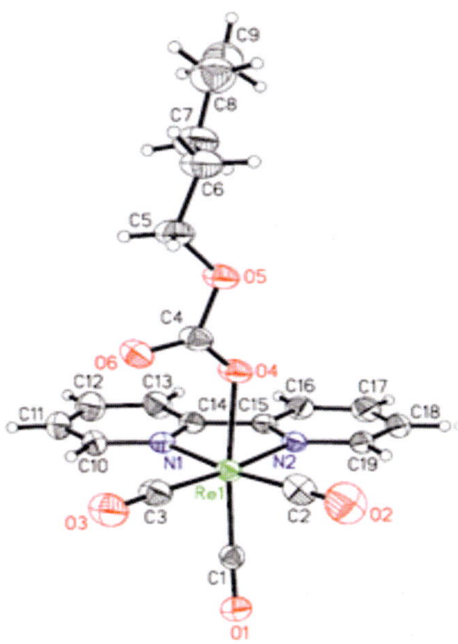

Figure 2: X-ray structure of PC1. (courtesy: Elsevier).

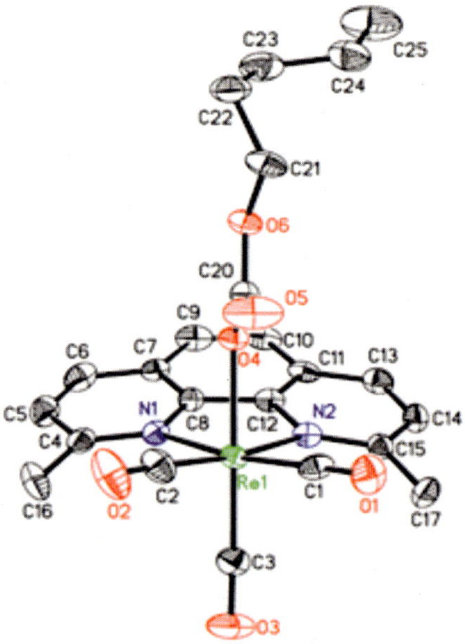

Figure 3: X-ray structure of PC4.

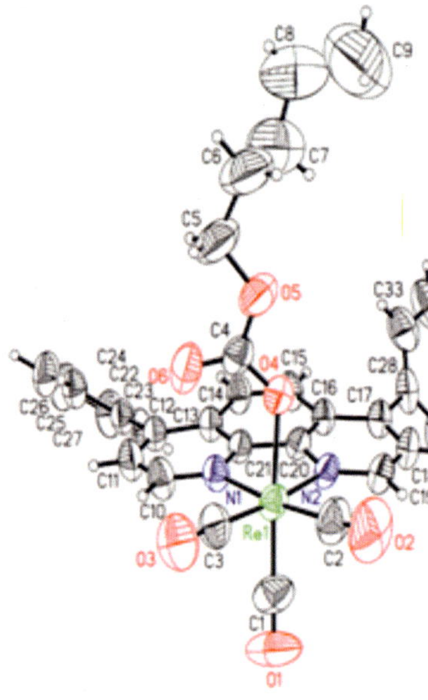

Figure 4: X-ray structure of PC6. (courtesy: Elsevier).

DNA binding studies are important because many anticancer agents work through binding to DNA causing DNA damage in cancer cells, interrupting replication and transcription processes, and inducing apoptosis. The electronic spectroscopy is the most common technique to find out the interaction of compounds with DNA [19]. When a compound binds to DNA through intercalation, the UV-Visible spectra usually shows hypochromism with a red shift. Each of the PC1-PC6 compounds is coordinated to a planar aromatic heterocyclic ligand (see, Figures 1-5). It is therefore possible that each compound will bind to DNA in an intercalative mode, involving a stacking interaction between the aromatic chromophores and DNA base pairs. Figure 6 depicts the UV-Visible spectra of the compound PC6 (20 μM in Tris buffer) in the absence and presence of calf thymus DNA. The intensity of the charge transfer band at 290 nm in PC6 decreases as the DNA concentration (range of 10-60 μM) increases. A small red shift of the band at 290 nm is observed concomitantly after each addition of DNA. A red shift to 295 nm is observed when 60 μM of DNA was added to fixed concentration (20 μM) of PC6. Based on the UV-visible spectra of the compound PC6, it may be concluded that PC6 binds to DNA in an intercalative fashion. Our conclusion may be corroborated by the fact that the phenanthroline rings are planar or near planar (see Figure 5) which facilitate their insertion into the DNA base pairs. Thus PC6 and may be other PC compounds PC2-PC5 except PC1 do not bind to DNA via covalent interaction through nitrogen bases (adenine, cytosine, guanine and thymine). PC1 might not follow intercalation mechanism because of the presence of nonintercalative 2,2-bipyridyl ligand. Cisplatin and other platinum metal drugs directly bind (form covalent bond through nitrogen bases of DNA) to the DNA. As a result normal cells are highly affected resulting in severe toxic side effects. It is therefore possible that many of the the PC-series of compounds might not exhibit severe side effects. It is worth mentioning that analogous rhenium compounds exhibited very negligible toxicity towards fibroblasts [15]. We have not extensively studied the mechanism of action using other techniques and with other PC-series of compounds. Therefore it is not impossible that the PC compounds will follow other modes of action too. 1, 10-phenanthroline (phen) is an intercalative ligand. It is expected that metal-phen complexes will follow intercalation mechanism. In fact, Barton group [20] proposed a major groove intercalation and a minor groove surface-bound interaction for the binding of $[Ru(phen)_3]^{2+}$ to DNA from equilibrium dialysis and photophysical methods. However, Satyanarayana, et. al. [21, 22] from viscosity measurements and Errikson, et. al. [23, 24] from

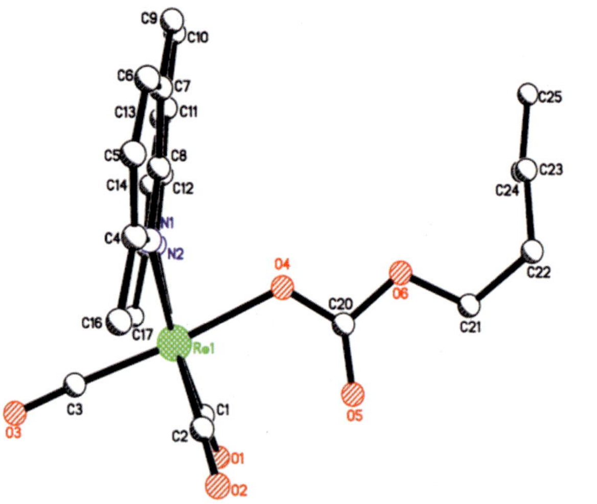

Figure 5: X-ray structure of PC4 showing a nearly planar phenanthroline ring.

Least Square Planes	Deviation (Å) from Planarity	Dihedral Angle (°)
N1-C4-C5-C6-C7-C8	0.0155	6.4(2)
C7-C8-C9-C10-C11-C12	0.0262	7.6(2)
N2-C11-C12-C13-C14-C15	0.0206	
N1- N2- Re1	0	
N1-N2-C4 thru C15	0.1166	19.7(1)
O4-O5-O6-C20-C21	0.0331	83.0(2)
C22-C23-C24-C25	0.0568	75.7(6)

Table 1: The least square planes and their deviations from planarity and dihedral angles for PC4.

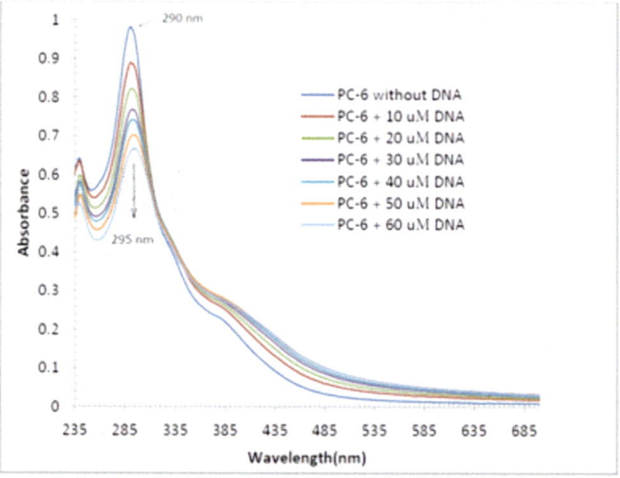

Figure 6: Electronic absorption spectra for the titration of 20 μM of compound PC6 in the absence and presence of varied amount of DNA.

NMR and CD spectra indicated no intercalative binding. We chose to study the DNA binding properties of PC6 because the planar aromatic heterocyclic ligand in PC6 has two additional planar aromatic (phenyl) rings which might facilitate PC6 to follow intercalaive mode of action. Although our prediction is right, the presence of planar aromatic ring is not the final arbiter of intercalation mechanism. For example, the mode of action of the complex $[Ru(bpy)_2(dppn)]Cl_2$ (where, dppn = benzo[*i*]dipyrido[3,2-a:2',3'-c]phenazine) seems to involve interactions with cell membranes, rather than interactions with DNA (in this case intercalation of the dppn ligand was anticipated) [25].

To determine the inhibitory potency of active compounds (GI50, 50% cell-growth inhibition), tests were extended for compounds PC1-PC6 to varying concentrations for different cancer cell lines. Most of the six drugs showed GI 50 around 3-4 μM (Table 2). Interestingly, the Re drugs (PC 1-PC6) showed less toxicity than cisplatin in the glomerular mesangial cells. Because nephrotoxicity is a major problem in cisplatin treatment, the titled Re compounds could be an ideal alternative. Significantly, all the Re compounds (PC 1-PC6) showed efficacy against all the cancer cells treated at a lower dose than the cisplatin.

Compounds	Lymphosarcoma	PC-3 Prostate Cancer Cells	Myeloid Leukemia Cancer Cells	Mesangial Cells
PC1	3. ± .5	4 ± .5	4 ± .9	15 ± .8
PC2	2. ± .6	2 ± .7	3 ± .5	13 ± .5
PC3	2 ± .7	4 ± .5	3 ± .6	12 ± .5
PC4	3 ± .6	2 ± .8	2 ± .5	14 ± .5
PC5	2 ± .8	3 ± .7	4 ± .7	11 ± .6
PC6	2 ± .5	2 ± .6	2 ± .5	13 ± .8
Cisplatin	6 ± .5	8 ± .5	10 ± .7	4 ± .4
DMSO	NA	NA	NA	NA

GI50 (mean ± S.D.) refers to the concentration required to have 50% cell-growth inhibition; (-) indicates that the isolated is inactive (negative at 100 or GI50>100 μM); N/A: no data available

Table 2: The effects of the PC-series of compounds (PC1-PC6) on lyphosarcoma, PC-3 prostate, and myeloid leukemia cancer cells and mesangial cells.

Acknowledgments

Partially supported by US-NIH-RISE, US-National Science Foundation LSAMP institutional support Award to Elizabeth City State University, NC, TMCF-DOE award to Dr.H.Banerjee and the work at MSU was partially supported by grants from the National Institutes of Health (Grant No. G11HD038439) and Nuclear Regulatory Commission (Grant No. NRC-HQ-12-G-27-0086). Crystallographic data were collected through the SCrALS (Service Crystallography at Advanced Light Source) program at Beamline 11.3.1 at the Advanced Light Source (ALS), Lawrence Berkeley National Laboratory. The ALS is supported by the U.S. Department of Energy, Office of Energy Sciences Materials Sciences Division, under contract DE-AC02-05CH11231.

References

1. Rosenberg B, Vancamp L, Krigas T (1965) Inhibition of Cell Division in Escherichia Coli by Electrolysis Products From a Platinum Electrode. Nature 205: 698-699.
2. Dyson PJ, Sava G (2006) Metal-Based Antitumour Drugs in the Post Genomic Era. Dalton Trans 16: 1929-1933.
3. Fricker SP (2007) Metal Based Drugs: From Serendipity to Design. Dalton Trans 43: 4903-4917.
4. Guidi F, Modesti A, Landini I, Nobili S, Mini E, Et. Al., (2013) The Molecular Mechanisms of Antimetastatic Ruthenium Compounds Explored through Dige Proteomics. J Inorg Biochem 118: 94-99.
5. Caruso F, Rossi M (2004) Antitumor Titanium Compounds. Mini-Rev Med Chem 4: 49-60.
6. Ott I (2009) On The Medicinal Chemistry of Gold Complexes as Anticancer Drugs. Coord Chem Rev 253: 1670-1681.
7. Hadjikakou SK, Hadjiliadis N (2009) Anti-Proliferative and Anti-Tumor Activity of Organotin Compounds. Coord Chem Rev 253: 235–249.
8. Reedijk J (2009) Platinum Anticancer Coordination Compounds: Study of DNA Binding Inspires New Drug Design. Eur J Inorg Chem 10: 1303-1312.
9. Leung C-H, Zhong H-J, Chan Ds-H, Ma D-L (2013) Bioactive Iridium and Rhodium Complexes As Therapeutic Agents. Coord Chem Rev 257: 1764-1776.
10. Tiekink Ert (2002) Antimony and Bismuth Compounds in Oncology. Crit Rev Oncol Hemat 42: 217-224.
11. Jakupec MA, Keppler BK (2004) Gallium in Cancer Treatment. Curr Top Med Chem 4: 1575-1583.
12. Yan Y-K, Cho Se, Shaffer K A, Rowell Je, Barnes Bj, et. al., (2000) Cytotoxicity of Rhenium(I) Alkoxo and Hydroxo Carbonyl Complexes in Murine and Human Tumor Cells. Pharmazie 55: 307-313.
13. Wang W, Yan YK, Hor TSA, Vittal JJ, Wheaton Jr, et.al., (2002) Synthesis, X-Ray Structures and Cytotoxicity of Rhenium(I) Carbonyl 2-(Dimethylamino) Ethoxide Complexes. Polyhedron 21: 1991-1999.
14. Zhang J, Vittal JJ, Henderson W, Wheaton Jr, Hall IH, et. al., (2002) Tricarbonylrhenium(I) Complexes of Phosphine-Derivatized Amines, Amino Acids and a Model Peptide: Structures, Solution Behavior And Cytotoxicity. J Organomet Chem 650: 123-132
15. Ho J, Lee WY, Koh KJT, Lee PPF, Yan YK (2013) Rhenium(I) Tricarbonyl Complexes of Salicylaldehyde Semicarbazones: Synthesis, Crystal Structures and Cytotoxicity. J Inorg Biochem 119: 10-20.
16. Mbagu Mk, Kebulu D, Winstead A, Pramanik Sk, Banerjee Hn, et. al., (2012) Fac-Tricarbonyl (Pentylcarbonato) (α-Diimine) Rhenium Complexes: One-Pot Synthesis, Characterization, Fluorescence Studies, And Cytotoxic Activity Against Human Mda-Mb-231 Breast, Ccl-227 Colon And Bxpc-3 Pancreatic Carcinoma Cell Lines. Inorg Chem Comm 21: 35-38.
17. Marmur J (1961) A Procedure for the Isolation of Deoxyribonucleic Acid from Micro-Organisms. J Mol Biol 3: 208-218.
18. Reichmann ME, Rice SA, Thomas CA, Doty PJ (1954) A Further Examination of the Molecular Weight and Size of Desoxypentose Nucleic Acid. J Am Chem Soc 76: 3047-3053.
19. Barton JK, Danishefsky A, Goldberg J (1984) Tris(Phenanthroline)Ruthenium(II): Stereoselectivity in Binding Dna. J Am Chem Soc 106: 2172-2176.
20. Barton JK, Goldberg JM, Kumer CV, Turro NJ (1986) Binding Modes and Base Specificity of Tris(phenanthroline)ruthenium(II) Enantiomers with Nucleic Acids: Tuning the Stereoselectivity. J Am Chem Soc 108: 2081-2088.
21. Satyanarayana S, Dabrowiak JC, Chaires JB (1992) Neither Delta- nor Lambda-tris(phenanthroline)ruthenium(II) binds to DNA by classical intercalation. Biochemistry 31: 9319-9324.

22. Satyanarayana S, Dabrowiak JC, Chaires JB (1993) Tris(phenanthroline) ruthenium(II) enantiomer interactions with DNA: Mode and specificity of binding. Biochemistry 32: 2573-2584.

23. Eriksson M, Leijon M, Hiort C, Norden B, Graeslund A (1992) Minor groove binding of [Ru(phen)3]2+ to [d(CGCGATCGCG)]2 evidenced by two-dimensional NMR. J Am Chem Soc 114: 4933-4934.

24. Eriksson M, Leijon M, Hiort C, Norden B, Graeslund A (1994) Binding of Delta- and Lambda-[Ru(phen)3]2+ to [d(CGCGATCGCG)]2 Studied by NMR. Biochemistry 33: 5031-5040.

25. Schatzschneider U, Niesel J, Ott I, Gust R, Alborzinia H, et. al., (2008) Cellular Uptake, Cytotoxicity and Metabolic Profiling of Human Cancer Cells Treated with Ruthenium(II) Polypyridyl Complexes [Ru(Bpy)2(N--N)]Cl2 With N--N=Bpy, Phen, Dpq, Dppz and Dppn. Chem Med Chem 3: 1104-1109.

9

Interaction of Whey Lactoferrin with Copper Immobilized in Polyacrylamide Cryogel

Carvalho BMA[1]*, Carvalho LM[2], Silva Jr WF[1], Minim LA[3] and Carvalho GGP[4]

[1]Department of Chemistry, Biotechnology and Bioprocess Engineering, Federal University of São João Del Rei, Ouro Branco, MG 36420-000, Brazil
[2]Department of Veterinary, Federal University of Viçosa, Viçosa, MG 36570-000, Brazil
[3]Department of Food Technology, Federal University of Viçosa, Viçosa, MG 36570-000, Brazil
[4]Department of Animal Science, Federal University of Bahia, Salvador, BA 40110-909, Brazil

Abstract

Design and optimization of Immobilized Metal Affinity Chromatography (IMAC) processes require deep knowledge of driving factors responsible for interaction between immobilized metal and biomolecules. Based on this requirement, interactions between lactoferrin from cheese whey and IDA-Cu^{2+}-cryogel system was investigated. Data from adsorption of lactoferrin in the system at pH 6, 7 and 8, as well as NaCl concentration from 200 to 1000 mmol L^{-1} were adjusted Langmuir, Freundlich, Temkin and Langmuir-Freundlich isotherm models. Although all models were able to explain the interaction lactoferrin-cryogel system, the Langmuir-Freundlich model was the most accurate one. In addition, it could explain quantitatively the cooperativity and heterogeneity of the bounds between protein and matrix. The methods used in this project are useful for both better understanding of the protein-immobilized metal interactions and developing preparative scale IMAC.

Keywords: IMAC; Adsorption of protein; Interactions; Separation

Introduction

Lactoferrin (LF) from cheese whey is a glycoprotein from the transferrin family, whose molecular mass is around 80 kDa. It has an iron atom in its inner and amino acid residues along its surface, such as histidine and tryptophan [1]. Given its wide application in medicine, biology, pharmaceutical and food industry, due to their biological properties, such as antiviral, antibacterial, antioxidant and immunomodulatory activities, several technologies of processing have been studied and developed for isolation of LF at high levels of purity. The majority of these processes are focused on chromatographic techniques, such as purification of lactoferrin by cation exchange [2], heparin-Sepharose affinity [3], ligand affinity [4] and heparin-poly(glycidyl methacrylate) affinity chromatography [5]. Because of the difficulties found in these biotechnological processes, such as the use of special columns and pieces of equipment, and the processing of complex solutions like cheese whey (whose particulates can reduce the resin porosity, leading to blockage of the column), the development of new techniques to separate and purify whey lactoferrin is necessary.

Cryogels are polymeric gels formed in moderately frozen conditions. They have large pores, which allow solutions containing cell fragments and particulates to be drained throughout chromatographic columns without causing their obstruction [1].

Immobilized Metal Affinity Chromatography (IMAC) is based on the differential affinity for chemical groups at the surface of some molecules shown by metals immobilized in solid matrix [6]. Such affinity arises from reversible coordination ligations between metal ion and amino acid residues (in the case of peptides), such as imidazole group in histidine, thiol in cysteine and indol in tryptophan, which donate electrons to the immobilized metal. In addition to the coordination binding, other driving forces, such as electrostatic, hydrophobic and Van der Waals, may be involved in the adsorption of protein in IMAC and it may not always be possible to determine their relative contributions [7,8].

Mathematical modeling is an essential part of the chromatographic theory for describing its dynamical process. It provides a procedure for predicting the dynamical behavior of solute in the column without extensive experiments. Due to different considerations of simplifications, several types of models have been established and applied to illustrate the behavior of chromatographic columns [9].

Then, the use of supermacroporous matrixes (e.g. cryogels), along with IMAC seems to be a promising alternative for capture of lactoferrin. Thus, the absorption of lactoferrin in cryogel supermacroporous matrix with immobilized Cu^{2+} at 30°C, different NaCl concentrations (200, 600, 800 and 1000 mmol L^{-1}) and pHs (6.0, 7.0 and 8.0) was analyzed in this work.

Material and Methods

Synthesis of IDA-Cu^{2+}-cryogel resin

Both the synthesis and the characterization of IDA-Cu^{2+}-cryogel resins were performed in accordance with [10] works [11]. Cryo-copolymerization reaction of solution containing acrylamide (AAm), N,N'-methylenebisacrylamide (MBAAm), Allyl Glycidyl Ether (AGE), N,N,N',N'-tetramethylethylenediamine (TEMED) and ammonium persulfate (APS) was performed. Iminodiacetic acid (IDA) was used to immobilize copper in the matrix.

Batch adsorption of lactoferrin and obtainment of equilibrium data

Static method was used to obtain absorption at each condition

***Corresponding author:** Carvalho BMA, Department of Chemistry, Biotechnology and Bioprocess Engineering, Federal University of São João Del Rei, Ouro Branco, MG 36420-000, Brazil

analyzed. Approximately 30 mg of cryogel matrix were weighed in eppendorf tubes and equilibrated with 400 μL of equilibration buffer (20 mmol L^{-1} HEPES, 0.2 mol L^{-1} NaCl, pH 7.0) for 2 h under mild stirring. After that, different volumes (0-300 μL) of 10 mg mL^{-1} lactoferrin solution were added to the tubes along with equilibration buffer enough to make the volume up to 1200 μL, leading to final concentrations of 0, 0.1, 0.2, 0.3, 0.4, 0.6, 0.9, 1.2, 1.5, 1.8 and 2.2 mg mL^{-1} lactoferrin. The tubes were kept under constant agitation for 24 h in a temperature controlled biological incubator (101 M Mod BOD/3 Eletrolab® Brazil) at 30°C up to establish the equilibrium. Then, the resins were taken away from tubes and the concentration of lactoferrin in the supernatant was measured according to the Bradford method [12] at 495 nm (Thermo Scientific Model BIOMATE 3) using calibration curve.

The amount of protein retained in the resin was obtained by mass balance (Equation 1).

$$q = \frac{V}{m}(C_o - C_b) \tag{1}$$

where q (mg g^{-1}) is the concentration of protein in the resin, V (mL) is the volume of the liquid phase, M (g) is the mass of the solid phase, C_0 (mg mL^{-1}) is the initial lactoferrin concentration in the liquid phase and C (mg mL^{-1}) is the final lactoferrin concentration in the supernatant after equilibrium was reached. The experiment was conducted using a fractional factorial design with three concentrations of salt (200 mmol L^{-1}, 600 mmol L^{-1} and 1000 mmol L^{-1}) and three pHs (6.0, 7.0 and 8.0).

Modeling of equilibrium data

Several researchers have shown that the relationship between free proteins in a solution and those absorbed in a matrix (at the equilibrium) may be described by the [13] isotherm. Chase [14] wrote a mass transfer equation which is coherent to the Langmuir isotherm. Protein mass transfer from solution to the matrix can be described by:

$$\frac{dq}{dt} = k_1 C(q_{m(L)} - q) - k_2 q \tag{2}$$

Where C is the concentration of protein in the liquid phase, q is the concentration of adsorbed protein, q_m is the resin maximum binding capacity and t is time. k_1 and k_2 are parameters which represent the rates of adsorption and desorption of protein to and from the immobilized metal ions. Supposing that all biding sites have same potential energy, regardless of their source, and that protein-biding site interactions are of type 1:1 (i.e. one molecule of protein is bound to one biding site), during the equilibrium, equation 2 can be reduced to:

$$q = \frac{q_{m(L)} C^*}{(k_d + C^*)} \tag{3}$$

Where q_m is the resin maximum binding capacity, k_d is a dissociation constant, q and C are protein concentration in the matrix and supernatant, respectively.

To validate the supposition of having the biding sites at the matrix surface, many linear transformations of equation 3 were performed, including the double and semi-reciprocal plots [15], as well as the Scatchard [16] plot:

Double reciprocal:

$$\frac{1}{q^*} = \frac{1}{q_{m(L)}} + \frac{K_d}{q_{m(L)}} \cdot \frac{1}{C} \tag{4}$$

Semi-reciprocal

$$\frac{C^*}{q^*} = \frac{K_d}{q_{m(L)}} + \frac{C^*}{q_{m(L)}} \tag{5}$$

Scatchard

$$\frac{q^*}{C^*} = \frac{q_{m(L)}}{K_d} - \frac{q^*}{K_d} \tag{6}$$

The shape of q^*/C^* vs q^* plots are particulary sensitive during independent, dependent non-identical or cooperative interaction process between protein and immobilized metal ion. If q^*/C^* vs q^* is convex, negative cooperation is observed. If its plot is linear, interactions between protein and ligand are independent. On the other hand, if q^*/C^* vs q^* plot is concave, there is positive cooperation in the adsorption process.

As protein-ligand interactions are frequently characterized by the participation of dependent binding sites, the Langmuir isotherm cannot describe experimental data satisfactorily. Another well-known isotherm, which is usually used to describe adsorption behavior, is the Freundlich [17] one. It correlates the concentration of adsorbed protein to the solute concentration:

$$q^* = KC^{\eta} \tag{7}$$

Where K and η are the equilibrium constant and capacity factor of Freundlich isotherm, respectively. K and η are constants for a particular protein-immobilized metal [17]. Although it is usually applied in a strictly empiric way, this equation may have interesting theoretical application in terms of adsorption on energetically heterogenic surface. One of the Freundlich isotherm limitations is that, according to it, amount of adsorbed protein increases indefinitely with concentration of protein in solution.

Reversible adsorption of proteins can also be described by Temkin's isotherm model, which was originally developed for adsorption of gases in heterogeneous surfaces. This model assumes an adsorption characterized by uniform distribution of ligation energy as result of the equation that predicts its [18]:

$$q^* = q_T \ln(1 + K_T C^*) \tag{8}$$

Where K_T (mol L^{-1}) is the equilibrium constant of ligation, which corresponds to the maximum ligation energy ($K_T = e^{-(\Delta G max/RT)}$), C^*(mol L^{-1}) is the concentration of protein in solution at the equilibrium and q_T is the surface differential adsorption capacity (in relation to the protein) per unity of energy.

The Temkin's isotherm has two main advantages on clarifying the heterogeneous nature of protein adsorption to ion exchanger and affinity resins when it is compared to the other ones (e.g. Freundlich and Langmuir) [19,20]: 1) it is described by two physically relevant parameters, whose values can be determined through adsorption equilibrium experiments; 2) it matches to Henry's law, which is an item necessary to explain the theory of local equilibrium in adsorption equilibrium for predictive chromatography.

In addition to Temkin's isotherm, Langmuir and Freundlich models were combined in order to overcome their inadequacies (e.g. indefinite increase of amount of adsorbed protein with concentration of protein in solution), leading to the Langmuir-Freundlich [21] model (Equation 9). This isotherm has been mentioned as an adequate way to describe the behavior of proteins in affinity resins.

$$q^* = \frac{q_{m(LF)}(C^*)^{\eta}}{K^*_d + (C^*)^{\eta}} \tag{9}$$

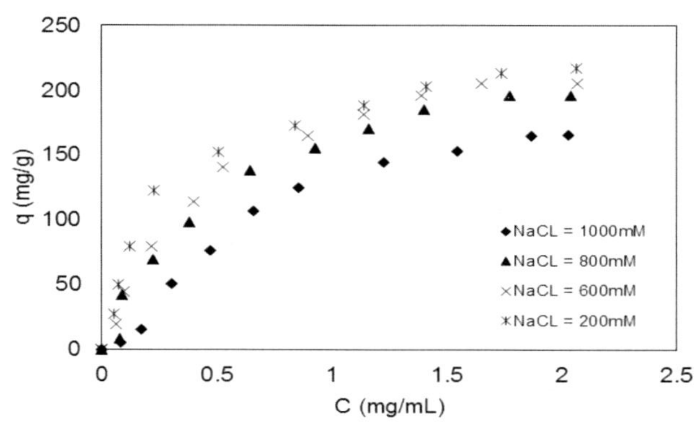

Figure 1: Adsorption of lactoferrin on IDA-Cu^{2+}-cryogel resin at different ionic strengths (HEPES buffer, 200-1000 mmol L^{-1} NaCl; pH 8.0, 30°C).

Where K^{*}_{d} is the constant of apparent dissociation, q_m is the maximum capacity of ligation and η is the Langmuir-Freundlich numeric coefficient. By analogy to multiple interactions made by proteins with ligands, it can be suggested that equation 9 can be used as model for cooperative adsorption. As this equation has three terms whose values must be found, it is better to explain adsorptions of heterogeneous nature and the adsorption cooperativity. For purely independent reasons, non-interative sites have $\eta = 1$. When $\eta > 1$, positive cooperation is suggested, while $0 < \eta < 1$ indicates negative cooperation in the binding process. Thus, value of η can be used as an empiric coefficient that represents what kind of cooperativity extension is present in the studied interaction.

Results and Discussions

Ionic strength effect

Figure 1 shows adsorption isotherms for lactoferrin- IDA-Cu^{2+}-criogel system. It can be observed that protein adsorption decreases with ionic strength increase. Similar behavior was obtained by [22] and [23] in adsorption of lysozyme on cooper ions immobilized in. Graham and co-authors [24], when the adsorption of lysozyme on CL-4B Sepharose with immobilized Cu^{2+} was analyzed, found diverse adsorption behaviors depending on the solution's pH and ion strength. In this matrix, the adsorption was positively influenced by ionic strength increase through all pH analyzed in their work. The TACN gel adsorption capacity was directly infuenced by pH and ionic strength, wherein at low values of pH (e.g. 5), the enzyme adsorption rised with NaCl concentration increase. However, at higher pH values (e.g. 9), adsorption behavior was reversed, wherein the maximum adsorption capacity was found in solutions without salt, which reinforces the complexity of phenomena behind IMAC and that the differential affinity of metal ions to biomolecules in solution is dependent on not only the chelate ion but also the composition of the mobile phase.

It is believed that protein adsorption in IMAC is result of cumulative effects of electrostatic, hydrophobic and/or coordination interactions. Though the addition of salt reduces electrostatic interactions, it also strengthens coordination and hydrophobic [6]. As consequence, different adsorption tendencies can be obtained for different proteins according to their hydrophobia. In this study, the increase of lactoferrin adsorption throughout ionic strength reduction indicates high effect of electrostatic interactions on the process. The resin adsorption capacities were significantly higher at 200 mmol L^{-1} NaCl; thus, it seems to be the best condition to achieve high protein biding capacity.

The higher adsorption capacity at low NaCl concentration may be due to non-specific interactions between protein and matrix produced by hydrophobic amino acid residues, such as cysteine and tryptophan. At pH 8, histidine is uncharged ($pK_a = 6.0$), which would provide additional hydrophobic force to the interaction -instead of electrostatic repulsion as if pH was lower than 6.0)-, besides more available electrons pairs to be donated to cooper ions. These amino acids are also considered to be responsible for affinity phenomenon in IMAC.

Data of lactoferrin adsorption on IDA-Cu^{2+}- cryogel at conditions discussed above were also adjusted to four isotherm models, that is; Langmuir, Freundlich, Temkin and Langmuir-Freundlich (Figure 2a-2d).

The Langmuir's model has been used to analyze adsorption data of protein on affinity resins [23,25]. In this study, Langmuir's model showed excellent accuracy, mainly at low salt concentrations. Its inability to describe experimental data was more visible at higher ionic strengths, though its coefficients of determination continued satisfactorily high. Estimated parameters from Langmuir, Freundlich, Temkin isotherms are shown in Table 1.

Although Freundlich's isotherm was adequate to explain experimental data, it was not fully explored in this work. Low binding capacity and high binding affinity are indicatives of larger deviations Temkin's isotherm (Table 1). Then, this model was not able to explain the behavior of protein adsorption. However, it has been successfully used to explain equilibrium data from heterogeneous protein adsorption in works where cooper was used as chelating metal [19,20].

Langmuir-Freundlich model [26] was the most appropriated one to describe experimental data at all analyzed conditions, yielding excellent coefficients of determination ($R^2 > 0.99$) (Figure 2a-2e). Besides, it could explain even the sigmoid (S) variation in the Langmuir's isotherm, observed at high salt concentrations. This change on isotherms shape with increase of ionic strength is an indicator of variation on relative contribution of electrostatic, hydrophobic and/or coordination interactions, as well as of change in the adsorption mechanism.

Values of $\eta > 1$ in all studied conditions, except for 200 mmol L^{-1}, indicate positive cooperation and adsorption of heterogeneous nature. This model is appropriate to explain the mechanism of protein adsorption at high salt concentration, in a quantitative way, in terms of η. Values of η increase with ionic strength rising, suggesting cooperation effect raise on protein-matrix binding. In despite of these facts, maximum adsorption values were obtained at 600mmol L^{-1} NaCl.

Nevertheless, experimental values of maximum protein adsorption (q_m) are much lower than the estimated (theoretical) ones, since they are based on 1:1 protein-ligand stoichiometry. As it is known, one protein molecule can interact with more than one metal ion, leading to lower actual q_m. It is also related to protein's tridimensional shape, and because its shape depends on the solution conditions (salt concentration, pH and temperature), the degree of heterogeneity of protein adsorption changes also varies due to these conditions.

Figure 3 shows the Scatchard's non-linear (q^*/C^* vs q^*) for experimental data. These plots' shape is particularly sensitive to presence of independent interactions, as well as to coordination bindings. Scatchard plot provides information about cooperation effects; its different maximum values through the studied NaCl concentrations imply change on adsorption mechanism through those conditions, reinforcing the cooperation effect on this process. Then, the hypotheses (i) all biding sites have same potential energy, (ii) interactions between protein and ligand are of type 1:1 stoichiometry, (iii) and binding sites

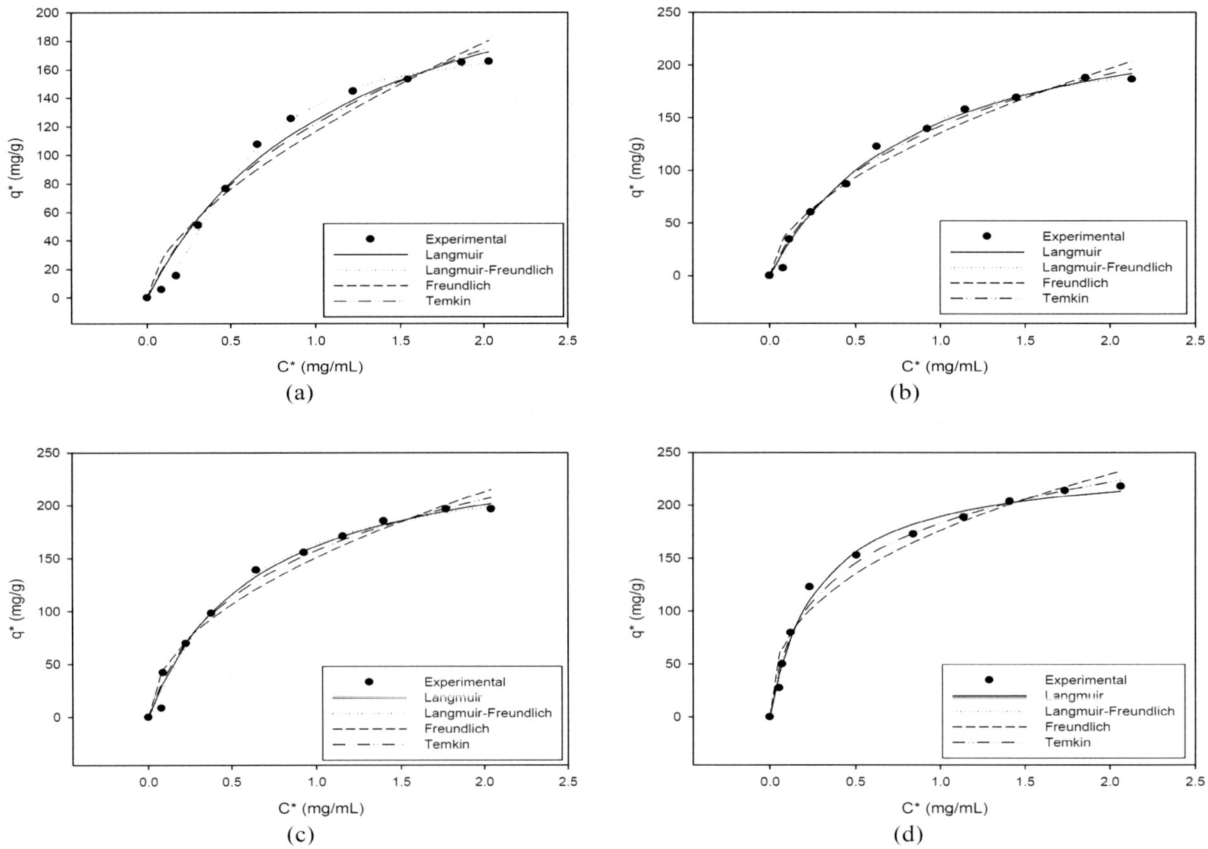

Figure 2: Comparison of accuracy of the Langmuir, Freundlich, Temkin and Langmuir-Freundlich isotherm models to predict lactoferrin adsorption IDA-Cu^{2+}-cryogel at different ionic strengths (temperature was 30°C and pH = 8.0); (a) 1000 mmol L^{-1}, (b) 800 mmol L^{-1}, (c) 600 mmol L^{-1}, (d) 200 mmol L^{-1} NaCl.

Theoretical Model	NaCl concentration (mmol L^{-1})			
	200	600	800	1000
Langmuir				
K_d (μmol L^{-1})	0.2751	0.6374	0.8472	1.1984
q_m (mg/mg gel)	274.40	264.55	263.59	241.61
Freundlich				
K (10^{-3}L/mg gel)	176.2	150.7	135.7	116.7
η	2.6207	2.0082	1.8562	1.6259
Temkin				
K_T (L mol^{-1})	22.5191	6.6284	4.5021	2.594
q_T (mg/mg gel)	57.9874	77.4595	83.3876	95.5819
Langmuir-Freundlich				
K^*_d (μM)	0.318	0.4502	0.3872	0.2785
q_m (mg/g gel)	242.2	244.7	237.8	178.5
η	1.85	1.16	1.12	0.99

[a] Buffer composition: HEPES 20mmol L^{-1}; 200-1000mmol L^{-1} NaCl; pH 8.0; 30°C.

Table 1: Estimated parameters of isotherm models for adsorption of whey lactoferrin on IDA-Cu^{2+}- cryogel system at different ionic strength conditions[a]

have independent nature, are no longer valid, invalidating the Langmuir model at these conditions.

Jiang and Hearn [27] observed significant changes in Scatchard's plots shapes with increase of salt concentration for adsorption of glair lysozyme on IDA-Cu^{2+}-resin. However, according to these authors, the intensity of deviation between these plots depends on the effective protein concentration at the immobilized ligand micro-environment.

pH effect

The influence of pH on the direction of lactoferrin adsorption on IDA-Cu^{2+}-cryogel (Figure 4) was investigated in the rage of 6 to 8, at 30°C and 200 mmol L^{-1} NaCl. It was observed that the larger the pH, the higher the adsorption capacity for this system. Similar results were noticed when obtaining lysozyme and nuclease adsorption isotherms from interaction CS-IDA-Cu^{2+} [22]. Indeed, optimum pH for this process depends on the used chelating chemical, the nature of the metal ion and the protein tridimensional shape.

Protons affect the net charge of both immobilized ion and protein. The pH range used in IMAC is usually 5-8 for immobilized ions, such as Cu^{2+}, Ni^{2+}, Co^{2+} e Zn^{2+}, where they coordinate mainly to hystidin residues at protein surface [6,28]. However, for metal ions, such as Ca^{2+} and Fe^{3+}, maximum adsorption occurs at low pHs (e.g. 5-6) [29,30]. Sulkowski [28] observed the rentention of Bovine Serum Albumin (BSA) in IDA-Ni^{2+}-sepharose was higher at pH 7.0 and negligible at pH 6.0 (Table 2).

Figure 5 shows Langmuir, Freundlich, Temkin and Langmuir-Freundlich isotherms. Though their coefficient of determination were relatively high, the three first models were not able to explain satisfactorily the adsorption of lactoferrin on IDA-Cu^{2+}-cryogel, which is seen by the deviations of simulated adsorption values from experimental data. Langmuir-Freundlich model yielded the best estimations at all investigated pH values (Figure 5a- 5c). Langmuir-Freundlich model parameters (Table 2) indicate that maximum adsorption capacity,

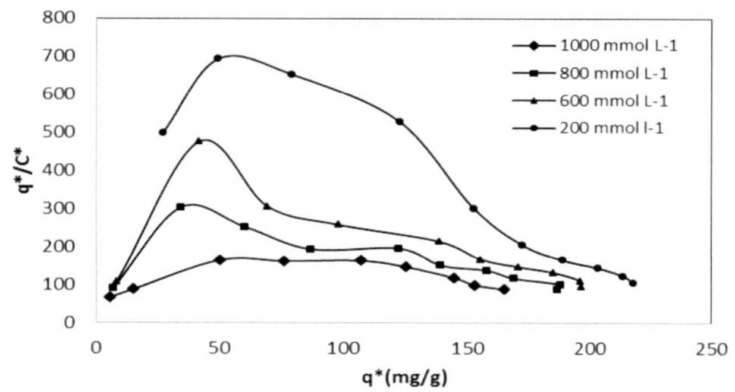

Figure 3: Scatchard's analysis for adsorption of lactoferrin on IDA-Cu^{2+}-cryogel in 20 mmol L^{-1} HEPES buffer, 200-1000 mmol L^{-1} NaCl, pH 8.0, 30°C.

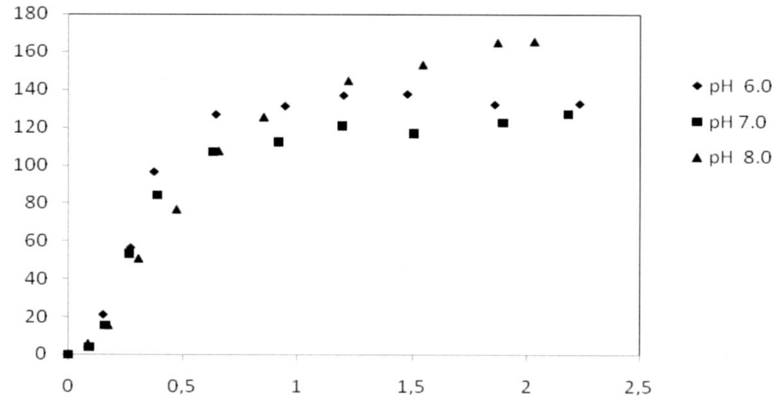

Figure 4: Adsorption of lactoferrin on IDA-Cu^{2+}-cryogel in HEPES buffer, pH 6.0-8.0, 200 mmol L^{-1} NaCl and 30°C).

Theoretical Model		**pH**	
	6.0	**7.0**	**8.0**
Langmuir			
K_d (μmol L^{-1})	0.2071	0.5264	0.2751
q_m (mg/g gel)	211.138	165.2006	274.4003
Freundlich			
K (10^{-3}L/g gel)	163.5	165.2	176.2
η	3.1062	2.4821	2.6207
Temkin			
K_T (L mol^{-1})	40.0969	15.6483	22.5191
q_T (mg/g gel)	40.495	60.9088	57.9874
Langmuir-Freundlich			
K^*_d (μmol L^{-1})	0.1552	0.0794	0.2785
q_m (mg/g gel)	203.1	196.2	242.2
η	1.1192	1.7007	0.9955

[a] 20mmol L^{-1} HEPES, pH 6.0-8.0, 200mmol L^{-1} NaCl, 30°C.

Table 2: Parameters of adsorption equilibrium for whey lactoferrin on IDA-Cu^{2+}-cryogel at different pH values [a]

q_m, was higher at pH 8. The elevation of this parameter may be due to the role of histidine (pK_a = 6) in protein- immobilized metal ion interactions. Studies performed by Andersson and Sulkowski [31] showed that residues of amino acids, such as lysine and arginine also significantly influence adsorption process at pH above 8. At pHs equal to or lower than 7, charged amino acids affect protein retention, mainly due to changes in driving forces between metal ion and amino acid residues [32]. At alkaline conditions, however, amino acid residues on the protein's surface will most likely be neutral or negatively charged, making them more bound to establish coordination bonds.

The binding affinity, that is, the inverse of the dissociation constant (K^*_d), initially increases; then, it reaches a maximum value at pH 7 and finally reduces. Similar tendencies were found for adsorption of ribonuclease A in Cu^{2+}- sepharose and for BSA in Cu^{2+}- superose at pH range of 5-7 [17,33]. Values of η >1 for all studied pHs, except for pH 8, suggest positive cooperation in the heterogeneous binding between lactoferrin and cryogel resin. Although η values varied through the three pHs, these changes were not significant, which implies that the adsorption mechanism remained the same.

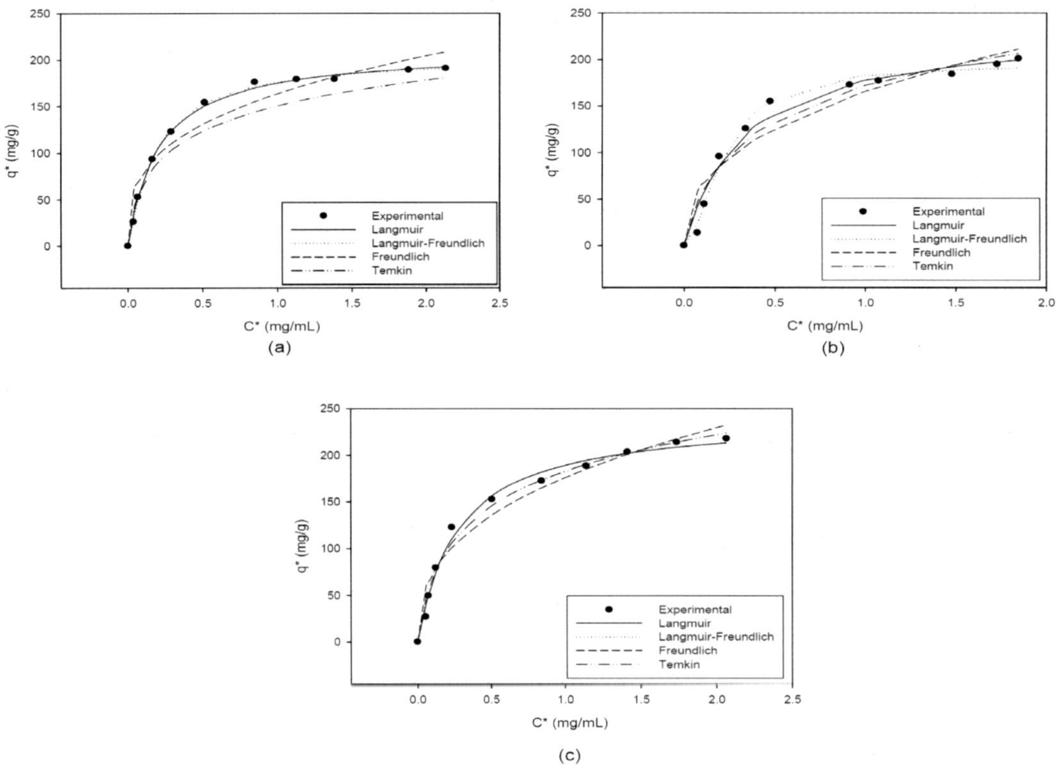

Figure 5: Comparison between several isotherm models for lactoferrin adsorption to IDA-Cu^{2+}- cryogel at three pH values: (a) 6.0, (b) 7.0 e (c) 8.0, temperature equal to 30°C and 200 mmol L^{-1} NaCl.

Conclusion

The results obtained from this work have important applications on immobilized metal affinity chromatography. Due to its high accuracy, Langmuir-Freundlich model parameters may be used to both describe interaction protein- immobilized ligand and simulate protein adsorption process at the conditions used to generate Langmuir-Freundlich isotherm, which is valuable for designing separation processes by IMAC in preparative scale and forecasting its performance. Thus, these results also show benefic information to generate process for capturing of lactoferrin in large scales and consecutively, to meet processing demands for this high value molecule.

References

1. Carvalho BM, Da Silva LH, Carvalho LM, Soares AM, Minim LA, et al. (2013) Microcalorimetric study of the adsorption of lactoferrin in supermacroporous continuous cryogel with immobilized Cu(2+) ions. J Chromatogr A 1312: 1-9.

2. Hahn R, Schulz PM, Schaupp C, Jungbauer A (1998) Bovine whey fractionation based on cation-exchange chromatography. J Chromatogr A 795: 277-287.

3. Bläckberg L, Hernell O (1980) Isolation of lactoferrin from human whey by a single chromatographic step. FEBS Lett 109: 180-183.

4. Noppe W, Plieva FM, Galaev IY, Vanhoorelbeke K, Mattiasson B, et al. (2006) Immobilised peptide displaying phages as affinity ligands. Purification of lactoferrin from defatted milk. J Chromatogr A 1101: 79-85.

5. Chen L, Guo C, Guan Y, Liu H (2007) Isolation of lactoferrin from acid whey by magnetic affinity separation. Separation and Purification Technology 56:168-174.

6. Bresolin ITL, Everson AM,Bueno SMA (2009) Cromatografia de afinidadeporíonsmetálicosimobilizados (IMAC) de biomoléculas: aspectosfundamentais e aplicações. Química Nova 32:1288-1296.

7. Ueda EK, Gout PW, Morganti L (2003) Current and prospective applications of metal ion-protein binding. J Chromatogr A 988: 1-23.

8. Sharma S, Agarwal GP (2002) Comparative studies on the metal sorption characteristics of chelating gels for immobilized metal ion affinity chromatography. Separation Science and Technology 37: 3491-3511.

9. Parveen S, Qamar S, Seidel-Morgenstern A (2015) Two-dimensional non-equilibrium model of liquid chromatography: Analytical solutions and moment analysis. Chemical Engineering Science 122: 64-77.

10. Arvidsson P, Plieva FM, Lozinsky VI, Galaev IY, Mattiasson B (2003) Direct chromatographic capture of enzyme from crude homogenate using immobilized metal affinity chromatography on a continuous supermacroporous adsorbent. J Chromatogr A 986: 275-290.

11. Carvalho BM, Carvalho LM, Silva WF Jr, Minim LA, Soares AM, et al. (2014) Direct capture of lactoferrin from cheese whey on supermacroporous column of polyacrylamide cryogel with copper ions. Food Chem 154: 308-314.

12. Bradford MM (1976) A rapid and sensitive method for the quantitation of microgram quantities of protein utilizing the principle of protein-dye binding. Anal Biochem 72: 248-254.

13. Langmuir I (1916) The constitution and fundamental properties of solids and liquids. J Am Chem Soc 30:2263-2295.

14. Chase HA (1984) Prediction of the performance of preparative affinity chromatography. J Chromatogr 297: 179-202.

15. Hanes CS (1932) Studies on plant amylases: The effect of starch concentration upon the velocity of hydrolysis by the amylase of germinated barley.Biochem J 26: 1406-1421.

16. Scatchard G (1949) The Attraction of Proteins for Small Molecules and Ions. Ann NY Acad Sci 51: 660-672.

17. Hutchens TW, Yip TT (1990) Protein interactions with immobilized transition metal ions: quantitative evaluations of variations in affinity and binding capacity. Anal Biochem 191: 160-168.

18. Sips R (1950) Advances in Chemical Engineering. Chemical Engineering 18: 1024.

19. Johnson RD, Arnold FH (1995) Review: Multipoint binding and heterogeneity in immobilized metal affinity chromatography. Biotechnol Bioeng 48: 437-443.

20. Johnson RD, Arnold FH (1995) The Temkin isotherm describes heterogeneous protein adsorption. BiochimBiophys Acta 1247: 293-297.

21. Andrade JD (1985)Surface and Interfacial Aspects of Biomedical Polymers. Volume 2, Plenum, New York, USA.

22. Chen WY, Lee JF, Wu CF, Tsao HK (1997) Microcalorimetric Studies of the Interactions of Lysozyme with Immobilized Cu(II): Effects of pH Value and Salt Concentration. J Colloid Interface Sci 190: 49-54.

23. Baydemir G, Andaç M, Derazshamshir A, Uygun DA, Özçaliskan, E, et al. (2013) Synthesis and characterization of amino acid containing Cu(II) chelated nanoparticles for lysozyme adsorption. Materials Science and Engineering: C 33: 532-536.

24. Graham B, Spiccia L, Hearn MT (2011) Comparison of the binding behavior of several histidine-containing proteins with immobilized copper(II) complexes of 1,4,7-triazacyclononane and 1,4-bis(1,4,7-triazacyclononan-1-yl)butane.J Chromatogr B Analyt Technol Biomed Life Sci 879: 844-852.

25. Góes LC, Miranda EA, Bueno SMA (2010) Interaction of histidine-tagged human proinsulin with immobilized nickel ion: Effect of chelating ligand and thermodynamics analysis. Colloids and Surfaces A: Physicochem. Eng. Aspects 369: 176-185.

26. James EA, Do DD (1991) Equilibria of biomolecules on ion-exchange adsorbents. Journal Chromatography A 542: 19-28.

27. Jiang W, Hearn MT (1996) Protein interaction with immobilized metal ion affinity ligands under high ionic strength conditions. Anal Biochem 242: 45-54.

28. Sharma S, Agarwal GP (2001) Interactions of Proteins with Immobilized Metal Ions: Role of Ionic Strength and pH. Journal of Colloid and Interface Science 243: 61-72.

29. Sulkowski E (1985) Purification of proteins by IMAC. Trends in Biotechnology3: 1-7.

30. Mantovaara T, Pertoft H,Porath J (1991) Purification of factor VIII:c coagulant activity from rat liver nonparenchymal cell culture medium by immobilized metal ion affinity chromatography. Biotechnol Appl Biochem 13: 120-126.

31. Andersson L, Sulkowski E (1992) Evaluation of the interaction of protein alpha-amino groups with M(II) by immobilized metal ion affinity chromatography. J Chromatogr 604: 13-17.

32. Arnold FH (1991) Metal-affinity separations: a new dimension in protein processing. Biotechnology (N Y) 9: 151-156.

33. Patwardhan AV, Ataai MM (1997) Site accessibility and the pH dependence of the saturation capacity of a highly cross-linked matrix. Immobilized metal affinity chromatography of bovine serum albumin on chelating Superose. J Chromatogr A 767: 11-23.

(Glycosylphosphatidylinositol-Based) Protein Chips and Biosensors for Biopharmaceutical Process Analytics

Günter Müller*

Department I, Genetics, Biocenter, Ludwig-Maximilians-University Munich, 82152 Martinsried near Munich, Germany

Abstract

Prior to the introduction of regulated process validation activities it was generally assumed that the production process for protein drugs *per se* is under control and appropriate for their generation, if the final product fulfils the criteria raised for the bioanalytical quality end control for therapeuticals. However nowadays, the production process is being considered in a much deeper and considerably more differentiated fashion. The use of in-process and on-/at-line controls and supervisions on a regular basis encompassing all critical parameters about the individual sub-processes and the emerging intermediary and side products is generally thought to significantly contribute to the demonstration that the production process is under control at the time point of the measurements and with high probability also thereafter. The deep understanding of the interrelationship between process and product can not completely eliminate but will considerably reduce the risk for the emergence of product variants, impurities and contaminants during critical sub-processes that may escape detection during final quality control for technical and/or economical reasons. The test systems required for elucidation of the multiple process-product interactions have to be chosen, validated and calibrated according to commonly accepted and approved criteria. In the near future novel platform technologies, such as protein chips and biosensors that are based on novel capturing/immobilising probes, e.g. glycosylphosphatidylinositol-anchored proteins, and detection probes, e.g. nanoparticles, will greatly contribute to the rapid and reliable measurement of many samples for multiple parameters in cell-free and cell-based assay configurations in parallel rather than consecutive fashion.

Keywords: Assay validation; Bioanalytical assays; Biosensors; Glycosylphosphatidylinositol; Nanoparticles; Protein chips; Protein drug production; Proteomics; Qualification and calibration; Quality control

Abbreviations: GPI: Glycosylphosphatidylinositol; GPI-proteins: GPI-anchored proteins; HCPs: Host cell proteins; NPs: Nanoparticles

Introduction

The scope of each process validation for the biopharmaceutical production of protein drugs is the demonstration that the manufacturing procedures envisaged will lead with high reproducibility and accuracy to a therapeutical that meets all patient's requirements with regard to safety and efficacy. Prior to the introduction of regulated process validation activities it was generally assumed that the production process per se is under control and appropriate for the generation of the protein therapeutical, if the final product fulfils the criteria requested for the bioanalytical quality end control. Accordingly, successful quality end control would be sufficient for approval of a protein therapeutical for the relevant purpose, e.g. phase I clinical studies. However nowadays, the production process is being considered in a much deeper and considerably more differentiated fashion, including its reproducibility and robustness during upstream and downstream processing in the industrial large-scale and under routine conditions as well as the possibility of the generation of undesired modifications of the protein therapeutical and of unexpected rare contaminations and impurities which might fail to be detected during routine quality end control. The use of in-process controls and supervisions on a regular basis encompassing all critical parameters about the individual sub-processes and the emerging intermediary and side products is generally thought to contribute to the demonstration that the process is under control at the time point of the measurements and with high probability also thereafter. Consequently, the development of production processes for protein drugs necessitates the rapid introduction of process analytics for both upstream and downstream processes [1-3].

Protein drugs must not only fulfil high-level product quality and safety requirements as requested by the regulatory agencies. In addition, they have to be introduced in the market in short time. As a consequence, at present the strategies for process development of protein drug purification are based on the scaling-up of procedures already implemented for the scale of laboratory or bench use and usually rely on the know-how of scientists and engineers upon successful experimentation at the lab and preclinical level. Nevertheless or for that reason this strategy often leads to sub-optimal individual steps regarding the technical equipment and auxiliary materials. Consequently, strategies and tools are urgently required that will generate optimal long-term solutions on the basis of a rational design in developing each step of the production process [4,5].

One strategy to approach this goal comprises the willingness to add to the currently used process development paradigm intensive efforts for the introduction of novel and attractive analytical tools, which will enable a more systematic and rationale design of the optimization strategy for the production of protein drugs. Examples for those tools encompass the so-called "high-throughput experimentation" (HTE) methods [6-8] and computer-aided process designing tools [9-11]. It is also of considerable importance to implement those novel designing tools during the development quite early since alterations in the

***Corresponding author:** Günter Müller, Sanofi Germany GmbH, R & D Diabetes, Industrial Park Höchst, Bldg. H821, 65926 Frankfurt am Main, Germany

production process and its analytical control may elicit considerable technological and regulatory risks [12]. The use of innovative designing and analytical tools for process monitoring and control is nowadays strongly encouraged by the US Food and Drug Administration (FDA) as delineated by its Process Analysis Technology (PAT) [13] and Quality by Design (QbD) [14] initiatives. The general mindset underlying these initiatives relies on the assumption that a deep understanding of the critical and rate-limiting process parameters will have the strongest positive influence on the quality of the protein drug produced.

In the early days of the biotechnological generation of protein drugs a thorough understanding of the production process was missing in most cases and the impact of the individual process parameters on the quality of the protein product could not be predicted in a reliable and adequate fashion. The ongoing improvement of bioanalytics and the underlying technologies as well as of the data evaluation caused an important shift in the analytical paradigm. On-/at-line sensors and measurement techniques allow the continuous determination of those process parameters critical for the adequate control of upstream and downstream processing. The introduction of statistical methods for data evaluation, like principal component analysis (PCA), facilitates the elucidation of the relationship between the process variables and the critical and rate-limiting process parameters. The availability of real-time data usually will have tremendous impact on the proper understanding of the production process. Thus the pro-active application of analytical data will create important novel opportunities within the production environment. A representative list of the molecular criteria for the quality assessment of the protein drug to be determined along its production and the analytical tools commonly used is given in Table 1.

Criteria for Process Analytical Methods Used for Protein Drug Production

The development of recombinant proteins as pharmaceutical drugs requires robust, sensitive and specific analytical assays to characterize the purified product with respect to its physicochemical as well as biological features, as well as bioactivity assays to quantify its protein amount and functionality in biological matrices. Well-established analytical assays are applied to determine the purity, absence of impurities and contaminants, identity, quantity, stability, specificity and potency of the purified recombinant protein during process development. The determination of the purity and identity of a protein drug is a particularly challenging task since recombinant proteins are produced from living systems that inherently lead to protein variants (e.g. post-translationally modified and/or fragmented

Protein Characteristic	Analysis Properties	(GPI-) Protein Chip	Alternative Methods
Physical and Chemical	identity	Yes	isoelectric focusing peptide mapping ion exchange HPLC
	purity	No	electrophoresis reverse phase HPLC size exclusion HPLC
	integrity/ molecular weight	yes/no	SDS electrophoresis mass spectrometry size exclusion HPLC light scatter
Potency and Activity	antigen binding	Yes	ELISA
	cell-based assays	Yes	cell proliferation cellular signaling cell metabolism cytotoxicity reporter-gene assays
Product-related Impurities	aggregation fragmentation conformation (secondary, tertiary structure)	(yes)	native electrophoresis Fourier-transformed infra-red (FTIR) size exclusion HPLC circular dichroism micro-rheometry differential scanning calorimetry raman spectroscopy dynamic/static light scatter analytical ultracentrifugation ultrasound
Process-related Contaminants	host cell protein	Yes	ELISA
	host cell DNA	No	DNA hybridisation qPCR DNA binding-threshold Fluorescent-picogreen
	protein A	Yes	ELISA
	medium proteins	Yes	ELISA
	viruses	Yes	qPCR electron microscopy in vivo/vitro assays
	microorganisms	Yes	"Bioburden" endotoxin-LAL test
	column/vessel leachates/extractables cell culture components reagents chemicals	(yes)	various e.g. reverse phase HPLC ion chromatography GC-MS

Table 1: Test methods for process analytics of protein drug production.

proteins) with altered characteristics, which may be difficult to separate from the authentic protein drug. In stability studies, those structural/functional features are evaluated that might be subject to change during storage/handling of the drug. Specificity measurements lead to a closer understanding of drug-target interaction(s), which might result in early hints about possible side effects in clinical trials. Finally, potency determinations are used to quantify the biological activity of the therapeutic protein.

On the other hand, bioanalytical assays are necessary to determine and quantify the protein drug in biological fluids. For example, validated bioanalytical assays are the key in the quantification of the protein drug in the course of pharmacokinetic studies. In particular, in the case of humanized/human monoclonal antibodies, bioanalytical assay development in human serum/plasma is challenging since the therapeutic concentration of antibodies can be very low (0.1 to 10 µg/ml or even lower), and because these antibodies are so similar to the native human antibodies that circulate in the blood at very high concentrations (up to 10 mg/ml).

Another major topic of scientific and regulatory consideration in the development of therapeutic proteins is the assessment of undesired immune responses toward the drug that may lead to a reduction in its efficacy and to adverse reactions. This assessment also requires validated bioanalytical assays, which allow to precisely measure the immune response. In future, there will be an even greater emphasis on the (bio) analytical description of biological substances because of an increase in numbers of protein therapeutics in clinical development, and of the advent of generics of biological drugs of the 1st generation (e.g. insulin and its variants) and the task of evaluating these compounds for clinical use. Developing methods and protocols for assessing bioequivalence of an original drug and its generics is a high priority for the FDA according to the current FDA Commissioner [15].

Taken together, the continuous measurement and control of all critical process parameters along the complete production process, which constitutes part of the so-called PAT initiative of the FDA, will significantly reduce the risk for the emergence of unexpected and non-predictable product variants as well as of contaminants not related to the product. Consequently, the deep understanding of the interrelationship between process and product can not eliminate but will considerably reduce the risk for the emergence of product variants and contaminants during critical sub-processes that escape detection during routine quality end control for technical or economical reasons.

Methods Currently Available for the Process Analytics of Protein Drug Production

During the last two decades a number of innovative analytical tools and methods has been introduced for the assessment of the quality and efficacy of the process for the recombinant production of protein drugs. They include chemometric sensors of high sensitivity and reliability [16] as well as on-line or at-line determinations of metabolites, nutrients, product and degradation products [17,18]. The use of those technologies will certainly narrow the gap usually existing between the understanding and the control of upstream processing. Protein chips and microarrays will become standard tools for the high-throughput analysis of protein expression. Chip technologies will also facilitate the fast characterization of the recombinant protein drug, even in brothes derived from process intermediates. Protein chips have already been successfully used for the purification and quantification of monoclonal antibodies and degradation products, such as half-antibodies, by capillary gel electrophoresis [19,20]. Chips specific for glycan structures have already been applied as tool to monitor glycosylated protein drugs. Moreover, the combination with sophisticated bioinformatical methods and relevant software tools will increasingly contribute to efficient clone screening and selection as well as to rational process development. A comparison of the major classes of the currently used methods for the identification of contaminations and impurities in protein drug preparations with host cell protein (HCP) during process development is given in Table 2.

Format	Direct Quantification	Identification	Limit of Detection	Use
ELISA	yes	no	1-100	lot release testing process control
Western Blotting	semi	yes	20-200	process characterisation
(GPI-) Protein Chip	yes	yes	0.1-200	process control and characterisation

Table 2: Comparison of the (GPI-) protein chip, ELISA and Western blotting formats for assaying HCP during process analytics of protein drug production.

Errors and Heterogeneity in the Production of Protein Drugs

Protein drugs, such as peptide hormones, growth factors, blood coagulation factors, enzymes and antibodies, are produced at large scale by recombinant microorganisms, such as bacteria (e.g. E. coli), funghi (e.g. yeast) or cultured mammalian cells (e.g. COS), predominantly, rather than by old-fashioned extraction from animal tissues or isolation from human plasma. Probably the best-known example represents the blood glucose-lowering polypeptide hormone, insulin, that is not available at adequate amounts in diabetic patients and therefore has to be substituted for by exogenous administration. Microorganisms become forced to synthesize the protein drug upon having received the corresponding information, i.e. the cDNA gene encoded by a circular

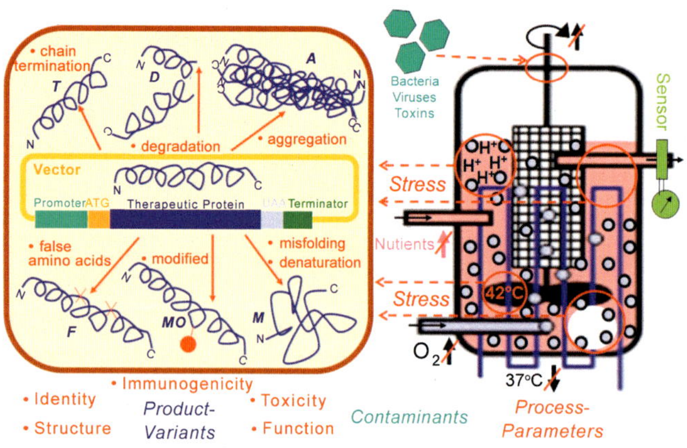

Figure 1: Errors and heterogeneity in the biotechnological production of protein drugs. Ectopic expression of protein drugs in recombinant host cells, such as bacterial, yeast and mammalian cells may lead to multiple product variants (*T, D, A, F, MO, M*) due to the operation of multiple physiological and stress response pathways under normal and sub-optimal growth conditions, respectively. Critical parameters for the quality of the resulting product variants may be differentially affected and become further impaired during fermentation in the large scale. The accompanying sub-optimal process parameters will lead to cellular stress due to local nutrient and oxygen depletion or pH and temperature increase. These deviations from the optimal process will not be recognized by sensors operating outside of the bioreactor. Furthermore, bioreactors are susceptible for infiltration by contaminants. *A*, aggregated protein; *D*, degraded protein; *F*, protein with false amino acids incorporated; *M*, misfolded denatured protein; MO, post-translationally modified protein; T, prematurely terminated protein.

DNA vector together with the appropriate upstream and downstream regulatory elements for transcription and translation. In many cases the microorganisms will produce the desired protein drug in functional state and in sufficient amounts for subsequent preparation and purification.

Microorganisms like human beings make errors. Moreover, populations of microorganisms like their human counterparts become heterogeneous with time. In consequence, the number and types of errors in protein drugs will increase with the scale and duration of the fermentation process. Errors occurring during protein synthesis in microorganisms include (Figure 1) the (i) incorporation of false or unphysiological (e.g. norleucine) amino acids leading to mutant proteins, (ii) premature termination of polypeptide elongation leading to carboxy-terminally shortened proteins, (iii) incorrect folding into the three-dimensional conformation leading to misfolded proteins, (iv) intracellular aggregation leading to protein precipitates or fibrilles and (v) intracellular proteolytic degradation leading to protein fragments [21-25]. These errors may have severe implications for the resulting protein variant with regard to its (i) molecular identity as reflected in the amino acid sequence (and already caused by a single amino acid exchange), (ii) structure as reflected in the three-dimensional conformation, (iii) physiological function as reflected in hypo- or hyperactivity (and putatively already caused by a single amino acid exchange), (iv) immunogenic potential as reflected in adverse antigenic response (and putatively already caused by a single amino acid exchange) and (v) toxicological potential as reflected for instance in proliferative activity with accompanying risk for cancer or protein aggregation with accompanying risk for embolic complications (and putatively already caused by a single amino acid exchange). The rate of errors in protein synthesis of non-recombinant and recombinant microorganisms under physiological growth conditions is rather low. However, this may change considerably upon their fermentation in bioreactors at large industrial scale. Microorganisms are usually grown in continuously operating bioreactors that are permanently fuelled with fresh glucose-containing medium under accompanying removal of the consumed glucose-deprived medium as well as with oxygen under concomitant precise control of the pH (by addition of buffering agent) and temperature (by cooling) of the medium (Figure 1).

The Origin of Contaminants and Impurities Generated during Recombinant Production

The 1994 draft, "Points to Consider in the Manufacturing and Testing of Monoclonal Antibody Products for Human Use" issued by the FDA [26] suggests that "wherever possible, contaminants should be below detectable levels using a sensitive assay capable of detecting 1 ppm expressed on a weight basis with respect to the monoclonal antibody". In practise, this often requires immunological assays, such as ELISAs, developed specifically to quantify known contaminants. This approach is likely to be required to measure the amounts of any cell culture medium additives, e.g. bovine serum albumin, transferrin, methotrexate, or any process-related contaminants, such as protein A or G molecules, leached from chromatography columns. In addition, HCPs also need to be measured. The development of such assays is not a trivial task as potentially a very wide range of molecules could contaminate the purified product [27]. Fortunately, with highly productive cell lines grown in serum-free or protein-free media, the purity target is usually not too difficult to achieve. This is particularly true when affinity chromatography is used in conjunction with one or more ion exchange separations.

The DNA specification is tighter than that of other contaminants with a recommended limit of no more than 100 pg cellular DNA per dose [26]. Often a target specification is set in the early stages of a product development program before the dose has been finalized and in these cases a specification is set that is based on the weight of the product, e.g. 1-10 pg DNA per mg antibody. It is not unusual for the specification to be set at the limit of quantification of the assay, as it is technically quite challenging to measure pg or sub-pg quantities of DNA in a concentrated protein solution. The level of DNA clearance required can be equivalent to 10 logs of deprivement or more.

Many mammalian cell lines are known to harbour viruses. In the past a number of biopharmaceutical products were used for therapy that have been contaminated by viruses and subsequently infected patients [28]. Therefore, viruses present a particular issue for the production of protein drugs when mammalian cell lines are used [29]. In addition to a thorough characterization of the cell line used for protein production and the testing of in-process samples, there are guidelines issued by the US and European regulatory authorities [26,30] relating to the incorporation of virus removal or inactivation steps in the purification process. In general, it is recommended that the purification schemes include at least one "robust" virus removal or inactivation step, such as solvent-/detergent- or low pH-treatment, or filtration steps. In addition, the chromatographic procedures used should also remove or inactivate viruses. The overall level of virus removal or clearance required is set by the number of virus-like particles detected in the unpurified bulk supernatant, which can be of the order of 10E8 or 10E9 per ml [31]. A safety margin of 3-6 logs is suggested [26], but it has been proposed that the safety margin should be variable and be dependent on the consequences of accidental exposure to the virus and the likelihood that a single virus particle will be pathogenic [28]. A similar strategy is based on the calculation of virus removal or clearance required on the probability that a single dose of product will contain a single virus particle. The figure that has been used for tissue plasminogen activator is a probability of less than 1 in 10E6 [32]. These types of calculations are often based on the assumption that a minimum of 15 logs of virus clearance/inactivation is required for the overall purification process.

In addition to the quantifiable targets in terms of the clearance of various contaminants, a key aspect of the purification of protein drugs is its "robustness" or tolerance to small variations in processing conditions. "Robustness" is directly related to the failure rate. With considerable time and money invested in the upstream production of a batch of protein drugs, the failure rate must be kept to an absolute minimum. It is essential to determine for each step of a purification procedure the key parameters that influence the performance of that step. For example ion echange chromatographical operations are sensitive to the pH and conductivity of the various equilibration, washing and elution buffers used. It is important to design those steps such that minor variations in the relevant parameters do not result in a decreased purification performance. As an additional precaution, some redundancy is frequently built into a purification process so that should one step perform poorly, subsequent steps can make up the deficiency and the final protein drug product will still meet the specifications. Thus the penultimate or the final step may be a polishing step that can remove process contaminants, but in most cases does not actually do so, as they have already been removed by preceding steps.

The absolute purity of a protein drug is often difficult, if at all possible, to determine. Regular and sometimes only subtle protein modifications, such as glycosylation, alternative disulfide bond formation, deamination, oxidation, phosphorylation, acetylation, sulfation, sulfoxidation, γ-carboxylation and pyroglutamate formation, lead to protein variants that may have more or less different characteristics. Also, truncated protein variants might be generated by the presence of cryptic or alternative start sites of transcription, by premature stop of the peptide chain elongation process, or by the action of host cell peptidases. Peptide mapping and mass spectrometry usually guarantee the detection of the majority of such protein variants. Aggregation is another modification of protein drugs, which can be the result of, for example, underglycosylation, oxidation, and/or deamination, and can be detected by size-exclusion chromatography. The amount of aggregated protein usually should account for less than 5%. It is highly recommended to investigate the nature and potential toxicity of such alterations. To analyse these variants is an essential yet challenging task, as their physicochemical features might not be very different from each other. Owing to the possible presence of highly related protein variants in the preparation, it is recommended to determine the purity of a protein drug by at least two independent methods, that is, methods that use different physicochemical principles, such as SDS-polyacrylamide gel electrophoresis and reverse-phase high pressure liquid chromatography.

Besides these protein variants, so-called process-related impurities have to be considered. Of major concern are residual antibiotics from fermentation, enzymes and antibodies from chromatography columns and other column leachates, endotoxin from bacterial hosts [33], (retro-) viruses [34], bacteria, fungi, mycoplasma, prions, various other media components, such as solvents, antifoam agents, heavy metal ions, as well as preservatives and HCPs. As for DNA contaminants, less than 10 to 100 pg per dose are allowed in the final drug product [35]. To check for the presence of antigenic expression host-related impurities, a polyclonal antiserum to the "empty" host, that is HCPs derived from host cells which are not expressing the product-encoding gene, is very helpful. Moreover, whenever possible, specific impurity standards should be used for impurity quantification and the limit of detection/quantification (LOD/LOQ) for impurity assays should be indicated. The acceptance limits should not be set higher than safety data justify, and it should not be lower than what is historically achievable by the production process and by reasonable analytical efforts. In some instances, the protein drug is conjugated to effector functions, such as radioisotopes, toxins, or other proteins, such as cytokines, that mediate the biological effect. Besides considering all aspects mentioned above for the individual components of the conjugate, special care has to be taken to determine the average coupling ratio as well as the amount of free components, if any, in the preparations.

Detection of Host Cell Proteins

Host cell proteins (HCPs) represent a complex set of analytes that have to be determined at very low concentrations (ppm or ng HCP per mg protein drug). The introduction of an adequate assay may be the most challenging task for process development. The assay system has to detect a wide array of analytes with high specificity and selectivity. Consequently, HCPs are commonly analysed with the use of immunoassay methods [27,36]. In most cases antibodies are raised against all the antigenic polypeptides expressed in the host cell line, which does not harbour the specific gene coding for the protein drug (so-called null cell line) [37,38]. Upon cultivation of the null cell line, the lysate proteins are extracted by procedures identical to those of the large-scale production process. Moreover, partially purified antigen preparations derived from selected steps of the large-scale purification process, may also be useful for generation of the antisera [39-41].

Three main formats of the immunoassays are currently in use, (i) conventional ELISA, (ii) Western blotting (involving electrophoretic separation of the HCPs and their electrotransfer onto a membrane, followed by decoration of the membrane with the anti-HCP antiserum) and (iii) protein chips (see below). Western blotting is rather effective for process characterization. It generates data about the efficacy of the individual purification steps for the elimination of particular HCPs. However, the technique of Western blotting is rather expensive and labour-intensive, that prevents it from application in automated and routine fashion, and therefore can not be used for online process control. Thus ELISA and protein chip technologies are the methods of choice for the overall quantitative evaluation of the levels of HCPs. The most challenging task for the development of both the ELISA and the protein chip formats is the accurate quantification of each individual HCP within a large concentration range by the same assay. For instance, as calibration standard the same mixed antigen preparation is commonly used that served for the preparation of the antisera for these assays. Upon immunization the range of HCPs in this preparation will trigger different immune responses and the generation of a variety of polyclonal antibodies with different affinities for each HCP. As a consequence, individual HCPs may yield different response factors in the ELISA and protein chip assays. This may impair the accuracy of both formats. Nevertheless, in practise the data obtained for samples derived from distinct process steps can be compared and will provide limits for appropriate process control. The principal procedures for the ELISA and protein chips represent standard immunoassay formats which can be automated and modified in straight-forward fashion to fit to at-/on-line process control where necessary. The relative merits of the three formats are given in Table 2.

For instance, protein A affinity chromatography represents a very efficient process step that is commonly used during purification for the production of monoclonal antibodies. It enables the highly efficient purification of the product in a single step. Importantly, any leaching of protein A from the column matrix will contaminate the therapeutic antibody and necessitate the unequivocal demonstration of removal of the protein A during the subsequent purification steps. Protein A is commonly detected by a conventional immunoassay. The method has to enable the dissociation of the protein A from the therapeutic antibody for accurate measurement, which in most cases is provoked by incubation at low pH (< 4) or detergent. The aim for both the ELISA and the protein chip technologies should be to approach limits of detection of less than 1 ppm (i.e. 1 ng protein per mg therapeutic antibody) to guarantee acceptable elimination of protein A [39].

Various sources for protein A are available and used for the commercial purification of therapeutic antibodies in pharmaceutical industry. The distinct protein A entities have varying size distribution and are heterogeneous in their accessibility to and extent of proteolytic degradation. Consequently, each species will trigger different response factors during the antibody generation for the immune assay. Thus, a specific assay standard is needed for each protein A species with the aim of its accurate quantitative evaluation relative to the standard in case of leaching from the purification matrix. Similar to the HCP assay, the ELISA and protein chip technologies represent standard immunoassay formats which can be readily automated and adapted for integration into at-/on-line process control if required.

Generation of Product Variants during Upstream Processing

Homogeneous and stable conditions regarding the most critical growth and fermentation process parameters, such as glucose, O_2, pH, temperature, inside the bioreactor vessel are often thought to be adequately maintained by rigorous and continuous mixing using a motor-driven propeller or bubbling of air. However, during large-scale operation of bioreactors, heterogeneities in the fermentation process parameters at certain local areas in the bioreactor vessel can not be avoided per se due to a number of technical and practical reasons. This may lead to the genesis of "hidden" spaces of poorly bubbled oxygen- and glucose-deprived and H^+-enriched medium "clouds" as well as of areas of elevated temperature at the surface of the propeller wings due to limited exchange with the surrounding medium of the heat generated by the propeller rotation (Figure 1). These local deviations from the optimal and homogeneous growth milieu for the microorganisms can not be prevented by simply increasing the influx of fresh medium or oxygen, the speed of the propeller or efficacy of the cooling. This would cause negative osmotic or toxic effects by high local concentrations of glucose or oxygen radicals, lead to undesired foam formation, further increase the temperature at the propeller surface or decrease the temperature in the immediate vicinity of the cooling device. The emergence of those local heterogeneities and deviations from the optimal fermentation parameters is favoured by the difficulties or even inability in recognizing them during the running process. For reasons of maintenance of sterile conditions and technical realization, the sensors for glucose, oxygen, pH and temperature are usually installed outside rather than inside the fermenter vessel at the critical positions. In consequence, the critical fermentation parameters are routinely measured at the efflux channel for the consumed medium. However, this mode of measurement provides average values for those process parameters, only. Nevertheless, these values will be used for the control and regulation of the fermentation process on basis of the simplifying assumption that they reflect the real conditions prevalent inside the vessel. Thus, the operators are prevented from proper recognition and, in consequence, from appropriate correction/counterregulation of local deviations from the optimal fermentation conditions.

In contrast, microorganisms will experience any sub-optimal growth conditions and try to overcome them by the initiation of stress-induced salvage pathways (Figure 1). In fact, the excessive production of ectopic "foreign" proteins in microorganisms per se exerts cellular stress that will induce physiological stress responses, which involve complex gene expression programmes for promoting cell survival and down regulation of functions that are not vital or even harmful for the microorganism. These include the (i) incorporation of unphysiological or false amino acids into (passenger) proteins in case of lack of certain amino acids in order to maintain the synthesis of vital proteins, (ii) premature chain termination of (passenger) proteins in order to save energy and (iii) degradation or aggregation of (passenger) proteins in order to avoid toxic effects due to their accumulation. Thus, each of these responses represents part of a physiological anti-stress strategy of the microorganism faced with the excessive production of a protein drug at large scale. But each of these responses may be responsible for the generation of undesired protein product variants of varying structure in varying amount. Furthermore, the growth of microorganisms in an industrial bioreactor always implies the danger of putatively contaminating the content of the fermenter, including the protein product, with foreign agents, such as bacteria and viruses (Figure 1). This is best exemplified with the most critical site of bioreactors of the "propeller" type, the bearing of the propeller axle in the vessel wall. It is technically difficult to create complete tightness for very small particles, such as viruses.

Product Variants and Contaminants during Downstream Processing

The possibility of the emergence of product variants and contaminants is not limited to the fermentation of the microorganisms, i.e. the upstream processing, but also encompasses the downstream processing for the preparation and purification of the final protein drug (Figure 2). This includes (i) the separation of the microorganisms from the medium by filtration or centrifugation in a separator, (ii) the homogenization of the collected microorganisms, e.g. by high pressure in the so-called "French Press", (iii) the conversion of immature, inactive and unfolded precursor polypeptide, that is often synthesized as the primary gene product in the microorganism, into the mature, active and correctly folded protein drug by specific processing steps, (iv) the purification of the mature protein from accompanying medium components and HCPs, including preservatives, anti-foam reagents, buffer substances, ions, nutrients, nucleic acids and lipids, by typical column chromatography procedures and finally (v) the concentration of the mature and purified but diluted protein drug by precipitation with high salt. This general scheme for downstream processing can be varied in multiple fashion, including the expression of the protein drug as insoluble inclusion bodies in the cytoplasm of bacteria or as secreted soluble entities in the growth medium of cultured mammalian cells. Each step specific for these expression systems, such as the refolding of the polypeptide from inclusion bodies, may contribute to the emergence of additional product variants and contaminants.

Furthermore, none of these steps during downstream processing operates with 100% efficacy, as is the case for the apparently simple separation of the medium from the host cells as well as the complex conversion of the unfolded precursor product into the mature protein drug. In consequence, medium components and precursor product putatively contaminate the product along all stages during the production process. Moreover, along the complete production process the protein drug is in contact to numerous "foreign" agents, such as column materials (e.g. silica beads) for the purification, which may not be stable but susceptible to aging, as well as the reagents used

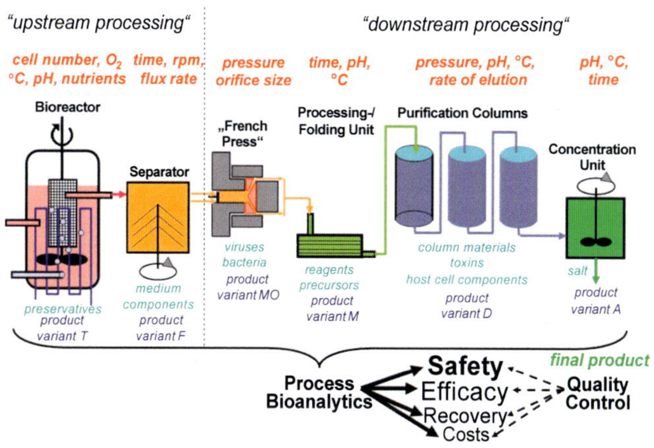

Figure 2: Process analytics as understanding of the relationship between process, product(-variants) and contaminants. During upstream and downstream processing, deviations from the optimal process parameters (in red) for each of the individual steps will lead to the emergence of the product variants *T, F, MO, M, D*, and *A* (in blue, symbols see Figure 1) and contaminants (in green) which are difficult to detect in the final product by routine quality control. These may affect the production costs, recovery, efficacy and safety of the protein drug in differential fashion with safety being of highest importance.

for the processing (e.g. enzymes) and precipitation (e.g. salt) steps (Figure 2). In consequence, at the end of the upstream and downstream processing the authentic protein product may often contain product variants and contaminants of varying nature and at variable amounts. In unfavourable cases the efficacy and safety of the protein drug may be negatively affected by the presence of product variants and contaminants in the final drug preparation.

Consequently, a deep understanding of the production process should encompass the relationship and interdependence of the relevant process parameters and the generation of the authentic protein, protein variants and contaminants under both optimal and sub-optimal running conditions (Figure 2). Thus, process analytics certainly exceeds the requirements for a mere analysis of the product, variants and contaminants with far-reaching consequences. In addition to the improvement of process control, this information will be required for the approval of a protein drug by the relevant health authorities, such as the EMA and FDA. For this, it is usually not sufficient to demonstrate the apparent purity of the final drug preparation. Instead, the manufacturer has to convince the authorities that (i) the production process envisaged will manage to generate a protein drug of the desired efficacy and safety in reproducible and reliable fashion, that (ii) any (critical) deviation of the production process from the optimal conditions will be recognized in time by the installed analytical devices and the data evaluation and that (iii) strategies are implemented that guarantee the rapid normalization of the sub-optimal process parameters for any critical step. Overall these demands for process validation, risk analysis and process control can be summarized with the statement "the process is the product". The consequences are far-reaching with regard to the official rules for production (GMP), storage (GSP) and documentation (SOPs), for which general recommendations and instructions (PAT) have been worked out by the FDA (see below). This requirement for an understanding of the production-product relationship is unique for protein drugs so far as is exemplified by the moderate or even missing rules and international standards for the production of human and animal food.

Hardware-Based Bioanalytics

The bioanalytical tests and assays required and selected for "state-of-the-art" process analytics have to fulfill a number of criteria which are summarized under the terms, qualification, calibration and validation. In a typical bioanalytical test the analyte contained in the sample induces a specific signal, the so-called read-out, e.g. light, current, fluorescence or luminescence impulses, most often elicited by a specific instrument. By nature of the test and instrumental configuration, even in absence of the analyte a signal will usually be generated. This "background" or "noise" should be as low as possible and is subtracted from each analyte-induced signal. As internationally accepted rule the signal-to-noise ratio has to exceed the factor of three, i.e. the analyte-induced signal has to be three-fold above the "noise" to be accepted as "specific" value for the analysis. This threshold defines the sensitivity or limit of detection (lod) of the test for this analyte without the scope for its quantification. With increasing amounts of sample, i.e. analyte, the analyte-induced signal will increase in non-linear fashion during a first phase, and then in linear fashion during a second phase. The lowest amount of analyte for which linearity holds true represents the lower limit of quantification (lloq). The slope of this line of direct proportionality should be high, resulting in pronounced increases in signal in response to rather low elevations in sample analyte and thereby defining the resolution of the test. At analyte concentrations exceeding the upper limit of quantification (uloq) any increase in the analyte will not result in a further linear elevation of the signal, but rather in gradually approaching a plateau phase. The start and end points of the linear phase between analyte concentration and signal define the dynamic "window", i.e. the range of measurable analyte concentrations, of the test. Certainly, this range should be as broad as possible for practical reasons, e.g. to avoid multiple sample dilutions prior to testing.

The test has to determine the analyte, e.g. human insulin, in very specific fashion, i.e. the ratio between the signals induced by the "active" (e.g. natively folded) sample analyte and the "inactive" (e.g. heat-denatured unfolded) reference analyte should be as high as possible. This ratio between the authentic "active" product and a more or less defective or interfering product variant defines the selectivity of the test. Furthermore, the reproducibility of the test is a critical factor and is affected by many different systematic and non-systematic errors, such as quality of the reagents, efficacy of the reactions involved, accuracy in pipetting, variation in instrumental data measurement, evaluation and calculation etc. By nature, distinct types of reproducibility can be discriminated, among them intra-assay variance, which encompasses measurements performed by the same operator with the same reagents, pipettes and instruments on the same day, and inter-assay variance, which characterizes measurements performed on different days by different operators using different pipettes, newly prepared reagents, independent reactions as well as distinct instrumentation, which may be switched off on and recalibrated. Inter-assay variance will always exceed intra-assay variance, but reflect the conditions for analysis of upstream and downstream processing products at the industrial scale in more realistic and reliable fashion.

Test Validation

The expenditure for and the scope of the validation of a test are related to its purpose and application. Furthermore, an in-house procedure requires a less exacting process than a method intended for multi-matrix and/or multi-laboratory use. For the latter tests, a full collaborative trial is necessary. However, for many purposes validation is limited to either demonstrating that method performance criteria established during development are met under routine laboratory conditions and/or showing test equivalence. The United States Pharmacopoeia identifies three categories of test: (i) Bioanalytical tests for quantitation of major components of bulk drug substances or active ingredients (including preservatives), (ii) bioanalytical methods for determination of impurities in bulk drug substances or degradation products in finished biopharmaceutical products, (iii) bioanalytical tests for determination of performance characteristics (e.g. dissolution, storage, drug release). Relevant parameters for all three types include accuracy, precision, selectivity/specificity, lod, lloq, uloq, linearity, dynamic range and robustness. One of the unfortunate previous choices of nomenclature is the use of "specificity" where what is actually required is "selectivity". Few bioanalytical tests are specific for a given analyte but generally can be made sufficiently selective for the purpose.

After quantitative determination of each of these criteria they have to be transformed and integrated into an "overall" parameter for the validity of the test reflecting all criteria in combination, the so-called z-factor. Calculated factors below or identical to 0.05 are regarded as indicative for a valid assay. However, it is of crucial importance to realize that there is no test validation as well as no validated assay per se, i.e. independent of the specific performance needs or purpose that have to be fulfilled by the test. For instance, validation of its resolution performance (see above) is critically dependent on the selection of the reference material(s) used and has to be justified toward the health authorities for drug approval. Furthermore, the reasons for the qualification of each test have to be presented to the authorities, including the arguments for the choice of the parameter analysed by the test, e.g. identity, structure, function. For instance, the validation of a test for the structure or function of a protein drug in comparison to a heat-inactivated material as reference may be useful if the emergence of the latter is conceivable in course of production of the drug.

Test Qualification

Analytical biochemists are by nature innovators and seekers of improvement. In the development area these qualities are invaluable in optimising the performance of a given method. However, far too often, this desire for continuous improvement spills over into the qualification of methods for process and quality control. Here consistency of application and rigorous control of processes and procedures have highest priority. These aspects are anathema for many practitioners of the "art of bioanalysis". Whilst this may be sustainable, albeit undesirable, for some applications within a single laboratory, discipline becomes a necessity when methods have to be transferred reliably between laboratories within an organisation or, even more critical, between distinct organisations. When the scope of operation encompasses different organisations, national boundaries, etc., uniformity of the approach is essential if comparable results have to be obtained. This discipline does not originate easily as it requires a control framework. The framework may be considered disturbing and unnecessary by some analytical biochemists, particularly those from a research environment. It is hoped to persuade those who doubt its necessity that the successful deployment of a method and its wide application rely heavily on such an approach and that flair, innovation and technical excellence alone are insufficient.

In general, the foundations for the confidence in a bioanalytical result require that (i) the sample is representative and homogeneous, (ii) the test selected is based upon sound scientific principles and has been shown to be robust and reliable for the sample analyte under test, (iii) the instrumentation used has been qualified and calibrated, (iv) a person who is both competent and adequately trained has carried out the analysis and (v) the data calculation is correct and statistically sound. Thus a control framework for the selection, development and validation of laboratory-based bioanalytical methods has to be established which includes both sample collection, preparation and storage (see above "i") and assay qualification (see above "ii-v"). Since many of the methods will be employed in generating data that could have profound legal or commercial impacts, the validity of bioanalytical results should be established beyond reasonable doubt. Importantly, validation of a bioanalytical test is not a single event. Rather it is a journey with a defined itinerary and stopping places as well as a final destination. The goal is a method that satisfies the original intent. A disciplined route is required which maps out both the qualification and validation process.

The key factors that need to be established for test qualification include (i) applicability of the bioanalytical principle(s) over the concentration range required, (ii) optimisation of the experimental conditions, (iii) selection of the calibration function, (iv) selection of the reference materials and standards, (v) evaluation of matrix effects and interferences, (vi) recovery experiments, (vii) robustness of the procedure toward changes in key parameters and (viii) generation of initial accuracy and precision data. The qualification procedure is likely to be an iterative one. However, it is essential that good written records are stored during this phase so that, in the event of problems at subsequent levels, investigations may be more readily carried out. However, far too often the excuse of "bioanalytical creativity" is cited for lack of such records. The most important outcome from this initial evaluation should be an assessment of the robustness of the developed test.

In general, it is critical if the test qualification is carried out by deciding to apply the method that is most popular or familiar. If a laboratory has expertise in a particular technique then is tempting it to use that expertise as the overriding factor for test qualification. Rarely is there a structured and rational approach for test qualification. Whilst it is often possible to make inappropriate methods work within a single laboratory, the impact on the reliable transfer between laboratories can be very large. In the past, the transferability of test methods has not been given the prominence it deserves. However, within the current climate of harmonisation and interchangeability, the technical requirements for test transfer and performance have been addressed in some detail. There are two areas which have received less attention and agreement, namely the inter-comparison of different tests for the same analytes in-house or within a few laboratories and the methods for describing and writing analytical tests. No test is "fit for purpose" unless there are clear and unambigous written instructions for carrying out the prescribed testing in accordance with the conditions laid down in the original test developmental cycle. The literature contains examples of collaborative efforts that only prove that the test was not fit for its intended purpose. The full IUPAC harmonised protocol is by its very nature an extensive and expensive exercise. From an economical perspective such trials should only be undertaken when there is well-documented evidence for sufficient robustness of the method under evaluation. Investment of time and intellectual effort for test qualification and the other aspects of the user requirements specification will pay great dividends. Prevention is better and almost always cheaper than cure.

Once the "User Requirements Specification" has been drawn up and the method performance criteria set, the test development process can begin. Quite often there are existing tests available within the literature or within trade and industry. On many occasions it is tempting to ignore the difficulties of a comprehensive literature search to save time. However, as a minimum, key word searches through the primary literature and abstracting journals, such as Analytical Abstracts, should be undertaken. For standard tests, it is essential to scan international standards from Europe and the USA as well as local sources and those deriving from statutory publications. Once already existing tests have been identified, it is good practice to compare them objectively. One way to do this is to list the performance criteria and relevant section of the User Requirements Specification and tabulate the corresponding data. An existing test may have a sufficiently good fit that adaptation is likely to lead to a suitable method. This relies upon professional knowledge and experience. For tests that are likely to be widely used, other aspects of suitability need to be considered. Some areas for consideration are listed below: (i) Can the method be written down sufficiently clearly and concisely to allow ease of the transfer? (ii) Can all the critical test parameters be identified and controlled, which is particularly important if automated systems are involved? (iii) Is the equipment readily available to all the likely participants? This assumes a special importance for internationally distributed tests and may involve questions of maintentance and support. (iv) Are all the reagents and solvents readily available in the appropriate quality? (v) Do the staff have the requisite skills and training to carry out the procedure? (vi) Are health and safety or environmental considerations likely to cause problems? (vii) Are standards and reference materials readily available to ensure that equipment and test systems are properly selected and calibrated?

In addition to validity, numerous other crucial criteria have to be considered for the qualification of bioanalytical tests, among them (i) the costs for the technical equipment, reagents and consumables as well as for the operators, (ii) the robustness, i.e. inertness of the measurement and data generation toward matrix components, (iii) the sample volume to save the product as well as reagents required for the analysis, and (iv) the possibility for process integration. For the efficient and rapid control of the production process on the basis of determination of the relevant product parameters, such as amount, product variants and contaminants, the data have to be available as soon as possible, i.e. on-line a short period after sample delivery. In case of delayed correction of and/or compensation for the sub-optimal process parameters, the product parameters would further deteriorate. This could lead to a vicious cycle within a single processing step and between successive steps if product variants and contaminants introduced by an upstream step cause deviations from the optimal parameters in downstream steps.

For integration into the production process, the bioanalytical test(s) have to be coupled intimately to the production process, at best at on-line and real-time mode. These requirements lead to severe limitations in the time available for eventual sample preparation, sample injection, actual analyte measurement and data analysis. Moreover, the bioanalytical test has to handle a considerable number of samples which have been collected along the whole production process at each critical step. Finally, since "a single test is no test", it is strongly recommended to test for more than a single product parameter, including identity, amount, structure, function, immunogenicity, toxicity and contaminants, in parallel at best or, if not feasible, successively within a short period.

The test criteria discussed above hold true irrespective of the product parameter analysed which include methods for demonstrating the (i) molecular identity via mass spectrometry (Figure 3) [22,42-44],

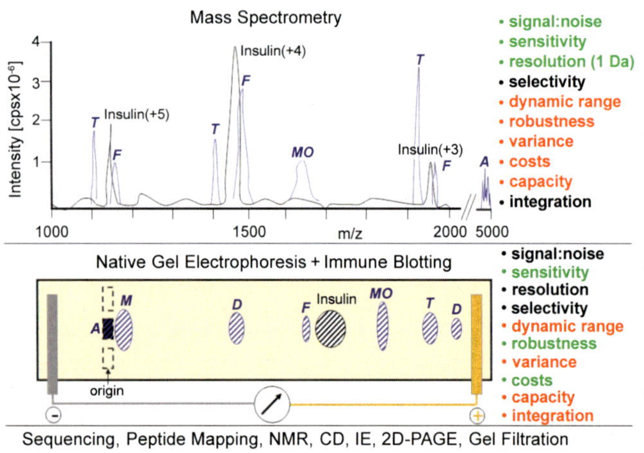

Figure 3: Bioanalytical tests for the parameters of molecular identity and structure. Advantages (in green), disadvantages (in red) and "neutral" criteria (in black) on the basis of the criteria for target qualification are given for mass spectrometry (nano-electrospray, MALDI-TOF, LC-MS/MS, 200 µM, 22°C, pH = 4.0, Zn2+, Cl-) and native gel electrophoresis (polyacrylamide gel, urea, 20 µg, pH = 6.8) combined with immunoblotting (polyclonal anti-insulin antiserum, 16 h) as exemplified for some insulin variants (symbols see Figure 1). Some alternative methods are listed at the bottom.

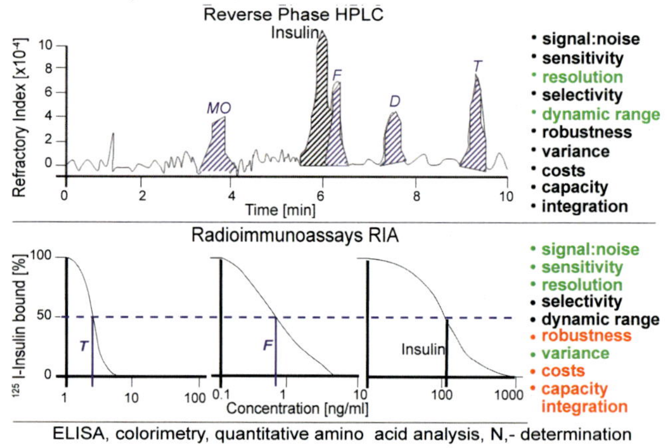

Figure 4 : Bioanalytical tests for the parameters of product amount on basis of physicochemical and immunological characteristics. Advantages (in green), disadvantages (in red) and "neutral" criteria on the basis of the criteria for target qualification are given for reverse phase HPLC (Eclipse XDB C8; 80A pore size, isokratic, 1 ml/min, 40 mM sodium phosphate, pH = 3.7, 85% phosphoric acid, 24% acetonitril, excitation 276 nm, emission 306) and radioimmunoassays (displacement curves with immobilised monoclonal antibody, [^{125}I] insulin, SDS-treated samples) as exemplified for some insulin variants (symbols see Figure 1). Some alternative methods are listed at the bottom.

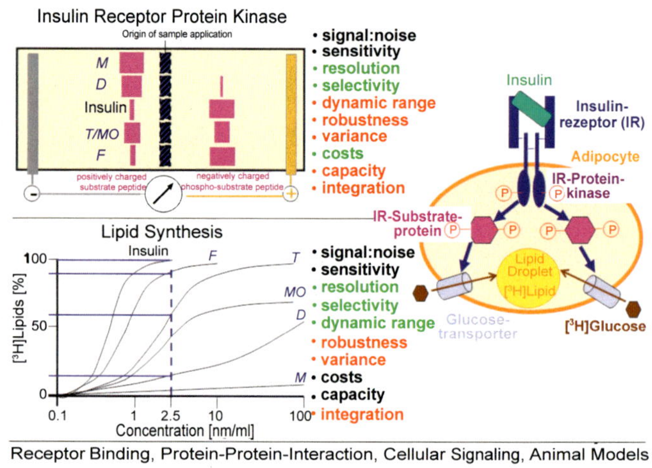

Figure 5: Bioanalytical tests for the parameters of molecular and cellular function. Advantages (in green), disadvantages (in red) and "neutral" criteria (in black) are given for assaying insulin receptor protein kinase activity (recombinant insulin receptor, fluorescent substrate peptide with consensus phosphorylation site, ATP, 0.8% agarose gel, 120 V, 55 mA, 20 min, extraction and fluorometric quantification of phosphorylated fluorescent peptide) and lipid synthesis (insulin concentration-response curves with response at 2.5 ng/ml for each variant of the incorporation of [3H]glucose into lipids in rat adipocytes, extraction and radiometric quantification of radiolabeled lipids) on the basis of the criteria for target qualification as exemplified for some insulin variants (symbols see Figure 1). Some alternative methods are listed at the bottom. The molecular basis for the activation of the insulin receptor protein kinase and lipid synthesis by insulin in adipocytes involving phosphorylation of the insulin receptor substrate protein in the cytoplasm and activation of the glucose transporter at the plasma membrane is depicted on the right.

(ii) two- and three-dimensional structure via native gel electrophoresis (Figure 3), Fourier-transformed infrared conformational analysis and circular dichroism [45-47], (iii) product amount via physicochemical (LC, 2D-NMR, RP-HPLC; Figure 4) [48,49] or immunological (radioimmunoassay; Figure 4) characteristics [1], (iv) function at the molecular (cell-free receptor or enzmye test; Figure 5) [50] or cellular

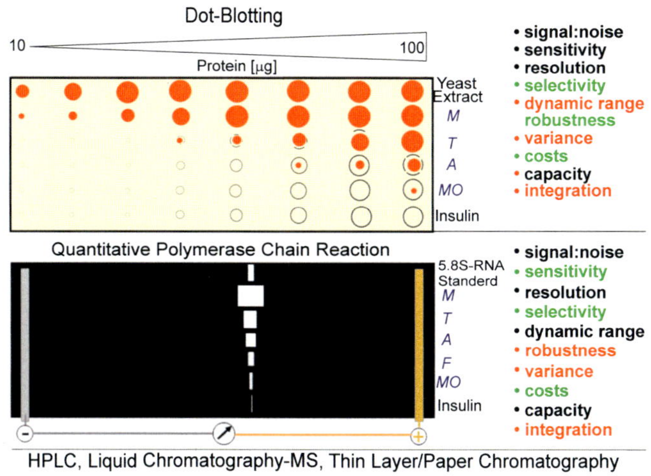

Figure 6: Bioanalytical tests for the parameters of endogenous contamination with HCPs and nucleic acids. Advantages (in green), disadvantages (in red) and "neutral" criteria on the basis of the criteria for target qualification are given for protein dot-blotting and quantitative polymerase chain reaction as exemplified for some insulin variants, which were produced in recombinant yeast cells (symbols see Figure 1), and extract from non-transfected yeast and 5.8S-RNA standard, respectively. Some alternative methods are listed at the bottom.

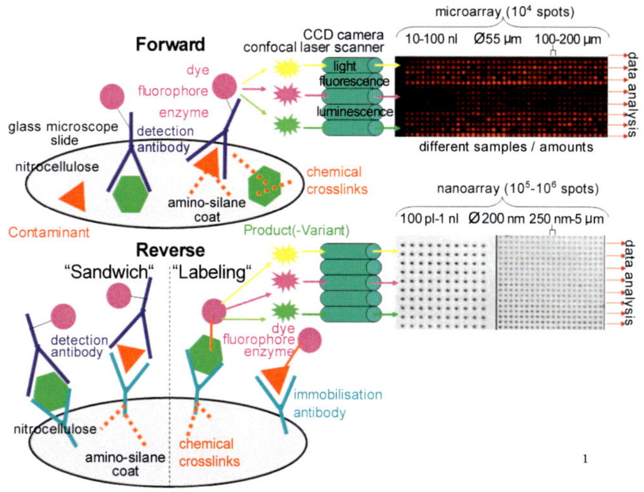

Figure 7: Principles of protein chip operation. The two distinct modes-of-action rely on either the direct (Forward) or indirect (Reverse) immobilisation of the protein analytes, such as products, product variants and contaminants, to the glass microscope slide by their non-covalent and chemical crosslinking or binding to immobilised antibodies, respectively. The protein analytes are subsequently detected by labeled antibodies in the "sandwich" or by labeling of the analytes in the "labeling" configuration. Read-out is identical for all modes and configurations and compatible with the micro- and nanoarray formats.

(cell-based physiological test; Figure 5) [50-53] level, (v) exogenous contaminants, such as toxins and antibiotics and (vi) endogenous contaminants, such as HCPs (dot blotting; Figure 6) or nucleic acids (RT-PCR; Figure 6) [54,55].

In conclusion, the qualification of tests to support validated, calibrated and approved process analytics along the whole biopharmaceutical production process necessitates high demands to be fulfilled by each test chosen as well as by each expert and operator involved [56,57]. It has to be accepted that most bioanalytical tests taken into consideration irrespective of the underlying principle will fail to meet all the requirements and represent compromises of numerous advantages and disadvantages [23]. The bioanalytics expert is responsible for the selection and creation of a test or test combination, which fits best to the process analytical task and, in most cases, represents an acceptable compromise. However, the design of any bioanalytical test is not "fixed in stone" and untouchable during subsequent periods of time. Rather the design has to be adapted to the current "state-of-the-analytical art" with regard to the technological progress as well as the gain in scientific knowledge about the putative pathophysiological relevance of the product variants and contaminants. This may necessitate the implementation of novel appropriate tests or the improvement of the sensitivity of already established ones in accordance with the rules and requests put forward by the health authorities.

Protein Chips – the Principle

During the last decade a large body of evidence has accumulated that protein chips may revolutionize the area of process analytics in biopharmaceutical production by their intrinsic capability of rapidly and simultaneously handling many samples under fulfillment of the criteria of validity, robustness, miniaturization and relatively low costs. Moreover, after special adaption they also enable multi-parameter analysis. Moreover, protein chips can not only be used for the evaluation of polypeptides of any size but also for the determination of small non-peptidergic analytes, such as lipids, carbohydrates and intermediary metabolites [56-64]. These characteristics considerably broaden the application profile of protein chips from mere biotechnological process analytics to novel biomedical research areas in diagnostics, drug discovery and biomaterial sciences, (i) personalized medicine with its scope of individualized diagnostics and therapy, (ii) systems biology with its aim of understanding the pathophysiology of common multifactorial diseases at the level of cells, tissues and the total organism including the interacting signaling cascades and metabolic networks, and (iii) tissue engineering with its potential to provide functional organs differentiated in vitro from (e.g. adipose tissue-derived) mesenchymal stem cells which had been isolated from the corresponding patient.

The mode of operation of protein chips, which are based on polypeptides rather than on nucleic acids as is the case for DNA/gene chips, can be easily explained (Figure 7). Protein chips, also known as protein microarrays, are miniaturized parallel assay systems which harbour minute amounts of highly purified proteins immobilised in a high-density format. They enable the simultaneous measurement of a variety of bioanalytes from limited amounts of complex sample mixtures in a single run. Initial approaches to simultaneously analyse large numbers of proteins for various parameters, such as molecular identity, amount, structure or function, were based on spotting down bacterial strains or bacterial total lysates of cDNA-driven protein expression libraries on nylon membranes [59-62,65-67]. A major progress was then the fabrication of protein chips consisting of 5,800 single yeast proteins on a specifically modified microscope glass slide [58]. Subsequently, a variety of useful applications, such as the identification of target proteins of small drug molecules, substantiated the practical value of protein chips.

Many protein chips are fabricated by the immobilisation of proteins onto a microscope glass slide with the help of a standard piezoelectric contact [68,69] or non-contact microarray printer [70-72]. Different slide surfaces are in use, including aldehyde- and epoxy-derivatized glass surfaces, so-called "Fullmoon" slides, or "Schott" NHS-derivatized slides for random linkage of the proteins through amines [69,73], nitrocellulose- [74,75] or gel- [76,77] coated slides for coupling through diffusion and absorption as well as nickel-coated slides for the non-covalent binding of His6-tagged proteins. In any case the immobilisation step has to operate in efficient, reliable and quantitative fashion. This means that each polypeptide contained in the sample has to be retained at the chip surface in quantitative fashion, irrespective of its nature and abundance. Furthermore, the nature of the interaction of the individual sample polypeptides with the chip surface, i.e. the type and number of the amino acids involved in cross-linking or secondary bond formation, is heterogeneous and can hardly be predicted. This may lead to variable and non-quantitative immobilisation. After successful immobilisation onto the slides, the proteins are evaluated for diverse parameters that classify the protein chips into two categories, analytical ones for the elucidation of the molecular identity, structure or amount and functional ones [76-84].

Forward Protein Chips

For the analysis of the identity, structure and/or amount of the protein drug or the detection of protein variants or contaminants, the protein chip becomes incubated with an appropriate well-characterized molecular probe which often is a specific antibody [85,86] (including single-chain or other variants), but nowadays instead may also consist of peptide-major histocompatibility complexes, carbohydrate-binding lectins, protein-interacting anti-/lipocalins, protein-nucleic acids or RNA aptamers. For the demonstration of binding of the molecular probes to the chip surface, they are coupled to a dye, a fluorophore or an enzyme which catalyzes a luminescent reaction. The light, fluorescence or luminescence signals are detected by a multi-channel laser scanner or CCD camera with high resolution power. The resulting typical patterns of (directly or indirectly) colored spots in regular arrangement (Figure 7) are evaluated for spot intensity, with the more intense, the higher the amount of the probe bound to the chip, the higher the amount of analyte immobilised onto the chip surface, the higher the analyte concentration of the sample. Data generation encompassing the incubation of the chip with the probe, the binding of the probe to the analyte, the washing of the chip for removal of unbound analyte, the reading-out of the chip for spot intensity and the computer-based data transformation and the calculation of the actual analyte concentration requires short periods of time only (e.g. 120 sec for a single cycle). The number of samples that can be processed in parallel depends on the power of the automatic chip printer used which generates spots of defined diameter and in defined distance from one another. With the currently available microarrays up to 10E4 spots can be applied onto a typical light microscope glass slide. Future nanoarrays will enable the spotting of up to 10E6 samples per slide with spot diameters as small as 250 nm, and will be limited only by the resolution of the available scanners. Thus, in principle, protein chips seem to meet the major bioanalytical demands, i.e. the measurement of many samples in parallel in short time with the option of multi-parameter analysis and automation.

With regard to functional protein chips, a large number of proteins in case of complex biological samples or the total proteome of a cell or organism in case of a systems biology approach are spotted. There is no need for extensive biochemical characterization of the proteins prior to immobilisation and chip analysis. The systematic screening for specific and divergent activities and functions may encompass protein-protein, protein-DNA, protein-carbohydrate, protein-lipid, protein-metabolite and protein-drug interactions as well as the identification of enzyme substrates or the detection of (undesired) immune and toxicological responses. The formats of the protein chips currently used vary considerably to match the diverse requirements from high-throughput analysis of routine parameters to the huge spectrum of low-throughput analysis of more specific functions.

Importantly, this type of forward protein chip critically depends on the quantitative immobilisation of the protein analytes by secondary bonds or covalent cross-links at the chip surface as well as their quantitative detection by the labeled molecular probes, such as antibodies. However, even in case of quantitative recovery of the analytes, it can not be excluded that unspecific interactions and modifications involved herein will lead to masking of certain protein epitopes which are recognized by the detecting probe. By nature, the amino acids involved in the interaction of the protein analyte with the chip surface as elicited by a selected procedure is difficult, and sometimes even impossible, to predict and may vary with each cycle of immobilisation. As a consequence, the detection of the analyte will not be quantitative leading to underestimation of the analyte concentration with considerable variation between distinct measurements.

Reverse Protein Chips

To circumvent these problems the so-called reverse protein chip has been introduced which relies on immobilisation of the protein analytes by binding to well-characterized immobilising probes (Figure 7). This type of analytical or functional protein chip represents the most convenient and powerful multiplexed detection platforms and is commonly used for determining protein expression, profiling cell surface markers, identifying biomarkers, and measuring diagnostic factors. The immobilising probes, most often antibodies, are themselves coupled onto the chip surface by secondary interactions with a (e.g. nitrocellulose) coat or by covalent cross-linking to functional (e.g. amino) groups of the glass slide [87-90]. In contrast to forward chips with their direct immobilisation of the protein analytes by the same means, the reverse chips can be controlled and normalized for the coupling efficacy of the immobilising antibodies under standard conditions to compensate for eventual non-quantitative recovery of the protein analytes. This solely depends on the number of functional immobilising antibodies coupled to the chip surface. For detection of the protein analyte, the so-called "sandwich" configuration relies on the specific binding of a 2nd antibody that recognizes an epitope distinct from the immobilising 1st antibody and is labeled with a dye, fluorophore or luminsecent enzyme. In consequence, "sandwich" protein chips have the advantage of exquisite selectivity for the protein analyte due to the simultaneous operation of two different antibodies. However, the identification of a pair of antibodies directed against the protein analyte, which are specific for distinct non-overlapping epitopes and do not mutually impair binding to the analyte due to steric hindrance, is often tedious, long-lasting and expensive and sometimes may even fail. However, the putative (partial) incompatibility of 1st and 2nd antibodies will be recognized in time during the chip development. Nevertheless, in some cases compensation for partial interference by normalization may be feasible and useful, since the same antibody pair can be used throughout the chip analysis. Thereby, underestimation and high variance in the determination of the analyte content would be minimized by the "reverse sandwich" chip. At variance, the "direct labeling" configuration circumvents the need for a 2nd detecting antibody (and, in consequence, may be of lower selectivity than the "sandwich" chip). For this the dye, fluorophore or enzyme is coupled to the protein analytes through secondary interactions or covalent cross-linking prior to incubation of the samples with the protein chip.

GPI-Protein Chips

Unfortunately, the technologies of both "forward" and "reverse" protein chips described above could lead to non-quantitative detection and immobilisation, respectively, of the (indirectly and directly tagged) analytes in course of (partial) masking of the epitopes recognized by the immobilising antibody/anticalin and detecting antibody, respectively. In consequence, these conventional protein chips are faced with formidable challenges in the quantitative detection and immobilisation, respectively, of the protein analytes. The problems may be (partially) overcome by a novel type of protein chip that is based on glycosylphosphatidylinositol- (GPI-) anchored proteins (GPI-proteins). These GPI-protein chips are currently being evaluated and validated for numerous applications in process analytics for the biopharmaceutical production, in "point-of-care-testing" for individualized diagnostics and health care as well as in drug discovery and compound optimisation.

Typical transmembrane proteins span the phospholipid bilayer of the cellular plasma membranes through a single or several stretch(es) of hydrophobic amino acids, the transmembrane domain(s), with large amino-terminal (carbohydrate- and disulfide bridge-harbouring) and carboxy-terminal polypeptide domains facing the cell surface and protruding into the cytoplasm, respectively. In contrast, GPI-proteins lack (a) transmembrane domain(s) but are embedded exclusively in the outer extracellular leaflet of the phospholipid bilayer by a covalently linked GPI structure [91-93]. This glycolipid anchor consists of phosphatidylinositol and several distinct and specifically linked carbohydrate moieties, the glycan portion, and is coupled via its terminal ethanolamine residue through an amide linkage to the carboxy-terminus of the extracellular protein domain. Thus, none of the amino acids of the GPI-protein moiety is in intimate contact to the plasma membrane. Rather, the GPI-protein is located at the cell surface solely through its GPI anchor, which can be specifically cleaved by phospholipase C leading to release and solubilisation of the protein moiety. GPI-proteins are expressed in all eucaryotic cells studied so far,

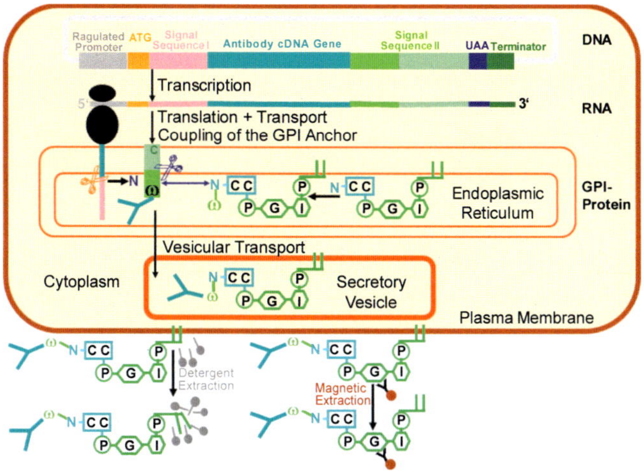

Figure 8: Recombinant production of GPI-proteins in eucaryotic cells. Soluble proteins, such as antibodies and protein drugs (e.g. insulin), are ectopically expressed as GPI-proteins in eucaryotic host cells in course of their plasmid-driven transcription, translation and co-translational translocation into the endoplasmic reticulum with concomitant coupling of the pre-fabricated GPI anchor to the carboxy-terminus of the protein domain and final transport to the outer surface of the plasma membrane along the classical secretory pathway. The GPI-proteins are then purified from the surface of the intact cells by detergent or "magnetic extraction".

from yeast to man, and fulfil diverse functions as cell surface receptors, enzymes, antigens, transporters and signaling and adhesion molecules [92].

For application in the protein chip technology, it is most critical that in principle each soluble passenger protein, such as a binding protein, receptor, anticalin or antibody, can be expressed ectopically as GPI-protein at the surface of eucaryotic host cells, such as the yeast S. cerevisiae, chinese hamster ovary cells or human embryonic kidney cells, using recombinant DNA technology (Figure 8). For this, the relevant host cells have to be transfected with a plasmid harbouring the cDNA derived from the corresponding protein gene and appropriate 5'- and 3'-regulatory elements for its inducible/repressible transcription and translation. In addition, for biogenesis of the passenger protein as GPI-protein two targeting signals, signal sequences I and II have to be placed at the 5'-/amino- and 3'-/carboxy-termini of the corresponding gene/protein constructs [94]. Signal sequence I directs the nascent GPI-protein through the endoplasmic reticulum and secretory vesicles to the plasma membrane along the typical secretory pathway. Simultaneously signal sequence II drives the covalent coupling of the GPI anchor, pre-fabricated in the endoplasmic reticulum by step-wise glycosylation of phosphatidylinositol, to the GPI-protein precursor in the course of removal of signal sequence II by a transamidase reaction occurring in the endoplasmic reticulum. The molecular machinery including the genes involved as well as the structural features of the signal sequences I and II have been elucidated during the past two decades [95-97]. This knowledge will be helpful for the ectopic expression of GPI-anchored versions of any soluble protein at the surface of host cells, in particular by engineering of optimised signal sequences as well as molecular components of the GPI-protein synthesis pathway [98].

Recombinant Production of GPI-Proteins/Antibodies

After the upstream processing, which includes fermentation of the recombinant host cells in a bioreactor and subsequent removal of the culture medium, total GPI-proteins will be extracted from the cell surface, preferably without solubilization of the cells. For this, non-ionic detergents, such as Triton X-100, Triton X-114, octylglucoside, are added at low concentration at low temperature. This results in disintegration of the host cell plasma membrane under concomitant aggregation of the GPI-proteins together with cholesterol and (glyco-) sphingolipids into so-called lipid rafts [99], that can be collected and enriched by sucrose gradient centrifugation on basis of their low buoyant density. Finally, the GPI-proteins are solubilized by high concentration of octylglucoside at room temperature and, if required, further purified by (several rounds of) conventional column (preferably affinity) chromatography.

An alternative to this controlled detergent extraction has recently been developed, the so-called "magnetic extraction" (Figure 9). It is based on antibodies which are directed against the glycan portion of the GPI-protein anchor and covalently coupled to metal beads with diameters in the µm-range [100]. Since the structure of the glycan portion is highly conserved and identical from yeast to man, these anti-GPI antibodies, once raised and demonstrated to be of high affinity and selectivity, are used for the "magnetic extraction" of any GPI-protein produced in any host cell. For this the recombinant host cells, such as yeast or cultured mammalian cells, have to be immobilised or adherent, respectively, to the bottom of the culture dish of the bioreactor. Upon successful plasmid-driven cell surface expression of the GPI-proteins, such as protein drugs or antibodies for protein chips, the anti-GPI antibodies coupled to metal beads are added to the culture medium.

Thereafter, an electromagnetic plate is positioned in the bioreactor in very close distance, however without direct contact, to the cell surface. Upon activation of the electromagnetic field and concomitant exposure of the cells to ultrasonic treatment, the magnetic plate is rapidly removed from the cell surface. This tightly controlled movement, which is performed by a high-precision robotic arm, causes the spontaneous displacement of the GPI-proteins, which are bound via the anti-GPI antibodies and the attached metal beads to the electromagnetic plate, from the lipid rafts of the host cell plasma membrane. Importantly the host cells will survive this procedure and can be re-used for the next fermentation cycle upon supplementation of fresh culture medium and re-induction of the GPI-protein expression system.

After washing of the electromagnetic plate with high concentrations of salt and low concentrations of non-ionic detergent to get rid of unspecifically associated medium and cellular components, the bound GPI-proteins are highly enriched toward plasma membrane (and total cellular) proteins since the major portion of them lacks the GPI anchor including its glycan moiety and therefore resists the "magnetic extraction". If required, the GPI-proteins bound to the anti-GPI antibodies can be purified to homogeneity, i.e. separated from endogenous host cell proteins, after release from the electromagnetic plate. For this, the well-established techniques of magnetic filtration and purification on the basis of the metal beads associated with the GPI-proteins alone or in combination with Triton-X114 partitioning and classical chromatographic procedures will be useful. The final desorption of the GPI-proteins from the anti-GPI antibodies and/or the electromagnetic plate are achieved by the addition of excess of carbohydrate, which competes for typical glycan components of the GPI anchor, such as mannose, or GPI anchor-cleaving bacterial phospholipase C, which leaves the passenger protein with a phosphoinositolglycan moiety attached (lacking diacylglycerol). The strategy preferred depends on the requirement for the passenger protein with either the intact GPI anchor (e.g. GPI-antibody/analyte for protein chips, protein drug for oral application [134,135]) or (partially) deleted GPI anchor (e.g. GPI-protein as protein drug for non-oral application). In the latter case the complete removal of the GPI anchor is desirable and this necessitates the cleavage off of the residual phosphoinositolglycan moiety by specific phosphatases and exoglycosidases (Figure 9). Meanwhile the "magnetic extraction" process has been successfully applied for the large-scale production of GPI-anchored antibodies for protein chips as well as protein drugs, such as human insulin. This pilot bioreactor uses an automatically and continuously operating assembly line for the successive fermentation, extraction, purification and desorption steps of the GPI-proteins and the integrated recycling of the anti-GPI antibodies [100].

Operation of GPI-Protein Chips

For the immobilisation of the GPI-antibodies at the chip surface, the microscope glass slide is coated with a monolayer of phospholipids, which is facilitated by hydrophobic interactions between the saturated long-chain fatty acids and the glass surface and mimics the structure of the extracellular leaflet of the plasma membrane phospholipid bilayer. Upon addition of the anti-analyte GPI-antibodies embedded in detergent micelles to the phospholipid-coated glass slide, the antibodies become spontaneously inserted into the phospholipid monolayers solely triggered by dilution. This can meanwhile be managed with sufficient reliability by adapted automated standard printers commercially available for the spotting of conventional protein chips and will result in defined orientation of the GPI-antibodies with the antigen-binding domain facing the chip surface at high density. The interaction resembles that of GPI-proteins reconstituted into the outer face of biological membranes upon their incubation with intact cells, liposomes or model membranes [101-104]. In comparison to covalent crosslinking, the GPI anchorage of the anti-analyte antibody to this special type of "reverse" protein chip has the advantage of being efficient and selective and causing lower background due to unspecific adsorption of the analyte. This often represents a problem with bifunctional chemical crosslinkers due to their hydrophobic nature and broad specificity. Moreover, the covalent coupling of the GPI anchor to the carboxy-terminus of the heavy chain of immunoglobulins or single-chain antibodies does not interfere with antigen recognition by the amino-terminal variable domains of their light (and heavy) chains, in general. This will guarantee high efficacy, robustness, reliability and reproducibility of the analyte immobilisation resulting in low variance between different measurements using the same chip or different chips, which have been prepared by different spotting procedures with different batches of anti-analyte GPI-antibodies.

Upon implementation of the required tools and equipment, the expenditure for the generic production of the anti-analyte GPI-antibodies and their immobilisation onto the chip using versatile cassette GPI-protein expression vectors (Figure 8), "magnetic extraction" (Figure 9) and an automated standard printer for embedding into the phospholipid monolayer is usually lower compared to covalent crosslinking with regard to both time and costs. Crosslinking requires intensive testing of a multitude of chemicals and reaction conditions and careful quality control for successful coupling of functional antibodies to the chip surface. Upon addition of the sample to the GPI-protein chip, the immobilised analyte is routinely detected by binding of an anti-analyte antibody labeled with a dye, fluorophore or luminescent enzyme, as is the case for conventional reverse protein chips of the "sandwich" mode. This directly leads to corresponding light, fluorescence or luminescence signals at "positive" spots of the array.

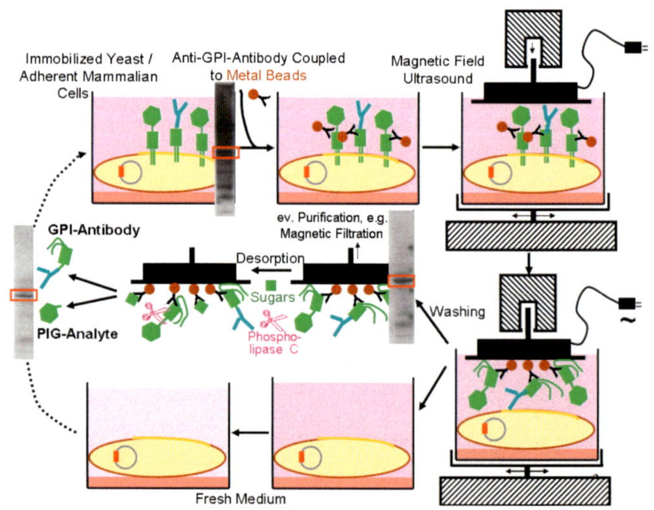

Figure 9: "Magnetic extraction" of GPI-proteins. GPI-proteins, such as antibodies and protein analytes, are extracted in the large scale by binding to anti-GPI-antibodies coupled to metal beads and subsequent removal of the complexes formed between them and the GPI-proteins from the host cell surface by attraction to an electromagnetic plate. Upon lifting of the electromagnetic plate, the highly enriched GPI-proteins are released with the intact GPI anchor by excess of sugar moieties or as phosphoinositolglycan (PIG)-analytes (with PIG remnant) by cleavage with phospholipase C.

Nanoparticle-Based (GPI-) Protein Chips

Faced with the problems of the identification of (pairs of) capturing/immobilising probes, most often monoclonal or polyclonal antibodies, as the most critical requirement for the development of sensitive and specific protein chips, intensive attempts have been made during the past decade to introduce alternative types of protein chips which (i) operate in solution rather than in solid phase, (ii) do not rely on specific capturing molecules, i.e. antibodies and (iii) do not require labeled detecting molecules, i.e. antibodies. These techniques take advantage of nanoparticles (NPs) and soluble reporter enzymes [105-110]. The reporter enzyme, such as bacterial ß-galactosidase, has to catalyze an easily detectable reaction and preferably should exhibit a large surface area with exposed polar as well as hydrophobic amino acid side chains and/or post-translational modifications of complex nature and be monitored by simple photometric or amperometric (e.g. ISFET) [111-114] measurement. ß-Galactosidase is characterized by a negative net surface charge at physiological pH. It was converted into its GPI-modified version in order to increase its apolar surface area without significantly reducing its high solubility or causing its aggregation by recombinant expression in mammalian host cells (e.g. HEK-293, CHO). Subsequently it was purified using "magnetic extraction" (see Figure 9). In contrast to the wildtype form, GPI-β-galactosidase displays amphiphilic character. This is caused by the carboxy-terminal GPI anchor harbouring two saturated long-chain fatty acids and the glycan moiety, which protrudes from the core of the polypeptide chain onto its surface, in combination with the negatively charged surface-exposed amino acids. Importantly, the GPI modification does not significantly impair the hydrolytic activity of β-galactosidase.

The NPs used will replace in a certain sense the antibodies and serve as "chemical noses or tongues". They should manage to "smell or taste" the surfaces of the reporter enzyme and protein analyte in a very differentiated fashion [109]. For this, the NPs consist of a gold core and an outer shell of covalently coupled organic molecules. These shell structures are built by a terminal positively charged ammonium residue with one variable functional group of different hydrophobicity as substituent. In consequence, they may bind with high selectivity and avidity to protein surfaces through electrostatic and hydrophobic/apolar interactions. In case of GPI-ß-galactosidase, its surface will be recognized by these cationic/hydrophobic NPs via salt bonds between the terminal ammonium groups and the negatively charged amino acids as well as via hydrophobic interactions between the functional groups and the GPI fatty acids. Coverage by the NPs of GPI-β-galactosidase causes its complete blockade, for instance through binding to its glucose substrate-binding site. Upon simultaneous incubation of protein analytes with the GPI-ß-galactosidase and the NPs, the analytes may fail to interact with the NPs thus leaving the reporter enzyme in the NP-bound and inhibited state. Alternatively, the analytes may succeed in interacting with the NPs and thereby displace them from the reporter enzyme thus relieving it from inhibition. The resulting activity signals will reflect the relative strength of the interactions between the NPs and the reporter enzyme vs. the protein analyte. They are positively and negatively correlated to the efficacy of the "smelling or tasting" of the surface characteristics of the protein analyte and GPI-β-galactosidase, respectively, by the NPs. Certainly a single type of NPs will enable only limited differentiation between different protein analytes contained in complex sample mixtures. It makes the combination of different types of NPs, each of them equipped with distinct functional groups that allow the unambiguous discrimination of a given analyte out of a complex biological sample. Taken together, protein chips based on NPs rather than antibodies seem to have the potential for sensitive, quantitative, selective and reliable determination of protein analytes from complex sample mixtures [110,115-117].

As a consequence, the generation of data signatures characteristic for the desired protein analyte in a complex sample, relies on multiple successive cycles with distinct types of NPs. The number of cycles will depend on the discrimination power of the NPs used, the nature of the analyte and sample as well as the specificity and resolution criteria desired for the assay. The multi-cycle handling of the microtiter plates, if performed manually, would considerably impede the rapid measurement of many samples with a multitude of different NPs. This delay would interfere with the requirements of process analytics during biotechnological production. To overcome this issue, the NP-based "smelling and "tasting" technology has been transferred from the microplate to the chip format that is based on a continuous flux of carrier fluid driving the components involved, such as buffers, samples, NPs, reporter enzyme and dye, from the corresponding reservoir chambers, which are integrated into a glass/silicon monolith, via microfluidic channels into the reactor. The basic principle including the electronic circuits and chips, which control the order of events through the regulation of microvalves are the same as described previously for the (GPI-) protein "chip-on-the-chip" [118]. However, here the reactor is just a temperature-controlled incubation chamber that becomes filled with binding buffer, reporter enzyme, NPs, sample and dye in defined order. The dye formed during the incubation is then measured in a distinct detection unit equipped with a photometer and installed at the efflux channel. Alternatively, the use of reporter enzymes leading to the formation of protons, such as glucose oxidase, can be simply monitored by a biosensor based on the technology of the ion-sensitive field effect transistor (ISFET) [111-114].

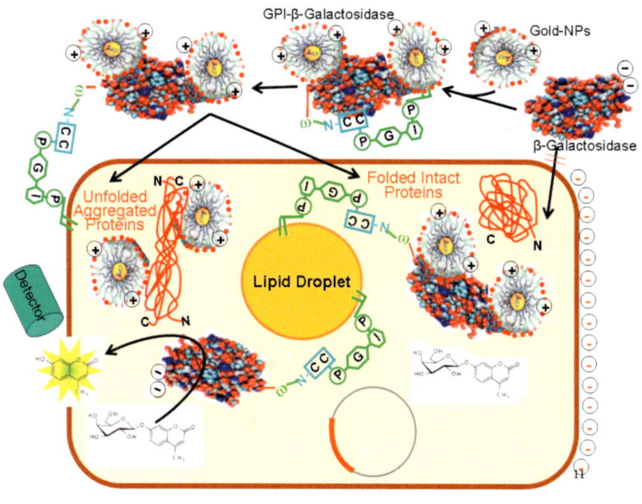

Figure 10: Principle of the NP-based cell chip. Upon interaction of the negatively charged reporter enzyme, GPI-β-galactosidase, with cationic hydrophobic gold-NPs, the positively charged complexes consisting of NPs and GPI-β-galactosidase come into contact with the outer leaflet of the plasma membrane bilayer and are subsequently translocated to the surface phospholipid monolayer of the cytoplasmic lipid droplets. Upon arrival at the cytoplasm with accumulated correctly folded or aggregated misfolded polypeptides, the NPs either remain bound or are relieved from binding to the GPI-β-galactosidase. The resulting inhibition or activation, respectively, of GPI-β-galactosidase is monitored as light generation and indicative for the amount of ectopically expressed passenger protein drug in the recombinant host cells that differs in structure from the authentic product.

Nanoparticle- and GPI-Protein-Based Cell-Based Assays

The "chemical noses and tongues" constituted by NPs can even "smell and taste" within intact cells (Figure 10). For instance misfolded and denatured proteins as might be generated in course of the biotechnological production of recombinant proteins in response to stress of the host cells during suboptimal upstream processing (see Figure 1) have to be detected by process analytics in sensitive fashion. However, the study of the synthesis of protein drugs in the host cell in vivo, for instance under stress conditions during the large-scale fermentation process, has turned out to represent a formidable challenge. For the corresponding application of the "chemical noses and tongues", the reporter enzyme has to be delivered into the cytosol of the cell to be monitored. Although some proteins can enter cells by receptor-mediated endocytosis, the majority is not recognized by appropriate cell surface receptors. However, some proteins lacking cell surface receptors have an intrinsic ability or are amenable for modifications to gain entry into the cytosol [119]. For instance, certain pancreatic-type ribonucleases readily undergo endocytosis and exhibit specific toxicity towards cancerous cells by degrading their RNA [120]. Binding of these ribonucleases to the cell surface is known to be unsaturable and mediated by non-specific ionic interactions. In particular, onconase, an 11.8 kDa pancreatic-type ribonuclease possessing endogenous anti-tumoral activity and cytotoxicity, becomes internalized at a rate being only slightly faster than fluid-phase uptake [121,122]. In contrast, natural cell-penetrating peptides, such as HIV-TAT or penetratin, are characterized by an abundance of cationic residues and endocytosed readily [119-125]. Interestingly, lysine and arginine residues are prerequisite for the internalization of those peptides which is presumably due to stable interactions with the carboxyl, phosphoryl and sulfuryl groups on the cell surface, thereby triggering endocytosis. In fact, increasing the net positive charge of a protein by either chemical modification or site-directed mutagenesis has been shown to increase its internalization [126]. For example, green fluorescent protein is a highly anionic protein that is not taken up by cells. Replacing five acidic residues with arginine on the surface of the folded green fluorescent protein endowed it with the ability to undergo endocytosis [120-122]. Installing even more cationic residues generated a variant that not only undergoes endocytosis, but also is capable to internalize passenger polypeptides. The efficacy of the internalization has been linked to (i) the net positive charge of the protein, (ii) the distribution of the charge over the protein surface and (iii) the type of positively charged amino acid involved, with arginine apparently being more potent than lysine residues [125,126].

Mode-of-Action

The reporter enzyme, β-galactosidase, with its negatively charged surface at physiological pH, is not amenable for transport across the eucaryotic plasma membrane at considerable efficacy, irrespective of whether being modified at its carboxy-terminus with a GPI structure or not. However, the negatively charged amino acids at the surface of GPI-ß-galactosidase can be (over-) compensated upon the introduction of positive charges by surface binding of NPs equipped with positively charged shell structures. Those GPI-β-galactosidase-NP complexes have been described to be efficiently taken up by eucaryotic cells [127,128]. The molecular mechanism involved in the rapid trafficking of GPI-ß-galactosidase from the extracellular medium via anchorage at the outer leaflet of the plasma membrane phospholipid bilayer to the cytosol is ill-defined. Apparently, it is based on a special type of clathrin-independent caveolin-mediated endocytosis which does not deliver its cargo to the endosomal/lysosomal degradation machineries.

The underlying trafficking pathway may be related to the recently described two-step transport of the GPI-proteins, Gce1 and CD73, which are both engaged in cAMP metabolism, from extracellular microvesicles onto the surface of cytoplasmic lipid droplets of adipocytes [129]. It has been demonstrated that in course of incubation of microvesicles harbouring Gce1 and CD73 with rat adipocytes, the GPI-proteins consecutively associate with the plasma membrane and then with the lipid droplets in spontaneous and regulated fashion [130,131]. This ultimately leads to expression of the cAMP-degrading activities of Gce1 and CD73 at the lipid droplet surface facing the cytoplasm [132,133]. Furthermore, GPI-proteins have been reported to be transported via transcytosis across rodent epithelial intestinal cells [100] and transferred between adipocytes of different size within the same adipose tissue depot via microvesicles [1334-135]. In analogy, the uptake of the GPI-ß-galactosidase-NP complex by the eucaryotic cell seems to be mediated by the GPI anchor, in addition to the positively charged surface. The GPI anchor efficiently triggers the interaction with the surface of the cell as well as with the cytosolic lipid droplets, which are present in almost each cell type, albeit to a varying degree [136,137]. Upon arrival of the GPI-ß-galactosidase-NP complexes at the lipid droplet surface, the NPs will either remain bound to or become displaced from the β-galactosidase dependent on the absence or presence, respectively, of competing interaction partners. As those denatured, aggregated or misfolded passenger polypeptides may operate that are generated by ectopic high-level expression. In consequence, these may induce the upregulation of ß-galactosidase activity and of the corresponding light signal (Figure 10). Thus, the NPs employed will "smell and taste" the surface landscape of the β-galactosidase and the incorrectly folded/assembled passenger protein, but not of its intact native version, in comparative fashion. In consequence, the proportion of native to misfolded protein drugs in the host cells along upstream processing can be followed by incubation of portions of them with appropriate GPI-reporter enzyme-NP complexes.

Automation

This cellular NP-based test for assaying the molecular and structural integrity of protein drugs in host cells can be easily transferred from

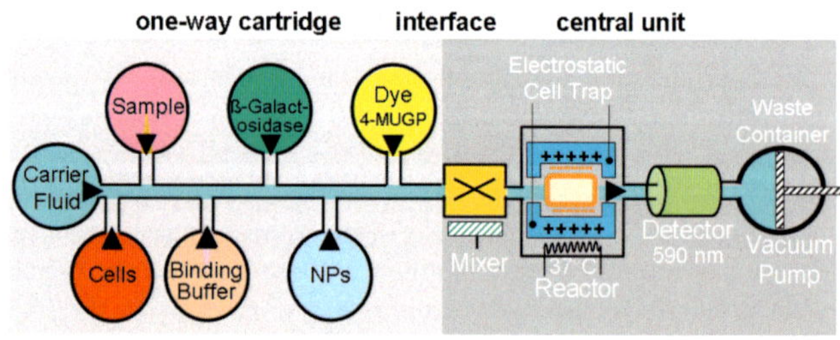

Figure 11: Automation of the NP-based cell chip. The individual reservoir chambers containing all the components required for the NP-based determination of the amount of misfolded protein drugs, including the host cells in suspension (see Figure 10), the carrier fluid, which mediates their regulated transfer via microfluidic channels, and the dye, 4-methylumbelliferyl-β-D-galactopyranoside (4-MUGP), are integrated into the one-way cartridge. The mixer, the reactor with the installed electrostatic cell trap, the photometric detector and the waste container are installed in a common central unit. The one-way cartridge and the central unit are connected via microfluidic channels and electrical interfaces. The reactor manages to trap and correctly position the host cells through electrostatic interactions between the cell surface and the trap wall.

the microtiter plate format to the automated "lab-on-the-chip" format as described above for the "chip-on-the-chip" configuration. For this, the NP-based (GPI-) protein chips have to be adapted and supplemented with (an) additional reservoir chamber(s) for the cells to be assessed (Figure 11) [138]. Confining cells in micrometer-sized reservoir chambers close to their intrinsic volume enables the handling and analysis of a small number of cells with minimized dilution [139]. Initial approaches for "lab-on-the-chip" cell analysis have been performed with silicon microfilter chambers, microcolumn separation techniques and capillary electrophoresis [140,141], including one- and two-dimensional resolving methods [141]. However, conventional capillary-based techniques lack the potential for high-throughput analysis of a few cells due to sophisticated cell-loading procedures. Microfluidic devices offer a versatile alternative to overcome this problem and enable the integration of the positioning and trapping of a limited number of cells on the micrometer scale. A typical workflow involves the transfer of a several-microliter sample of the cell suspension from an off-chip culture to the reservoir chamber of the microfluidic device [143]. A selection step for the cells of interest follows together with a navigation step in order to immobilise the cells at a certain trapping position in a downstream channel of the microfluidic device [144].

Cell Trapping

A variety of methods for the selection, navigation and trapping steps during the cell manipulation has been employed [144-146]. They commonly are based on mechanical, optical, magnetic field or electric field principles. For example, the mechanical trapping of cells using microfabricated filters or other physical barrier designs has proven successful, albeit, in general, it does not offer the versatility of other methods. Laser tweezers, magnetic fields and droplet-based microfluidics have been used to select, trap and move cells [147-150]. However, approaches involving the use of electrical fields remain the most popular, since they combine the ease of field generation and regulation with speed and flexibility. In particular, both electrophoresis and electroosmosis have been used extensively for sorting, transportation and navigation of cells in microchannels [151-154].

In case of the cellular NP-based (GPI-) protein chip, an electrostatic cell trap has been installed. It relies on multiple ionic interactions between the negatively charged surface of the yeast or eucaryotic cells, which are formed by the cell wall glucans and mannans or the plasma membrane phospholipids and transmembrane proteins, respectively, and the positively charged wall of the cell trap chamber (Figure 11). The positioning of the cells transported via the carrier fluid into the cell trap for the analysis and the release of the cells from the cell trap into the carrier fluid for delivery to the waste and initiation of the next cycle of measurement (with a distinct type of NPs or cell sample) after the analysis is under control of an electronic chip that regulates the electric field in the trap. However, it is known that a (pulsed) electric field applied onto a cell induces an extra transmembrane potential (TMP) across the cellular membrane, which can compromise its integrity [155,156]. The applied electric field may trigger the emergence of pores in the plasma membrane, which could increase in diameter and eventually become hydrophilic at a threshold TMP of 0.5-1 V. Below the threshold TMP, reversible electroporation occurs, in which pores are not generated at all or are formed but can reseal. The small distance between the electrodes installed in the chamber walls (Figure 11) allows the formation of electric field strengths (< 10 kV cm-1) sufficient for transient trapping and correct and stable positioning of the cells despite the low voltages applied. These are sufficiently low to avoid irreversible electroporation, mechanical breakdown of the plasma membrane, cell lysis and cell inactivation as well as electrolysis and problems associated with local gas generation and pH shifts.

An additional problem pertains to intrinsic cell-to-cell variability, which is particularly prominent in case of analysing a few cells, only [157]. Biochemical studies are routinely based on millions of cells for a single experiment and as a result record the average across this population. However, increasing evidence suggests that individual cells vary in their responses due to noise and by design. For example, protein levels fluctuate between different cells of the same population and stress signaling pathways are often designed to respond switch-like rather than with classical Michaelis-Menten type kinetics. However, a switch-like response of individual cells will be monitored as a graded response across a cell population. If stress conditions during fermentation increase the number of host cells accumulating misfolded proteins, they become indistinguishable at the population level from a genuinely graded response of individual cells [158]. It is of importance to discriminate these response types as the effects of stress during fermentation from cellular variability. In hypersensitive (switch-like) systems it is sufficient to induce the stress response above a threshold value in order to achieve a maximal amount of misfolded polypeptides, while an analogous system with a graded stress response will show increasing misfolding proportional to the increase in unfavourable fermentation conditions. Thus, it is of general importance to deconvolute cell population vs. few-cell responses [159]. It is assumed that in future applied physiology, biochemistry and cell biology will increasingly tackle this problem for microengineering of "lab-on-the-chips" as well as for chip-based life cell analysis [160,161].

Software-Based Bioanalytics

The measurement of product parameters critical for the quantity and quality of the protein drug, such as amount, identity, structure, function, non-immunogenicity and non-toxicity, in the course of process validation, risk analysis and process control is admittedly accompanied by considerable technical difficulties and expenditure. Therefore, many attempts have been initiated to predict product parameters from relevant process parameters that are seemingly more easily to follow, such as glucose, oxygen, pH and temperature for the fermentation process. For this, a specific software called the "softsensor" was developed that is fuelled with data obtained from the simple online-measurement of the process parameters at the efflux channel of the bioreactor (Figure 12). As an advantage, there is no need for additional sample collection and analytical equipment for the generation of "softsensor"-specific experimental data. Instead the "softsensor" relies on model calculations and algorithms which are driven by "historical" experimental data. These have been derived from previous systematic variation of the corresponding process parameters, such as for the fermentation step. The relevant product parameters, which are directly or indirectly dependent on the process parameters, will then be calculated by transformation of these process parameters upon their online measurement. These "actual" values are then forwarded to the regulatory control unit for comparison with the corresponding "ideal" values. In case of a calculated difference, the control unit, which is connected to electric valves for the regulation of the supply with medium, oxygen, buffer etc., will induce increases or decreases of the corresponding influxes for compensation of eventual deficiencies or excesses, respectively.

In combining the most critical parameters for a specific process, such as fermentation, this configuration is designed for the automated

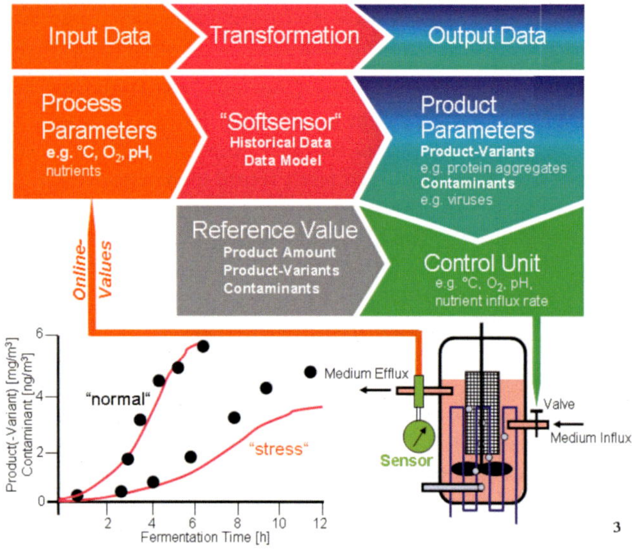

Figure 12: Software-based process bioanalytics. Output data on product parameters are derived from the input data on process parameters that are actually measured online and then transformed on the basis of historical data and a data model by the "softsensor". The process, e.g. fermentation, will be regulated by the control unit which regulates the influx of medium, oxygen and buffer as well as the operation of heating and cooling devices in comparison to reference values for the corresponding product parameter. The diagram shows the change of product parameters along the fermentation period under normal and "stress conditions". Black circles, measured data; pink line, data calculated by the "softsensor".

process control on the basis of estimated and predicted product parameters. In fact, under standard conditions the "softsensor" will predict the critical product parameters, such as amount of the authentic protein drug or the emergence of product variants, e.g. modified protein drug, from the relevant process parameters, such as glucose or oxygen concentration in the fermentation broth, with remarkable accuracy. This is concluded from the almost perfect match of the calculated and the experimentally determined values (Figure 12). Thus, standard conditions during "regular" operations are compatible with the capability of the "softsensor" to delineate the causal relationship between process and product parameters. However, "non-standard" conditions during "irregular" operations that may occur rarely, but can never be excluded, are not and per se can not be predicted in the data model that forms the basis of the "softsensor". In consequence, they can not be calculated and estimated by the "softsensor".

Taken together, there are considerable advantages for the application of the "softsensor" in process analytics that encompass (i) the simple continuous real-time observation of all critical steps in upstream and downstream processing, (ii) the use of already available online-data rather than expensive instrumentation, (iii) the estimation of the "true" product parameters with reasonable accuracy as long as the corresponding process operates within the "normal" range, and (iv) the valuable supplementation of modern and intelligent process control and automation. Certainly, the "softsensor" will be installed for future process analytics on a routine basis (Figure 12). However, the "softsensor" does not replace the need for the generation of experimental data regarding relevant product parameters, such as the product amount, product variants and contaminants, and this is true for both seemingly normally and, in particular, obviously irregularly operating processes. Thus, the "softsensor" will not substitute for the "hardware analytics" that relies on bioanalytical tests and assays including the appropriate instrumentations and data analyses.

Conclusions and Perspectives

The following general consequences have to be drawn for the state-of-the-art process analytics in biopharmaceutical production: (i) Large-scale production of protein drugs is subject to errors and heterogeneity. (ii) Analysis of the production process relies on the deep understanding of the relationship between the individual production steps (under normal and irregular conditions) and the authentic product, product variants and contaminants. (iii) "Software-based" analytics will supplement, but not substitute for "hardware-based" analytics. (iv) Test qualification and validation critically depend on the production steps, product variants and contaminants to be followed. (v) Combinations of several tests evaluating distinct product parameters, such as identity, structure and function, are preferred vs. a single test system. (vi) Specific adaptation of the test (combination) for the corresponding production step and relevant parameter is usually required. (vii) For any given production step and parameter the tests are not "ideal" and "fixed" in most cases, but rather represent acceptable compromises at the time point of implementation. They have to be improved according to current state-of-the-art both with regard to the regulatory and technological progress [162-171]. Examples for recent breakthroughs in the parallel multi-parameter process analytics represent cell-free and cell-based protein chips and biosensors relying on GPI-proteins as capturing/immobilising probes and NPs as detection probes which can be integrated into chip-on-the-chip and lab-on-the-chip formats for full automatic routine handling.

This kind of relativism reflected in "non-ideal" and "non-fixed" tests apparently underlies each bioanalytical procedure, in general, and test qualification, in particular. The relativism of process bioanalytics displays some interesting parallels with other analytical processes, for instance the interpretation of music pieces. Starting with the compository material, adequate musical instruments and knowledge about the process of its composing, including eventually the biography of the composer, are prerequisites. Nevertheless, there will be enough space left for creativity and inspiration for the musician as well as for the bioanalyst. This apparent relative freedom, by no means to mix up with randomness, in the interpretation of bioanalytical data represents the most interesting aspect in the field of process analytics.

References

1. Kelley B, Greg B, Lee A (2009) Downstream processing of monoclonal antibodies: current practices and future opportunities. (Gottschalk U, edn), In Process Scale Purification of Antibodies. John Wiley & Sons.
2. Low D, O'Leary R, Pujar NS (2007) Future of antibody purification. J Chromatogr B 848: 48-63.
3. Sommerfeld S, Strube J (2005) Challenges in biotechnology production – generic processes and process optimization for monoclonal antibodies. Chem Eng Process 44: 1123-1137.
4. Nfor BK, Ahamed T, van Dedem GW, van der Wielen LA, van de Sandt EJ, et al. (2008) Design strategies for integrated protein purification processes: challenges, progress and outlook. J Chem Technol Biotechnol 83: 124-132.
5. Asenjo JA, Andrews BA (2004) Is there a rational method to purify proteins? From expert systems to proteomics. J Mol Recognit 17: 236-247.
6. Bensch M, Schulze Wierling P, von Lieres E, Hubbuch J (2005) High throughput screening of chromatographic phases for rapid process development. Chem Eng Technol 28: 1274-1284.
7. Rege K, Pepsin M, Falcon B, Steele L, Heng M (2006) High-throughput process development for recombinant protein purification. Biotechnol Bioeng 93: 618-630.

8. Wiendahl M, Schulze Wierling P, Nielsen J, Fomsgaard Christensen D, Krarup J, et al.(2008) High throughput screening for the design and optimization of chromatographic processes – miniaturization, automation and parallelization of breakthrough and elution studies. Chem Eng Technol 31: 893-903.

9. Gosling I (2005) Process simulation and modeling for industrial bioprocessing: tools and techniques. Indust Biotechnol 1: 106-109.

10. Lu Y, Clarkson A, Titchener-Hooker N, Pantelides C, Bogle D (1994) Simulation as a tool in process designing and management for production of intracellular enzymes. Chem Eng Res Des 72: 371-375.

11. Petrides DP (1994) BioPro Designer: an advanced computing environment for modeling and design of integrated biochemical processes. Comp Chem Eng 18: S621-S625.

12. Hodge G (2004) Investing in process development. Bioprocess J 31-35.

13. FDA (2004) Guidance for industry PAT – A framework for innovative pharmaceutical development, manufacturing and quality assurance, FDA.

14. FDA (2009) International committee on harmonization, Q8: Pharmaceutical Development, FDA.

15. Schuppenhauer M (2003) BioCentury A5.

16. Clementschitsch F, Kern J, Pötschacher F, Bayer K (2005) Sensor combination and chemometric modeling for improved process monitoring in recombinant *E. coli* fed-batch cultivations. J Biotechnol 120: 183-196.

17. Baker KN, Rendall MH, Patel A, Boyd P, Hoare M, et al. (2002) Rapid monitoring of recombinant protein products: a comparison of current technologies. Trends Biotechnol 20: 149-156.

18. Peuker T, Riedel M, Kaiser C, Ellert A, Lenz K, et al. (2004) At-line determination of glucose, ammonia, and acetate in high cell density cultivations of *Escherichia coli*. Eng Life Sci 4: 138-143.

19. Forrer K, Hammer S, Helk B (2004) Chip-based gel electrophoresis method for the quantification of half-antibody species in IgG4 and their by- and degradation products. Anal Biochem 334: 81-88.

20. Vasilyeva E, Woodard J, Taylor FR, Kretschmer M, Fajardo H, et al. (2004) Development of a chip-based capillary gel electrophoresis method for quantification of a half-antibody in immunoglobulin G_4 samples. Electrophoresis 25: 3890-3896.

21. Müllner S, Karbe-Thönges B, Tripier D (1993) Charge heterogeneity of insulin fusion proteins expressed in *Escherichia coli* is not due to proteolytic degradation. Anal Biochem 210: 366-373.

22. Nettleton EJ, Tito P, Sunde M, Bouchard M, Dobson CM, et al. (2000) Characterization of the oligomeric states of insulin in self-assembly and amyloid fibril formation by mass spectrometry. Biophys J 79: 1053-1065.

23. Staack RF, Stracke JO, Stubenrauch K, Vogel R, Schleypen J, et al. (2011) Quality requirements for critical assay reagents used in bioanalysis of therapeutic proteins: what bioanalysts should know about their reagents. Bioanalysis 3: 523-534.

24. Kuster B, Mann M (1998) Identifying proteins and post-translational modifications by mass spectrometry. Curr Opin Struct Biol 8: 393-400.

25. Sluzky V, Tamada JA, Klibaonov AM, Langer R (1991) Kinetics of insulin aggregation in aqueous solutions upon agitation in the presence of hydrophobic surfaces. Proc Natl Acad Sci USA 88: 9377-9381.

26. CBER (Center for Biologics Evaluation and Research)(1994) Points to consider in the manufacturing and testing of monoclonal antibody products for human use. Rockville: Food and Drug Administration.

27. Eaton LC (1995) Host cell contaminant protein assay development for recombinant biopharmaceuticals. J Chromatogr A 705: 105-114.

28. Shadle PJ, McAllister PR, Smith TM, Lubiniecki AS (1995) Viral validation strategy for recombinant products derived from established animal cell lines. Cytotechnol 18: 21-25.

29. Entzeroth M (2003) Emerging trends in high-throughput screening. Curr Opin Pharmacol 3: 522-529.

30. CPMP Biotechnology Working Party (1996) Note for guidance on virus validation studies: the design, contribution and interpretation of studies validation, the inactivation and removal of viruses. London: The European Agency for the Evaluation of Medical Products, Canary Wharf.

31. Poiley JA, Bierley ST, Hillesund T, Nelson RE, Monticello TM, et al. (1994) Methods for estimating retroviral burden. Pharmaceutic Technol 5: 32-35.

32. Lubiniecki AS, Wiebe ME, Builder SE (1990) Process validation for cell culture-derived pharmaceutical proteins. Bioprocess Technol 10: 515-541.

33. CDER/CBER/CDRH/CVR-FDA (1987) Guidelines on validation of the limulus amebocyte lysate test as an end-product endotoxin test for human and animal parenteral drugs, biological products, and medical devices.

34. ICH, Q5A (1998) Guidance for industry: viral safety evaluation of biotechnology products derived from cell lines of human or animal origin.

35. Werz W, Werner RG (1998) Recombinant tissue plasminogen activator (rt-PA) in fibrinolytic therapy. J Biotechnol 61: 157-158.

36. Eaton LC (1989) Quantification of putative glycoprotein X in bioengineered pseudorabies vaccine virus culture medium by ELISA. J Immunol Methods 118: 125-128.

37. Hammerl P, Hartl A, Thalhamer J (1993) Particulate nitrocellulose as a solid phase for protein immobilization in immuno-affinity chromatography. J Immunol Methods 165: 59-66.

38. Hammerl P, Hartl A, Thalhamer J (1993) Improvement of antisera raised against complex antigen mixtures by the use of heterologous sources of antigen for immunization. J Immunol Methods 160: 155-161.

39. Flatman S, Alam I, Gerard J, Mussa N (2007) Process analytics for purification of monoclonal antibodies. J Chromatogr B 848: 79-87.

40. Nfor BK, Verhaert PDEM, van der Wielen LAM, Hubbuch J, Ottens M (2009) Rational and systematic protein purification process development: the next generation. Trends Biotechnol 27: 673-679.

41. Graumann K, Premstaller A (2006) Manufacturing of recombinant therapeutic proteins in microbial systems. Biotechnol J 1: 164-186.

42. Thomas A, Thevis M, Delahaut P, Bosseloir A, Schänzer W (2007) Mass spectrometric identification of degradation products of insulin and its long-acting analogues in human urine for doping control purposes. Anal Chem 79: 2518-2524.

43. Fabris D, Fenselau C (1999) Characterization of allosteric insulin hexamers by electrospray ionization mass spectrometry. Anal Chem 71: 384-387.

44. Kuster B, Mann M (1998) Identifying proteins and post-translational modifications by mass spectrometry. Curr Opin Struct Biol 8: 393-400.

45. Goldman J, Carpenter FH (1974) Zinc binding, circular dichroism, and equilibrium sedimentation studies on insulin (bovine) and several of its derivatives. Biochemistry 13: 4566-4574.

46. Vecchio G, Bossi A, Pasta P, Carrea G (1996) Fourier-transform infrared conformational study of bovine insulin in surfactant solutions. Int J Pep Prot Res 48: 113-117.

47. Wei J, Lin YZ, Zhou JM, Tsou CL (1991) FT-IR studies of secondary structures of bovine insulin and its derivatives. Biochem Biophys Acta 1080: 29-33.

48. Mercolini L, Musenga A, Saladini B, Bigucci F, Luppi B, Zecchi V, Raggi MA (2008) Determination of insulin in innovative formulations by means of LC coupled to fluorescence detection. J Pharmaceut Biomed Analy 48: 1303-1309.

49. Hua Q, Weiss MA (1991) Comparative 2D-NMR studies of human insulin and des-pentapeptide insulin-sequential resonance assignment and implications for protein dynamics and receptor recognition. Biochemistry 30: 5505-5515.

50. Sommerfeld M, Müller G, Tschank G, Seipke G, Habermann P, et al. (2010) *In vitro* metabolic and mitogenic signaling of insulin glargine and its metabolites. PLoS One 5: e9540.

51. Müller G, Ertl J, Gerl M, Preibisch G (1997) Leptin impairs metabolic actions of insulin in isolated rat adipocytes. J Biol Chem 272: 10585-10593.

52. Müller G, Herling A, Sandow J (2008) Antidiabetic activity in drug discovery and evaluation–Pharmacological assays .(Vogel HG Edn) Berlin, Heidelberg, New York.

53. Müller G, Jordan H, Petry S, Wetekam E-M, Schindler P (1997) Analysis of lipid metabolism in adipocytes using fluorescent fatty acids. I. Insulin stimulation of lipogenesis. Biochim Biophys Acta 1347: 23-39.

54. IABs international meeting (2001) Immunogenicity testing. Immunogenicity of therapeutic biological products.

55. CDER/CBER/CDRH/CVR-FDA (1997) Guidelines on validation of the limulus amebocyte lysate test as an end-product endotoxin test for human and animal parenteral drugs, biological products, and medical devices.

56. Flatman S, Alam I, Gerard J, Mussa N (2010) Process analytics for purification of monoclonal antibodies. J Chromatogr B Analyte Technol Biomed Life Sci 848: 79-87.

57. Lottspeich F (2009) Introduction to proteomics. Methods Mol Biol 564: 3-10.

58. Chen R, Snyder M (2010) Yeast proteomics and protein microassays. J Proteom 73: 2147-2157.

59. Lynch M, Mosher C, Huff J, Nettikadan S, Johnson J, Henderson E (2004) Functional protein nanoarrays for biomarker profiling. Proteomics 4: 1695-1702.

60. Lin JC-H (2010) Protein microarrays for cancer diagnostics and therapy. Med Princ Pract 19: 247-254.

61. Malinowsky K, Wolff C, Ergin B, Berg D, Becker KF (2010) Deciphering signaling pathways in clinical tissues for personalized medicine using protein microarrays. J Cell Physiol 225: 364-370.

62. Mueller C, Liotta LA, Espina V (2010) Reverse phase protein microarrays advance to use in clinical trials. Mol Oncol 4: 461-481.

63. Yang L, Guo S, Li Y, Zhou S, Tao S (2011) Protein microarrays for systems biology. Acta Biochim Biophys Sin 43: 161-171.

64. Merbl Y, Kirschner MW (2011) Protein microarrays for genome-wide posttranslational modification analysis. WIREs Syst Biol Med 3: 347-356.

65. Park J, Kurosawa S, Takai M, Ishihara K (2007) Antibody immobilization to phospholipid polymer layer on gold substrate of quartz crystal microbalance immunosensor. Colloid Surf B Biointerf 55: 164-172.

66. Hu S, Xie Z, Qian J, Blackshaw S, Zhu H (2011) Functional protein microarray technology. WIREs Syst Biol Med 3: 255-268.

67. Templin MF, Stoll D, Schrenk M, Traub PC, Vöhringer CF, et al. (2002) Protein microarray technology. Trends Biotechnol 20: 160-166.

68. Zhu H, Bilgin M, Bangham R, Hall D, Casamayor A, Bertone P, Lan D (2001) Global analysis of protein activities using proteome chips. Science 293: 2101-2105.

69. MacBeath G, Schreiber SL (2000) Printing proteins as microarrays for high-troughput function determination. Science 289: 1760-1763.

70. Delehanty JB, Ligler FS (2003) Method for printing functional protein microarrays. Biotechniques 34: 380-385.

71. Delehanty JB (2004) Printing functional protein microarrays using piezoelectric capillaries. Methods Mol Biol 264: 135-143.

72. Jones VW, Kenseth JR, Porter MD, Mosher CL, Henderson E (1998) Microminiaturized immunoassays using atomic force microscopy and compositionally patterned antigen arrays. Anal Chem 70: 1233-1241.

73. Kusnezow W, Jacob A, Walijew A, Diehl F, Hoheisel JD (2003) Antibody microarrays: an evaluation of production parameters. Proteomics 3: 254-264.

74. Kramer A, Feilner T, Possling A, Radchuk V, Weschke W, et al. (2004) Identification of barley CK2alpha targets by using the protein microarray technology. Phytochemistry 65: 1777-1784.

75. Stillman BA, Tonkinson JL (2000) FAST slides: a novel surface for microarrays. Biotechniques 29: 630-635.

76. Angenendt P, Glokler J, Murphy D, Lehrach H, Cahill DJ (2002) Toward optimized antibody microarrays. a comparison of current microarray support materials. Anal Biochem 309: 253-260.

77. Charles PT, Goldmn ER, Rangasammy JG, Schauer CL, Chen MS, et al. (2004) Fabrication and characterization of 3D hydrogel microarrays to measure antigenicity and antibody functionality for biosensor applications. Biosens Bioelectron 20: 753-764.

78. Feilner T, Hultschig C, Lee J, Meyer S, Immink RG, et al. (2005) High throughput identification of potential Arabidopsis mitogen-activated protein kinase substrates. Mol Cell Proteomics 4: 1558-1568.

79. Chen CS, Korobkova E, Chen H, Zhu J, Jian X, et al. (2008) A proteome chip approach reveals new DNA damage recognition activities in *Escherichia coli*. Nat Methods 5: 69-74.

80. Alhamdani MS, Schroder C, Hoheisel JD (2010) Analysis conditions for proteomic profiling of mammalian tissue and cell extracts with antibody microarrays. Proteomics 10: 3203-3207.

81. Gulmann C, Sheehan KM, Conroy RM, Wulfkuhle JD, Espina V, et al. (2009) Quantitative cell signalling analysis reveals down-regulation of MAPK pathway activation in colorectal cancer. J Pathol 218: 514-519.

82. Jones RB, Gordus A, Krall JA, MacBeath GA (2006) A quantitative protein interaction network for the ErbB receptors using protein microarrays. Nature 439: 168-174.

83. Hsu KL, Mahal LK (2006) A lectin microarray approach for the rapid analysis of bacterial glycans. Nat Protoc 1: 543-549.

84. Tao SC, Li Y, Zhou J, Qian J, Schnaar RL, et al. (2008) Lectin microarrays identify cell-specific and functionally significant cell surface glycan markers. Glycobiology 18: 761-769.

85. Lueking A, Horn M, Eickhoff H, Bussow K, Lehrach H, et al. (1999) Protein microarrays for gene expression and antibody screening. Anal Biochem 270: 103-111.

86. SC, Chen CS, Zhu H (2007) Applications of protein microarray technology. Comb Chem High Throughput Screen 10: 706-718.

87. Hamelinck D, Zhou H, Li L, Verweij C, Dillon D, et al. (2005) Optimized normalization for antibody microarrays and application to serum protein profiling. Mol Cell Proteomics 4: 773-784.

88. Wingren C, Borrebaeck CA (2008) Antibody microarray analysis of directly labelled complex proteosomes. Curr Opin Biotechnol 19: 55-61.

89. Wingren C, Ingvarsson J, Dexlin L, Szul D, Borrebaeck CA (2007) Design of recombinant antibody microarrays for complex proteome analysis: choice of sample labeling-tag and solid support. Proteomics 7: 3055-3065.

90. Chen CS, Zhu H (2006) Protein microarrays. Biotechniques 40: 423-425.

91. Müller G , Wetekam E-M, Jung C, Bandlow W (1994) Membrane association of lipoprotein lipase and a cAMP-binding ectoprotein in rat adipocytes. Biochemistry 33: 12149-12159.

92. Nosjean O, Briolay A, Roux B (1997) Mammalian GPI proteins: sorting, membrane residence and functions. Biochim Biophys Acta 1331: 153-186.

93. Ikezawa H (2002) Glycosylphosphatidylinositol (GPI)-anchored proteins. Biol Pharm Bull 25: 409-417.

94. Caras IW, Weddell GN (1989) Signal peptide for protein secretion directing glycophospholipid membrane anchor attachment. Science 243: 1196-1198.

95. Orlean P, Menon AK (2007) Theamtic review series: lipid post-translational modifications. GPI anchoring of protein in yeast and mammalian cells, or: How we learned to stop worrying and love glycophospholipids. J Lipid Res 48: 993-1011.

96. Pittet A, Conzelmann A (2007) Biosynthesis and function of GPI proteins in the yeast *Saccharomyces cerevisiae*. Biochim Biophys Acta 1771: 405-420.

97. Tiede A, Bastisch I, Schubert J, Orlean P, Schmidt RE (1999) Biosynthesis of glycosylphosphatidylinositols in mammals and unicellular microbes. Biol Chem 380: 503-523.

98. Kinoshita T, Fujita M, Maeda Y (2008) Biosynthesis, remodelling and functions of mammalian GPI-anchored proteins: Recent progress. J Biochem 144: 287-294.

99. Varma R, Mayor S (1998) GPI-anchored proteins are organized in submicron domains at the cell surface. Nature 394: 798-801.

100. G (2011) Novel applications for glycosylphosphatidylinositol-anchored proteins in pharmaceutical and industrial biotechnology. Mol Membr Biol 28: 187-205.

101. Nosjean O, Roux B (1999) Ectoplasmic insertion of a glycosylphosphatidylinositol-anchored protein in glycosphingolipid- and cholesterol-containing phosphatidylcholine vesicles. Eur J Biochem 263: 865-870.

102. Nosjean O, Roux B (2003) Anchoring of glycosylphosphatidylinositol-proteins to liposomes. Methods Enzymol 372: 216-232.

103. Sesana S, Re F, Bulbarelli A, Salerno D, Cazzaniga E, et al. (2008) Membrane features and activity of GPI-anchored enzymes: alkaline phosphatase reconstituted in model membranes. Biochemistry 47: 5433-5440.

104. Ilangumaran S, Robinson PJ, Hoessli DC (1996) Transfer of exogenous

glycosylinositol (GPI)-linked molecules to plasma membranes. Trends Cell Biol 6: 163-167.

105. Miranda OR, Creran B, Rotello VM (2010) Array-based sensing nanoparticles: 'Chemical noses' for sensing biomolecules and cell surfaces. Curr Opin Chem Biol 14: 728-736.

106. Miranda OR, Chen HT, Chang-Cheng Y, Mortenson DE, Yang XC, et al. (2010) Enzyme-amplified array sensing of proteins in solution and in biofluids. J Am Chem Soc 132: 5285-5289.

107. Picollet-D'hahan N (2011) Live cell analysis: When electric detection interfaces microfluidics. J Biochip Tissue chip S1-001.

108. Mallick P, Kuster B (2010) Proteomics: a pragmatic perspective Nat Biotechnol 28: 695-709.

109. Anslyn EV, Rotello VM (2010) Chemosensory models: approaches and applications of differential sensing. Curr Opin Chem Biol 14: 683-684.

110. Baldini L, Wilson AJ, Hong J, Hamilton AD (2004) Pattern-based detection of different proteins using an array of fluorescent protein surface receptors. J Am Chem Soc 126: 5656-5657.

111. Sannomiya T, Vörös J (2011) Single plasmonic nanoparticles for biosensing. Trends Biotechnol 29: 343-351.

112. Lei J, Ju H (2010) Nanotubes in biosensing. Nanomed Nanobiotechnol 2: 496-509.

113. Nambiar S, Yeow JT (2011) Conductive polymer-based sensors for biomedical applications. Biosens Bioelectron 26: 1825-1832.

114. Ray S, Chandra H, Srivastava S (2010) Nanotechniques in proteomics: Current status, promises and challenges. Biosens Bioelectron 25: 2389-2401.

115. M, Rana S, Akpinar H, Miranda OR, Arvizo RR, et al. (2009) Sensing of proteins in human serum using nanoparticle-green fluorescent protein conjugates. Nat Chem 1: 461-465.

116. Rana SS, Yeh YC, Rotello VM (2010) Engineering the nanoparticle-protein interface: applications and possibilities. Curr Opin Chem Biol 14: 828-834.

117. Wright AT, Griffin MJ, Zhong Z, McCleskey SC, Anslyn EV, et al. (2005) Differential receptors create patterns that distinguish various proteins. Angew Chem Int Ed Engl 44: 6375-6378.

118. Müller G (2011) Glycosylphosphatidylinositol-Anchored Protein Chips for Patient-Tailored Multi-Parameter Proteomics. J Biochip Tissue chip S3: 001.

119. Fuchs SM, Raines RT (2004) Pathway for polyarginine entry into mammalian cell. Biochemistry 43: 2438-2444.

120. Futami J, Yamada H (2008) Design of cytotoxic ribonucleases by cationization to enhance intracellular protein delivery. Curr Pharm Biotechnol 9: 180-184.

121. Turcotte RF, Lavis LD, Raines RT (2009) Onconase cytotoxicity relies on the distribution of its positive charge. FEBS J 276: 3846-3857.

122. Futami J, Maeda T, Kitazoe M, Nukui E, Tada H, et al. (2001) Preparation of potent cytotoxic ribonucleases by cationization. Enhanced cellular uptake and decreased interaction with ribonuclease inhibitor by chemical modification of carboxyl groups. Biochemistry 26: 7518-7524.

123. Gosh P, Yang X, Arvizo R, Zhu ZJ, Agasti SS, et al. (2010) Intracellular delivery of a membrane-impermeable enzyme in active using functionalized gold nanoparticles. J Am Chem Soc 132: 2642-2645.

124. Kim CK, Ghosh P, Rotello VM (2009) Multimodal drug delivery using gold nanoparticles. Nanoscale 1: 61-67.

125. Amand HL, Fant K, Norden B, Esbjorner EK (2008) Stimulated endocytosis in penetratin uptake: Effect of arginine and lysine. Biochem Biophys Res Commun 371: 621-625.

126. Fuchs SM, Raines RT (2007) Arginine grafting to endow cell permeability. ACS Chem Biol 2: 167-170.

127. Müller G, Over S, Wied S, Frick W (2008) Association of (c)AMP-degrading glycosylphosphatiylinositol-anchored proteins with lipid droplets is induced by palmitate, H2O2 and the sulfonylurea drug, glimepiride, in rat adipocytes. Biochemistry 47: 1274-1287.

128. Müller G, Jung C, Wied S, Biemer-Daub G, Frick W (2010) Transfer of the glycosylphosphatidylinositol-anchored 5'-nucleotidase CD73 from adiposomes into rat adipocytes stimulates lipid synthesis. Brit J Pharmacol 160: 878-891.

129. Müller G, Jung C, Wied S (2008) Translocation of glycosylphosphatidylinositol-anchored proteins from plasma membrane microdomains to lipid droplets in rat adipocytes is induced by palmitate, H2O2 and the sulfonylurea drug, glimepiride. Mol Pharmacol 73: 1513-1529.

130. Müller G, Wied S, Jung C, Over S (2008) Hydrogen peroxide-induced translocation of glycolipid-anchored (c)AMP-hydrolases to lipid droplets mediates inhibition of lipolysis in rat adipocytes. Brit J Pharmacol 154: 901-913.

131. Müller G, Jung C, Straub J, Wied S, Kramer W (2009) Induced release of membrane vesicles from rat adipocytes containing glycosylphosphatidylinositol-anchored microdomain and lipid droplet signalling proteins. Cell Signal 21: 324-338.

132. Müller G, Schneider M, Biemer-Daub G, Wied S (2011) Microvesicles released from rat adipocytes and harboring glycosylphosphatidylinositol-anchored proteins transfer RNA stimulating lipid synthesis. Cell Signal 23: 1207-1223.

133. Müller G (2011) Take-over – Multiple mechanisms of inter-adipocyte communication. J Mol Cell Biol 3: 81-90.

134. Müller G (2010) Oral protein therapy for the future? Transport of glycolipid-modified therapeutic proteins:vision or fiction? Pharmacology 86: 92-116.

135. Müller G (2011) Oral delivery of protein drugs – Driver of personalized medicine ? Curr Issues Mol Biol 13: 13-24.

136. Ducharme NA, Bickel PE (2008) Lipid droplets in lipogenesis and lipolysis. Endocrinology 149: 942-949.

137. Fuimoto T, Ohsaki Y, Cheng J, Suzuki M, Shinohara Y (2008) Lipid droplets: a classic organelle with new outfits. Histochem Cell Biol 130: 263-279.

138. Andersson H, van der Berg A (2003) Microfluidic devices for cellomics: a review. Sens Actuators B 92: 315-325.

139. Andersson H, van den Berg A (2004) Microtechnologies and nanotechnologies for single-cell analysis. Curr Opin Biotechnol 15: 44-49.

140. Preckel T, Luedke G, Chan SDH, Wang B, Dubrow R, Buhlmann C (2002) Detection of cellular parameters using a microfluidic chip-based system. J Assoc Lab Autom 7: 85-89.

141. Lundqvist JA, Sahlin F, Aberg MA, Strömberg A, Eriksson PS, et al. (1998) Altering the biochemical state of individual cultured cells and organelles with ultramicroelectrodes. Proc Natl Acad Sci USA 95: 10356-10360.

142. Wallingford RA, Ewing AG (1988) Capillary zone electrophoresis with electrochemical detection in 12.7 microns diameter columns. Anal Chem 60: 1972-1975.

143. Toriello NM, Douglas ES, Mathies RA (2005) Microfluidic device for electric field-driven single-cell capture and activation. Anal Chem 77: 6935-6941.

144. Wheeler AR, Throndset WR, Whelan RJ, Leach AM, Zare RN, et al. (2003) Microfluidic device for single-cell analysis. Anal Chem 75: 3581-3586.

145. Li PCH, Harrison DJ (1997) Transport, manipulation and reaction of biological cells on-chip using electrokinetic effects. Anal Chem 69: 1564-1568.

146. Price AK, Culbertson CT (2007) Chemical analysis of single mammalian cells with microfluidics. Anal Chem 79: 2614-2621.

147. Kennedy RT, Oates MD, Cooper BR, Nickerson B, Jorgenson JW (1989) Microcolumn separations and the analysis of single cells. Science 246: 57-63.

148. Hellmich W, Pelargus C, Leffhalm K, Ros A, Anselmetti D (2005) Single cell manipulation, analytics, and label-free protein detection in microfluidic devices for systems nanobiology. Electrophoresis 26: 3689-3696.

149. Buhlmann C, Preckel T, Chan S, Luedke G, Valer M (2003) A new tool for routine testing of cellular protein expression: Integration of cell staining and analysis of protein expression on a microfluidic chip-based system. J Biomol Tech 14: 119-127.

150. Cai X, Klauke N, Glidle A, Cobbold P, Smith G, et al. (2002) Ultra-low-volume, real-time measurement of lactate from the single heart cell using microsystems technology. Anal Chem 74: 908-914.

151. Carlson R, Gabel C, Chan S, Austin R (1997) Self-sorting of white blood cells in a lattice. Phys Rev Lett 79: 2149-2152.

152. Chao T-C, Ros A (2008) Microfluidic single-cell analysis of intracellular compounds. J R Soc Interface 5: S139-S150.

153. Sedgwick H, Caron F, Monaghan PB, Kolch W, Cooper JM (2008) Lab-on-a-chip technologies for proteomic analysis from isolated cells. J R Soc Interface 5: S123-S130.

154. Wong PK, Wang TH, Deval JH, Ho CM (2004) Electrokinetics in micro devices for biotechnological applications. IEEE/ASME Trans Mechatron 9: 366-376.

155. Yi C, Zhang Q, Li CW, Yang J, Zhao J, Yang M (2006) Optical and electrochemical detection techniques for cell-based microfluidic systems. Anal Bioanal Chem 384: 1259-1268.

156. Yi C, Li C-W, Ji S, Yang M (2006) Microfluidics technology for manipulation and analysis of biological cells. Anal Chim Acta 560: 1-23.

157. Sigal A, Milo R , Cohen A , Geva-Zatorsky N, Klein Y,et al. (2006) Variability and memory of protein levels in human cells. Nature 444: 643-646.

158. Ferrel JE, Xiong W (2001) Bistability in cell signaling: how to make continuous processes discontinuous, and reversible processes irreversible. Chaos 11: 227-236.

159. Wilding P, Kricka LJ, Cheng J, Hvichia G, Shoffner MA, et al. (1998) Integrated cell isolation and polymerase chain reaction analysis using silicon microfilter chambers. Anal Biochem 257: 95-100.

160. Di Carlo D, Lee LP (2006) Dynamic single-cell analysis for quantitative biology. Anal Chem 78: 7918-7925.

161. Picollet-D'hahan N (2011) Live cell analysis: When electric detection interfaces microfluidics. J Biochip Tissue chip S1:001.

162. ICH (1998) Q5A Guidance for industry: viral safety evaluation of biotechnology products derived from cell lines of human or animal origin.

163. FDA (1988) Guideline for submitting documentation for the stability of human drugs and biologics.

164. ICH (1996) Q5C Guideline for industry: quality of biotechnological products: stability testing of biotechnological/biological products.

165. ICH (2002) Q1A (R1) Stability testing of new drug substances and products.

166. ICH (2002) Q1E Evaluation of stability data.

167. US Pharmacopeia (2003) USP 26-NF 21.

168. European Pharmacopeia (2002) Sections 2.6.1, 5.1, (4th Edn), European Directorate for the Quality of Medicines (EDQM), Strasbourg.

169. FDA-CBER (1997) Points to consider in the manufacture and testing of monoclonal antibody products for human use.

170. FDA (2001) Guidance for industry: bioanalytical method validation.

171. AAPS (2003) Workshop on bioanalytical method validation for macromolecules in pharmacokinetic profiling.

11

Characterization of a High Activity (S)-Aminotransferase for Substituted (S)-Aminotetralin Production: Properties and Kinetics

Abraham R. Martin[1]*, David Shonnard[1], Sachin Pannuri[1, 2] and Sanjay Kamat[2]

[1] Dept. of Chemical Engineering, Michigan Technological University, Houghton, MI 49931, USA
[2] Cambrex North Brunswick, 661 Highway One, North Brunswick, NJ 08902, USA

Abstract

The production of substituted (S)-Aminotetralins requires biocatalyst that have high activities at high temperatures and considerable tolerance to amine donors to shift the reaction towards products. The biocatalyst used in this process are (S)-Aminotransferases. An (S)-aminotransferase of a high activity derived from Athrobacter citreous by directed evolution for the production of substituted (S)-aminotetralin was characterized in the form of whole cells. Its optimum conditions were pH 7 and 55 °C. Maximum activity was 0.21 mM/min per gram of whole cells. Substrate affinities were 750 mM for isopropylamine and 10.4 mM for substituted tetralone. A kinetic study to describe the production of substituted (S)-aminotetralin showed that two major reactions were involved: one enzyme catalyzed –by (S)-aminotransferase- that is the production of substituted (S)-aminotetralin from isopropylamine and substituted tetralone; the other a non-enzyme catalyzed reaction that forms a byproduct consisting in the imine formed by substituted tetralone and substituted (S)-aminotetralin.

Keywords: (S)-aminotransferase; Substituted (S)-aminotetralin

Introduction

The production of substituted (S)-Aminotetralins requires biocatalyst that have high activities at high temperatures and considerable tolerance to amine donors to shift the reaction towards products. The biocatalyst used in this process are (S)-Aminotransferases. When the correspondent parochial amine and an amine acceptor are mixed with the biocatalyst at the proper conditions of pH and temperature, among other factors, it is desired that the majority of the prochiral ketone be converted to the corresponding amine. From the economical point of view, the prochiral ketone is the most expensive, meanwhile the amine donor is the cheapest. Thermodynamically, it is possible to achieve in a reversible reaction high conversion of one of the reactants if another is added in excess; but in order to do that, the biocatalyst must tolerate such conditions. Especially, in this study was selected the biocatalyst that allows the production of Subtituted (S)-Aminotetralins from (S)-Aminotetralone. Substituted (S)-aminotetralins are substances with potential for the treatment of Parkinson's disease [1-2], depression [3], and cardiovascular problems [4-5]. Due to this several synthesis procedures have been developed. These procedures consist in a sequence of organic reactions [6-10], use of catalysts [10-12], and use of biocatalyst [13-18]. In this study a characterization of the properties, and kinetics of an (S)-aminotransferase (S-AT) developed for production of substituted (S)-aminotetralins (SS-AT) was done.

Materials and Methods

S-AT

The S-AT used in this study was derived from Athrobacter citreous after five rounds of mutagenesis by directed evolution (Mutant CNB05-01, International Patent WO 2006063336 [18]. It was produced by the microorganism Escherichia coli MG1655 (Cambrex) at an expression level of about 65 mg per gram of whole cells. Its molecular weight is about 65 kDa. S-AT for production of SS-AT was used in the form of whole cells, since the S-AT in this form does not requires purification neither it is required by the specifications of the desired product.

Source: The aminotransferase used in this study was obtained from a naturally occurring S-AT found in Athrobacter citreus in the fifth round of mutagenesis and selection. This S-AT as well as its predecessors was coded by a gene of 1.5 kbp, its molecular weight was 65 kDa and was composed by 480 amino acids. The corresponding genes were isolated, and then sub cloned into pSE420 and transformed into Escherichia coli MG1655. Once the selected gene was expressed in the host cell, it was subjected to random mutagenesis, and selected to improve the properties of the enzyme.

Production: The S-AT used in this study (CNB05-01, gene sequence given on International Patent WO 2006063336 [18] was produced by Escherichia coli MG1655 as follows. To a 15-liter fermenter (New Brunswick®) was added 8 liters of water, 55 grams of KH_2PO_4 (Fisher Scientific), 18 grams of $MgSO_4{\cdot}7H_2O$ (Fisher Scientific), 2 ml of Antifoam (Sigma), and 80 ml of a solution whose composition per liter was 6 grams of Fe (III) citrate hydrate (Aldrich), 1.6 grams of EDTA free acid (Aldrich), 1.5 grams of $MnCl_2{\cdot}2H_2O$ (Aldrich), 0.8 grams of $Zn(CH_3COO)_2{\cdot}2H_2O$ (Aldrich), 0.3 grams of H_3BO_3 (Aldrich), 0.25 grams of $Na_2MoO_4{\cdot}2H_2O$ (Aldrich), 0.25 grams of $CoCl_2{\cdot}6H_2O$ (Aldrich), and 0.118 grams $CuCl_2$ (Aldrich). After that, the fermenter was sterilized at 121 °C for 30 minutes. The system was cooled down to 30 °C and then a sterile solution of 300 ml with 150 grams of glucose was added and the pH was adjusted to 7 by addition of NH_4OH (Sigma). After 18 hours of growth, the glucose supplied at inoculation time had been consumed. Then a sterile solution of glucose 50% w/w and $MgSO_4{\cdot}7H_2O$ 2% w/w (Fisher Scientific), was supplied by a peristaltic pump -equipped with sterilized tubes sterilized controlled by the fermenter concentration of dissolved oxygen. Ten hours later 0.6 grams

***Corresponding author:** Abraham R. Mártin, Dept. of Chemical Engineering, Michigan Technological University, Houghton, MI 49931, USA

of isopropyl-β-D-thiogalactopyranoside (IPTG) from Sigma was added to the fermenter, and then 12 hours later the cells were harvested by spray drying. Typical yields were 250-350 grams of whole cells per 8 liters of fermentation. S-AT expression level was about 65 mg per gram of whole cells.

Synthesis of SS-AT

The synthesis of SS-AT occurs upon reaction of isopropylamine (Sigma) and substituted tetralone (Cambrex) catalyzed by S-AT in the presence of pyridoxal-5-phosphate (PLP) as the enzyme cofactor (Schweizer Hall) and a buffer (Description of starting material and product available in International Patent WO 2006063336 [18]. The synthesis reaction is shown in Figure 1.

In addition to the reaction shown in Figure 1, during production of the desired product there is a side reaction. This undesirable process consists on the reaction of SS-AT and substituted tetralone to form an imine (Figure 2). Its formation was important later on the system, once that there was a considerable amount of the desired product, being negligible early on the reaction.

Analysis of SS-AT by HPLC

Enzyme properties and enzyme kinetics were determined by monitoring the concentration of SS-AT in the reaction system. Periodically samples were withdrawn from the system. Samples were diluted, centrifuged and then analyzed using an HPLC (Hewlett-Packard Model 1100 Series). Absorbance readings were taken at 254 nm after injecting a volume of 15-25 µl. Flow rate was set to 1.0-1.5 ml/min for 20 minutes through a Nova-Pak Phenyl (Waters) column. The mobile phase was comprised of isopropanol: aqueous buffer at 30%:70% (v/v), where the aqueous buffer was comprised of 0.24% (w/v) 1-Octanesulfonic acid (Janssen Chimica) and 0.15% (v/v) H3PO4 (Sigma).

Enzyme properties

Optimum pH: The effect of pH on initial rate was investigated incubating 50 ml of reaction mixture at specific pH and 55°C, at a concentration of free whole cells of S-AT of 5 grams per liter. The composition of the corresponding reaction mixture was 130 mM substituted tetralone (Cambrex), 750 mM isopropylamine (Sigma), 2 mM PLP (Schweizer Hall), and 200 mM

sodium acetate as a buffer (Sigma). Samples of 0.5 ml were withdrawn from the system and mixed with 9.5 ml of HCl 0.1 N (Sigma); at 20, 40, 60, 80, and 100 min.

Optimum temperature: The effect of temperature on initial rate was investigated incubating 50 of ml reaction mixture at specific temperature and pH 7, at a concentration of free whole cells of S-AT of 5 grams per liter. The composition of the corresponding reaction mixture was 130 mM substituted tetralone (Cambrex), 750 mM isopropylamine (Sigma), 2 mM PLP (Schweizer Hall), and 200 mM sodium acetate as a buffer (Sigma). Samples of 0.5 ml were withdrawn from the system and mixed with 9.5 ml of HCl 0.1 N (Sigma); at 20, 40, 60, 80, and 100 min.

Substrate affinity: The affinity of the enzyme for isopropylamine was investigated incubating 50 of ml reaction mixture of specific isopropylamine concentration at 55°C and pH 7, at a concentration of free whole cells of S-AT of 5 grams per liter. The composition of the corresponding reaction mixture was 130 mM substituted tetralone (Cambrex), 2 mM PLP (Schweizer Hall), and 200 mM sodium acetate as a buffer (Sigma). Samples of 0.5 ml were withdrawn from the system and mixed with 9.5 ml of HCl 0.1 N (Sigma); at 20, 40, 60, 80, and 100 min.

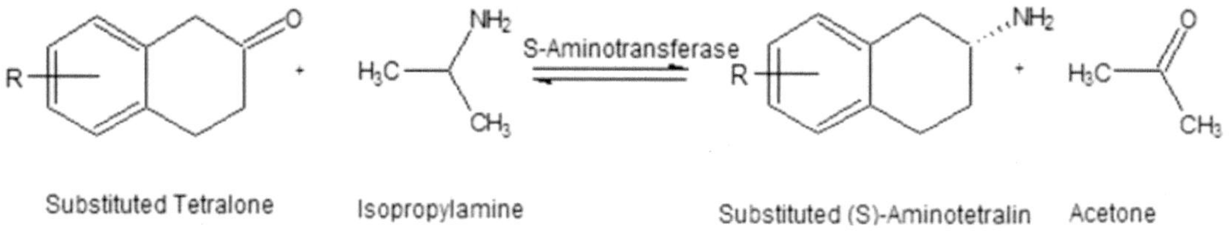

Figure 1: Reaction of substituted tetralone and isopropylamine catalyzed by S-AT to produce a SS-AT.

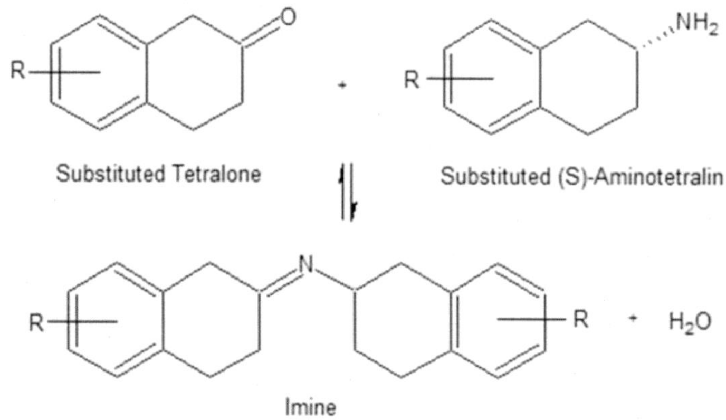

Figure 2: Reaction of imine formation.

Following the same methodology, the affinity for substituted tetralone was determined, withdrawing samples at 5, 10, 15, and 20 min.

Kinetics

Enzyme kinetic: Enzyme kinetic was studied at 55 °C, optimum pH, 2 mM concentration of PLP and at one gram of whole cells per liter. The kinetic studies consisted on the effect of substituted tetralone and isopropylamine in the concentration ranges from 2 to 30 mM and 750 to 2000 mM, respectively; the rate of reverse reaction (reaction between SS-AT and acetone); and the effect of product concentration (SS-AT and acetone). The reaction systems described previously for studying of enzyme kinetic were analyzed following concentration of SS-AT withdrawing 0.5 ml samples at 5, 10, 15, and 20 minutes. After withdrawing the sample it was mixed with 2 ml of water and centrifuged. Under the range of time chose to sample the system imine formation was considered to be negligible.

Kinetic of Imine formation: Imine formation was monitored during SS-AT production at 55°C, and pH 6 and 7. The reaction system was composed of 130 mM substituted tetralone, 750 mM isopropylamine, 2 mM PLP, 200 mM sodium acetate at an enzyme concentration of 5 grams of whole cells per liter; in a reaction volume of 500 ml. At the desired sampling time two samples of 0.5 ml were withdrawn. One of these samples was added to 9.5 ml of distilled water centrifuged and analyzed; the other was added to 9.5 ml 0.1N HCl. The sample added to HCl solution was incubated at 80°C for 30 minutes, then brought to room temperature, centrifuged and analyzed by HPLC. This sample would provide information about the total SS-AT in the system –since HCl and heat hydrolyzes the imine formed-, whereas the sample added water would provide the amount of free SS-AT in the system; then by difference imine concentration was obtained.

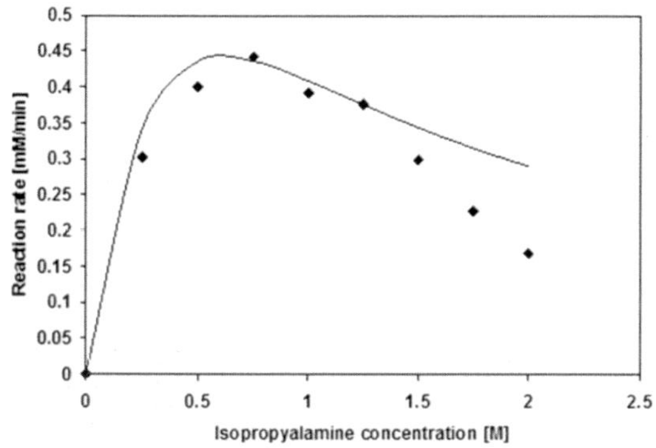

Figure 3: Isopropylamine affinity of S-AT for SS-AT production at 55°C, pH 7, and 5 grams per liter of whole cells. The continuous trend correspond to Equation (4) evaluated at V_m = 1.50 mM/min, K_m = 0.83 M, and K_S = 0.45 M.

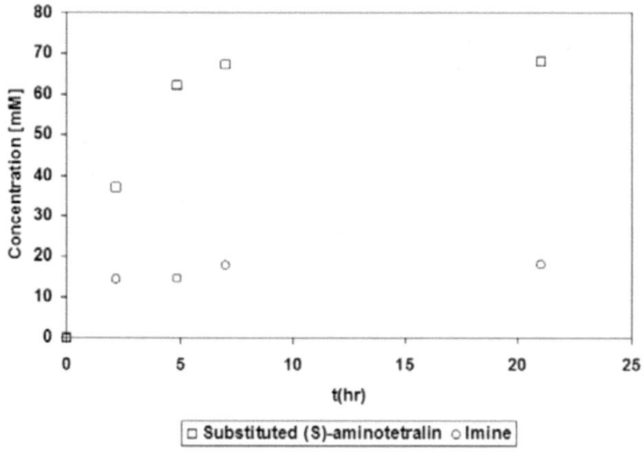

Figure 4: Profile concentration of SS-AT and Imine at 55°C, pH 7, at a concentration of 5 grams per liter of whole cells of S-AT.

S-AT model

The data collected during the experiments were fitted to a modified ping-pong mechanism [19-21] that included substrate and product inhibition by isopropylamine and SS-AT, respectively. Enzyme inhibition by substituted tetralone was not present in the range of solubility of this substance at the temperatures of experimentation. Acetone inhibition was not considered since its boiling point is 56.53°C, and the temperature of the system was 55°C, therefore acetone evaporation rate was high, which did not allow acetone accumulation in the system to cause strong enzyme inhibition. The model used is presented below.

$$v = \frac{v^f_{max}v^r_{max}\left((A)(B)-\frac{(P)(Q)}{Keq}\right)}{v^r_{max}K_{mB}(A)\left(1+\frac{(A)}{K_{iA}}\right)+v^r_{max}K_{mA}(B)+v^f_{max}\frac{K_{mQ}}{Keq}(P)+v^f_{max}\frac{K_{mP}}{Keq}(Q)\left(1+\frac{(Q)}{K_{iQ}}\right)+v^r_{max}(A)(B)+\frac{v^f_{max}K_{mQ}}{KeqK_{iA}}(A)(P)+\frac{v^f_{max}}{Keq}(P)(Q)+\frac{v^r_{max}K_{mA}}{K_{iQ}}(B)(Q)} \quad (1)$$

Where v^f_{max} is the forward rate of reaction, v^r_{max} is the reverse rate of reaction, Keq is the equilibrium constant, K_{iA} is inhibition constant of isopropylamine [A], K_{iQ} is the inhibition constant of SS-AT, and K_{mA}, K_{mB}, K_{mP} and K_{mQ} are the affinities for isopropylamine [A], substituted tetralone [B], acetone [P] and SS-AT [Q], respectively.

SS-AT production model

Equation 1 describes the main reaction occurring in the system early on, when imine formation is negligible. A complete description of the reaction system considering imine formation was done. Imine [I] formation occurs parallel to synthesis of SS-AT [Q]. Substituted tetralone [B] reacts with isopropylamine [A] in the presence of S-AT, rendering as products SS-AT [Q] and acetone [P]. SS-AT [Q] reacts further with substituted tetralone [B], in a reversible fashion, given as product imine [I]. The models that describe imine formation [I] and SS-AT [Q] are shown below.

$$A+B \underset{\text{Enzyme catalyzed process}}{\leftrightarrow} Q+P$$

$$B+Q \underset{k_2}{\overset{k_1}{\rightleftarrows}} I$$

$$\frac{dQ}{dt} = \frac{v^f_{max}v^r_{max}\left((A)(B)-\frac{(P)(Q)}{Keq}\right)}{v^r_{max}K_{mB}(A)\left(1+\frac{(A)}{K_{iA}}\right)+v^r_{max}K_{mA}(B)+v^f_{max}\frac{K_{mQ}}{Keq}(P)+v^f_{max}\frac{K_{mP}}{Keq}(Q)\left(1+\frac{(Q)}{K_{iQ}}\right)+v^r_{max}(A)(B)+\frac{v^f_{max}K_{mQ}}{KeqK_{iA}}(A)(P)+\frac{v^f_{max}}{Keq}(P)(Q)+\frac{v^r_{max}K_{mA}}{K_{iQ}}(B)(Q)} + k_2(I) - k_1(Q)(B) \quad (2)$$

$$\frac{dI}{dt} = k_1(Q)(B) - k_2(I) \quad (3)$$

In these equations the relation between substrates and products is done by mass balance, monitoring the concentration of SS-AT [Q] and imine [I], A=[Ao]-([Q]+[I]), P=([Q]+[I]), and B=[Bo]-[Q]-2[I]. If total concentration of B is higher than the solubility of it at 55°C (33 mM), then B concentration is considered as its solubility.

Effect of PLP

The effect of PLP concentration, the enzyme cofactor, on initial rate of S-AT used in this study was investigated at 55 °C. It was done in systems of 50 ml of solution of composition 130 mM substituted tetralone, 750 mM isopropylamine, 200 mM sodium acetate, and variable concentration of the cofactor. Enzyme concentration was 1 gram per liter. Samples were withdrawn from the systems at 5, 10, 15 and 20 minutes and then analyzed by HPLC to measure concentration of SS-AT.

Results

Enzyme properties and kinetic

Enzyme properties of the S-AT used for production of SS-AT are shown in Table 1. During the experiments done to determine substrate affinity it was noticed that isopropylamine causes enzyme inactivation as shown in Figure 3, while substituted tetralone does not in its soluble range. Substituted tetralone is a compound that has limited solubility (around 33 mM at 55°C). Experimentally, optimum isopropylamine concentration was 750 mM, but Equation (1) predicts an optimum concentration of of 0.61 M, obtained as the square root of the product of *Km KS*. The values of the parameters shown Equation (4) are *Vm* = 1.50 mM/min, *Km* = 830 mM, and *KS* = 450 mM for the data shown in Figure 3.

$$v = \frac{V_m[S]}{K_m + [S] + \frac{[S]^2}{K_S}} \quad (4)$$

where v is reaction rate, *Vm* is maximum reaction rate, S is substrate concentration, *Km* is the Michaelis-Menten constant, and *KS* is the inhibition constant of the substrate [19].

Imine formation

Imine formation in the system together with SS-AT in solution is presented in Figure 4 and 5. Imine formation is believed to be a reversible process where substituted tetralone and SS-AT combines to for the imine, at the same time imine decomposes to form substituted tetralone and SS-AT. Maximum imine and SS-AT was at about 7 hours, and remained practically constant towards the rest of the biotransformation. SS-AT reached a concentration about 70 mM, whereas imine concentration reached a maximum concentration of about 20 mM.

During the production of SS-AT at pH 6 and 7, at a 500 ml scale, the same overall conversion was achieved. Twenty-one percent of the SS-AT produced at pH 7 was obtained after hydrolysis of imine, meanwhile 3.5% of the SS-AT produced at pH 6 was originated after acid hydrolysis of imine. Therefore, low pH appears to decrease overall imine formation, but at the pH of 6 enzyme activity is lower than the activity of the enzyme at pH 7, therefore pH is selected as the optimum for this S-AT.

Effect of pyridoxal-5-phosphate

The data shown in Figure 6 were correlated to a Michaelis-Menten-like model of saturation kinetics.

$$r = \frac{V_m S}{K_m + S} \quad (5)$$

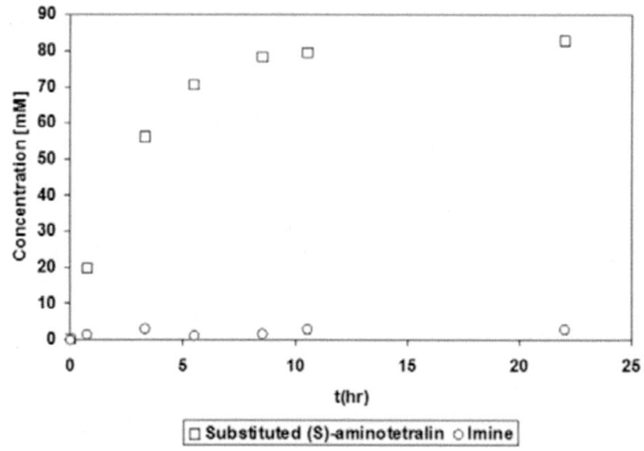

Figure 5: Profile concentration of SS-AT and Imine at 55°C, pH 6, at a concentration of 5 grams per liter of whole cells of S-AT.

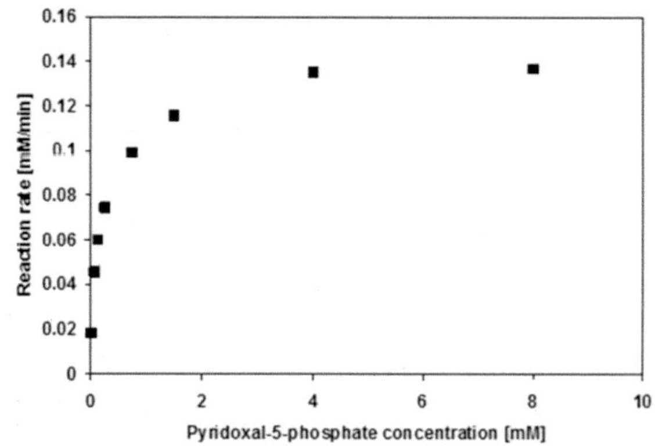

Figure 6: Effect of Pyridoxal-5-phosphate (PLP) concentration on reaction rate of free S-AT catalyzing the reaction system of substituted tetralone at 55 °C, pH 7.

Property	Value
Optimum pH	7
Optimum Temperature (°C)	55
Activation Energy (kcal/mol)	13.0
Deactivation Energy (kcal/mol)	43.9
Maximum activity per gram of whole cells per liter (mM/h)	12-18
Apparent Michaelis-Menten constant for Substituted Tetralone (mM)	10.4
Optimum Isopropylamine concentration (mM)	750

Table 1: S-AT properties.

The model gave as a result a maximum value of rate Vm of 0.14 mM/min, and a Km of 0.22 mM by means of non-linear regression using the solver tool from Microsoft® Office Excel 2003. Figure 6 also shows a reaction rate of 0.02 mM/min at a cofactor concentration of 0 mM, obtained when the enzyme was used as it was after obtaining the whole cells by spray drying. Adding cofactor up to a concentration of 4 mM increases the activity (reaction rate) by a factor of about 7 times. From these data the need is shown of adding the cofactor, PLP, to all the reactions mediated by S-AT expressed in *Escherichia coli* MG1655, since it appears that the amount of PLP inside of the cell is not enough to form the complex enzyme-cofactor with all the S-AT produced.

Discussion

The experiments regarding to enzyme properties and kinetic showed clearly that S-AT presents substrate inhibition by isopropylamine, and suggested that SS-AT might cause product inhibition. The data collected during the experiments mentioned above allowed modeling the data by a modified ping-pong mechanism [20-22] that included substrate and product inhibition by isopropylamine and SS-AT, respectively. The parameters that best describe the experimental data are shown in Table 2. These parameters were estimated by non-linear regression, using as guessing parameters the values of substrate affinity constants and maximum activity shown in Table 1.

Individual parameter	Value
v^{f}_{max}	0.21 (mM/min) [12.6 mM/hr]
v^{r}_{max}	0.02 (mM/min)
Keq	0.33
K_{mB}	1.72 (mM)
K_{iA}	200.10 (mM)
K_{mA}	42.59 (mM)
K_{mQ}	0.45 (mM)
K_{mP}	14.95 (mM)
K_{iQ}	6.93 (mM)

Table 2: Apparent kinetic parameters of whole cells of S-AT.

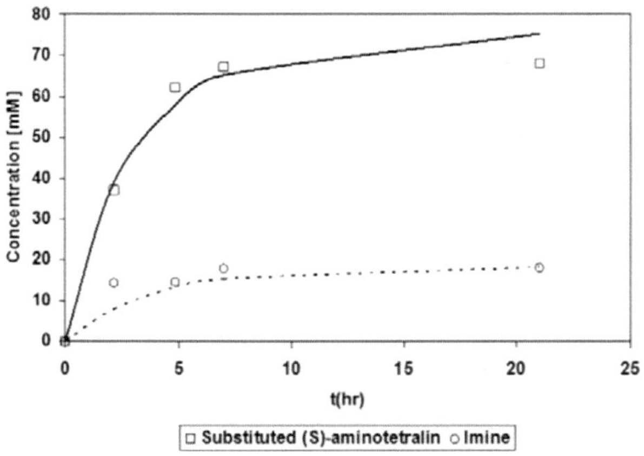

Figure 7: Description of profile concentration of SS-AT and Imine at 55°C, pH 7, at a concentration of 5 grams per liter of whole cells of S-AT by Equation 3 and 4 evaluated with the parameters shown in Table 3.

Parameters	Values
v^{f}_{max} (mM/hr)	59.29 [11.9 mM/hr/gram whole cells per liter]
v^{r}_{max} (mM/hr)	59.23
Keq	0.33
K_{mB} (mM)	1.71
K_{iA} (mM)	200.10
K_{mA}(mM)	42.59
K_{mQ} (mM)	0.45
K_{mP}(mM)	14.95
K_{iQ}(mM)	6.93
k_1 (mM*hr)$^{-1}$	0.02
k_2 (hr)$^{-1}$	2.04
k_{eq} for Q	0.33
k_{eq} for I	0.01

Equilibrium constant were estimated using the values of final system composition. Where, $k_{eq}=\frac{(P)(Q)}{(B)(A)}$ for Q and $k_{eq}=\frac{(I)}{(B)(Q)}$ for I.

Table 3: Apparent kinetic parameters of whole biotransformation at 55°C, pH 7, 5 grams of whole cells per liter

The whole process of SS-AT production is described by Equations 2 and 3 with the parameters shown in Table 2 were used to model both SS-AT and imine formation during the biotransformation. Equation 2 and 3 were solved numerically by Runge-Kutta method of fourth order and the imine formation parameters obtained by non-linear regression. The parameters that describe the whole biotransformation process, at 55°C, pH 7, 5 grams of whole cells per liter are shown in Table 3; and their prediction of the biotransformation is shown in Figure 7.

Figure 7 shows the biotransformation description by Equation 2 and 3 evaluated with the parameters shown in Table 3. Overall SS-AT formation is described fairly well, at least to the point where it reaches it maximum concentration, after that the model over predicts its concentration (by 17% more). Imine formation apparently is not well described early on the biotransformation, but its final concentration is predicted accurately.

Acknowledgments

Cambrex Corporation for financial support, laboratory facilities, and expertise, to Michigan Technological University, and to the National Council of Science and Technology of Mexico (Conacyt).

References

1. Scheller D, Dressen F (2005) Use of substituted 2-aminotetralins for the preventative treatment of Parkinson's disease. PCT Int Appl 35 pp.
2. Warawa EJ, Migler B (1998) Preparation of aminotetralins for the treatment or prevention of movement disorders. US 16 pp.
3. Scheller D, Breidenbach A, Selve N (2005) Substituted 2-aminotetralins for the treatment of depression. PCT Int Appl 34 pp.
4. Montanari S, Cavalleri P, Santangelo F, Marchini F (1998) Preparation of 6-hydroxy-5-hydroxymethyl-2-aminotetralins as cardiovascular agents. PCT Int Appl 19 pp.
5. Chiesi P, Bongrani S., Razetti R, Civelli M, Umile A (1996) Aminotetralin derivative for the therapy of cardiovascular diseases. PCT Int Appl 18 pp.
6. Nair V, Balagopal L, Rajan R, Mathew J (2004) Recent Advances in Synthetic Transformations Mediated by Cerium(IV) Ammonium Nitrate. Acc Chem Res 37: 21-30.
7. Youngman MA, Willard NM, Dax SL, McNally JJ (2003)The synthesis of novel cis-a-substituted-b-aminotetralins. Synthetic Communications 33: 2215-2227.
8. Yamaura M (2002) Preparation of optically active 2-aminotetralins from racemates. Jpn Kokai Tokkyo Koho 5 pp.
9. Yamada M, Nagashima N (1999) Preparation of 7-methoxy-2-aminotetralin carbamates and improvement of optical purity of 7-methoxy-2-aminotetralin using them. Jpn Kokai Tokkyo Koho 8 pp.
10. Lennon IC (2001) Speeding up synthesis of chiral chemicals. Chemical Engineering 108: 70-72.
11. Dupau P, Hay AE, Bruneau C, Dixneuf PH (2001) Synthesis of optically active 2-aminotetralin derivatives via enantioselective ruthenium-catalyzed hydrogenation of ene carbamates. Tetrahedron: Asymmetry 12: 863-867.
12. Kanai Y, Saito K (2000) Preparation of optically active 2-amino-substituted tetralins. Jpn. Kokai Tokkyo Koho 8 pp.
13. Honda T, Fujii A, Inoue K, Yasohara Y, Itagaki Y, et al.(2003) Process for preparation of 2-aminotetralin derivatives and intermediates thereof. PCT Int Appl 47 pp.

14. Orsini F, Sello G, Travaini E, Di Gennaro P(2002) A chemoenzymic synthesis of (2R)-8-substituted 2-aminotetralins. Tetrahedron: Asymmetry 13: 253-259.

15. Khumtaveeporn K, Cosway TC, Yeh WL (2001) Enzymatic resolution of aminotetralins. U.S 5 pp.

16. Takashima H (2001) Manufacture of optically-active aminotetralins with Pichia. Jpn. Kokai Tokkyo Koho 7 pp.

17. Pannuri S, DiSanto R, Kamat S (2002) Encyclopedia of Catalysis. 1, JohnWiley 610 & Sons Inc 722–737.

18. Pannuri S, Kamat SV , Garcia ARM (2006)Methods for engineering Arthrobacter citreus w-transaminase variants with improved thermostability for use in enantiomeric enrichment and stereoselective synthesis. PCT Int Appl 54 pp.

19. Shuler ML, Kargi F (2002) Bioprocess Engineering Basic Concepts. Second Edition. Prentice Hall International Series in the Physical and Chemical Engineering Sciences. Upper Saddle River, New Jersey.

20. Cleland WW (1963) The kinetics of enzyme-catalyzed reactions with two or more substrates or products. I. Nomenclature and rate equations. Biochimica et biophysica acta 67: 104-137.

21. Cleland WW (1963) The kinetics of enzyme-catalyzed reactions with two or more substrates or products. II. Inhibition: nomenclature and theory. Biochimica et biophysica acta 67: 173-187.

22. Cleland WW (1963) The kinetics of enzyme-catalyzed reactions with two or more substrates or products. III. Prediction of initial velocity and inhibition patterns by inspection. Biochimica et biophysica acta 67: 188-96.

12

Marine Oils as Potential Feedstock for Biodiesel Production: Physicochemical Characterization

Deepika Dave*, Vegneshwaran V Ramakrishnan, Sheila Trenholm, Heather Manuel, Julia Pohling and Wade Murphy

Centre for Aquaculture and Seafood Development, Marine Institute of Memorial University of Newfoundland, St. John's, Newfoundland, Canada

Abstract

Physico-chemical characteristics of four crude marine oils including farmed salmon, seal, cod liver and wild salmon are compared and interpreted with regard to their suitability as biodiesel feedstock. The physico-chemical properties including specific gravity, pH, ash content, acid value, iodine value, saponification value, p-anisidine value, peroxide value, TOTOX value, free fatty acid, flash point, kinematic viscosity, refractive index, lipid classes and fatty acid classification of all four marine oils were evaluated. The characterized marine oils were pale yellow to orange in color and stable in the liquid state at room temperature. The pH (6.5-6.8) values of all oils were neutral. The specific gravity (0.921-0.924 g/cm^3), water content (179-325 ppm), ash content (0.0027-0.00455%), free fatty acids (0.03-1.23%), acid value (0.057-0.771 mg KOH/g), peroxide value (5.13-9.17 meq O_2/kg oil) and p-anisidine value (3.36-9.67) of all oils were within recommended limits except for higher water content in the seal oil (689 ppm), higher acid value in farmed salmon (2.441 mg KOH/g) and seal oil (0.958 mg KOH/g) and higher iodine value (116-139.15 g I_2/100 g). A drying step needs to be implemented for the removal of water as it can lead to corrosion of internal combustion engine components. Due to higher iodine values, all the oils were drying oils except farmed salmon oil which was semidrying oil and susceptible to become rancid which causes reduction of pour point of biodiesel produced in the absence of antioxidant. All four marine oils were more likely to polymerize in the heat of the engine if used directly without transesterification. Flash point of all marine oils were above 200°C so, there is no risk of fire outbreaks in case of accidents. Due to higher triacylglycerol (81-93%) content, all the marine oils are suitable as a feedstock for biodiesel production via transesterification. Cod liver (14.72%) and wild salmon oil (9.92%) were rich in polar lipids while the farmed salmon (2.43%) and wild salmon (2.43%) were low in polar lipids. The phospholipids (1.21-1.67%) were higher than the recommended limit of ≤10 ppm so degumming process is required prior to biodiesel production. All the marine oils in this study have a high degree of unsaturation and polyunsaturated fatty acids and therefore the biodiesel produced from all oils will have less oxidation stability and result in the precipitation of the biodiesel components in a fuel feeding system or combustion chamber. Therefore, it is essential to stabilize the oil using an antioxidant immediately after extraction/production to obtain a high quality biofuel.

Keywords: Marine oil; Biofuel; Feed stock

Introduction

The continuously increasing demand for energy has been translated into increased cost of crude oils, shortage of fossil fuels and intensified emission of greenhouse gases worldwide. Adverse environmental concerns due to use of non-renewable fossil fuel and the rapid decline of their reserves are the two largest incentives to find energy alternatives that are renewable and unlimited [1-4]. Renewable energy resources of biological origin (biofuels) have smaller net greenhouse gas emissions. Currently, biodiesel and bioethanol production are gaining momentum all across the globe due to the shrinking supply of oil reserves, security of source, cost of production and the impending threat of global warming [4-6]. However, sustainable production of biofuels will require a resourceful biomass conversion process.

Biodiesel is a biofuel that is obtained from plant and marine oils or animal fats. Biodiesel, as a diesel-equivalent, has a potential share among biofuels of about three quarter of all refinery distillate fuel oils [7-9]. One of the biggest challenges in biodiesel production is the availability of feed-stock. There is a concern about using plant derived oils and fats since the crops used for biodiesel production are also needed for food, feed and oleochemical industries [10,11]. Biodiesel factories must compete with food, cosmetic, chemical and livestock feed demands. There is also an environmental concern because an increased demand for vegetable oils requires an increase in the use of fertilizers which contribute to greenhouse gas emissions. In fact, biodiesel production from heavily fertilized crops could result in a 70% increase (from the current value) in greenhouse gas emissions [11-13].

Canada's commercial fishing industry is valued at approximately $3 billion a year with its aquaculture industry worth $845 million. Of the 799,567 tons marine landings and 163, 036 tons aquaculture in 2012, 34% and 14%, respectively came from Newfoundland and Labrador [14]. There are 187 registered fish processing facilities in Newfoundland, ranging in size from feeder plants (processing fish to the fillet) to large year-round plants (processing fish into various fresh and frozen products including secondary processing) [15]. Of the 145 licensed aquaculture plants, 84 produced salmonids [16]. In 2012, salmonid, cod and seal production accounted 16,831 tons (79.2%), 8334 tons (3.2%) and 67,567 (26.3%), respectively and was valued at $90.1 million [17]. Processing of fish generates large amounts of solid wastes, up to 30-80% of the body weight of the processed fish. Currently, most of the fish processing waste is dumped at sea or in landfills [9].

Sealing has thrived in Newfoundland and Labrador for hundreds of

***Corresponding author:** Deepika Dave, Centre for Aquaculture and Seafood Development, Marine Institute of Memorial University of Newfoundland, St. John's, Newfoundland, Canada

years and the sealing industry has been worth over $55 million per year to the province's economy [18]. Only relatively unrefined seal products (raw seal skins, seal oil and seal meat) can be identified in Canada's export statistics [19]. While the focus is directed on the method used for killing seal, much of the negativity arises from the utter waste of the resource and environmental impact of discarding seal carcasses into the ocean. This is largely due to the lack of economic feasibility to convert these waste materials into high value added products.

Newfoundland and Labrador's seafood industry generates on the average 102,850 tons (25% of Canada's fish waste) of marine wastes (processing discards). Currently, most of the processing waste is bulk packaged and sold to mink producers and/or pet food manufacturers in eastern Canada. A small percentage of the processing discards (gut material) and morts are bulk shipped to a composting operation in the Corner Brook area. There is a planned expansion of Newfoundland salmon aquaculture up to 50,000 metric tons in the next 5 years which will increase the volume of waste generated from aquaculture processing operations and mortalities. Currently, NL has limited capacity to handle this waste in terms of composting, mink feed, rendering and landfilling [15]. Thus, it is important to develop by-product applications that demand large volumes of marine waste, in order to make these industries viable, greener and environmentally friendly. Newfoundland and Labrador has the potential to produce 2600 tons of marine oils extracted from seal harvesting and fish processing waste as the largest potential source of biodiesel feedstock [20]. Converting marine oil into biodiesel would benefit the marine industry sectors in reducing the disposal cost of these wastes to landfills and utilize biodiesel for operating feed barges, marine vessels and generators located at their remote locations.

The properties of marine oils are less uniform compared to fresh vegetable oils because of the physical and chemical changes mainly due to oxidative and hydrolytic reactions that take place during handling, stabilization, storage and oil extraction process [21]. The initial evaluation of the physical and chemical composition of the feed stock oil is very essential for qualitative identification prior to utilization in biodiesel production. Feed stock for biodiesel production are given priority selection and pre-treatment according to the level of free fatty acids, impurities, water content, ash content, acid value, iodine value, saponification value, p-anisidine value, peroxide value, free fatty acid, flash point, kinematic viscosity and refractive index [22-25].

In the present study, the physico-chemical characteristics of four crude marine oils including farmed salmon, seal, cod liver and wild salmon are compared and interpreted with regard to their suitability as biodiesel feedstock.

Materials and Methods

This study was carried out at the Marine Bioprocessing Facility of the Centre for Aquaculture and Seafood Development, Marine Institute of Memorial University of Newfoundland in St. John's, Newfoundland, Canada. Fatty Acid profiling was performed at the Oceans Sciences Centre of Memorial University of Newfoundland, Logy Bay, Newfoundland, Canada.

Chemicals and reagents

Reagent 'A' (Calcium Hydride) and Reagent 'B' were obtained from Sandy Brae Laboratories (Wilmington, Delaware, USA). Buffer Solutions (pH 4.0, 7.0 and 10.0), sodium thiosulfate, cyclohexane and potassium hydroxide were obtained from Sigma-Aldrich (Oakville, Ontario, Canada). Wijs iodine solution, iso-octane (2,2,4-trimethylpentane), glacial acetic acid (Ricca), concentrated sulfuric acid (Acros Organic), para-anisidine (Acros Organic), hydrochloric acid (concentrated), soluble starch and chloroform were obtained from Fisher Scientific (Ottawa, Ontario, Canada). Potassium iodide (BDH), potassium dichromate (BDH), and formic acid (BDH) were obtained from VWR International (Mississauga, ON, Canada). Ethanol (95%) was obtained from East Chem (Paradise, NL, Canada).

Marine oils

Crude cod (*Gadus morhua)* liver and pacific salmon *(Oncorhynchus)* oil were purchased from J. Edwards International Inc. (a bulk oil provider, Quincy, MA, USA). Crude Atlantic salmon *(Salmo salar)* and Harp seal *(Pagophilus groenlandicus)* oils were purchased from an oil extraction plant in Newfoundland, operated by Barry Group Inc. Salmon and seal oil were extracted from the waste stream from salmon guts and seal blubber, respectively that were obtained from local processing plants Both oils were extracted using a meat grinder, Contherm™ scraped-surface heat exchanger, 2-phase decanter centrifuge and Westfalia polishing centrifuge. The extracted oils were stabilized using a food grade industrial antioxidant ('Dadex Toro', Caldic Canada Inc, Mississauga Ontario, Canada) immediately after extraction and no further oil refining was performed. All four crude oils were stored in a dark place at room temperature (18-20°C) in tightly sealed containers.

Physical properties

Smell, colour and physical state: Odour, colour and physical state of the oils were assessed by sensory evaluation.

Specific gravity: Specific gravity was determined using a hydrometer (Model # 11582, Fisher Scientific, Ottawa, Ontario, Canada) at 15.5°C (60°F).

Water content: The water content was determined using a water test kit (01-WTK-DELUXE, Sandy Brae Laboratories Inc., Wilmington, Delaware, USA) which measures the pressure produced after the reaction of trace water with calcium hydride in a pressure chamber [26].

Chemical properties

pH: The pH was measured using a pH pen (Model #850050, Super Scientific, Scottsdale, AZ, USA).

Ash content: The ash content was determined gravimetrically using a muffle furnace (Thermolyne Type F6000, Thermo Scientific, Asheville, North Carolina, USA) at 550°C following BS ISO 6884:2008 (British Standards Institution International Organization for Standardization). Successive portions of oil were ashed after initial ashing until a sufficient quantity was obtained to calculate the percent ash.

Saponification value: The saponification value was determined by following ASTM procedure D5558-95. To determine the saponification value, 4-5 g of oil sample was accurately weighed and filtered through a Whatman No. 40 filter paper to remove moisture and impurities. The oil samples after filtration were collected in a round bottom flask. An aliquot of 50 mL of alcoholic potassium hydroxide was added to the oil sample using a pipette and the pipette was allowed to drain for a definitive period of time. The alcoholic potassium hydroxide was prepared by adding a few grams (5 to 10 g) of potassium hydroxide to 1.5 L of 95% ethyl alcohol and boiled for 30-60 min under a reflex condenser on heating mantle. The boiled ethyl alcohol was distilled in a rotary evaporator and collected. Potassium hydroxide (40 g) (low in carbonate) was dissolved in 1 L of distilled ethyl alcohol at 15.5°C until the solution was clear. A blank solution without oil was also prepared and the experiment was carried out simultaneously. The oil sample and the blank sample were gently boiled until the sample completely saponified. Complete saponification took place approximately in 1 h and it was generally determined by the clarity and homogeneity of the test solution. After 1 h, the samples were cooled down but not sufficiently to gel the content of the sample and the condenser was washed with a little amount of distilled water. To the samples, 1 mL of phenolphthalein indicator was added and titrated against 0.5 N HCl until the pink color disappeared completely. The saponification value of the oil sample was calculated by Equation 1 [96]

$$Saponification\,Value\,(mg\,KOH\,/\,g) = \frac{56.1 \times N(A-B)}{wt.of\;oil\,(\mathrm{g})} \quad (1)$$

N	=	Normality of HCl
A	=	Titration of blank (mL)
B	=	Titration of sample (mL)
56.1	=	Molecular weight of potassium hydroxide (KOH)

Free fatty acids and Acid number: The free fatty acid content (%FFA) and Acid number (AN) were determined according to AOCS Official method Ca 5a-40 [97]. A well-mixed oil sample (7.05 ± 0.05 g) was accurately weighed into a 250 mL Erlenmeyer flask and 75 mL of hot neutralized 95% ethanol and 2 mL of 1% phenolphthalein indicator solution was then added to the oil sample. The hot neutralized 95% ethanol was prepared by heating 75 mL of 95% ethanol with 2 mL of 1% phenolphthalein indicator solution to incipient boiling and then ethanol was neutralized by adding 0.25N sodium hydroxide solution until a faint permanent pink color appeared. The oil samples were then titrated against 0.25N sodium hydroxide until the appearance of the first permanent pink color of the same intensity as that of the neutralized ethanol before the addition of sample. The permanent pink color must persist for at least 30 seconds during titration. The free fatty acids content (%FFA) and acid number were calculated using Equations 2 and 3.

$$FFA(\%) = \frac{mL\,of\;alkali \times N \times 28.2}{w} \quad (2)$$

Where:

N	=	Normality of NaOH solution
W	=	Weight of oil (g)

$$Acid\;number\;(mg\;KOH/g) = 1.99 \times FFA\;(\%) \quad (3)$$

Iodine value: The iodine value was determined by following AOAC official method 993.20. The test oil sample was filtered through a Whatman No. 40 filter paper to remove moisture and impurities. A 0.16 g sample of filtered fish oil was accurately weighed to 0.2 mg in a 250 mL Erlenmeyer flask. To the fish oil sample 15 mL of cyclohexane-acetic acid solvent was added and swirled to mix with a magnetic stirrer. Two blank solutions were also prepared simultaneously without the addition of fish oil. Wijs solution (25 mL) was added and the flask was then stored in a dry dark place for 2 h at 25 ± 5°C. After 2 h, the reaction was terminated by adding 20 mL of potassium iodide (KI) and and swirling to mix and then adding 150 mL of distilled water within 3 min. The samples were titrated against standardized 0.086 M (0.1M) standard sodium thiosulfate ($Na_2S_2O_3$) until the yellow colour of the solution has disappeared. A 1-2 mL of starch indicator was added to the samples and the titration was continued until the blue color of the solution disappeared. The iodine value of the oil sample was calculated using Equation 4 [27]:

$$Iodine\,Value = \frac{(\mathrm{B-S}) \times M \times 12.96}{wt.of\;oil\,(\mathrm{g})} \quad (4)$$

Where:

B	=	Titration of blank (mL)
S	=	Titration of test solution (mL)
M	=	Molarity of $Na_2S_2O_3$

Peroxide value (PV): The peroxide value was determined by AOCS official method Cd 8-53. The test oil sample was filtered through a Whatman No. 40 filter paper to remove moisture and impurities. A 5 g sample of filtered oil was accurately weighed to 0.05 g in a 250 mL Erlenmeyer flask and 30 mL of 3:2 acetic acid-chloroform was added and swirled to mix well using a magnetic stirrer. Two blank samples were simultaneously prepared without addition of fish oil. To the samples, 0.5 mL of saturated potassium iodide solution was added and allowed to stand for 1 min. Saturated potassium iodide solution was prepared by adding 10 g potassium iodide to 6 mL boiled distilled water so that undissolved potassium iodide crystals were present during analysis. Then 30 mL of distilled water was added to the oil samples and swirled to mix with a magnetic stirrer. The samples were titrated against 0.1 N sodium thiosulfate until the yellow iodine color disappeared. Starch indicator (2 mL) was added and the titration was continued against 0.1 N sodium thiosulfate until the blue color disappeared. The blank titration value must not exceed 0.1 mL and the peroxide value was calculated by using Equation 5. Preliminary results showed a titration value less than 0.5 mL and therefore the peroxide value determination was carried out using 0.01 N sodium thiosulfate [28].

$$Peroxide\,Value\,(\text{milliequivalents/ 1000 g sample}) = \frac{(\mathrm{S-B}) \times N \times 1000}{W} \quad (5)$$

Where:

S	=	Volume of titrated sample (mL)
B	=	Volume of titrated blank (mL)
N	=	Normality of sodium thiosulfate solution
W	=	Weight of oil (g)

***p*-Anisidine value:** The *p*-anisidine value was determined by AOCS official method Cd 18-90. The test oil sample was filtered through a Whatman No. 40 filter paper to remove moisture and impurities. A 0.5-4 ± 0.001g sample of oil was accurately weighed in a 25 mL volumetric flask. The oil samples were dissolved and diluted with 25 mL iso-octane. The absorbance (A_B) of the oil sample was measured at 350 nm using a spectrophotometer (Jenway 6400/6405, Jenway Incorporated, Stone, Staffordshire, UK). A 5 mL sample of oil was pipetted into one test tube and 1 mL of *p*-anisidine reagent was added. A 5 mL of iso-octane was added to another test tube and 1 mL of p-anisidine reagent was added to it and used as a blank. The p-anisidine reagent was prepared by adding 0.25 g p-anisidine to 100 mL of glacial acetic acid. After 10 minutes, the absorbance (A_S) of the oil sample with the p-anisidine reagent was measured at 350 nm using a spectrophotometer. The *p*-anisidine value was calculated by using following Equation 6 [29].

$$p-anisidine\,value = \frac{25 \times (1.2\,\mathrm{A_S} - \mathrm{A_B})}{W} \quad (6)$$

Where:

A_S	=	Absorbance of the fat solution after reaction with

the *p*-anisidine reagent

A_B = Absorbance of the fat solution

W = Weight of oil (g)

TOTOX value: TOTOX means "Total Oxidation", calculated as twice the Peroxide value plus Anisidine value.

Flash point: The flash point was measured using a Pensky-Marten closed cup tester (K162XX, Koehler Instruments, Bohemia, New York, USA) [98] according to Procedure A in ASTM D93-12. The test cup was filled with 75 mL of oil sample and the cup was closed with a test cover and placed in the assembly, ensuring that the locating groove was engaged. The temperature of the test cup and test specimen should be at least 18°C below the expected flash point. The test flame was switched on and the oil was heated at a rate of 5-6°C/minute. The oil sample was stirred in a downward direction at 90-120 rpm. The observed flash point was recorded at the temperature when a distinct flash occurred in the interior of the cup. The oil sample was deemed to have flashed when a large flame appeared and instantaneously propagated itself over the entire surface of the oil sample. The observed flash point was corrected for barometric pressure.

Viscosity: Dynamic viscosity (cP) was measured with a Brookfield viscometer (DV-I, Brookfield Engineering Laboratories, Inc., Middleboro, MA, USA) with a S61 spindle at 50 rpm at 40°C. Kinematic viscosity (cSt) was calculated using equation 7:

$$v = \frac{\mu}{\rho} \tag{7}$$

Where:

v = Kinematic viscosity (mm^2/s)

μ = Dynamic viscosity (N_s/m^2)

ρ = Density (kg/m^3)

Refractive index: The refractive index was measured with a pocket ATAGO refractometer (PAL-S, ATAGO Co. Ltd., Tokyo, Japan) per manufacturer's instructions. 2-3 drops of the crude oil was applied to the prism surface. The start key was pressed and the BRIX value was displayed and recorded. The Brix value was converted to refractive index.

Cetane number: The Cetane Number was calculated using equation (8) [30].

$$CN = 46.3 + \frac{5458}{SN} - .225 \times IV \tag{8}$$

Where:

CN = Cetane number

SN = Saponification number of oil (mg KOH/g)

IV = Iodine value of oil

Higher heating value: The higher heating value (HHV) was calculated using equation (9) [31].

$$HHV = 0.0317VS + 38.053 \tag{9}$$

Where:

HHV = Higher Heating Value (kJ/g)

VS = Kinematic Viscosity (mm^2/s) .

Lipid class determination and fatty acid analysis

Lipid extraction: Lipid samples were extracted according to Parrish [32]. An aliquot of 250 µl of oil (between 170 and 215 mg) sample was weighed in a test tube containing 2 mL of chloroform. Previous to addition of the oil sample, the test tubes and Teflon® lined caps were rinsed 3 times with methanol and chloroform, respectively. 1 mL of ice-cold methanol, 1 mL of 2:1 chloroform:methanol and 0.5 mL of chloroform extracted water were added to the test tube. Chloroform extracted water was prepared by adding 1L of distilled water and 30 mL of chloroform to a separating funnel. The funnel was manually shaken for 2 minutes and the chloroform was allowed to settle and removed from the bottom of the funnel. This procedure was repeated twice to remove any lipids present in the distilled water. The test tube was then recapped and sonicated for 10 minutes followed by centrifugation for 2-3 minutes at 3000 rpm using an international clinical centrifuge (model CL, International Equipment Co, Needham, Mass). The entire lower organic lipid layer was removed by a double pipetting technique and transferred to a 15 mL vial that was cleaned 3 times with methanol and chloroform, respectively [99]. The double pipetting technique was performed in three steps. Firstly, an ashed 14 cm pipette was passed through the top aqueous layer in the test tube, by bubbling air with the pipette bulb to prevent the aqueous layer from entering the 14 cm pipette until it touched the bottom of the test tube. Secondly, the pipette bulb was removed and a 27 cm pipette was placed inside the shorter pipette until it touched the bottom of the test tube. Thirdly, the lipid layer was removed using the long pipette and transferred to a second cleaned 15 mL vial. Each of the short and long pipettes were washed with 3 mL ice-cold chloroform and the wash was collected, subsequently. The samples were again resonicated, centrifuged, double pipetted and the pipettes were rinsed three times as previously described and all the organic layers were pooled together. The extracted lipid was then evaporated under a gentle stream of nitrogen, sealed with Teflon® tape and stored in the freezer at -20°C until use.

Lipid class composition: Lipid class composition was determined using an Iatroscan Mark VI TLC-FID (Iatron Laboratories Inc., Tokyo, Japan), silica coated Chromarods and a three-step development method [33]. The Chromarods were calibrated (0.2–20 µg) using lipid standards including: n-nonadecane (HC-aliphatic hydrocarbon), cholesteryl palmitate (WE/SE-wax esters/steryl ester), n-hexdecan-3-one (KET-ketone), glyceryl tripalmitate (TAG-triacylglycerol), glyceryl-1,2-dihexadecanoate (DG-diglyceride), 1-hexadecanol (ALC-free aliphatic alcohol), cholesterol (ST-free sterol), 1-monopalmitoyl-rac-glycerol (AMPL-acetone mobile phase lipids) and 1,2, di-0-hexadecyl-sn-glycerol-3-phosphatidylcholine (PL-phospholipids) that were obtained from Sigma–Aldrich from Sigma Chemicals (Sigma Chemicals, St. Louis, Mo., USA).

The lipid extracts and standards were applied to the Chromarods and focused to a narrow band using 100% acetone. During standard or sample application, the frame holding the chromarods were placed on a warm hot-plate with the lower edge extending beyond the end of the hot plate so that the bottom of the rods were not directly over the heat source.

Four different solvent systems were used to obtain three chromatograms per rod. The first development system was hexane: diethyl ether: formic acid (99.95:1:0.05). The rods were developed for 25 minutes, removed from the system for 5 minutes and placed again in the system for 20 minutes for double development. The first chromatograms were obtained by scanning each rod to the lowest point behind the ketone (KET) peak. The second development was for

40 minutes in hexane:diethyl ether:formic acid (79:20:1). The second chromatogram was obtained by scanning each rod to the lowest point behind the diglyceride (DG) peak. The final development was carried out in two steps, in the first step the lipid extracted was developed using 100% acetone for two 15 minute time periods, followed by two 10 minute periods in chloroform:methanol:chloroform-extracted water (5:4:1). The third chromatogram was obtained as the complete scan after two double developments. Before each solvent system the rods were dried in a constant humidity chamber. After each development system the rods were scanned in the Iatroscan and the data collected using Peak Simple software (ver 3.67, SRI Inc).

Preparation of fatty acid methyl esters (FAME) with H_2SO_4 in MeOH: An aliquot of 40 μl of lipid extract was transferred to a lipid cleaned (rinsed 3 times with methanol and chloroform, respectively) vial and 1.5 mL of methylene chloride and 3.0 mL Hilditch reagent were added, subsequently. The Hilditch reagent was prepared by adding 1.5 mL of concentrated H_2SO_4 to 100 mL of dry methanol (100 mL methanol was transferred to a volumetric flask and sufficient amount of Na_2SO_4 was added to the methanol to cover the bottom of the flask and mixed manually by inverting the flask and left for 10 minutes and then decanted). The sample was capped and vortexed for approximately 5 seconds followed by sonication for 4 minutes. The tube was then flushed with nitrogen, capped, sealed with Teflon® tape and heated at 100°C for 1 hour in a VWR drying oven (VWR international, Mississauga, Ontario, Canada). The vials were then cooled to room temperature. Approximately 0.5 mL of saturated sodium bicarbonate solution (9 g/100 mL of chloroform extracted water) was slowly and carefully added to the vial, followed by addition of 1.5 mL of hexane and vortexing for 5-10 seconds. The top organic layer was carefully removed to a new vial without disturbing the bottom layer and the hexane was evaporated with a gentle stream of nitrogen. The fatty acids were re-suspended by adding approximately 0.5 mL of hexane, capping the vial with nitrogen, and Teflon® tape and sonicating for an additional 4 minutes.

FAME analysis: An aliquot of 10 μL of the mixture was separated by fatty acid class based on the carbon atom by a gas chromatography system (HP6890 Series II, Agilent Technologies, Mississauga, Ontario, Canada), coupled with flame ionization detector (FID) and 7683 auto sampler. A ZB wax + polar capillary column 30 m in length, 0.32 mm of internal diameter and 0.25 μm film thicknesses (Phenomenex, Torrance, CA, USA) was used for analyses. The separated samples were injected directly into the column with the initial oven temperature of 65°C for 5 minutes, followed by ramped temperature of 195°C at a rate of 40°C/min for 15 minutes and again ramped to a final temperature of 220°C at a rate of 2°C/min. A final temperature of 220°C was held for 0.75 minutes. The detection system was equipped with a flame ionization detector (FID) operating at 260°C with hydrogen as a carrier gas at a flow rate of 2 mL/min. The injector temperature was started at 150°C and ramped to a final temperature of 250°C at a rate of 120°C/minute. Peaks were identified using retention times from standards purchased from Supelco: 37 component FAME mix (Product number 47885-U), Bacterial acid methyl ester mix (product number 47080-U), PUFA 1 (product number 47033) and PUFA 3 (product number 47085-U). Chromatograms were integrated using the Varian Galaxie Chromatography Data System, version 1.9.3.2. The total run time was 32 minutes.

Results and Discussion

Physical properties

Physical properties of four crude marine oils including farmed salmon, seal, cod liver and wild salmon are shown in Table 1.

Smell, colour and physical state: All four marine oils had an agreeably oily smell while cod liver and wild salmon also had a slightly fishy smell. Seal and cod liver oils were pale yellow in color while farmed Atlantic salmon and wild Pacific salmon had an orange and light orange colours, respectively. The crude marine oils were in a clear liquid state at room temperature except wild pacific salmon oil which was a cloudy liquid with small solid particles in suspension at room temperature.

Specific gravity: Specific gravity is the ratio of the density of the substance to that of water (1 g/cm^3) at 15.6°C. It can be used to determine the purity of oil to match desired standards. Specific gravity of farmed salmon, seal, cod liver and wild salmon oil was 0.921, 0.923, 0.924 and 0.922 g/cm^3, respectively, which were close to the biodiesel standard range of 0.87–0.90 g/cm^3 [34,35]. The higher viscosity of oils tends to cause problems including incomplete combustion and particulate matter emissions when used directly in diesel engines [36,37]. Trans-esterification of oils and fats using short-chain alcohols result in monoesters having viscosities closer to petroleum- based diesel fuel [38,39]. The specific gravity of biodiesel will depend on the type of feed stock, fatty acid composition of the mixed esters and their purity. It is very essential to monitor specific gravity of diesel fuel and correlate against performance indicators, such as cetane number and heating value. However, the specific gravities of hydrocarbons are strongly affected by temperature [40].

Water content: The water content of farmed salmon, seal, cod liver and wild salmon oil was 325, 689, 179 and 312 ppm, respectively. Obtained water content values were within the ASTM biodiesel standards (<500 ppm) except for seal oil. The water content in oil must be within ASTM standards to prevent excessive soap formation during the chemical trans-esterification reaction. The produced soap also increases the viscosity of the reaction mixture, sometimes causing gel formation which can traps the resulting ester and glycerin together thereby making the separation of glycerol from ester difficult [41]. Most vegetable and marine oils are degummed in an extraction plant using

S.No	Properties	Crude Farmed Salmon Oil (Atlantic)	Crude Seal Oil	Crude Cod Liver Oil	Crude Wild Salmon Oil (Pacific)	ASTM specification for Biodiesel
1	Odour	Fresh oil odour, not fishy	Fresh oil odour, not fishy	Fresh oil odour, slightly fishy	Fresh oil odour, slightly fishy	---
2	Color	Orange	Pale yellow	Pale yellow	Pale orange	---
3	Specific gravity	0.921	0.923	0.924	0.922	---
4	Water content (ppm)	325	689	179	312	< 500
5	Physical state at room temperature	Clear liquid	Clear liquid	Clear liquid	Cloudy liquid, small solid particles suspended in otherwise clear oil.	---

Table 1: Physical properties of four crude marine oils.

low amounts of water and the used water is removed subsequently during centrifugation. Most vegetable oils have a clear appearance even though they hold 200-1000 ppm water in the dissolved state. Cloudy or hazy oil/FAME indicates water above 1500 ppm (0.15%) at room temperature. Generally it is recommended that feedstock oil should not contain more than 1500 ppm of water to ensure successful trans-esterification. If it does, a drying step has to be conducted prior to trans-esterification. Free water can be removed by centrifugation or settling out at elevated temperature. Dissolved water in contrast has to be removed by steaming off (i.e. heating to near the boiling point of water) [42,43]. The higher water content in biodiesel can cause corrosion of internal combustion engine components such as pumps, injectors and fuel line tubes and affect heat of combustion which results in greater power consumption [44]. The presence of higher amounts of water also allows microbes to grow during storage and may affect the engine components such as filters and pumps. Formation of gelling and nucleation of oil/biodiesel can take place as water may freeze at low temperatures [45,46].

Chemical properties

Chemical properties of the four crude marine oils including farmed salmon, seal, cod liver and wild salmon are shown in Table 2.

pH: The reported pH values for all four marine oils were neutral in the range of 6.5-6.8. The pH of feedstock oil is a major factor in both base/acid catalyzed and enzyme catalyzed trans esterification reaction, because the catalyst activity may be severely affected at lower or higher pH values. Devanesan et al., [47] studied the effect of pH while producing biodiesel from Jatropha oil using immobilized *Pseudomonas fluorescence* and reported an optimum pH value of 7 for biodiesel production. Yee et al., [48] reported that the activity and stability of the enzyme during enzymatic trans-esterification depends upon the pH of the reaction mixture. Different enzymes have different pH ranges and the optimal pH range not only depends on the nature of the enzyme but also on the substrate, substrate concentration and stability of the enzyme. Kulkarni and Pandit [49] reported an optimum pH of 7.0 for enzymatic trans-esterification of castor oil for biodiesel production.

Ash content: Ash consists of the residue left after the fuel is heated to a sufficiently high temperature when combustible material burns and leaves as CO_2 and H_2O. The ash content of farmed salmon, seal, cod liver and wild salmon oil was 0.0045, 0.0027, 0.0037 and 0.0034%, respectively. The ash content of all four marine oils was much lower than the ASTM biodiesel standards (0.02%). Contaminants including abrasive solids, soluble metallic soaps, inorganic materials and unremoved catalyst present in the fuel may produce ash during combustion that can be abrasive and contribute to wear in fuel injector, fuel pump, piston and ring wear. Sodium and potassium metals are likely to be the main sources for ash in biodiesel. Higher amounts of ash content in the feedstock oil tends to produce rejected quality of biodiesel according to ASTM standards [50,51].

Saponification value: When oil and fat reacts with alkali their long chain fatty acid salts results in soap formation, glycerols and fatty acids. Soaps, which are the salts of longer chain fatty acids, are produced by treating a fat with alkali [52,53]. The saponification value, which is defined as the number of milligrams of potassium hydroxide required to saponify 1 g of fat, is an indicator of the average molecular weight of the triacylglycerols into the fat sample. Dividing the mean molecular weight by 3 gives an approximate mean molecular weight for the fatty acids present in the fat sample [52]. Fat that is composed of short chain fatty acids will have a greater number of fatty acids compared to fats containing long chain fatty acids [54]. Boran et al., [24] studied the effect of temperature and time on saponification value of garfish, golden mullet, shad and horse mackerel oil. The fish oils were stored at 4 and -18°C prior to the analysis. The results from this study indicated that the saponification value of all the fish oils increased at both temperatures. The highest increase in the saponification value was found in fish oil stored at 4°C. During the storage of fish oil, hydrolysis and oxidation causes lipid breakdown forming free fatty acids, aldehydes and ketones. These end products can contributes to an increase in saponification value. The results also indicated that the saponification values of garfish, golden mullet, shad and horse mackerel oil were in the range of 181-195, 176-187, 191-200 and 192-202 mg KOH/g, respectively. Fish oils stored at -18°C had a lower saponification value and twice the longer shelf life than samples stored at 4°C. Garfish and shad oil showed the greatest stability against oxidation and preserved their acceptability properties for at least 150 days at -18°C. In the present study, the saponification value of farmed salmon, seal, cod liver and wild salmon oil was 185.85, 184.61, 179.55, 176.19 mg KOH/g, respectively. These

S. No	Properties	Crude Farmed Salmon Oil (Atlantic)	Crude Seal Oil	Crude Cod liver Oil	Crude Wild Salmon Oil (Pacific)
1	pH	6.8	6.5	6.8	6.7
2	Ash content (%)	0.0045	0.0027	0.0037	0.0034
3	Acid value (mg KOH/g)	2.441	0.958	0.057	0.771
4	Iodine value	116.79	138.67	139.15	138.79
5	Saponification value (mg KOH/g)	185.85	184.61	179.55	176.19
6	p-Anisidine value	3.36	4.21	6.20	9.67
7	Peroxide value (meq/kg)	9.17	7.04	6.92	5.13
8	TOTOX Value	21.69	18.29	20.03	19.92
9	Free fatty acid (%)	1.23	0.48	0.03	0.39
10	Flash point (°C)	267.5	284.0	251.0	232.0
11	Kinematic viscosity (mm^2/s, 40°C)	29.35	26.08	27.46	26.79
12	Refractive index	1.47	1.48	1.48	1.48
13	Cetane number	49.39	44.67	45.39	46.05
14	Higher heating value (kJ/g)	39.00	38.88	38.92	38.90

Table 2: Chemical properties.

values were within the standard range of 176-195 [55]. This indicates a high proportion of fatty acids of low molecular weight. This shows a low tendency towards soap formation and less difficulties in the separation of products if utilized as feedstock for biodiesel production.

Free fatty acids: It is important to determine free fatty acids (FFA) content as it is one of the main criteria for the quality determination of fats and oils. Also, FFA was employed to assess fish deterioration during frozen storage [56] and FFA content increases with increase in storage time. According to Miyashita and Takagi [57], FFA have shown a catalytic effect on lipid oxidation development on soybean, safflower, olive and linseed lipids. Higher amounts of FFA in feedstock oils limits their use for biodiesel production by acid-catalyzed and alkali-catalyzed methods. In the case of biodiesel production by the alkali-catalyzed method, the maximum allowable is below 2.5 wt.% FFA. A pretreatment step becomes necessary if the oil or fat feedstock has a FFA content over 2.5 wt.% [58]. Aryee et al., [59] reported that lipids extracted from fish tissue are highly prone to both lipolysis and oxidation. The presence of more than 0.5% FFA in the oil can cause soap formation and reduce the efficiency of the catalyst during trans-esterification. The author studied free fatty acid formation in stored crude fish oil at -20°C over a period of 120 days. In this study, the free fatty acids (FFA) content in all four oils were relatively low compared to allowable limits and indicates that esterification prior to trans-esterification may not be necessary and a one-step reaction can complete the trans-esterification process and result in a higher yield of biodiesel. However, Math et al., [60] reported that the most effective conversion of triglycerides into esters during the alkaline catalyzed trans-esterification process can take place when the free fatty acid level is less than 1%. The acid value (AV) is approximately 2 mg KOH/g oil at 1% FFA concentration.

Acid value: Oil acidity is an important quality parameter determining the presence of free fatty acid (FFA) and other non-lipid acid compounds [61]. FFA are mostly generated by a hydrolysis reaction of triacylglycerides. Oil acidity depends on several factors such as oil composition, extraction procedures and raw material freshness. The acid value quantifies the amount of acid present in the sample. It is the mass of potassium hydroxide in mg that is required to neutralize 1g of chemical substance [23]. As oil goes rancid, triacylglyceride (TAG) coverts to fatty acid (FA) and glycerol which increases the acid number. Kai et al., [62] reported that the acid value in feedstock oil should be less than 2.5 mg KOH/g oil for the base-catalyzed transesterification process. However, Ding et al., [63] reported that acid value of oil should be less than 1 mg KOH/g oil to meet the alkaline catalyzed trans-esterification conditions. The recommended ASTM standard for acid value of biodiesel is 0.8 mg KOH/g [58]. Freedman et al., [64] obtained maximum ester yield using moisture free alcohol and vegetable oil with an acid value less than 1 mg KOH/g. In the present study, the acid value of farmed salmon, seal, cod liver and wild salmon oil were 2.441, 0.958, 0.057 and 0.771 mg KOH/g, respectively. The acid values of cod liver and wild salmon oils were below the ASTM biodiesel standards (0.8 mg KOH/g) except for farmed salmon and seal oil.

Iodine value: The iodine number gives an indication about the amount of unsaturated fatty compounds (number of double bonds) in the oil and thereby indicates the ease of oxidation or drying capacity of the oil. However, it does not give any information on the nature of the unsaturated and saturated compounds [50]. As the number of double bonds does not change during trans-esterification, measuring the iodine value in feedstock oil will give an indication of the stability of the biodiesel produced from the same feedstock. Europe's EN14214 specification allows a maximum of 120 for the Iodine number (g I_2/100 g) in biodiesel [65]. In this present study, the iodine value of farmed salmon, seal, cod liver and wild salmon oil were 116.79, 138.67, 139.15 and 138.79 g I_2/100 g, respectively. Oils are classified as drying, semi drying and non-drying on the basis of iodine value. Oils with an iodine value above 125 g I_2/100 g are classified as drying oils; those with an iodine value 110–140 g I_2/100 g are classified as semidrying oils and those with iodine value less than 110 g I_2/100 g are considered as non-drying oils [34]. All the characterised marine oils were drying oils except farmed salmon oil which was a semidrying oil. Drying oils are susceptible to become rancid and cannot be preserved for long without the addition of an antioxidant. They can cause the lowering of the pour point of biodiesel produced. All four marine oils were more likely to polymerize in the heat of the engine if used directly without trans-esterification [66].

Peroxide value (PV): The peroxide value is the measurement of the primary oxidation product hydroperoxide and is a widely used chemical test for the determination of fats and oil quality [67]. The oxidative process of oils and fats is one of the main causes of the deterioration of the principal organoleptic and nutritional characteristics of foodstuffs. The number of peroxides present in the oil is an index of their primary oxidative level and its tendency to go rancid. The complex oxidation process can be summarized into two phases: in the first phase, fatty acids react with oxygen and determine odorless compounds as peroxides; during the second phase the peroxides degrade into many substances as volatile aldehydes, responsible for the rancid odor and flavor and in the non-volatile portion. The lower the peroxide value the better the oil quality and its state of preservation [25]. In the present study, the peroxide values of farmed salmon, seal, cod liver and wild salmon oil were 9.17, 7.04, 6.92 and 5.13 meq O_2/kg oil, respectively. Bimbo [68] reported acceptable PV values for fish oil were between 3 and 20. In this study, PVs of all examined fish oil samples did not exceed 20 meq O_2/kg oil. However, according to Boran [24], the acceptability limit for PV of crude fish oil is 7–8 meq O_2/kg oil and all examined oil samples did not exceed this limit and showed oxidative stability except farmed salmon oil. If peroxide values are higher, oils will not remain stable and become rancid easily and should be utilized immediately after extraction. Fritsche and Johnston [69] reported that peroxide value of the purified diet containing menhaden oil without antioxidant had elevated 5-6 folds within 24 h and 12 fold within 48 h when exposed to air at room temperature.

***p*-Anisidine value (AV):** The *p*-anisidine value is used to measure the secondary products of oxidation and determines the aldehyde in the lipid, primarily 2-alkene present in the fat. Aldehyde present in the oil and the p-anisidine reagent react under acidic condition [25,100]. The color obtained not only depends on the aldehydes present, but also their structure. Further degradation of lipids generates off-flavours and off-odours. On the contrary, other tests consider the volatile portion of aldehydes and, due to their intrinsic variable nature, result in less reliable data [22]. In the present study, the *p*-Anisidine values of farmed salmon, seal, cod liver and wild salmon oil were 3.36, 4.21, 6.20 and 9.67, respectively. All studied marine oils *p*-Anisidine values are within the recommended range of ≤20 for crude fish oil [71]. Pak [72] studied the stability and quality of fish oil and found the p-anisidine value to be 19.8. Che Man and Setiowaty [73] studied the *p*-anisidine value of oxidized palm olein using FTIR and reported the p-anisidine value of 8.684.

The formation of primary and secondary degradation products can be prevented by adding a stabilizing agent (antioxidant) immediately after oil extraction. However, it is very much essential to perform peroxide and p-anisidine value tests prior to the alkaline catalyzed trans-esterification process for biodiesel production.

Totox value: The p-anisidine value is often used in conjunction with the peroxide value to calculate the total oxidation value or Totox value. The Totox value is calculated by the formula AV + 2PV to determine the overall oxidation state of the oil. The peroxide value can decrease over time so p-anisidine value and/or Totox calculation are essential to get an understanding of total oxidation. FAO and WHO [71] recommended Totox value is ≤ 26 for fish oil. In the present study, Totox values of all the examined four marine oils were within the recommended range.

Flash point: The flash point is the lowest temperature at which a product of petroleum gives off sufficient flammable vapours to ignite or momentarily flash. The value of the flash point is used for the classification of flammable and combustible materials needed for safety and shipping regulations. The flash point is determined by heating a sample of the fuel in a stirred container and passing a flame over the surface of the liquid. If the temperature is at or above the flash point, the vapor will ignite and an easily detectable flash can be observed [74]. In this study, the flash points of farmed salmon, seal, cod liver and wild salmon oils were 267.5, 284.0, 251.0 and 232.0°C, respectively. The flash point of all four marine oil samples were above the recommended ASTM D6751 value of 130°C and therefore pose no risk of fire outbreaks in case of accidents. They were all also above the 120°C minimum EN 14214 (European) recommended range. It is very essential to measure the alcohol content and flash point after each batch of biodiesel produced as very small quantities of residual alcohol present in biodiesel can significantly decrease the flash point [75].

Kinematic viscosity [cSt]: Viscosity is a measure of a fluid's resistance to flow. The greater the viscosity, the less readily the liquid flows. It is one of the most important parameters required in the design of combustion processes. A direct relationship exists between the oil viscosity and some chemical characteristics of the lipids including the degree of unsaturation and chain length of the fatty acids that constitute the triacylglycerols. Viscosity slightly decreases with increased degree of unsaturation and rapidly increases with polymerisation [76,77]. Abromovič and Klofutar [76] reported that viscosity linearly decreases as the iodine value increases and there was no dependence of viscosity on saponification value, which characterises the chain length of fatty acids. Gunstone [78] reported that the viscosity of oil is highly temperature dependent and each oil has a specific viscosity curve, which depends on its chemical composition and presence and quantity of impurities. Dubey et al., [79] found the reduction (7.4 times) in viscosity while the temperature was raised from 10-70°C and reported a viscosity of 47.3cSt at 20°C. In the present study, all crude marine oils showed a viscosity between 26-30 cSt at 40°C. The viscosity of biodiesel is highly dependent on the type of feedstock oil. The range for biodiesel is between 1.9 and 6.0 cSt as specified by ASTM standard D D6751-12. The major concern about biodiesel is that its viscosity tends to be higher than diesel fuel. Biodiesel consisting of the methyl esters of soybean oil has a viscosity between 4.2 and 4.6 cSt [80] while ethyl esters of rapeseed oil, have been reported to be as high as 6.0 cSt [81].

Refractive index: The refractive index is used to measure the increase in autoxidation of fats and oils. The refractive index of oils depends on their molecular weight, fatty acid chain length, degree of unsaturation, and degree of conjugation. The refractive index of an oil increases (nonlinearly) with chain length and unsaturation of fatty acids. The refractive indices and peroxide values of the oils and fats can significantly increase while exposed to light and heat [82]. Generally oils' refractive index values vary between 1.447 and 1.482. The degree of unsaturation of an oil greatly influences its oxidative stability and the stability of the biodiesel derived from it [83]. The fatty acid analysis shows a high degree of unsaturation in all four marine oils tested in this study. Hence, the stabilization of the oil as well as the biodiesel immediately after extraction/production is crucial for obtaining a high quality biofuel. Sadrolhosseini et al., [84] reported that the state of biodiesel can be evaluated based on change of refractive index. Biodiesel turns into cloudy state when the biodiesel temperature is near to the cloud point. In this study, refractive index of farmed salmon, seal, cod liver and wild salmon oils were 1.47, 1.48, 1.48 and 1.48, respectively.

Cetane number: The cetane number is used to measure the combustion quality of biodiesel and diesel fuels during compression ignition. In this study, the calculated cetane numbers of farmed salmon, seal, cod liver and wild salmon oils were 49.39, 44.67, 45.39 and 46.05, respectively. Except for farmed salmon oil, these values are lower than the ASTM D6751-12 recommended cetane number minimum of 47. Bello et al., [85] found the cetane number of cashew nut, egusi melon and rubber seed oils increased 10-40% after trans-esterification and the increase depended on the degree of unsaturation. Longer fatty acid carbon chains and more saturated molecules leads to higher cetane numbers [86]. It is expected that trans-esterification of the studied crude marine oils would result in an increase in the cetane numbers.

Higher heating value: Higher Heating Value is an important property defining the thermal energy released per unit quantity of fuel after complete combustion of fuel to CO_2 and H_2O at the initial temperature and pressure. It is an important property that describes the energy content of fuels, such as oils and biodiesel. HHV can be determined by use of a bomb calorimeter or calculated from fuel characteristics and physical properties. Demirbas [31] found high regression between viscosity and HHV of vegetable oil and biodiesel and suggested that the correlations may be used to estimate the HHV of mixtures of biodiesels obtained from vegetable oils. Sardine oil has a HHV of 39.6, whereas, sardine oil biodiesel has a HHV of 39.7 [87]. In this study, the calculated higher heating values of farmed salmon, seal, cod liver and wild salmon oils were 39.00, 38.88, 38.92 and 38.90, respectively. These values are lower than the ASTM D6751 recommended higher heating value minimum of 39.2.

Lipid classification

Lipid classes of the farmed salmon, seal, cod liver and wild salmon oil are presented in Table 3. Among the lipid classes, triacylglycerols, free fatty acids, phospholipids, sterols and polar lipids were the predominant constituents, while hydrocarbons were minor components in all the marine oils studied. Hydrocarbons, sterol esters and triacylglycerol were the major components of the neutral lipids in the four marine oils. The presence of a higher percentage (≥ 80%) of triacylglycerols was found in all four marine oils including farmed salmon (92.83%), seal (93%), cod liver (85.26%) and wild salmon (81.20%); which is essential for the trans-esterification of feedstock oil and higher biodiesel production yield. Hydrocarbon content was absent in the farmed and wild salmon oil, while 0.11 and 0.33% was present in the crude seal and cod liver oil, respectively. Farmed salmon, seal, cod liver and wild salmon oil also contained 3.74, 2.33, 3.24 and 2.02% of sterol esters, respectively. Cod liver (14.72%) and wild salmon oil (9.94%) were rich in polar lipids while the farmed salmon (2.43%) and wild salmon (2.44%) were low in polar lipids. The phospholipids present in the farmed salmon, seal, cod liver and wild salmon oil were 1.43, 1.53,

S. NO.	Parameter	Crude Farmed Salmon Oil (Atlantic)	Crude Seal Oil	Crude Cod liver Oil	Crude Wild Salmon Oil (Pacific)
1	Hydrocarbons	0.00	0.11	0.33	0.00
2	Steryl Esters/Wax Esters	0.00	0.12	0.00	0.00
3	Ethyl Esters	0.00	0.00	0.00	0.00
4	Methyl Esters	0.00	0.00	0.00	0.00
5	Ethyl Ketones	0.00	0.00	0.00	0.00
6	Methyl Ketones	0.00	0.00	0.00	0.00
7	Glycerol Ethers	0.00	0.00	0.00	0.00
8	Triacylglycerols	92.83	93	85.26	81.20
9	Free Fatty Acids	1.23	0.48	0.03	0.39 permanent pink
10	Alcohols	0.00	0.00	0.00	0.00
11	Sterols	3.74	2.33	3.24	2.02
12	Diacylglycerols	0.00	0.00	0.00	0.00
13	Acetone Mobile Polar Lipids	2.43	9.94	14.72	2.44
14	Phospholipids	1.43	1.51	1.23	1.67

Table 3: Composition (wt%) of marine oils lipid classes.

1.21 and 1.67%, respectively which are higher than the ASTM D6751-12 recommended limit of ≤10 ppm. Phospholipids (gums) in oils refer to hydratable and non-hydratable phosphatides, lecithin, sugars, trace metals and other impurities [88,89]. For biodiesel producers, gums are a concern for the following reasons: (a) inhibit the catalyst during the trans-esterification reaction (b) difficult to separate biodiesel and glycerol due to the emulsifying effect that occurs after the trans-esterification reaction (c) metals contamination in the final product (P and Ca specifically), and (d) yield loss. [43,64,90]. In the present study, all the characterized marine oils contain higher amounts of phospholipids and will require a degumming process prior to biodiesel production. Liu et al., [91] studied the removal of phospholipids from Jatropha oil through a conventional degumming process combined with ultrafiltration membrane separation in a small-scale batch system and reported that the phospholipid content of jatropha oil was reduced from 1200 ppm to 60 ppm and was further reduced to less than 20 ppm by subjecting the oil to ultrafiltration membrane separation. Fan et al., [43] reported a 99.7% reduction of phosphorus content in the crude soybean oil after acid and water degumming. A relatively low free fatty acid content in all four marine oils was observed compared to allowable limits of <2.5%. Math et al., [60] reported that the most effective conversion of triglycerides into esters during alkaline catalyzed trans-esterification process can take place when the free fatty acid level is less than 1% which indicates that esterification prior to trans-esterification may be necessary for Atlantic farmed salmon oil. El-Mashad et al., [92] concluded that only alkaline-catalysed trans-esterification was not an effective method for producing biodiesel from salmon oil and implemented a two-step process including a sulphuric acid-catalysed pretreatment followed by alkaline-catalysed transesterification and observed a reduction of the acid value from 12.0 to 3 mg KOH/g oil in the first step of acid catalysed pretreatment.

Fatty acid composition

There were about 78 fatty acids identified in each examined marine oil using the gas chromatography analysis however, only 12 fatty acids had values >0.5%. Fatty acid composition including saturated, monounsaturated and polyunsaturated fatty acids of marine oils are shown in Table 4. Farmed salmon, seal, cod liver and wild salmon oil have consisted of 19.83, 13.53, 16.05, and 22.52% of saturated fatty acids including myristic acid (14:0), palmitic Acid (16:0), stearic acid (18:0), respectively. Farmed salmon, seal, cod liver and wild salmon oil also contains 47.38, 58.88, 53.79 and 44.97% of monounsaturated fatty acids including palmitoleic acid (16:1n-7), oleic acid (18:1n-9), vaccenic acid (18:1n-7), respectively. The total polyunsaturated fatty acids including linoleic acid (18:2n-6), alpha-linolenic acid (18:3n-3), arachidonic (20:4n-6), EPA (20:5n-3), DPA (22:5n-3) and DHA (22:6n-3) were also present in the farmed salmon, seal, cod liver and wild salmon oil comprising of 32.23, 26.64, 29.05 and 31.42%, respectively. Biodiesel produced from saturated fat have a higher cetane number and improve oxidative stability however, they also have higher cloud point and poor temperature properties which can lead to gel formation at ambient temperature. If more polyunsaturated fatty acids are present in the feedstock oil then a reduced cloud point, cetane number and stability can be observed in the biodiesel. In the present study, all four studied marine oils have higher polyunsaturated fatty acids so biodiesel produced will have less oxidative stability and may results in the precipitation of biodiesel components in a fuel feeding system or combustion chamber [93-95]

Conclusion

In the present study, farmed salmon, seal, cod liver and wild salmon marine oils were characterized to identify their physical and chemical properties and determine their feasibility as a feedstock for biodiesel production. All oils were pale yellow to orange in color and were stable in the liquid state at room temperature. The specific gravity of all oils was significantly closer to the recommended parameters. The water content of all oils were within the recommended limit (<500 ppm) except for seal oil (689 ppm). A drying step has to be implemented to remove the water present in the feedstock seal oil as water in biodiesel can cause corrosion of internal combustion engine components and results in greater power consumption. The pH and ash content values of all oils were in the recommended range. All oils studied were comprised mainly of short chain fatty acids due to their high saponification values. The free fatty acids (FFA) of all oils were relatively low compared to the allowable limits of 2.5%. The acid values of all oils were within the ASTM biodiesel standards (0.8 mg KOH/g) except for farmed salmon and seal oil. All the characterised marine oils were identified as drying oils except farmed salmon oil which was a semidrying oil because of the higher iodine value (116.79-139.15 g I2/100 g) and are susceptible to become rancid which causes reduction of pour point of biodiesel produced in the absence of antioxidant. All four marine oils were more likely to polymerize in the heat of the engine if used directly without trans-esterification. The peroxide value and p-anisidine value of all oils were in the recommended range. The flash point of all marine oils

Type of Fatty Acid	Crude Farmed Salmon Oil (Atlantic)	Crude Seal Oil	Crude Cod liver Oil	Crude Wild Salmon Oil (Pacific)
Saturated Fatty Acid				
Myristic acid (14:0)	3.20	4.23	3.11	4.52
Palmitic Acid (16:0)	12.79	7.69	9.60	13.88
Stearic Acid (18:0)	3.29	1.02	2.23	2.54
Other saturated fatty acids	0.55	0.59	1.10	1.59
Subtotal	**19.83**	**13.53**	**16.05**	**22.52**
Monounsaturated Fatty Acid				
Palmitoleic Acid (16:1n-7)	7.37	16.11	8.17	5.32
Oleic Acid (18:1n-9)	31.19	21.20	17.48	14.06
Vaccenic acid (18:1n-7)	3.44	4.76	5.36	3.15
Other monounsaturated fatty acids	5.38	16.81	22.78	22.44
Subtotal	**47.38**	**58.88**	**53.79**	**44.97**
Polyunsaturated Fatty Acid				
Linoleic acid (18:2n-6)	14.50	1.54	1.91	1.63
Alpha-linolenic acid (18:3n-3)	1.69	0.61	0.78	0.97
Arachidonic (20:4n-6)	0.61	0.49	0.38	0.63
EPA (20:5n-3)	4.63	7.12	8.52	9.54
DPA (22:5n-3)	0.13	0.09	0.11	0.12
DHA (22:6n-3)	3.48	8.45	11.36	10.55
Other polyunsaturated fatty acids	7.20	8.33	5.99	7.97
Subtotal	**32.23**	**26.64**	**29.05**	**31.42**

Table 4: Fatty acid composition (wt%).

were well above 200°C and therefore poses no risk of fire outbreaks in case of accidents. Slightly lower in cetane number was found after calculation in all marine oils except farmed salmon oil. However, they would increase after oil to biodiesel conversion.

All the marine oils have significant potential for use as feedstock due to higher (81%) triacylglycerol. A degumming process prior to biodiesel production will be required as the phospholipids (1.21-1.67%) were higher than the recommended limit of ≤10 ppm. All four marine oils have saturated (13.53-22.52%), monounsaturated (44.97-58.88%) and polyunsaturated (26.64-32.23%) fatty acids. All the studied marine oils have higher polyunsaturated fatty acids so biodiesel produced will have less oxidative stability and may result in the precipitation of biodiesel components in a fuel feeding system or combustion chamber. Therefore, it is crucial to stabilize oils and biodiesel immediately after extraction/production to obtain a high quality biofuel.

Acknowledgement

Financial support under the Harris Centre–MMSB Waste Management Applied Research Fund is gratefully acknowledged. We are also grateful to our supporting partners including the Newfoundland Aquaculture Industry Association (NAIA) and Cook Aquaculture.

References

1. Fukuda H, Kondo A, Noda H (2001) Biodiesel fuel production by transesterification of oils. Journal of Bioscience and Bioengineering 92: 405-416.
2. Akoh CC, Chang S, Lee G, Shaw J (2007) Enzymatic approach to biodiesel production. Journal of Agricultural and Food Chemistry 55: 8995-9005.
3. Fukuda H, Hama S, Tamalampudi S, Noda H (2008) Whole-cell biocatalysts for biodiesel fuel production. Trends in Biotechnology 26: 668-673.
4. Shafiee S, Topal E (2009) When will fossil fuel reserves be diminished? Energy Policy 37: 181-189.
5. Demirbas A (2007) Progress and recent trends in biofuels. Prog Energy Combustion Sci 33: 1-18.
6. Basha SA, Raja Gopal K, Jebaraj S (2009) A review on biodiesel production, combustion, emissions and performance. Renewable and Sustainable Energy Reviews 13: 1628-1634.
7. Hinton D, Masterson N, Lippert A, King R (1999) Petroleum an energy profile. EIA Publication, Pittsburgh, PA, USA.
8. Srivastava A, Prasad R (2000) Triglycerides-based diesel fuels. Renewable and Sustainable Energy Reviews 4: 111-133.
9. Murugesan A, Umarani C, Chinnusamy TR, Krishnan M, Subramanian R, et al., (2009) Production and analysis of bio-diesel from non-edible oils-A review. Renewable and Sustainable Energy Reviews 13: 825-834.
10. Gui MM, Lee KT, Bhatia S (2008) Feasibility of edible vs. non-edible oil vs. waste edible oil as biodiesel feedstock. Energy 32: 1646-1653.
11. Jegannathan KR, Abang S, Poncelet D, Chan ES, Ravindra P (2008) Production of biodiesel using immobilized lipase-A critical review. Critical Reviews in Biotechnology 28: 253-264.
12. McNeff CV, McNeff LC, Yan B, Nowlan DT, Rasmussen M, et al., (2008) A continuous catalytic system for biodiesel production. Applied Catalysis A: General 343: 39-48.
13. Ranganathan SV, Narasimhan SL, Muthukumar K (2008) An overview of enzymatic production of biodiesel. Bioresource Technology 99: 3975-3981.
14. http: //www.dfo-mpo.gc.ca/stats/commercial/land-debarq-eng.htm
15. http: //www.fishaq.gov.nl.ca/stats/aquaculture_2011-2012%20factsheet.pdf
16. http: //www.fishaq.gov.nl.ca/licensing/processing.html
17. http: //www.fishaq.gov.nl.ca/sealing/index.html
18. http: //www.fishaq.gov.nl.ca/stats/landings/landings_2012_prel.pdf
19. http://www.dfo-mpo.gc.ca/seal-phoque/reports-rapports/mgtplan-plangest0610/mgtplan-plangest0610_e.htm#top.
20. Manuel H, Buckle K, Onodenlaore C (2006) Inventory and assessment of marine sources of feedstock for biodiesel in Newfoundland and Labrador. Centre for Aquaculture and Seafood Development, Canada.
21. http: //lipidlibrary.aocs.org/processing/marine/index.htm
22. St. Angelo AJ (1996) Lipid oxidation in foods. Critical reviews Food Science and Nutrition 36: 175-224.
23. Wrolstad RE, Decker EA, Schwartz SJ, Sporns P (2004) Handbook of food analytical chemistry, water, proteins, enzymes, lipids, and carbohydrates.
24. Boran G, Karacam H, Boran M (2006) Changes in the quality of fish oils due to storage temperature and time. Food Chemistry 98: 693-698.
25. O'Brien RD (2009) Fats and oils: Formulating and processing for applications.

2nd edtn, CRC press, London, UK.

26. Sandy Brae Laboratories (2009) Water test kit instructions (Version 1-4-02). Newark, Delaware, USA.

27. AOAC Official Method 993.20 (2000) Iodine value of fats and oils, Wijs (cyclohexane-acetic acid solvent) method. American Oil Chemists Society, Champaign, Illinois, USA.

28. AOCS Official Method Cd 8-53 (2003) Peroxides in fats and oils. American Oil Chemists Society, Champaign, Illinois, USA.

29. AOCS Official Method Cd 18-90 (1997) p-Anisidine value. Sampling and analysis of commercial fats and oils. American Oil Chemists Society, Champaign, Illinois, USA.

30. Bose PK (2009) Empirical approach for predicting the cetane number of biodiesel. International Journal of Automotive Technology 10: 421-429.

31. Demirbas A (2008) Relationships derived from physical properties of vegetable oil and biodiesel fuels. Fuel 87: 1743-1748.

32. Hooper T, Parrish C (2009) Profiling neutral lipids in individual fish larvae by using short column gas chromatography with flame ionization detection. Limnology and Oceanography: Methods 7: 411-418.

33. Parrish CC (1987) Separation of aquatic lipid classes by Chromarod thin-layer chromatography with measurement by latroscan flame ionization detection. Canadian Journal of Fisheries and Aquatic Sciences 44: 722-731.

34. Ibeto CN, Okoye COB, Ofoefule AU (2012) Comparative study of the physicochemical characterization of some oils as potential feedstock for biodiesel production. ISRN Renewable Energy.

35. Kywe TT, Oo MM (2009) Production of biodiesel from Jatropha oil (Jatropha curcas) in pilot plant. World Academy of Science, Engineering and Technology 3: 387-393.

36. Fort EF, Blumberg PN (1982) Performance and durability of a turbocharged diesel fueled with cottonseed oil blends. International Conference on Plant and Vegetable Oils as Fuel, Fargo, ND, USA.

37. Ziejewski M, Kaufman K, Pratt G, Walter J (1993) "Flower Power" Tractor Fuel Testing Program 1981-1987. SAE Technical Paper: 932476.

38. Graboski MS, McCormick RL (1998) Combustion of fat and vegetable oil derived fuels in diesel engines. Progress in Energy and Combustion Science 24: 125-164.

39. Noureddini H, Teoh BC, Clements DL (1992) Viscosities of vegetable oils and fatty acids. Journal of American Oil Chemists Society 69: 1189-1191.

40. Tat ME, Van Gerpen JH (2000) The specific gravity of biodiesel and its blends with diesel fuel. Journal of the American Oil Chemists' Society 77: 115-119.

41. Wright HJ, Segur JB, Clark HV, Coburn SK, Langdon EE, et al., (1944) A report on ester interchange. Oil Soap 21: 145-148.

42. Djikstra AJ (1998) Degumming Revisited. In: Oléagineux, Corps Gras, Lipides. Dossier : Innovations technologiques 5: 367-370.

43. Fan X, Burton R, Austic G (2010) Conversion of degummed soybean oil to biodiesel: Optimization of degumming methods and evaluation of fuel properties. International Journal of Green Energy 7: 594-599.

44. Jones DA (1996) Principles and prevention of corrosion. 2nd edtn, Prentice Hall, Michigan, USA.

45. Young RG, Lovell PA (1991) Introduction to Polymers. 2nd edtn, CRC Press, New York, USA.

46. Mendez-Villuendas E, Bowles R (2007) Surface nucleation in the freezing of gold nanoparticles. Physical Review Letters 98: 185503

47. Devanesan MG, Viruthagiri T, Sugumar N (2007) Transesterification of Jatropha oil using immobilized Pseudomonas fluorescens. African Journal of Biotechnology 6: 2497-2501.

48. Yee LN, Akoh CC, Philips RS (1997) Lipase PS-catalyzed transesterification of citronellyl butyrate and geranyl caproate: effect of reaction parameters. Journal of American Oil Chemists Society 74: 255-260.

49. Kulkarni SR, Pandit AB (2005) Enzymatic hydrolysis of castor oil: An approach for rate enhancement and enzyme recovery. Indian Journal of Biotechnology 4: 241-245.

50. Van Gerpen J, Shanks B, Pruszko R, Clements D, Knothe G (2004) Biodiesel Production Technology. National Renewable Energy Laboratory, Colorado, USA.

51. Ferrari RA, Leticia A, Pighinelli MT, Park KJ (2011) Biodiesel Production and Quality.

52. Nielsen SS (1998) Food analysis. 2nd Edtn, Aspen Publishers, Gaithersburg, Maryland, USA.

53. Ockerman HW, Hansen CL (2000) Animal by-product processing and utilization. CRC Press, London, UK.

54. Potter N, Hotchkiss J (1995) Food science. 5th Edtn, Chapman & Hall, New York, NY, USA.

55. Abdulkadir M, Abubakar GI, Mohammed A (2010) Production and characterization of oil from fishes. ARPN Journal of Engineering and Applied Sciences 5: 1-5.

56. de Koning AJ, Mol TH (1991) Quantitative quality tests for frozen fish: Soluble protein and free fatty acid content as quality criteria for hake (Merluccius capensis) stored at −18°C. Journal of the Science of Food and Agriculture 54: 449-458.

57. Miyashita K, Takagi T (1986) Study on the oxidative rate and prooxidant activity of free fatty acids. Journal of the American Oil Chemists Society 63: 1380-1384.

58. Leung DYC, Wu X, Leung MKH (2010) A review on biodiesel production using catalyzed transesterification. Applied Energy 87: 1083-1095.

59. Aryee ANA, van de Voort FR, Simpson BK (2009) FTIR determination of free fatty acids in fish oils intended for biodiesel production. Process Biochemistry 44: 401-405.

60. Math MC, Kumar SP, Chetty SV (2010) Technologies for biodiesel production from used cooking oil-A review. Energy for Sustainable Development 14: 339-345.

61. Rubio-Rodriguez N, de Diego SM, Beltran S, Jaime I, Sanz MT, et al., (2008) Supercritical fluid extraction of the omega-3 rich oil contained in hake (Merluccius capensis-Merluccius paradoxus) by-products: Study of the influence of process parameters on the extraction yield and oil quality. The Journal of Supercritical Fluids 47: 215-226.

62. Kai T, Kubo A, Nakazato T, Takanashi H, Uemura Y (2012) Influence of the acid value on biodiesel fuel production using a two-step batch process with a homogeneous catalyst. International Journal of Biomass & Renewables 1: 15-20.

63. Ding J, Xia Z, Lu J (2012) Esterification and deacidification of a waste cooking oil (TAN 68.81 mg KOH/g) for biodiesel production. Energies 5: 2683-2691.

64. Freedman B, Pryde EH, Mounts TL (1984) Variables affecting the yields of fatty esters from transesterified vegetable oils. Journal of American Oil Chemists Society 61: 1638-1643.

65. Son SM, Kusakabe K, Guan G (2010) Biodiesel synthesis and properties from sunflower and waste cooking oils using CaO catalyst under reflux conditions. Journal of applied Sciences 10: 3191-3198.

66. Sulistyo H, Suardjaja IM, Rahayu SS (2006) Transesterification of candlenut oil with ethanol to biodiesel. Regional Symposium on Chemical Engineering, Singapore.

67. Aidos I, Luten JB, Boom RM, Padt AV (2001) Upgrading of Maatjes herring by-products. Production of crude fish oil. Journal of Agricultural Food Chemistry 49: 3697-3704.

68. Bimbo PA (1998) Guidelines for characterizing food grade fish oil: International Fish meal and Oil Manufacturers Association. Research report 9: 473-483.

69. Fritsche KL, Johnston PV (1988) Rapid autoxidation of fish oil in diets without added antioxidants. Journal of Nutrition 118: 425-426.

70. Dieffenbacher A, Pocklington WD (1987) Standard methods for the analysis of oils, fats and derivatives. Oxford, Blackwell, UK.

71. FAO, WHO (2013) Codex Alimentarius Commission joint FAO/WHO food standards programme codex committee on fats and oils Twenty-third Session, Langkawi, Malaysia.

72. Pak CS (2005) Stability and quality of fish oil during typical domestic application. The United Nations university, Fisheries Training Programme, Iceland.

73. Che Man YB, Setiowathy G (1999) Determination of anisidine value in thermally

oxidized palm olein by fourier transform infrared spectroscopy. Journal of the American Oil Chemists Society 76: 243-247.

74. Gmehling J, Rasmussen P (1982) Flash points of flammable liquid mixtures using UNIFAC. Industrial & Engineering Chemistry Fundamentals 21: 186-188.

75. Faber Boog JH, Silveira ELC, de Caland LB, Tubino M (2011) Determining the residual alcohol in biodiesel through its flash point. Fuel 90: 905-907.

76. Abromovic H, Klofutar C (1998) The temperature dependence of dynamic viscosity for some vegetable oils. Acta Chimica Slovenica 45: 69.

77. Stanciu I (2011) Modeling the temperature dependence of dynamic viscosity for rapeseed oil. Journal of Science and Arts 1: 55-58.

78. Gunstone F (2008) Oils and Fats in the Food Industry. John Wiley & Sons Ltd., West Sussex, United Kingdom.

79. Dubey AK, Sarviya, Rehman A (2011) Characterization of Processed Jatropha Oil for use as Engine Fuel. Current World Environment 6: 101-107.

80. Tat ME, Van Gerpen JH (1999) The Kinematic Viscosity of Biodiesel and Its Blends with Diesel Fuel. Journal of the American Oil Chemists' Society 76: 1511-1513.

81. Peterson CL, Wagner GL, Auld DL (1983) Vegetable oil substitutes for diesel fuel. Trans. ASAE 26: 322-332.

82. Arya SS, Ramanujam S, Vijayaraghavan PK (1969) Refractive index as an objective method for evaluation of rancidity in edible oils and fats. The Journal of the American Oil Chemists Society 48: 28-30.

83. Shahidi F (2005) Quality Assurance of Fats and Oils. John Wiley & Sons, West Sussex, United Kingdom.

84. Sadrolhosseini AR, Moksin MM, Nang HLL, Norozi M, Yunus MMW, et al., (2011) Physical properties of normal grade biodiesel and winter grade biodiesel. International Journal of Molecular Sciences 12: 2100-2111.

85. Bello EI, Out F, Osasona A (2012) Cetane number of three vegetable oils, their biodiesels and blends with diesel fuel. Journal of Petroleum Technology and Alternative Fuels 3: 52-57.

86. http: //www.biodiesel.org/reports/19960901_gen-187.pdf

87. Vilela L, Mata T, Caetano NS (2010) Biodiesel Production from Fish Oil with High Acidity. International Symposium on Energy from Biomass and Waste, Venice, Italy.

88. McDonnell KP, Ward SM, Timoney DJ (1995) Hot water degummed rapeseed oil as a fuel for diesel engines. Journal of Agricultural Engineering Research 60: 7-14.

89. Indira TN, Hemavathy J, Khatoon S, Gopala Krishna AG, Bhattacharya S (2000) Water degumming of rice bran oil: a response surface approach. Journal of Food Engineering 43: 83-90.

90. Xiaohu F, Burton R, Austic G (2010) Conversion of degummed soybean oil to biodiesel: optimization of degumming methods and evaluation of fuel properties. International Journal of Green Energy 7: 593-599.

91. Liu KT, Gaoa S, Chunga TW, Huang CM, Lin YS (2012) Effect of process conditions on the removal of phospholipids from Jatropha curcas oil during the degumming process. Chemical Engineering Research and Design 90: 1381-1386.

92. El-Mashad HM, Zhang R, Avena-Bustillos RJ (2008) A two-step process for biodiesel production from salmon oil. Biosystems Engineering 99: 220-227.

93. Tanwar D, Tanwar A, Sharma D, Mathur YP, Khatri KK, et al., (2013) Production and Characterization of Fish Oil Methyl Ester. International Journal of Innovative Technology and Research 1: 209-217.

94. Pinyaphong P, Sriburi P, Phutrakul S (2011) Biodiesel Fuel Production by Methanolysis of Fish Oil Derived from the Discarded Parts of Fish Catalyzed by Carica papaya Lipase. World Academy of Science, Engineering and Technology52: 466-472.

95. Ramadhas AS, Jayaraj S, Muraleedharan C (2005) Biodiesel production from high FFA rubber seed oil. Fuel 84: 335-340.

96. 96. ASTM D5558-95 (2011) Standard test method for determination of the saponification value of fats and oils. ASTM International, West Conshohocken, Pennsylvania, USA.

97. AOCS Official Method Ca 5a-40 (1997) Free Fatty Acids. American Oil Chemists Society, Champaign, Illinois, USA.

98. ASTM D93-12 (2012) Standard test methods for flash point by Pensky-Martens closed cup tester. ASTM International, West Conshohocken, Pennsylvania, USA.

99. Hooper T, Parrish C (2009) Profiling neutral lipids in individual fish larvae by using short column gas chromatography with flame ionization detection. Limnology and Oceanography: Methods 7: 411-418.

100. IUPAC (1987) Standard Method, Preparation of fatty acid methyl ester, In: Standard methods for the analysis of oils, fats and derivatives. Oxford, Blackwell, UK.

Study on Interactions between Fed-Batch and Batch Operating Parameters for the Development of Monoclonal Antibody Fed-Batch using Design of Experiments

Junaid Muneer Raja[1,2], Nurina Anuar[1*], Badarulhisam Abdul Rahman[2] and Jamaliah Md Jahim[1]

[1]*Department of Chemical and Process Engineering, Faculty of Engineering and Built Environment, University Kebangsaan Malaysia, 43600 UKM, Bangi, Selangor, Malaysia*

[2]*InnoBiologics Sdn. Bhd., Lot 1, Persiaran Negeri, 71800 Nilai, Negeri Sembilan, Malaysia*

Abstract

In conventional fed batch process development approaches, batch operating parameters (such as pH, temperature, seeding density, dissolved oxygen concentration) are kept constant and only feeding parameters such as feeding time, post-feed concentration are manipulated. The batch and fed batch operating parameters are assumed to be independent of each other. This approach to process development ignores any interactions that might exist between the batch and fed-batch operating parameters are therefore not evaluated. However in a complex bioprocess, none of the factors affecting the process can be assumed to be independent of each other and mutually exclusive. Therefore in this study an attempt was made to study the interaction between fed-batch operating parameter-post feed glucose concentration (A) and batch operating parameters- seeding density (B), temperature (C), and dissolved oxygen concentration (D) by their simultaneous manipulation, as well as the effect of these interactions on cell growth and monoclonal antibody (mAb) production. NS0 cell line producing the mAb's against carcino-embryogenic antigen (Anti-CEA) was used. The final mAb concentration, viable cell density and integral of viable cell concentration (IVCC) were the responses evaluated. Statistical analysis experimental data showed that parameter (A) and its interaction with parameter (B) were the main factors affecting both the response variables. In comparison to the batch run which yielded 5.21 mg/L mAb, the developed fed batch process increased the mAb titer by 10 fold (59.40 mg/mL), and the IVCC was increased by 7 fold. The maximum VCD value (3.46×10^6 cells/mL) of the developed fed batch process was 1.25 over fold the value for batch.

Keywords: Fed-batch; Design of experiments; Monoclonal antibody; Anti-CEA; NS0

Introduction

Following the pioneering development of hybridoma technology by Kohler and Milstein in 1975, the usage of mAbs as therapeutic and diagnostic agents has seen a rapid increase during the past decade. There are several approved mAb based therapeutic and diagnostic agents in the market today. The prime advantage of mAbs as therapeutic and diagnostic agents is their high target specificity, resulting in a low side effect profile. Due to this high therapeutic potential, the process development and production of monoclonal antibodies is being pursued by biopharmaceutical companies globally. Although in the beginning, monoclonal antibodies were exclusively produced in hybridoma cells, currently various cell lines are employed for their production of which NS0 and CHO cell lines are the most common, others including murine hybridomas, hamster BHK21 and the human PER.C6 cell line. However, for the choice of the appropriate cell line, certain key criteria have to be considered which include: the ability to provide a close to human glycosylation pattern, the capability to produce high mAb concentrations in the chosen production system, the ability to consistently produce a product of uniform characteristics (stability), and the speed with which a high yielding cell line can be obtained [1,2].

NS0, a murine myeloma cell line, has been adopted by a number of biotechnology companies for the expression of therapeutic antibodies [3-7]. The successful use of NS0 cells in several fusion, transfection, selection and production approaches are among the few properties of this cell line that make it a perfect candidate for the expression of the desired product [8]. However, NS0 cells are known to be susceptible to apoptosis especially under environmental and/or nutritional stress [9]. Lack of Hsp70 expression potential and variability in the membrane cholesterol concentration are among the several factors that are responsible of this particular susceptibility of these cells [10].

A common method for the industrial manufacture of cell culture based recombinant therapeutics is the fed-batch culture. Fed batch cultures have been developed with the objective of achieving maximal increase in the culture viable cell concentrations and prolonging the culture lifetime for increased product concentrations [3,5,11-15] Strategies such as development of the feeding solutions, feeding rate control, maintaining low residual glucose and/or glutamine concentrations, employing dialysis membranes to remove low molecular weight components from the culture, feeding based on Oxygen Uptake Rate (OUR) are employed to achieve the aforementioned objectives. However, the batch operating parameters such as pH, DO, seeding density, temperature are kept constant (or optimized separately) during the fed batch process development. Such approach does not evaluate the interactions between the process parameters (batch and fed-batch) that influence the overall outcome of the process (Figure 1).

***Corresponding author:** Nurina Anuar, Department of Chemical and Process Engineering, Faculty of Engineering and Built Environment, University Kebangsaan Malaysia, 43600 UKM, Bangi, Selangor, Malaysia

Hence, based on the hypothesis that none of the factors affecting the production process can really be independent, fractional factorial design using design expert software was carried out to find the interaction between the batch (Dissolved Oxygen, Temperature and Seeding Density) and fed-batch (post-feed Glucose concentration) process parameters. The usage of design of experiments also facilitates shorter development time, hence more cost effective. This study was carried out to assess the effect of control parameters (post-feed Glucose concentration, temperature, dissolved oxygen, seeding density) on culture response parameters (IVCC, final mAb concentration). The interaction between control parameters and their effects on the production of Humanized-Anti CEA mAb production by NS0 cells in the bioreactor was also evaluated.

Materials and Methods

Cell line and culture medium

The NS0 cell line (Hu2nA33) producing humanized anti- CEA (carcino-embryogenic antigen) was obtained from the Protein Science Department of Inno Biologics Sdn. Bhd. The cells were maintained in cryo vials and stored in liquid nitrogen tanks until usage. The basal medium used to support cell growth was Dulbecco's Modified Eagle Medium (DMEM, Product # T-043, Biochrom AG).

The supplements added to the basal medium were 2% (v/v) Ultra Low IgG FBS (Product # 16250-078, Gibco), 2 mM L-glutamine (Product # G8540, Sigma Aldrich) and 0.2% (v/v) Pluuronic F-68 (Product # P1300, Sigma Aldrich).

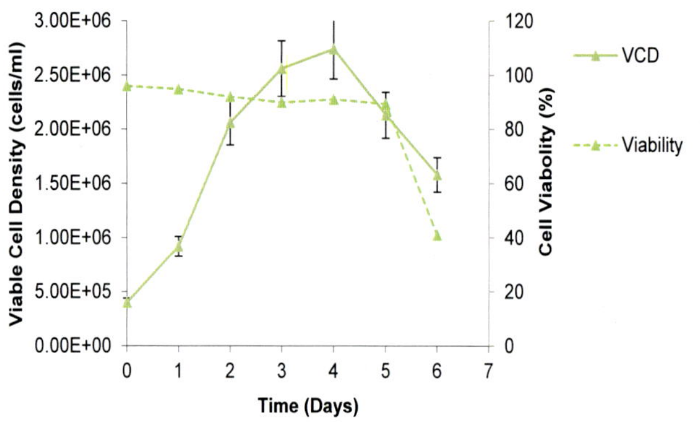

Figure 1: Mayonnaise sauces pH changes containing *L.acidophilus* in Free State, encapsulated by Hi maize-alginate and alginate 4% during storage.

The feeding medium was a 10X concentrated DMEM basal medium (Product # F0455, Biochrom AG) with 4.5 g/L D-glucose and 8 mg/L Phenol red.

Inoculum preparation

Cell culture was initiated by thawing a cryovial followed by subsequent expansion for 1 to 2 weeks using a sequence of spinner flasks (Techne, UK) prior to inoculating into the bioreactor as outlined in Table 1. Throughout the expansion process, cells were maintained in exponential growth phase. Cells were maintained at 37°C in a humid atmosphere of 5% CO_2 in an incubator (Galaxy R+, RS Biotech, UK). In order to ensure that all the fed batch runs were inoculated with cells in same physiological state, for each run a new cryopreserved vial was revived. All the cells in the cryopreserved vials were from the same culture flask.

Experimental design

An experimental design was created on DESIGN EXPERT 6.0.8 software (STAT-EASE Inc., US), and half-fractional factorial design (Resolution IV), was used for fed-batch culture incorporating only 12 of the 20 possible combinations (if using full factorial) of 4 selected control parameters at high and low levels. The experiment was designed to be carried out in four blocks (labeled 1, 2, 3 and 4). The blocking was done to ensure that all the experiments were run under homogeneous conditions and shun bias (Table 2). Four centre points was added to the design in order to avoid the risk of missing a curvilinear relationship. Since the experiment design was a Resolution IV design, which meant that no main effect was aliased with any other main effect or with any two-factor interactions, but two-factor interactions are aliased with each other [16]. Thus it was assumed that three factor interactions do not exist or have no significant effect on the output.

Day	Procedure
0	Cell Revival in 125 mL spinner flask (50 mL working volume)
1	First Medium Change
2	Second Medium Change
3	Subculture in 125 mL spinner flask (50 mL working volume)
5	Scale upto 250 mL spinner flask (125 mL working volume)
7	Scale upto 500 mL spinner flask (250 mL working volume)
9	Scale upto 1000 mL spinner flask (500 mL working volume)
11	Inoculation into the bioreactor

Table 1: General Procedure followed to inoculate the 2L bioreactors.

Run No.	Block	A: Post-Feed Glucose concentration (g/L)	B: Seeding Density (x 106 cells /mL)	C: Temperature (°C)	D: Dissolved Oxygen (DO)
1	1	8	0.3	35	60
2	1	5	0.4	36.5	45
3	1	8	0.5	35	30
4	2	8	0.3	38	30
5	2	2	0.3	38	60
6	2	5	0.4	36.5	45
7	3	2	0.3	35	30
8	3	2	0.5	35	60
9	3	8	0.5	35	60
10	4	5	0.4	36.5	45
11	4	5	0.4	36.5	45
12	4	2	0.5	38	30

Table 2: Half-fractional factorial design for the validation of process control parameters for fed batch.

The high and low levels of parameters were as follows: Post-feed Glucose concentration: 2 g/L(11.11 mmol/L) and 8 g/L (44.44 mmol/L) [13] dissolved oxygen tension (DOT) with respect to medium saturation with air (20% and 60%), seeding density (0.3×10^6 cells/mL and 0.5×10^6 cells/mL), Temperature (34°C and 38°C). pH value was set at 7.20 for all the runs.

Bioreactor equipment

All the runs were carried out in 2L double wall glass bioreactors (Univessel, B. Braun Biotech, AG). Agitation in the vessel was achieved by standard 65 mm diameter, 45° pitch-blade impeller. The process control parameter set points were maintained using a Digital Control Tower (B.Braun Biotech, AG). The pH was measured using a gas-filled electrode (Mettler-Toledo) and the dissolved oxygen concentration (DO) was measured using a polarographic electrode (Mettler-Toledo). The adjustment of pH was carried out using CO_2 gas and 0.5M NaOH. The bioreactor was configured with two mass-flow controllers (Sierra Instruments, Monterey, CA), one for air and one for oxygen. The gas flow rate was held constant while the ratio of air and oxygen was adjusted to maintain the DO set point.

Fed-batch runs

Following cell revival, fed-batch experiments were performed in the bioreactor inoculated with an initial cell concentration of 0.3×10^6 cells/mL in 1.5L working volume with 4 mmol/L glutamine and 25 mmol/L glucose. The operating parameters were set according to the experimental design. The experiment duration varied between 5-7 days.

During the initial 2 days of cultivation, bioreactor sampling was performed once daily, taking samples of 5-7 mL. Each day thereafter, sampling was performed twice daily; one before feeding and one within one hour after the feeding. These samples were analyzed for viable cell concentration, percent viability, pH, and biochemical analysis.

Feeding strategy and feeding medium: The initial addition of feed medium was performed 2 days after inoculation, once the glucose concentration in the bioreactor was depleted than the intended target post feed glucose concentration values (Figure 3). The feeding strategy was based on maintaining a desired post-feed glucose concentration according to the experimental design. Post-feed target glucose concentrations ranged from 2-8 g/l. The volume of feed medium (*Vfeed*) to be added was calculated using the following equation, as described by Sauer et al., [13], whereas the post feed L-glutamine concentration was maintained at 2 mM [13].

$$V_{feed} = \left(\frac{glc_{tar} - glc}{glc_{feed} - glc_{tar}} \right) . V$$

where glc_{tar} was the post-feed target glucose concentration, glc was the culture's glucose concentration prior to feeding, glc_{feed} was the feed medium's glucose concentration, and V was the culture volume prior to addition.

Analytical methods

Viable cell density (VCD) and viability were determined by the tryphan blue exclusion method using a haematocytometer. Prior to cell counting, samples were diluted 2 to 10-fold with 0.4% trypan blue solution (GIBCO, USA) depending on the cell density. From the VCD measurement, IVCC was calculated in the same manner as described by Sauer et al. [13].

For biochemical analysis, 5 to 7 ml of the sample were centrifuged for 5 minutes at 3000 rpm to remove cells and further aliquoted and stored at -20°C for metabolites analysis. Glucose, Lactate, and Glutamine concentration were determined enzymatically by the amperometric biosensors of Nova Bioprofile 100 Plus Analyzer (Nova Biomedical Corp., Waltham, MA). Ammonia concentration was determined potentiometrically by the potentiometric sensor of of Nova Bioprofile 100 Plus Analyzer (Nova Biomedical Corp., Waltham, MA)

Total humanized anti-CEA IgG titer or monoclonal antibody concentration was determined using the Enzyme Linked Immuno-Sorbent Assay (ELISA) in 96-well micro titer plates as described by Xie and Wang [14]. Human IgG, whole molecule, unconjugated (Pierce, Product # 31154) was used as a standard. First, 100 µL capture antibody solution with a dilution of 1:1000 (anti-human IgG-Fc specific, produced in goat; Sigma, Product # I2136) was placed in a 96-well plate and incubated at 2-8°C overnight. Next, after rinsing with 1×200 µL washing buffer (0.05% Tween-20 in 0.9% NaCl), 200 µL of blocking buffer (0.01% w/v BSA and 0.02% v/v Tween-20 in 1XPBS) was added and incubated for another 2 hours at 37°C followed by washing 4X with washing buffer. A 100 µL sample or standard solution was added to each well and incubated for 2 hours at 37°C. After rinsing with 4X200 µL washing buffer, 100 µL labeled secondary antibody, 1:2500 dilution (anti-human gamma chain specific peroxidase conjugate, produced in goat; Sigma, Product # A6029) was added and incubated for another 2 hours at 37°C. The plate was then rinsed with 4X200 µL washing buffer. Then, 100 µL of enzyme substrate working solution (ABTS tablets, Roche, Product # 11 112 442 001, dissolved in ABTS buffer, Roche, Product # 11 112 597 001) was added and incubated for 1 hour in the dark. Absorbance was measured at a wavelength of 405 nm via a kinetic micro plate reader (Bio-Tek Instruments Inc., Highland Park, Winooski). Samples were diluted 1000 to 10,000-fold with blocking buffer prior to assay depending on the antibody concentrations.

Results and Discussion

Batch run

A batch run was conducted in order to be used as a control and basis for the comparison to fed-batch runs. A maximum cell density of 2.74x106 cells/mL was achieved on day 4 of the batch run which resulted in the maximum antibody concentration of 5.21 mg/L on the same day. A biochemical analysis for Glucose and Glutamine revealed that they were depleted from the medium, suggesting that the cells stopped replicating due to nutrient limitation, hence the need for fed-batch process development.

Fed-batch runs

Twelve fed-batch runs were conducted according to the experimental design (Table 2), and the summarized data was presented in Table 3 for IVCC, maximum viable cell concentration, cell viability, final mAb concentration, average specific consumption and production for metabolites (glucose, glutamine, lactate, ammonia). Three post-feed target glucose concentrations (2, 5 and 8 g/L) was investigated according to the experimental design (Table 2). These different values resulted in different volumetric feed rates which in turn result in different final working volumes (Table 3) and also affected the cell growth. Four among the 12 fed-batch runs were run at same parameter setting since they were the four centre points in the design. The addition of centre points was done to test for curvature without adding a large number of experimental runs.

The maximum VCD for the 12 runs conducted, ranged from 0.28 to 3.43×10^6 cells/mL with the final mAb titer in the range of 4.96 to

Run # 12	Run # 11	Run # 10	Run # 9	Run # 8	Run # 7	Run # 6	Run # 5	Run # 4	Run # 3	Run # 2	Run # 1	Parameter
107.96	115.43	115.08	86.15	119.1	184.13	136.75	160.55	101.65	121	116.23	96.24	Culture Time (hrs)
22.47	21.45	22.66	15.59	16.18	22.1	32.35	59.4	8.54	12.63	23.68	4.96	Final Antibody Concentration (mg/L)
1.28	1.43	1.35	1.43	1.28	1.23	1.34	1.33	1.41	1.47	1.38	1.53	Final Working Volume (L)
2.56	3.03	3.09	3.03	2.88	3.08	3.46	3.43	1.06	1.75	3.14	1.19	Peak Viable Cell Concentration (10^6 cells/mL)
1.4	1.25	1.3	1.21	1.97	2.09	1.96	2.94	0.4	1.02	1.32	0.45	IVCD (x 10^8 cells day/L)
0.15	0.16	0.17	0.16	0.08	0.14	0.15	0.19	0.13	0.11	0.14	0.09	Average Specific Antibody Production Rate (mg/10^9 cells per day)
11.1	7.79	12.5	7.42	12.57	8.31	1.4	9.88	26.05	40.49	5.9	20.63	Average Specific Glucose Consumption Rate (mg/10^9 cells per day)
3.67	3.17	2.69	3.02	3.36	1.95	2.42	2.17	12.87	3.04	3.02	10.79	Average Specific Glutamine Consumption Rate (mg/10^9 cells per day)
20.23	20.31	17.3	19.34	20.58	5.37	14.67	6.66	79.67	32.84	18.89	38.23	Average Specific Lactate Production Rate (mg/10^9 cells per day)
0.78	0.59	0.5	0.56	0.39	0.24	0.55	0.22	1.63	0.54	0.61	0.98	Average Specific Ammonia Production Rate (mg/10^9 cells per day)
35.1	33.8	32.1	35.55	49.1	25.75	35	26.6	38.4	40.3	36.2	27.3	Final lactate concentration (g/L)
7.03	4.77	4.7	4.77	5.23	4.64	7.15	4.31	4.52	4.08	6.3	3.18	Final ammonia concentration (mmol/L)
0.52	0.82	0.62	0.82	0.61	0.84	0.43	1.52	0.41	0.41	0.23	0.18	Viable cell concentration at harvest (10^6 cells/mL)
41.08	36.6	39.96	36.6	36.06	36.72	31.12	38.51	36.24	37.72	38.05	34.28	Viability at harvest (%)

Table 3: Data for the fed-batch runs conducted according to the experimental design.

59.40 mg/L. Meanwhile, calculated IVCC values ranged from 0.45 to 2.94×10^8 cells day. The different values of the response variables are attributed to the fact that the operating parameters were different for the runs. The highest mAb titer of 59.40 mg/L was recorded from Run # 5 with the highest VCD and IVCC of 3.43×10^6 cells/mL and 2.94×10^8 cells day/L, respectively. Also, the highest average specific mAb production rate and lower average specific lactate and ammonia production rates were observed from Run # 5, indicating an inverse correlation between them.

In comparison to the batch run, Run # 5 improved the mAb yield by 10 fold whereas the maximum cell concentration was increased by 1.25 fold (Figure 2). The improvement can be attributed to the fact that post-feed glucose concentration was maintained at a target value for the fed-batch process development. Sauer et al., (2000) reported 7.9 fold increase in average final mAb concentration developed the fed-batch process based on post-feed glucose concentration. According to Sauer et al. [13], different post-feed glucose concentration result in producing different final mAb product titer, which is in accordance with our findings. The viability dropped significantly from day 5 to day 6 for Run #5, in spite of the feeding performed on day 5, indicating that the cells have entered the death phase of the cell cycle. However the cumulative antibody titer at day 6 was 59.40mg/L, the highest observed. This increase of the specific mAb productivity in the late cultivation phase is likely due to the higher osmolality (data not shown) and in agreement with the findings of Zhou et al. [15]. The increase in osmolality towards the end of the culture can be attributed to feeding of feeding medium and the addition of NaOH for pH adjustment.

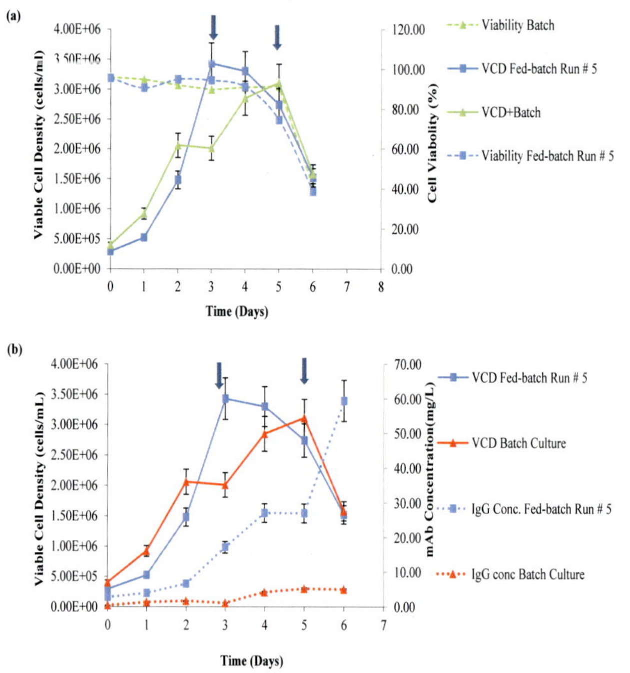

Figure 2: Comparison of the cell growth (a) and mAb production (b) between Fed-batch Run # 5 and the batch. The arrows (↓) indicate the feeding time for the Fed-batch.

The usage of medium concentrates is another factor that is responsible of the improvement in the final mAb titer. Feeding medium concentrates has been reported to dramatically increase cell densities and final product titers in recombinant CHO [17] and NS0 cell cultures [3].

Figure 3 shows the consumption and production profiles for the most important biochemical components in the growth medium namely Glucose, Lactate, Glutamine and Ammonia. The feeding was initiated on day 3 and thus the post feed glucose concentration was maintained at 2 g/L=11.11 mM (according to the experimental design for Run # 5). The post feed glutamine concentration for all fed-batch runs was maintained at 2 mM. Limiting the availability of glutamine at low concentrations throughout the fed-batch cultivation helps to reduce metabolite accumulation [18,19].

ANOVA approach for determining the interactions

The analysis was performed using the Analysis of Variance (ANOVA) feature of the design expert software. The response variables (IVCC and mAb concentration) of the culture data were modeled with a mathematical function indicating the dependence of these response variables on the various control parameters (Figure 4). The development of such models helps to establish the interactions/main

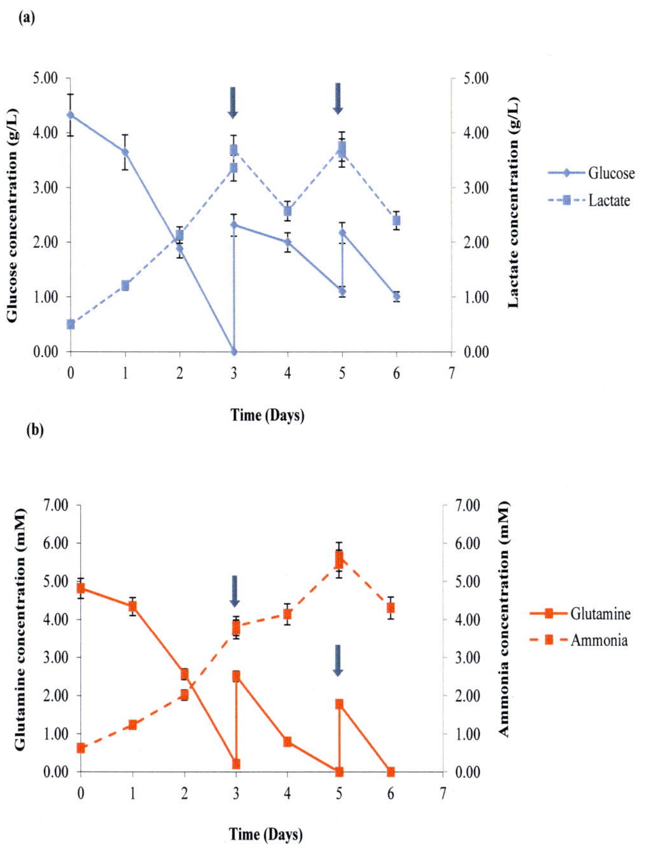

Figure 3: Glucose consumption Vs Lactate production (a) and Glutamine consumption Vs Ammonia production (b) profiles for fed-batch Run # 5.

effects responsible for the process responses. The ANOVA statistic, the F-value or F-ratio, indicates the ratio of the variability in a response caused by changing a control parameter, and the variability caused by random error alone. Larger F-values (>4) indicate the variation explained by the model is greater than would be expected by chance. Values of "Prob>F" less than 0.0500 indicate model terms are significant, whereas values greater than 0.1000 indicate the model terms are not significant. The statistical models relating culture responses to control parameter variation showed highly significant outcomes (F-value>4, p<0.05).

Statistical model for viable cell density (IVCC): The ANOVA F-Value and p-value data for IVCC is shown in Table 4, the F-value and p-values for the model were 17.16 (>4) and 0.0017(<0.01) respectively indicating a good fit for the model.

From Table 4, it is clear that interaction exists between fed-batch operating parameter-post feed glucose concentration (A) and batch operating parameter seeding density (B) and has a significant (F-value=18.98) effect on IVCC. Although the interaction AB could be aliased BC, since the experimental design was Resolution IV fractional factorial design, AB was assumed to be the true interaction since factor

Source	F-Value	p -Value Prob > F
Model	17.16	0.0017
A: Post feed Glucose concentration	58.14	0.0003
B: Seeding Density	0.16	0.7021
D: Dissolved Oxygen	5.66	0.0548
AB	18.98	0.0048
AD	2.86	0.1418
Lack of Fit	0.08	0.9687
Curvature	0.02	0.8970

Table 4: ANOVA F-Value and p-value for IVCC.

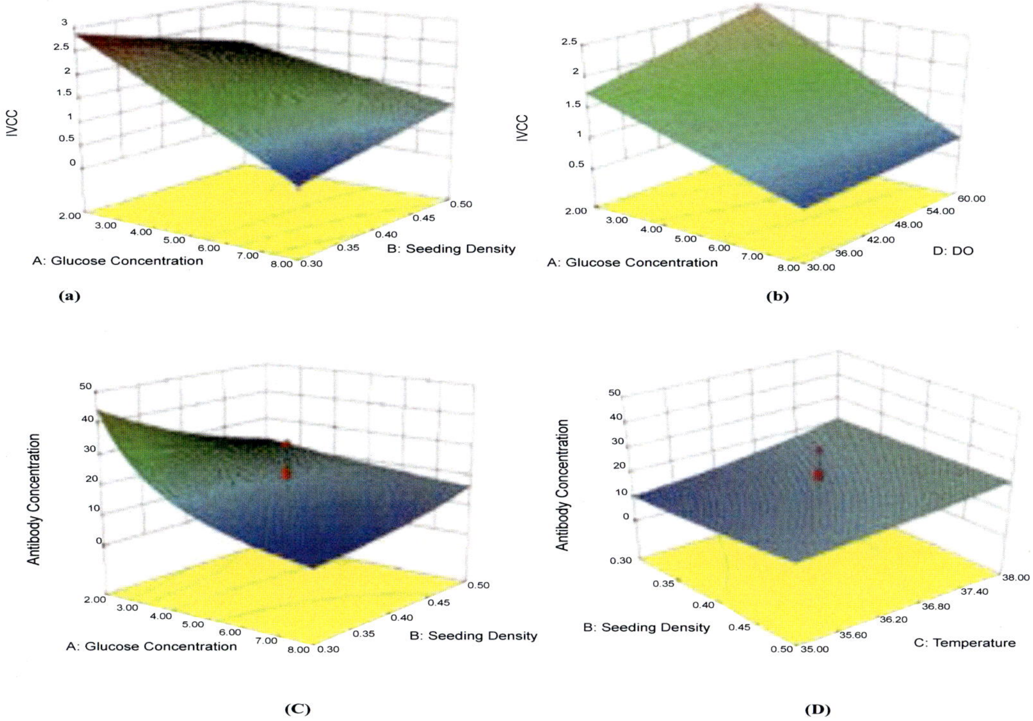

Figure 4: Effect of process control parameters on IVCC (a and b) and mAb production (c and d).

A had a major effect in the model with an F-value of 58.14 and also the factor C didn't have any effect on IVCC, thus interaction BC was unlikely to be true. At A and B low and D high and the IVCC was at its maximum.

The final equation governing the ANOVA model for IVCC in terms of coded factors was:

IVCC=+ 1.44 - 0.64 * A - 0.035 * B + 0.21 * D + 0.38 * A * B-0.15 * A * D eq.1

Statistical model for mAb concentration: The ANOVA F-Value and p-value data for mAb concentration is shown in Table 5. An inverse square root transformation was performed as recommended by the software and the ANOVA model exhibited a significant curvature (F-value being 44.40 (>4) and the p-value of 0.0026 (<0.05), indicating a good fit for the model.

From Table 5, it is clear that interaction exists between fed-batch operating parameter-post feed glucose concentration (A) and batch operating parameter seeding density (B) has a significant (F-value=18.98) effect on the responses mAb concentration. The final equation governing the ANOVA model for mAb in terms of coded factors was:

mAb = + 0.24 - 0.065 * A - 0.017 * B - 0.032 * C + 4.173 * 10-3 * D - 0.047 * A * B + 0.016 * B * C eq.2

Effect of interactions on mAb concentration and IVCC: The significant interactions AB involved for IVCC and mAb concentrations are shown in Figure 5. The intersection of lines in Figure 5 indicates that factors A and B are involved in the interaction, in case of no interaction, parallel lines would have been observed. The summarized effect of process control variables on these response variables of the fed-batch process is presented in Table 6.

Source	F-Value	p -Value Prob > F
Model	44.40	0.0013
A: Post feed Glucose concentration	139.03	0.0003
B: Seeding Density	9.85	0.0349
C: Temperature	33.09	0.0045
D: Dissolved Oxygen	0.56	0.4942
AB	70.74	0.0011
BC	7.79	0.0492
Lack of Fit	0.08	0.7956
Curvature	44.40	0.0026

Table 5: ANOVA F-Value and p-value for mAb concentration.

It is evident from Figure 4 that for both the response variables (IVCC and mAb concentration), factor A: glucose concentration is the most significant factor and AB interactions are the main driving factors.

The studentised residual versus run number shows random scatter (Figure 6) implying the lack of the effect of run number on culture response variables (IVCC and MAb concentration). The studentised residual is basically a dimensionless statistic used to measure the difference between the actual value of the response observed in the experiment and the predicted value for the response by the experimental model. The studentised residual is used to estimate for the block effect of any of the measured response of a blocked factorial experiment.

The interaction between fed-batch operating parameter-post feed glucose concentration (A) and batch operating parameter seeding density (B) was established and found to be significant for both the culture responses evaluated (IVCC and mAb concentration). The existence of such interactions between fed-batch and batch operating parameters prove the hypothesis that none of the factors affecting the responses of a fed-batch process can be considered independent. Therefore conventional approach to keep batch operating parameters constant while varying only the fed-batch operating parameters needs to be carefully evaluated, since the interactions between the batch and fed-batch parameters are ignored. The knowledge of such interactions during early stages of fed batch process development makes it easier to fully optimize the process. The usage of design expert as a tool for experimental design shortens the early process development and also allows establishing the main factors driving the process response and

Culture Response	A: Glucose Concentration	B: Seeding Density	C: Temperature	D: DO
IVCC	Low	Low	High/Low	High
mAb Concentration	Low	Low	Low	High/Low

Table 6: Summarized of the effects of the investigated control parameters on culture responses.

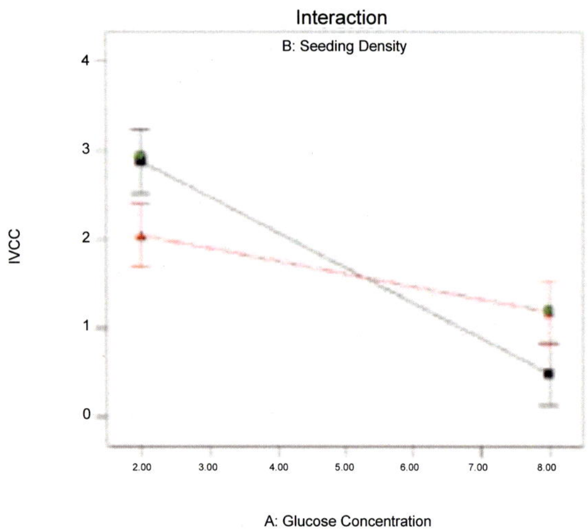

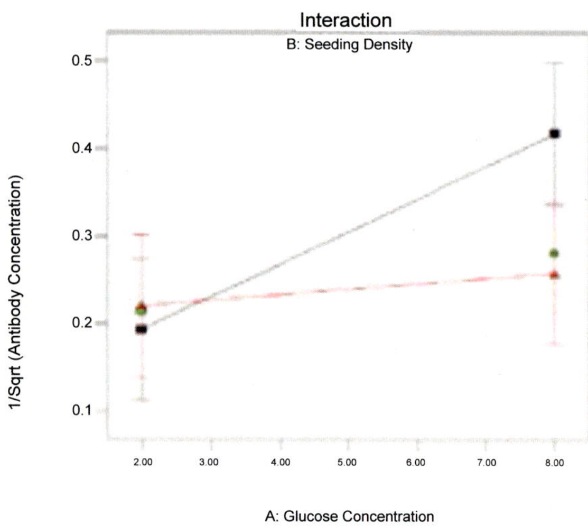

Figure 5: Interaction between A: Post-feed Glucose concentration and B: seeding density for IVCC (a) and mAb concentration (b).

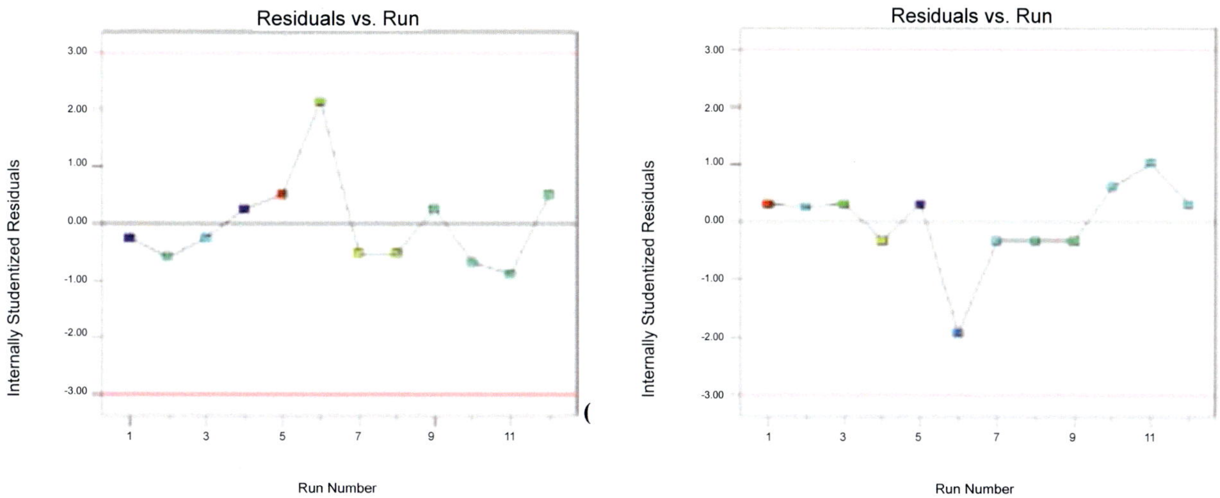

Figure 6: Studentized residuals vs. run order for IVCC (a) and mAb concentration (b).

the interactions between them. Resolution IV Experimental design designed using design expert proved an efficient tool for analyzing the fed-batch process, since the number of runs needed to be carried out were 12 (including 4 center points) in comparison to that of a full factorial design (20, incuding 4 center points). The best among the fed-batch runs conducted increased the final mAb titer by 10 fold in comparison to a batch culture. The maximum VCD value of the developed fed batch process was 1.25 over fold the value for batch.

Conclusion

Fed batch process for the production of anti carcino embryogenic antigen from NS0 hybridoma cells was developed with feeding based on post feed glucose concentration. Design of experiments was used to study simultaneous variation of four different parameters of post feed glucose concentration, seeding density, dissolved oxygen concentration and temperature on cell growth and mAb production. Resolution IV Experimental design designed using design expert proved an efficient tool for analyzing the fed-batch process, since the number of runs needed to be carried out were 12 (including 4 center points) in comparison to that of a full factorial design (20, including 4 center points). The best among the fed-batch runs conducted increased the final mAb titer by 10 fold in comparison to a batch culture. The maximum VCD value of the developed fed batch process was 1.25 over fold the value for batch.

Also the interaction between fed-batch operating parameter-post feed glucose concentration (A) and batch operating parameter seeding density (B) was established and found to be significant for both the culture responses evaluated (IVCC and mAb concentration). The existence of such interactions between fed-batch and batch operating parameters proved the hypothesis that none of the factors affecting the responses of a fed-batch process can be considered independent. Therefore conventional approach to keep batch operating parameters constant while varying only the fed-batch operating parameters needs to be carefully evaluated, since the interactions between the batch and fed-batch parameters are ignored. The knowledge of such interactions during early stages of fed batch process development makes it easier to fully optimize the process.

Acknowledgment

I want to extend my sincere gratitude and appreciation towards Ministry of Science and Technology for funding this research. I am also indebted to InnoBiologics Sdn Bhd and National University of Malaysia (UKM) for letting me uses their facility for conducting my research and providing necessary support. I would like to thank Dr. Nurina Anuar, Dr. Badarulhisam Abdul Rahman and Dr. Jamaliah Md Jahim for their constant guidance, unwavering support and an objective criticism of my work that helped in shaping up my subsequent research efforts.

References

1. Birch JR, Racher JA (2006) Antibody production. Advanced Drug Delivery Reviews 58: 671-685.
2. Sommerfeld S, Strube J (2005) Challenges in biotechnology production-generic processes and process optimization for monoclonal antibodies. Chemical Engineering and Processing 44: 1123-1137.
3. Bibila TA, Ranucci CS, Glazomitsky K, Buckland BC, Aunins JG (1994) Monoclonal antibody process development using medium concentrates. Biotechol Prog 10: 87-96.
4. Dempsy J, Ruddock S, Osborne M, Ridley A, Sturt S, et al., (2003) Improved fermentation processes for NS0 cells expressing human antibodies and glutamine synthetase. Biotechnol Prog 19: 175-178.
5. DeZengotita VM, Miller WM, Aunins JG, Zhou W (2000) Phosphate feeding improves high-cell-concentration NS0 myeloma culture performance for monoclonal antibody production. Biotechnol Bioeng 69: 566-576.
6. Whitford W (2003) NS0 serum-free culture and applications. Bio Process Int 1: 36-47.
7. Zhou W, Chen C, Buckland B, Aunins J (1997) Fed-batch culture of recombinant NS0 myeloma cells with high monoclonal antibody production. Biotechnol Bioeng 55: 783-792.
8. Manwaring J, Barnett B, Pence B, Whitford W (2005) NS0 Derivatives: MAB Production in Largescale SFM Formats. Animal Cell Technology Meets Genomics & ESACT Proceedings 2: 581-584.
9. Arden N, Betenbaugh MJ (2004) Life and death in mammalian cell culture: strategies for apoptosis inhibition. Trends Biotechnol 22: 174-180.
10. Lasunskaia EB, Fridlianskaia II, Darieva ZA, da Silva MS, Kanashiro MM, et al., (2003) Transfection of NS0 myeloma fusion partner cells with HSP70 gene results in higher hybridoma yield by improving cellular resistance to apoptosis. Biotechnol Bioeng 81: 496-504.
11. Bibila TA, Robinson DK (1995) In pursuit of the optimal fed-batch process for monoclonal antibody production. Biotechnol Prog 11: 1-13.
12. Ramirez OT, Mutharasan R (1990) Cell cycle- and growth phase dependent variations in size distribution, antibody productivity, and oxygen demand in hybridoma cultures. Biotechnol Bioeng 36: 839-848.
13. Sauer PW, Burky JE, Wesson MC, Sternard HD, Qu L (2000) A High-Yielding,

Generic Fed-Batch Cell Culture Process for Production of Recombinant Antibodies. Biotechnol Bioeng 67: 585-597.

14. Xie L, Wang DIC (1996) High cell density and high monoclonal antibody production through medium design and rational control in a bioreactor. Biotechnol Bioeng 51: 725-729.

15. Zhou W, Rehm J, Europa A, Hu WS (1997) Alteration of mammalian cell metabolism by dynamic nutrient feeding. Cytotechnology 24: 99-108.

16. Meyers RH, Montgomery DC, Anderson-Cook CM (2002) Response Surface Methodology; Process and Product Optimization Using Designed Experiments.3rd Edtn, John Wiley & Sons, New Delhi, India.

17. Hettwer DJ, Escobar E, Fieschko J (1991) Development of a serum free suspension process for recombinant CHO cells. Presented at the American Institute of Chemical Engineers Annual Meeting, Los Angeles CA, USA.

18. Xie L, Daniel ICV (1997) Integrated approaches to the design of media and feeding strategies for fed-batch cultures of animal cells. Trends Biotechnol 15: 109-113.

19. Xie LZ, Wang DIC (1994) Fed-Batch Cultivation of Animal Cells Using Different Medium Design Concepts and Feeding Strategies. Biotechnol Bioeng 43: 1175-1189.

14

Enhanced Saccharification of Steam-Pretreated Rice Straw by Commercial Cellulases Supplemented with Xylanase

Jairam Choudhary, Saritha M, Lata Nain and Anju Arora*

Division of Microbiology, Indian Agricultural Research Institute, New Delhi 110 012, India

Abstract

For successful commercialization of biomass to ethanol process, the techno- economic hurdles have to be overcome. Complete substrate utilization is the key for making economics favorable. Existing enzyme preparations do not give high saccharification efficiency, thus, a broader suite of hydrolases is required. Xylanase supplementation in enzyme cocktail is an important strategy to increase sugar yields. In our study, hyperxylanolytic *Aspergillus awamori* F18 was used to produce xylanase. *A. awamori* F18 expressed high levels of xylanase during Solid state fermentation of corncob and also, Carboxy Methyl Cellulase (CMCase), filter paper activity (FP lyase) and β-glucosidase (524.43, 43.95, 8.64 and 29.81 IU mg^{-1} proteins respectively). A 10.6-12.5 fold concentration of these activities was achieved by single-step acetone precipitation. Supplementing concentrated xylanase to Accellerase®1500 resulted in 69.5% enhancement in sugars released after 72 h of saccharification of steam-pretreated rice straw. HPLC analysis of hydrolysates showed higher glucose levels along with the presence of xylose and arabinose. There was a 12.4% increase in the amounts of glucose as well as total sugars released when xylanase was supplemented to commercial cellulase and β-glucosidase.

Keywords: Bioethanol; Lignocellulosic biomass; Enzymatic saccharification; Xylanase; Cellulase

Introduction

Depleting fossil fuel reserves, dwindling fuel prices, increased concerns over climate change, and most importantly the search for energy security have led the global research community to look for renewable, environment friendly and sustainable energy sources. In this perspective, ethanol derived from biomass can be used to supplement conventional fossil fuels and cut down on petroleum consumption. For first generation bioethanol, industry mostly depended on sugar and starchy feed stocks. Thus, this option is limited to the countries where actual and sustainable surplus of these crops is available. Moreover, since these feed stocks are a part of food, this resulted in a sudden increase in food prices all over the globe. This led to the concept of 2nd generation bioethanol, which is derived from lignocellulosic biomass and do not threaten food security. In developing countries like India, the possible competition with food is one of the risky factors when using agricultural crops for ethanol production. In this regard, lignocellulosic biomass, which is not a part of human food chain, is envisaged to supply a significant portion of the feed-stocks for bioethanol production in the medium and long term due to their low cost and abundance. Lignocellulosic biomass constitutes cellulose and hemicellulose (carbohydrate polymers), and lignin. Cellulose is a linear, crystalline homopolymer made up of repeated units of glucose strung together with β-glycosidic bonds whereas hemicellulose is a highly branched heteropolymer consisting of xylose, glucose, galactose, mannose, and arabinose. The exact sugar composition of hemicellulose varies depending on the plant type and origin.

Biomass conversion to ethanol proceeds in a step wise manner where pretreatment and saccharification are the two most important steps which determine the overall cost and efficiency. Saccharification can be carried out by chemical or biological means by employing glycosyl hydrolases. Bacteria and fungi are good sources of cellulases and xylanases that cause the breakdown of cellulose and hemicellulose chains in order to produce monosaccharides for the subsequent fermentation step. Hydrolysis of cellulose is brought about by cellulase complex while hydrolysis of hemicelluloses involves enzymes like glycoside hydrolases, carbohydrate esterases, polysaccharide lyases, endo-hemicellulases and others, the concerted action of which hydrolyse glycosidic bonds, ester bonds and remove the chain's substituents or side chains [1]. These include endo-1, 4-β-xylanase, β-xylosidase, β-mannanase, β-mannosidase, α-glucuronidase, α- L-arabinofuranosidase, acetyl xylan esterase and other enzymes. Although the enzymatic saccharification is inherently slow process, it demonstrates better results using pretreated lignocellulosic biomass because no degradation products are formed which inhibit fermentation step. The other prominent advantages of enzymatic hydrolysis are the mild process conditions which do not require expensive materials of construction and low process energy requirements. Thus, enzymatic hydrolysis of biomass carbohydrates is the technology of choice. However, existing technologies and available cellulase cocktails do not allow complete saccharification of plant polymers as there are many factors hindering the action of cellulases. Currently employed cellulases are sourced from saprophytic fungi [2] and these cellulase preparations are subject to tight induction and regulation systems and also inhibition from various end products [3,4]. Therefore, more potent and efficient enzyme preparations need to be developed for the enzymatic saccharification process to be more economical. To overcome the limitation of enzymes and to get higher saccharification efficiency, a broader suite of enzymes is required for hydrolysis of cellulose and hemicelluloses to fermentable sugars [2,5]. Additional critical enzymes need to be present in commercial cocktails at optimal levels for improving the efficiency of these enzyme mixtures [6]. Also, the xylan and xylan fragments present in the biomass have been suggested to be one of the causes of the

***Corresponding author:** Anju Arora, Division of Microbiology, Indian Agricultural Research Institute, New Delhi 110 012, India

reduced activity of cellobiohydrolases [7]. Therefore, supplementation of enzyme cocktails with xylanases can alleviate the problem of inhibition of cellulase enzyme action caused by xylan oligosaccharides. Xylanase supplementation in combination with other pretreatments and cellulase improves sugar yields [8,9]. The reactivity of component enzymes differs with the enzymatic activity and biomass, and for higher saccharification ability, cellulases need high hemicellulase activity as well as high β-glucosidase activity [10]. However, there are very few reports on supplementation of cellulase cocktails with xylanases for increasing saccharification yields. The present study was conducted to produce xylanase from a hyperxylanolytic fungal strain, *Aspergillus awamori* F18 and assess the impact of xylanase supplementation to standard cellulases (Accellerase®1500, commercial cellulase and β-glucosidase) on enhancement of sugar yields during saccharification.

Materials and Methods

Organism

The organism used in the study was *Aspergillus awamori* F18 obtained from the culture collection, Division of Microbiology, Indian Agricultural Research Institute, New Delhi. The fungus was grown on Potato Dextrose Agar (PDA) medium at 30°C. After one week of incubation, the slant cultures were stored in refrigerator at 8°C until used.

Qualitative assays of xylanolytic and cellulolytic abilities

Xylanolytic potential of *A. awamori* F18 was evaluated on the basis of its ability to grow on Reese's mineral medium with 0.1% xylan as sole carbon source. The ability of *A. awamori* F18 to produce endoglucanases was assessed by its ability to produce zones of hydrolysis on CMC (Carboxy Methyl Cellulose) agar (Congo-red test), as described earlier [11].

Xylanase production using A. awamori F18

Xylanase was produced by *A. awamori* F18 under both submerged and solid state fermentation by using different substrates. For submerged fermentation, 0.1% xylan or 1% rice straw was used as substrate and for solid state fermentation, corn cob was used.

Production of xylanase by *A. awamori* F18 under submerged fermentation

Submerged fermentation was carried out in 250 ml Erlenmeyer flasks containing 50 ml Reese's mineral medium supplemented with 1% rice straw / 0.1% xylan as carbon source. The flasks were autoclaved at 15 psi pressure and 121°C temperature for three consecutive days. They were inoculated with three agar plugs (6 mm) of fungal mycelia and incubated at 28°C for 7 days in stationary condition. After incubation, the contents of the flasks were filtered through Whatman No. 42 filter paper and the filtrate was used for estimation of different enzyme activities after 3, 5 and 7 days of incubation.

Production of xylanase by *A. awamori* F18 under solid state fermentation

Solid state fermentation was carried out using corn cob as substrate. The substrate was dried and ground to pass through a 2 mm sieve. Erlenmeyer flasks (500 ml) containing 5 g substrate with 25 ml Reese's mineral medium was autoclaved at 121°C temperature and 15 psi pressure for three consecutive days and inoculated with 1 ml spore suspension (4×10^6 spore/ml) of *A. awamori* F18. Flasks were incubated at a temperature of 30°C for 7 days. Treatments were carried out in triplicates. After 7 days of incubation, 70 ml of sodium citrate buffer (0.05 M, pH 4.8) was added to each flask and contents were mixed on a shaker at 200 rpm for 2 hours for extraction and vacuum filtered. The filtrate was used as source of enzymes and assayed for xylanase, cellulases and proteins.

Enzyme was also produced on large scale by solid state fermentation of 100 g of corncob in glass trays. All the parameters regarding inoculation, incubation and extraction were kept the same.

Xylanase and cellulase activity assays

Xylanase activity was assayed by the method described by Ghose and Bisaria [12]; filter paper (exo-β-glucanase) activity and CMCase activity (endo-β-1, 4-glucanase) were assayed as described by Ghose [13]. The reducing sugars released were measured by the DNSA method [14]. One International Unit of xylanase was expressed as 1 μmole of xylose formed per minute and one International Unit of filter paper activity or CMCase corresponded to 1 μmole of glucose formed per minute during hydrolysis. β-glucosidase assay was performed using p-nitrophenyl-β-D-glucopyranoside as substrate [15] and the activity was calculated in terms of μmoles of p-nitrophenol produced per ml of culture filtrate per minute.

Concentration of enzyme

Crude enzyme extract produced by solid state fermentation was concentrated by acetone precipitation. The enzyme extract was centrifuged at 8000 rpm for 10 min to remove the impurities present in crude extract. This was then mixed with chilled acetone (stored at -20°C) in a ratio of 1:5 and kept at 4°C overnight. The protein precipitate was collected by centrifugation at 10,000 rpm for 10 minutes. Pellet was air dried and dissolved in sodium citrate buffer of pH 4.8. Concentrated xylanase enzyme was used for enzymatic saccharification of pretreated rice straw.

Biochemical characterization of enzyme

Temperature and pH optima of enzyme activities were determined. To study the effect of temperature on the activity, reaction mixture was incubated at different temperatures i.e.35, 45, 50, 55, 65 and 75°C. Enzyme assays were carried out as described earlier. To determine optimum pH for xylanase, cellulase, FPase and β- glucosidase activities, respective substrates were dissolved in 0.05 M sodium citrate buffer of pH ranging from 3-6, Sorensen phosphate buffer of pH 7-8 and glycine NaOH buffer of pH 9. The respective buffers were used for enzyme assays at the particular pH. Activities were measured after incubation with enzyme.

Saccharification of rice straw with commercial cellulases supplemented with xylanase

To assess the effect of xylanase supplementation on hydrolysis of rice straw and sugar yields, the concentrated xylanase preparation from *A. awamori* in combination with Accelerase®1500 (Genencore) or cellulase (Sigma) + 0.1 ml β glucosidase (Sigma) were used.

Pretreatment of rice straw for saccharification

Chopped paddy straw (1 cm pieces) having 35% cellulose, 22.2% hemicellulose and 3% moisture was used as substrate for saccharification was pretreated with steam. Erlenmeyer flasks, containing 5 g rice straw, were autoclaved at 15 psi pressure and 121°C temperature for 30 min followed by extraction with 0.1 N NaOH (in a 1:10 ratio of biomass sample to NaOH). It was followed by two washings with distilled water in the same ratio to bring down the pH and the biomass was vacuum filtered. The washed biomass was then used as the substrate for saccharification.

Compositional analysis of pretreated rice straw

For compositional analysis the rice straw was dried to constant weight and moisture content was calculated. Hemicellulose content in the substrate was estimated by the TAPPI standard method [16] and the cellulose content of dried biomass samples was estimated by the procedure given by Updegraff [17].

Enzymatic hydrolysis of pretreated rice straw using standard cellulases supplemented with concentrated xylanase produced by *A. awamori* F18

Saccharification of steam-pretreated rice straw was carried out following the procedure given by Saritha et al. [11]. The substrate in wet form (1 g) with 85% moisture was taken in 50 ml plastic bottles to which different enzyme mixtures were added and volume was made up to 10 ml by adding 0.05 M citrate buffer (pH 4.8). The reaction mixture in bottles was incubated in a shaker water bath at 50°C and 0.5 ml aliquots were withdrawn periodically and reaction stopped by boiling the mixtures for 2 minutes. Total reducing sugars released during saccharification were estimated by the DNSA method and HPLC. All the treatments were run in triplicates. The different enzyme combinations used for saccharification of pretreated rice straw were:

- 0.5 ml Accelerase®1500 (Genencore)
- 0.1 ml Accelerase®1500 (Genencore)
- 1.0 ml concentrated xylanase
- 0.1 ml Accelerase®1500 (Genencore) + 1.0 ml concentrated xylanase
- 0.1 ml cellulase (Sigma)
- 0.1 ml cellulase (Sigma) + 1.0 ml concentrated xylanase
- 0.1 ml cellulase (Sigma) + 0.1 ml β glucosidase (Sigma)
- 0.1 ml cellulase (Sigma) + 0.1 ml β glucosidase (Sigma) + 1.0 ml concentrated xylanase

HPLC analysis of enzyme hydrolysate

The hydrolysates were analyzed for the presence of sugars using High Performance Liquid Chromatography (Waters 515 (Waters Corporation, Milford, MA, USA) equipped with a Waters 2414 Refractive Index Detector (RID)). The Aminex HPX-87H column was operated with 5 mM H_2SO_4 as mobile phase at a flow rate of 0.5 ml min^{-1} and the oven temperature was kept at 40°C.

Xylanase	CMCase	FP activity	β-D-glucosidase
2.22 ± 0.109[x]	0.63 ± 0.04	0.09 ± 0.001	1.88 ± 0.10
(183.64)[y]	(46.96)	(6.91)	(140.98)

[x]Data represents the means ± SD, n = 3
[y]Values in brackets show the specific activity (IU/mg proteins)

Table 1: Enzyme activities (IU/ml) produced by *A. awamori* F18 on Reese's medium with paddy straw (1%) as sole carbon source under submerged fermentation for 7days.

Results and Discussion

Xylanase production by hyperxylanolytic *A. awamori* F18 strain

Xylanolytic ability of *A. awamori* F18 strain was evident by its growth on Reese's mineral medium containing xylan as sole carbon source. It also produced hydrolysis zone on CMC agar indicating cellulase production. High levels of xylanase and cellulases were produced by *A. awamori* F18 under submerged culture with rice straw as substrate (Table 1). With 0.1% xylan, high xylanase (25.5 IU/ml), FPase and β-glucosidase activities were expressed after 5 days incubation and maximum CMCase was obtained after 7 days (Table 2). Most fungi are reported to produce a wide array of enzymes capable of depolymerizing plant cell wall polysaccharides, extracellularly [18]. Lignocellulosic substrates induce the production of cell wall degrading enzymes by bacteria and fungi under submerged and solid state conditions [19-21]. Most of the commercial cellulases are produced by the filamentous fungi - *Trichoderma reesei* or *Aspergillus niger* under submerged fermentation. As filamentous fungi have the tendency to grow adhered to surfaces and considering that submerged free floating fungal growth is not natural, favorable physiological aspects are associated with solid-state fermentation [22]. *A. awamori* F18 produced very high levels of xylanase on corncob substrate under solid state fermentation. Activities (IU g^{-1} of substrate) of different enzymes were 1432.09, 121.69, 19.06 and 86.80, for xylanase, CMCase, FPase and β-glucosidase, respectively (Table 3). Solid state fermentation technology has obvious advantages for production of commercial enzymes using fungi and lignocellulosic substrates as the cultivation conditions are suitable for growth of fungi which are able to grow at low water activities. Therefore, enzyme productivity in solid state fermentation is higher than that of submerged fermentation [23]. Many researchers have used agricultural and other wastes as substrates for production of cellulases and xylanases with this technology [20,24,25].

The enzyme activities were concentrated by precipitation with acetone and 11.5, 12.5 and 10.6 fold concentration of xylanase, CMCase and FPase, respectively, was achieved, with respective activity recoveries of 52.39%, 61.31% and 36.09% (Table 4). Simkhada et al. [26] produced crude xylanase enzyme in corn cob medium and obtained 4.7 fold purification and 23.6% activity recovery using a single step size exclusion chromatography.

Characterization of xylanase produced by *A. awamori* F18

To use different enzymes in combination, it is important that they have compatible activity parameters. Xylanase activity had an optimum temperature of 45°C and optimum pH of 6.0. Xylanases produced from *Aspergillus* strains were found to be active in a pH range of 5.0-7.0 with an optimum temperature from 40-70°C [27]. Most of the cellulases tried or recommended for bioconversion perform better at pH range 4-6 and at temperature ranging from 50 to 65°C; though at lower ends of the temperature range, the effective life time of enzyme is higher [22].

Xylanase			CMCase			FP activity			β-D-glucosidase		
3 days	5 days	7 days	3 days	5 days	7 days	3 days	5 days	7 days	3 days	5 days	7 days
11.6 ± 4.28[x]	25.5 ±1.59	24.8 ± 2.10	0.3 ± 0.05	2.09 ± 0.2	2.37 ± 0.1	0.05 ± 0.001	0.07 ± 0.001	0.07 ± 0.001	0.14 ± 0.015	0.35 ± 0.04	0.35 ± 0.03

[x]Data represents the means ± SD, n = 3

Table 2: Enzyme activities (IU/ml) produced by *A. awamori* F18 on Reese's medium with xylan (0.1%) as sole carbon source under submerged fermentation.

Enzymatic saccharification

Enzymatic saccharification studies were carried out on steam-pretreated rice straw with different combinations of enzymes. Pretreated rice straw showed enriched cellulose (46%) and hemicellulose (27.5%) contents as compared to raw rice straw which had 35% and 22.2% cellulose and hemicelluloses, respectively on dry weight basis. It had moisture content of 85%. This enrichment in holocellulose is because of lignin removal during steam pretreatment. Different pretreatments result in enrichment of holocellulose by removing lignins/ hemicelluloses. Depending on severity of applied conditions, steam pretreatment typically results in significant depolymerisation of lignin component and extraction subsequently carried out after steam pretreatment will likely solubilize a significant fraction of this depolymerised lignin [28]. Saritha et al. [11] achieved enrichment of cellulose content resulting from delignification by biological pretreatment followed by alkali wash.

Substrate	Activity (IU/g substrate)			
	Xylanase	**CMCase**	**FP activity**	**β-D-glucosidase**
5 g corn cob	1432.09 ± 17.17[x]	121.69 ± 3.23	19.06 ± 2.47	86.80 ± 11.52
	(456.19)[l]	(38.57)	(7.14)	(27.88)
100 g corn cob	856.03 ± 12.19	61.67 ± 4.01	8.12 ± 1.87	44.32 ± 3.05
	(524.43)	(43.95)	(8.64)	(29.81)

[x]Data represents the means ± SD, n = 3

[l]Values in brackets shows the specific activity of enzymes

Table 3: Enzyme activities produced by *A. awamori* F18 under solid state fermentation of corn cob for 7 days.

	Enzyme activity (IU/ml)			Proteins (mg/ml)	Specific activity (IU/mg proteins)		
	Xylanase	**CMCase**	**FPase**		**Xylanase**	**CMCase**	**FPase**
Crude extract	96.8	8.1	1.5	0.21	456.19	30.57	7.14
Extract after acetone precipitation	1110.9	101.38	15.9	0.42	2645	241.38	37.85
	% Recovery				**Fold concentration**		
	52.39	61.31	36.09		11.5	12.5	10.6

Table 4: Concentration of enzymes produced by *A. awamori* F18 under solid state fermentation on corn cob substrate.

Enzymatic saccharification of pretreated rice straw resulted in release of sugars in high amounts in all the treatments. Treatments in which concentrated *Aspergillus* xylanase was supplemented to standard cellulases resulted in significantly higher sugar yields as compared to respective controls. Accelerase®1500 (0.1 ml) supplemented with 1ml concentrated xylanase yielded 538.7 mg of sugars g^{-1} dry substrate (Figure 1a) which was 69.5% more than when Accelerase®1500 (0.1 ml) was used alone. However, the combination of concentrated xylanase, cellulase (Sigma) and β-glucosidase (Sigma) yielded the maximum sugars (629.7 mg g^{-1} dry substrate) after 72 hours (Figure 1b). Enzyme hydrolysate analyzed by HPLC showed significantly higher glucose release, and also release of xylose and arabinose in treatments receiving concentrated xylanase (Table 5). A 12.4% enhancement in the release of glucose and total sugars was evident from the results. Enhanced levels of total sugars as detected with DNSA method and also higher levels of different sugars as observed by HPLC showed that enhancement was there in degradation of cellulose and also hemicellulose was hydrolyzed as xylose was also present. This indicated that addition of xylanase enhanced total sugar recovery not only by degrading hemicelluloses but also facilitated enhanced hydrolysis of cellulose by cellulase enzymes. Saccharification of different pretreated agro residues by crude xylanase produced from *Aspergillus foetidus* MTCC4898 was attempted by Chapla et al. [29]. The maximum reducing sugars (mg g^{-1}) produced ranged from 151.6-193.86 and HPLC analysis of hydrolysates showed

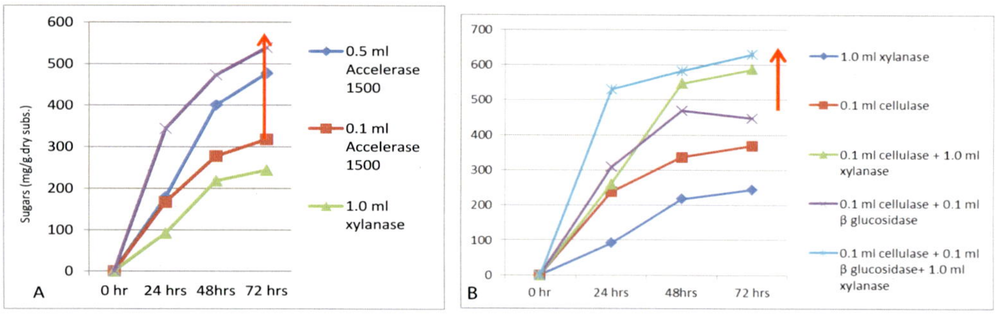

Figure 1: Enhancement of sugar yields by xylanase supplementation to
(a) Accellerase®1500 (Genencore)
(b) Cellulase (Sigma) and β-glucosidase (Sigma)

Treatment	Sugars(mg/g dry substrate)			Total sugars (mg/g dry substrate)
	Glucose	**Xylose**	**Arabinose**	
0.1 ml Accelerase®1500	105.33	14	0	119.33
0.1 ml cellulase (Sigma) + 0.1 ml β-glucosidase (Sigma)	256.66	58	178	492.66
0.1 ml cellulase (Sigma) + 0.1 ml β-glucosidase (Sigma) + 1.0 ml conc. Xylanase	288.66	68	196.66	553.33
SE(m) ±				1.702
CD@5%				9.45

Table 5: Different sugars present in hydrolysates detected by HPLC.

xylose to be the major end product with only traces of glucose. Hu et al. [30] also reported an enhancement in enzyme hydrolysis of steam pretreated corn stover when xylanase were added to minimum amount of cellulase used for hydrolysis. The co-hydrolysis of xylan increased accessibility of cellulose to cellulases. Apparently supplementing the cellulases with xylanases increased both the cellulose and hemicelluloses hydrolysis. This effect holds true for all pretreatments with the benefit of xylanase supplementation depending on type of pretreatment with total sugar yield boost between 40-100% [8].

Hemicelluloses may add to the recalcitrance of cellulose and enzymes such as xylanase are common in industrial enzyme cocktails for lignocellulose processing [31]. Rice straw pretreated with a GH11 endoxylanase from *Chaetomium globosum* showed a decrease in hemicellulose content from 25.8 to 11.4% and an increased saccharification rate [9]. This was due to the hydrolysis of hemicelluloses, thereby exposing a more accessible surface area of cellulose to cellulase. While role and synergistic actions of classical endoglucanases and cellobiohydrolases in degradation of cellulose are well understood, conversion yields with these are well below 100% [32]. Presence of other enzymes in the cocktails and their role in increasing the efficiency of cellulases by increasing accessibility of cellulose is important. One of the main beneficial effects of cellulase supplementation with xylanase during biomass saccharification is thought to be the result of improved cellulose accessibility as a result of xylan solubilization [8]. To use lignocellulosic biomass industrially, it is important that cellulase and hemicellulase work in concert. The hemicellulose-degrading enzyme activities in most of the commercial cellulase preparations are low to achieve significant conversion of the residual hemicellulose [33]. Therefore, supplementation of cellulase with the so-called accessory enzymes, including xylanases, has been the most common approach to increase the overall fermentable sugar yields [30]. There are definite reports that xylanase supplementation in combination with pretreatment or cellulases during saccharification enhance the yields of sugars [8,34]. An analysis of the saccharification ability of high-functional cellulase for various pretreated biomass by Kawai et al. [10] showed that commercial enzymes with enhanced cellobiase activity gave low glucose yields as the remaining xylan inhibited the degradation of cellulose. Therefore, they found that addition of a hemicellulase to Accelerase®1500 increases not only xylose yield but also glucose yield. Benefits of xylanase addition during pretreatment of wheat straw and rice straw have resulted in enhanced sugar yields [8,9]. In the present study, xylanase was supplemented to commercial biomass hydrolyzing cellulases during saccharification step and there was 12.4% enhancement in total sugar yields. The results of the study, therefore, confirmed that accessory xylanase improved saccharification of pretreated rice straw by cellulases and enhanced sugar yields. Accessory enzymes that facilitate more complete utilization of plant biomass could be used to develop less energetically and chemically intensive processing and allow for greater fermentable sugar recovery. Thus, enzyme mixtures of the future comprising accessory enzymes need to be tailored for essential lignocellulolytic activities to achieve improved overall performance yields.

Conclusions

In coming years bio refineries will form key components of sustainable bio based economy and conversion of polysaccharides present in biomass with enzymes will be an important technology. The present study showed that to realize higher sugar yields and conversion of both the constituent polysaccharides of lignocellulosic biomass, a mixture of enzymes is required and is more efficient. Results of this study demonstrate the role of accessory enzyme xylanase in enhancing saccharification efficiency of lignocellulosic biomass.

Authors' contributions

Jairam Choudhary conducted the experiments. Jairam Choudhary and Saritha M wrote the manuscript. Anju Arora and Lata Nain conceptualized and critically reviewed the manuscript. All authors read and approved the final manuscript.

Acknowledgement

Authors are thankful to Dr. S. Singh for HPLC analysis of hydrolysates. Jairam Choudhary and Saritha M are thankful to the Post Graduate School and Director, Indian Agricultural Research Institute, New Delhi, India and the Department of Science and Technology, GOI for providing for fellowship. We are also grateful to the Division of Microbiology, IARI, New Delhi for providing necessary facilities for undertaking this study.

References

1. Sweeney MD, Xu F (2012) Biomass converting enzymes as industrial biocatalysts for fuels and chemicals: Recent developments. Catalysts 2: 244-263.
2. King BC, Waxman KD, Nenni NV, Walker LP, Bergstrom GC, et al. (2011) Arsenal of plant cell wall degrading enzymes reflects host preference among plant pathogenic fungi. Biotechnol Biofuels 4: 4.
3. Panagiotou G, Olsson L (2007) Effect of compounds released during pretreatment of wheat straw on microbial growth and enzymatic hydrolysis rates. Biotechnol Bioeng 96: 250-258.
4. Ximenes E, Kim Y, Mosier N, Dien B, Ladisch M (2010) Inhibition of cellulases by phenols. Enzy Microb Tech 46: 170-176.
5. McMillan JD, Jennings EW, Mohagheghi A, Zuccarello M (2011) Comparative performance of precommercial cellulases hydrolyzing pretreated corn stover. Biotechnol Biofuels 4: 29.
6. Banerjee G, Scott-Craig JS, Walton JD (2010) Improving enzymes for biomass conversion: A basic research perspective. Bioenerg Res 3: 82-92.
7. Baumann MJ, Borch K, Westh P (2011) Xylan oligosaccharides and cellobiohydrolase I (TrCel7A) interaction and effect on activity. Biotechnol Biofuels 4: 45.
8. Kumar R, Wyman CE (2009) Effects of cellulase and xylanase enzymes on the deconstruction of solids from pretreatment of poplar by leading technologies. Biotechnol Prog 25: 302-314.
9. Singh RK, Tiwari MK, Kim D, Kang YC, Ramachandran P, et al. (2013) Molecular cloning and characterization of a GH11 endoxylanase from *Chaetomium globosum*, and its use in enzymatic pretreatment of biomass. Appl Microbiol Biotechnol 97: 7205-7214.
10. Kawai T, Nakazawa H, Ida N, Okada H, Tani S, et al. (2012) Analysis of the saccharification capability of high-functional cellulase JN11 for various pretreated biomasses through a comparison with commercially available counterparts. J Ind Microbiol Biotechnol 39: 1741-1749.
11. Saritha M, Arora A, Nain L (2012) Pretreatment of paddy straw with *Trametes hirsuta* for improved enzymatic saccharification. Bioresour Technol 104: 459-465.
12. Ghose TK, Bisaria VS (1987) Measurement of hemicellulase activities part 1: xylanases. Pure Appl Chem 59: 1739-1752.
13. Ghose TK (1987) Measurement of cellulase activities. Pure Appl Chem 59: 257-268.
14. Miller GL (1959) Use of dinitrosalicylic acid reagent for determination of reducing sugar. Anal Chem 31: 426-429.
15. Wood TM, Bhat KM (1988) Methods for measuring cellulase activities. Methods Enzymol 160: 87-112.
16. Wilson WK, Mandel J (1960) Determination of pentosan. Tappi 43: 998.
17. Updegraff DM (1969) Semimicro determination of cellulose in biological materials. Anal Biochem 32: 420-424.

18. Tribak M, Ocampo JA, García-Romera I (2002) Production of xyloglucanolytic enzymes by *Trichoderma viride, Paecilomyces farinosus, Wardomyces inflatus, and Pleurotus ostreatus*. Mycologia 94: 404-410.

19. Masui DC, Zimbardi AL, Souza FH, Guimarães LH, Furriel RP (2012) Production of a xylose-stimulated β-glucosidase and a cellulase-free thermostable xylanase by the thermophilic fungus *Humicola brevis var. thermoidea* under solid state fermentation. World J Microbiol Biotechnol 28: 2689-2701.

20. Sherief AA, El-Tanash AB, Atia N (2010) Cellulase production by *Aspergillus fumigatus* grown on mixed substrate of rice straw and wheat bran. Resear J Microbiol 5: 199-211.

21. Song JM, Wei DZ (2010) Production and characterization of cellulases and xylanases of *Cellulomicrobium cellulans* grown in pretreated and extracted bagasse and mineral nutrient medium M9. Biomass Bioenerg 34: 1930-1934.

22. Singhania RR, Sukumaran RK, Patel AK, Larroche C, Pandey A (2010) Advancement and comparative profiles in the production technologies using solid-state and submerged fermentation for microbial cellulases. Enzyme Microb Tech 46: 541-549.

23. Haltrich D, Nidetzky B, Kulbe KD, Steiner W, Zupancic S (1996) Production of fungal xylanases. Bioresour Technol 58: 137-161.

24. Betini JH, Michelin M, Peixoto-Nogueira SC, Jorge JA, Terenzi HF, et al. (2009) Xylanases from *Aspergillus niger, Aspergillus niveus* and *Aspergillus ochraceus* produced under solid-state fermentation and their application in cellulose pulp bleaching. Bioprocess Biosyst Eng 32: 819-824.

25. Nair SG, Sindhu R, Shankar S (2008) Fungal xylanase production under solid state and submerged fermentation conditions. African J Microbiol Res 2: 82-86.

26. Simkhada JR, Yoo HY, Choi YH, Kim SW, Yoo JC (2012) An extremely alkaline novel xylanase from a newly isolated *Streptomyces* strain cultivated in corncob medium. Appl Biochem Biotechnol 160: 2017-2027.

27. Zheng J, Guo N, Wu L, Tian J, Zhou H (2013) Characterization and constitutive expression of a novel endo-1,4-β-D-xylanohydrolase from *Aspergillus niger* in *Pichia pastoris*. Biotechnol Lett 35: 1433-1440.

28. Kumar L, Chandra R, Saddler J (2011) Influence of steam pretreatment severity on post-treatments used to enhance the enzymatic hydrolysis of pretreated softwoods at low enzyme loadings. Biotechnol Bioeng 108: 2300-2311.

29. Chapla D, Divecha J, Madamwar D, Shah A (2010) Utilization of agro-industrial waste for xylanase production by *Aspergillus foetidus* MTCC4898 under solid state fermentation and its application in saccharification. Biochem Eng J 49: 361-369.

30. Hu J, Arantes V, Saddler JN (2011) The enhancement of enzymatic hydrolysis of lignocellulosic substrates by the addition of accessory enzymes such as xylanase: is it an additive or synergistic effect? Biotechnol Biofuels 4: 36.

31. Várnai A, Huikko L, Pere J, Siika-Aho M, Viikari L (2011) Synergistic action of xylanase and mannanase improves the total hydrolysis of softwood. Bioresour Technol 102: 9096-9104.

32. Horn SJ, Vaaje-Kolstad G, Westereng B, Eijsink VG (2012) Novel enzymes for the degradation of cellulose. Biotechnol Biofuels 5: 45.

33. Berlin A, Maximenko V, Gilkes N, Saddler J (2007) Optimization of enzyme complexes for lignocellulose hydrolysis. Biotechnol Bioeng 97: 287-296.

34. Rémond C, Aubry N, Crônier D, Noël S, Martel F, et al. (2010) Combination of ammonia and xylanase pretreatments: impact on enzymatic xylan and cellulose recovery from wheat straw. Bioresour Technol 101: 6712-6717.

Bacterial Cellulose Production and its Industrial Applications

Sherif MAS Keshk[1,2*]

[1]*King Khalid University, Faculty of Science, Chemistry Department, P.O. Box 9004, Abha 61413- Saudi Arabia*
[2]*Ain-Shams University, Faculty of Environmental Studies and Research, Basic Science Department, Abbassia, Cairo 11566, Egypt*

Abstract

Microbial cellulose has proven to be a remarkably versatile biomaterial and can be used in wide variety of applied scientific endeavors, such as paper product, electronics and biomedical devices. In fact, biomedical devices recently have gained a significant attention owing to increase in tissue-engineering products for both wound care and the regeneration of demand or diseased organs. So, microbial cellulose production, characterization and influence of additives during its production are important to understand the microbial pathway and create new application.

Keywords: Bacterial Cellulose; Fiber Structure; Production Enhancement; Medical Application

Introduction

Cellulose is the earth's major biopolymer and is of tremendous economic importance globally. Cellulose is the major constituent of cotton (over 94%) and wood (over 50%). Together, cotton and wood are the major resources for all cellulose products such as paper, textiles, construction materials, and cardboard, in addition to cellulose derivatives such as cellophane, rayon, and cellulose acetate. Cellulose from major land and forest plants and cotton is assembled from glucose, which is produced in the living plant cell from photosynthesis. In the oceans, however, most cellulose is produced by unicellular plankton or algae using the same type of carbon dioxide fixation found in photosynthesis of land plants. In fact, it is believed that these organisms, the first in the vast food chain, represent nature's largest resource for cellulose production. Without photosynthetic microbes, all animal life in the oceans would cease to exist. Several animals, fungi, and bacteria can assemble cellulose. However, these organisms are devoid of photosynthetic capacity and usually require glucose or some organic substrate synthesized by a photosynthetic organism to assemble their cellulose. Several excellent reviews and papers concerning the nature of cellulose structure and biosynthesis have appeared and some of the unifying features are recounted [1-4]. This review offers synopsis of the many recent developments reported for nature of its fibril structure, enhancement of bacteria cellulose production and influence of several organic additives on the crystal structure of bacterial cellulose as well as its applications in several industries.

Microbial cellulose: A source of pure cellulose

Among the bacteria, one of the most advanced types of purple bacteria is the common vinegar bacterium, *Acetobacter*. This non-photosynthetic organism can procure glucose, sugar, glycerol or other organic substrates and convert them into pure cellulose [5,6]. Unlike the cellulose from wood pulp, cellulose produced by an *Acetobacter* strain is devoid of other contaminating polysaccharides and its isolation and purification are relatively simple, not requiring energy- or chemical-intensive processes. Further more, environmental problems due to by-products of wood pulping given an added impetus to study unexplored sources of cellulose [7]. There is no single system, which has emerged as an ideal system for the study of cellulose biosynthesis. For those new to the field, it is helpful to describe briefly some that have proven particularly useful. Very few genera of bacteria can synthesize cellulose, but the gram-negative bacterium *Gluconacetobacter xylinus* (Formerly known as *Acetobacter xylinum*) secretes large quantities of cellulose as microfibrils from a row of synthetic sites along the longitudinal axis of the cell [8,9]. The microfibrils from each synthetic site merge to form a large ribbon of cellulose; in the growth medium. These ribbons and the associated cells tangle form a floating pellicle that allows the non-motile, strictly aerobic bacteria to grow in the higher oxygen tension at the surface of the growth medium. Upon contact with host plant cells, the tumor-forming bacterium *Agrobacterium tumefaciens* secrets cellulose fibrils from all sides of the cell, a process that aids in cell attachment and promotes virulence [10,11]. Both *Gluconacetobacter xylinus* and *Agrobacterium tumefaciens* can be grown in large quantities and can be transformed. Mutants of both can be selected that are impaired in cellulose biosynthesis and genes and enzymes involved in this process have been identified in both bacteria. The cellulosic algae have proven quite useful for freeze-fracture studies in which putative synthase complexes were first visualized in the plasma membrane, but they have proven difficult to study at the biochemical and molecular levels [5]. Algal cellulose from different origins has been analyzed with special reference to the crystalline features such as allomorphism, dimension [12]. The slim mold *Dictyostelium discoideum* synthesizes cellulose at various stages in its life cycle. A cellulose synthase activity has been demonstrated in this organism [13] and many sophisticated genetic approaches are available that make it very attractive for developmental studies of cellulose biosynthesis. The water mold *Saprolegnia* contains both 1,4-β-D-glucan and β-D-1,3-glucan in its cell walls and separable enzymes have been detected that synthesize these glucans in vitro, but genetic studies are not advanced with this organism [14]. In higher plants, cotton fibers represent an interesting object of study [15]. These are single cells that elongate from the epidermal layer of the ovule and they elongate synchronously within the boll. There is no experimental evidence that bacterial cellulose biosynthesis differs from those of green plants, but Uridine Diphosphoglucose (UDPG) is necessary intermediate in the synthesis of bacterial cellulose while Guanidine Diphosphoglucose (GDPG) is involved in the biosynthesis of green plant cellulose [7]. A good model system to study the mechanism of cellulose biogenesis is the bacterium *Gluconacetobacter xylinus*

***Corresponding author:** Sherif Mohamed Abdel Salam Keshk, King Khalid University, Faculty of Science, Chemistry Department, P.O. Box 9004, Abha 61413- Saudi Arabia

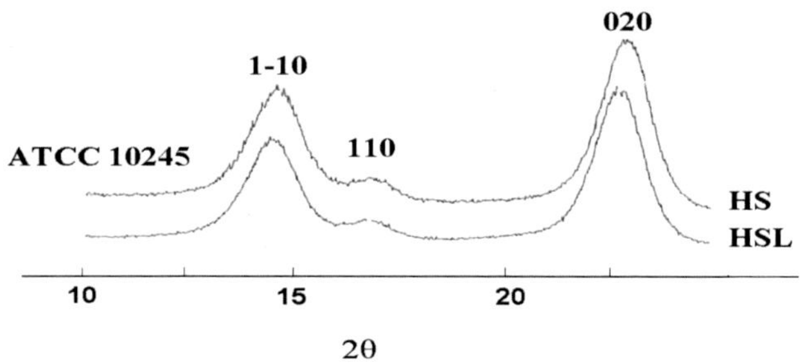

Figure 1: X-ray pattern of BC produced from HS and HSL media in presence of *Gluconacetobacter xylinus* ATCC 10245.

that produces pure cellulose as an extracellular product. Figure 1 explains the proposed model for regulation of cellulose synthesis in *Gluconacetobacter xylinus* [16-21].

Bacterial cellulose fibril structure

Application of solid state 13C NMR analysis to cellulose from 1980 taken a new turn for structural analysis of cellulose I, which indicate that native cellulose consist of two different crystal structures, cellulose I_α and I_β [22,23]. The primary difference between cellulose I_α and I_β appears at the pattern of C1 resonance around 106 ppm, singlet resonance for I_α and doublet one for I_β. Hirai et al. [24] reported that almost pure cellulose I_β could be prepared from cellulose I_α -rich samples by hydrothermal treatment under slightly alkaline conditions at 260°C for 30 min. Since cellulose I_β is formed irreversibly from cellulose I_α, the structure of cellulose I_β must be thermodynamically more stable than that of cellulose I_α. Cellulose I_α is crystallize in larger-size microfibrils, whereas cellulose I_β is formed in smaller-size microfibrils. Cellulose I_α must be crystallized in the higher energy state compared to the case of cellulose I_β, because the crystal transformation is induced from cellulose I_α to I_β at higher temperatures. The unit cell of cellulose I_α is a triclinic and that for cellulose I_β is monoclinic unit cell [25].It seems almost acceptable that native cellulose crystals are composites of two allomorphs, cellulose I_α and I_β [26], although there is some significant exception in nature [25,27]. The ratio of cellulose I_α and I_β differs greatly from species to species [25]. Why do two allomorphs exist and why does their ratio vary in nature. One approach to answering these questions is to characterize the crystallization process for a cellulose-producing bacterium, *Gluconacetobacter xylinus,* as a model system. The biosynthesis of cellulose by this bacterium has been relatively well studied and their ratio is changed depending on the culture conditions, e.g. additives and temperature [28,29]. According to the recent structural model [30], several tens of cellulose synthesizing sites are located in the cytoplasmic membrane, being aligned parallel with the longitudinal axis of the bacterial cell. Cellulose synthases in the respective sites produce 12-15 cellulose chains and extrude them into the culture medium as fine fibrils with a lateral width of about 1.5 nm (such fine fibrils are often called supplementary fibrils) through small pores in the outer membrane. These supplementary fibrils aggregate to form microfibrils 3-6 nm in width, and the resulting microfibrils further aggregate to produce a typical ribbon assembly with a lateral width of 40-60 nm. The term of microfibril has been proposed to denote the thinnest fibril structures visible in the field of the electron microscope. Crystallization of cellulose will be induced during the aggregation processes because the supplementary fibrils may be too fine to crystallize. Sea algal cellulose and bacterial cellulose, which is produced by *Gluconacetobacter xylinus,* are found to be rich in I_α; the average mass fraction of cellulose I_α is about 0.63. For example, the mass fractions of cellulose I_α are 0.64 for *Valonia macrophysa;* 0.60 for *Valonia aegarropila*, 0.67 for *Chaetomorpha*. In the case of bacterial cellulose, the mass fraction also depends on strains and culture temperature, ranging from 0.64 to 0.71. Careful purification with aqueous alkaline solution will also induce the reduction of the content of cellulose I_α by high percent in bacterial cellulose. On the other hand, cellulose forming cell walls for higher plants such as cotton and ramie is rich in cellulose I_β, the mass fraction being about 0.8, as for the characterization of wood [25]. Bacterial cellulose can be produced by two kinds of culture method, namely static and agitated [31]. Under static culture conditions a typical gelatinous membrane of bacterial cellulose is accumulated on the culture surface whereas, under agitated culture conditions bacterial cellulose is produced in well-dispersed slurry as irregular masses such as granule, stellate and fibrous strand. The agitated culture method seems to be suitable for the industrial production of bacterial cellulose and serves commercial application in various fields. The static culture method has been adopted mainly for the investigation of cellulose production because cellulose deficient mutants sometimes appear in agitated cultures; these mutants were believed to interfere with the fermentative production of bacterial cellulose [31]. Several researchers, however, have succeeded in producing bacterial cellulose using stable cellulose-producing strains in agitated culture [31,32]. Nevertheless, a patent claimed that a specific reticulated structure observed by scanning electron microscopy was generated in bacterial cellulose produced in agitated culture, but not in that produced in static culture. Watanabe et al. [31] revealed that bacterial cellulose produced in agitated culture had a lower degree of polymerization (DP) and crystallinity compared with that produced in static culture. These finding suggest that the smaller particle size of agitated bacterial cellulose lead to a higher water holding capacity (WHC) than static bacterial cellulose. ACP/ MAS ^{13}C NMR analysis revealed that the cellulose Iα content of the cellulose produced in agitated culture was lower than that of the cellulose produced in static culture. Therefore it has a lower Young's modulus of sheet, higher water holding capacity and a higher suspension viscosity in the disintegrated form than produced in static culture [32].

Biosynthesis of cellulose from various carbon sources

In general, glucose has been used as a carbon source for cellulose production by *A. xylinum*. It has been reported, however, that cellulose was also synthesized from other carbon sources, such as 5- or 6-carbon monosaccharides, oligosaccharides, starch, alcohol and organic acids [33]. Fructose and glycerol gave almost the same cellulose yield as glucose [34]. The yield of cellulose, relative to the glucose consumed, decreased with an increase in initial glucose concentration, and gluconic acid accumulated at a high initial glucose concentration. The decrease in cellulose yield could be due to some glucose being metabolized to gluconic acid. The optimum pH for cellulose production is between 4.0 to 6.0. Production of cellulose from D-arabitol by *Gluconacetobacter xylinus* KU-1 (it was obtained from the type culture collection of the laboratory of applied biological chemistry, Yamaguchi University, Japan) was studied [35]. The amount of cellulose from D-arabitol was more than 6 times as much as that from D-glucose. In D-arabitol medium, the final pH did not decrease and D-gluconic acid was not detected, this seems to be one of the reasons for high productivity of cellulose from D-arabitol. Romano measured the yield of cellulose from glucose, galactose or xylose as a carbon source. Galactose and xylose gave smaller yields, mostly because of slower growth rates. The microfibrils obtained from xylose are also less regular than those obtained from glucose. Gas chromatography analysis showed that the composition of sugars in the polymers (cellulose and other by-product) produced from xylose as a carbon source were still 80% glucose and 20% other sugars, and there were no differences in the degree of polymerization. Times initiation of production was also different for the different carbon sources. The use of the cheaper carbon sources might be able to reduce the production cost. Bacterial cellulose yield from sucrose is only half those from glucose, it is due to the low activity of sucrase in *A. xylinum*. If sucrose is efficiently hydrolyzed, it will increase bacterial cellulose productivity. Tajima et al. [36] have succeeded in enhancing bacterial cellulose productivity by the co-cultivation of two different types of *Gluconacetobacter xylinus* strains (NCI 1051 or ATCC 10245 and NCI 1005). The bacterial cellulose productivity of *Gluconacetobacter xylinus* by co-cultivation was greater than that by monocultivation. This seems to be due to the formation of glucose and fructose through the hydrolysis of sucrose by sucrase secreted from *Gluconacetobacter xylinus* NCI 1005. These researchers used *Gluconacetobacter xylinus* strains (IFO, ATCC or NCI) while other studies employed *Gluconacetobacter xylinus,* the nata organism, which was isolated directly from a nata. Nata is a favorite dessert delicacy native to the Philippines and produced mainly in the coconut regions. The cellulose "gel" or nata is synthesized by a strain of *Gluconacetobacter xylinus* grown in coconut water or fruit juices (pineapple, tomato, etc.) supplemented with sucrose [37,38]. In this case, fructose gave the best yield, followed by a combination of fructose + lactose. Sucrose and a combination of fructose + lactose also gave good yields. These four carbohydrate sources gave significantly better yields than any of the other sources. The presence of glucose in the medium alone or in combination with other sugars produced lower amounts of cellulose. The cellulose yields from sucrose, although lower than those from fructose.

Enhancement of bacterial cellulose production

Bacterial cellulose production has been reported to be enhanced by the addition of a small amount of a cellulase complex (cellulase complex has endoglucanase, exocellobioglucosidase and β-glucosidase activities and cellulose binding) to the culture broth [39]. Tonouchi et al. [40] examined cellulose production by *Gluconacetobacter xylinus* using a single species of β-1,4-endoglucanase and its active site mutant, which still had cellulose-binding ability, but was devoid of the enzyme activity, to identify the effect of the cellulose-binding ability on cellulose production. They found that the endoglucanase activity itself, not the cellulose-binding ability, enhanced bacterial cellulose production. The addition of small amount cellulase does not affect the degree of polymerization of the bacterial cellulose but enhance the production rate [40]. It has been found that the addition of a commercial sulfite pulping waste fraction (named CP powder) into culture medium remarkably enhanced the efficiency for cellulose production of *Gluconacetobacter xylinus* [41]. The CP powder was fractionated by gel filtration with Sephadex G-25 into the high and low molecular fractions (named A and B) and it was clarified that the high molecular lignosulfonate fraction (A) was essential for the remarkable yield increase in the cellulose production of *Gluconacetobacter xylinus.* Premjet et al. [42] found a simpler separation method as follow; the larger amount of CP powder was treated by single ethanol precipitation and divided into a precipitate (high molar fraction, C) and non-precipitate (low molar fraction, D). Fraction C consists mainly of high molecular lignosulfonate almost equivalent to fraction A (which separate by Sephadex G-25) and shows low sugar contents and strong UV absorption at 280 nm. Fraction D, which shows high sugar contents and lower UV absorption is almost equivalent to fraction B (which separate by Sephadex G-25) and is deduced to consist mainly of the low molecular weight carbohydrates and the low molecular weight lignosulfonate. Fraction A differs from fraction C in that the former contains UV absorbing substance in the D region. Fraction B also has a little UV absorption in C region. Both fraction A and B greatly affected the cellulose production [41]. In which, fraction A is much more effective than fraction B. The remarkable yield increase by the addition of fraction A means that the ethanol precipitation method is equally effective as the gel-filtration method to separate the substance which contribute to the remarkable yield increase from the whole CP powder. Fraction A was further purified by the repetition of ethanol precipitation to give precipitate (fraction A1) and non-precipitate (fraction A2) then fraction A1 was further purified to give precipitate (fraction A3) and non-precipitate (fraction A4). It was observed that high molecular fraction A3 gave the highest efficiency and the low molecular fraction A2 and A4 were followed [41]. This indicates that the fraction A3 is the highly purified high molecular lignosulfonate with the lowest sugar content. The efficiency of fraction A3 is about two times larger than those of fractions C and A. From these results the carbohydrates in CP powder do not play any important role on the cellulose production. The more the high molecular lignosulfonate fraction was purified the more the efficiency for cellulose production increased. This means that the purification to eliminate both carbohydrates and low molecular fraction is effective. The prominent cellulose productivity was also observed by the addition of either sodium or calcium lignosulfonates on the market, which indicates that the high molecular lignosulfonate in CP powder is the dominant contributor for the high cellulose productivity [41]. Influence of Vitamin C (Ascorbic acid) on bacterial cellulose (BC) production and crystal structure was studied using four strains of *Gluconacetobacter xylinus (*ATCC 10245, IFO 13693, 13772 and 13773) [43]. BC productivity of all strains was increased in presence of Vitamin C (0.5% w/w), the average BC production reached to 0.47g/30ml compared with 0.25g/30ml without Vitamin C. Enhanced productivity is associated with a decrease in gluconic acid concentration that produced from *Gluconacetobacter xylinus* during BC production. X-rays results, showed, the crystallinity index of BC produced in presence of ascorbic acid was the lowest with remarkable change in d-spacing. These results confirmed by using solid state ^{13}CNMR. The increase in BC yield in presence of Vitamin C is due to its antioxidant behavior and confirms our past work on lignosulfonate influence on BC.

Influence of the additives on the crystal structure of BC

Lignosulfonate: The influence of lignosulfonate on BC structure and productivity was studied using six strains of *G. xylinus* (ATCC 10245, IFO 13693, 13772, 13773, 14815 and 15237) [44]. BC productivity of all strains was improved almost 57% in the presence of (1% w/v) lignosulfonate. Ft-IR result showed that BC produced in the presence of lignosulfonate displayed the higher crystallinity idex and I_{α}-rich cellulose. Furthermore, X-ray diffractogram confirmed the FT-IR results and showed no remarkable change in d-spacing (Figure 1). From CP/MAS ^{13}C NMR spectra (Figure 2), the amorphous region in the presence of lignosulfonate was relatively lower, that means the lignosulfonate enhanced the crystalinity of BC. These results indicate that the increase in BC yield is due to the inhibition of gluconic acid formation in the presence of the antioxidant, polyphenolic compounds in the lignosulfonate. Furthermore, a scanning electron micrograph showed the ribbons that were produced in presence of lignosulfonate are coarse and less entangled compared with the ribbons that were produced by the control medium [44].

Moreover, the average Young's modulus was 633.00 Mpa compared with 450.76 Mpa in absence of lignosulfonate (Figure 3). It was also

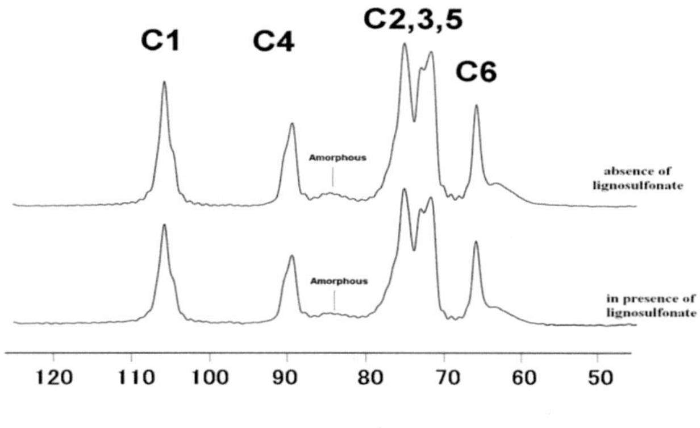

Figure 2: Solid state ^{13}C NMR spectra of BC produced from HS and HSL media in Presence of *G. xylinus* ATCC 10245.

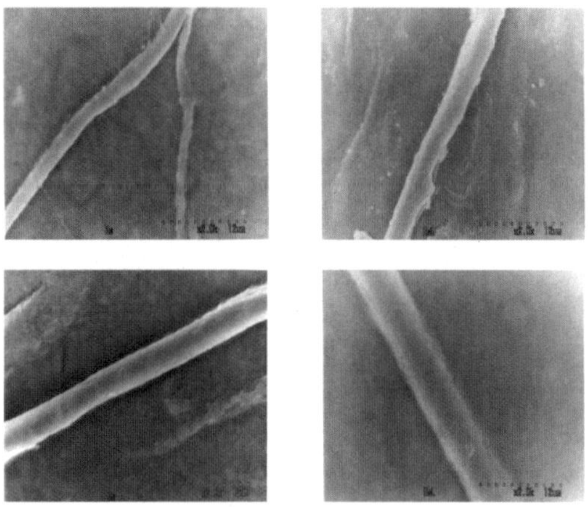

Figure 3: Scanning Electron Micrographs (12 μm) of (Left) Bacterial cellulose in control medium. (Right) Bacterial cellulose in lignosulfoante medium.

Sample	BC* Yield mg/ml	Tensile Strength (N)	Young's Modulus (MPa)	BC Viscosity (cP)	DP
HS 10245	4.4	10.0	331.75	34.87	1747.71
HSL 10245	7.2	18.28	606.45	70.05	2240.61
HS 13772	8.7	15.51	514.55	37.12	1791.40
HSL13772	11.4	18.69	620.05	84.80	2377.81
HS 13773	10.1	15.03	498.62	37.05	1790.08
HSL 13773	16.2	19.23	637.95	84.25	2372.80
HS 13693	7.9	13.81	458.15	36.88	1786.86
HSL 13693	16.3	21.53	714.27	68.82	2227.89

*Based on the volume of the medium
RV= relative value= yiled in HS/ yiled in H

Table 1: Characterization of Bacterial Cellulose and the Sheets Produced from Four Strains of *Gluconacetobacter xylinus.*

found that the average viscosity values of the produced sheets from lignosulfonate medium was 76.98 cP compared with 36.48 cP from control medium, which indicates a high degree of polymerization.

The ribbon morphology was estimated by electron microscopy from both HS and HSL media [44]. The electron micrographs of the two samples are presented in Figure 3. The morphology of these two samples was not the same as observed in Figures in which the morphology of BC produced from HS comprises thin ribbon compared with thick ribbon that were produced from HSL medium. The ribbons produced in the presence of lignosulfonate appeared coarser and less entangled than the ribbons produced by the control medium [44]. Since 1976, Brown et al, have interesting cytological observations on the mode of ribbon production at cell division. In which, the long row of microfibril synthesizing sites is longitudinally duplicated prior to division and that both parent and daughter sites are active just before division. The two sites are intercepted at cell fission, and each daughter cell receives a comparable set of synthesizing complexes, which does not extend in length and which synthesizes a microfibrillar ribbon at constant dimensions.

The two sites are intercepted at cell fission, and each daughter cell receives a comparable set of synthesizing complexes, which does not extend in length and which synthesizes a microfibrillar ribbon at constant dimensions.

This means that, the fibrils of BC is, perhaps, formed by the bunching of microfibrils, which are excreted from pores aligned on the surface of the cells in a row along their longitudinal axis. If the excretion continues beyond cell generation, the microfibrils of parent cells will inherit one-half by each of the daughter cells, forming a branching point on the fibril, so the diameter of fibril may be narrower after cell division, although it must recover as the new cell matures. Alternatively, the normal number of pores for excreting microfibrils will be provided in the new cells at the stage of division as suggested by Brown et al. [5]. In the presence of lignosulfoante, cell division appears to be inhibited, and the ribbons do not split and adjacent ribbons could to merge to form wider ribbons or aggregates of ribbons [44]. Thus adding lignosulfonate produced ribbons which appear wider than the ribbon in the control medium, because several ribbons from a cluster of cells whose dividing process was inhibited. The effects of the lignosulfoante on the tensile strength and Young's modulus for BC sheets are listed in Table 1.

The sheets made from BC cultivated for 30 days at 28°C on adding lignosulfonate gave 20-56% higher Young's modulus than those from the control sheets. Furthermore, the viscosity and the degree of polymerization of BC sheets produced from HSL were higher than those from the control medium (Table 1). These results are attributed not only to the increase in the effective cross-sectional area that produces an increase in cross-sectional momentum but also the increase in the number of 1, 4 covalent bond. Hence, the wider ribbons gave a better uniplanar orientation, which produced a higher Tensile strength and Young's modulus. One factor that gave a higher Young's modulus of sheets made from BC is considered to be its ribbon form, because ribbons are more easily oriented in one plane when they are pressed [45]. These results matched with Huang et al., results. Where they studied the effect of lignosulfonate on properties of soy protein plastics and showed that, the introduction of lignosulfonate enhanced the tensile strength and Young's modulus of soy protein plastics.

Hemicellulose: Hemicellulose is also one of the main components of liginified cell wall. Monosaccharide residue constituting hemicelluloses are arabinose and xylose (pentose), glucose, galactose and mannose (hexose), rhamnose and fucose (6-deoxyhexose) and galacturonic, glucuronic and 4-O-methylglucuronic acids (Uronic acids). The dominating hemicellulose in hard wood is an O-acetyl (4-O-methylglucurono) xylan (so called hard wood xylan) accompanied by a small amount of glucomannan. On the other hand, softwood contains an O-acetylgalactomannan (so-called softwood glucomannan) as the major hemicellulose with a small amount of an arabino (4-O-methylglucurono) xylan (so-called softwood xylan).

Cellulose produced in presence of hemicellulosic polysaccharides altered crystallinities and were found to be more like the I_β-type crystalline structure. Mannan is believed to lower the degree of crystallinity of cellulose by decreasing the crystallite size, while xylan and xyloglucan are likely to co-crystallize with the cellulose and change the cellulose lattice structure by creating lattice defects [46,47]. The high affinity of xyloglucan for bacterial cellulose corresponds with xyloglucan-cellulose networks in the primary cell wall. In the case of secondary cell wall, however, the situation is not simple, because hemicellulsoe is chemically associated with lignin. It is recommended to use native hemicellulose present in woody plants for further investigation of the arrangement of cellulose-hemicellulose composites in the secondary cell walls of higher plant. For revelation of the affinity between cellulose and hemicellulose in secondary cell wall of higher plant, *Gluconacetobacter xylinus* produced the cellulose composites in the presence of glucuronoxylan glucomanan, O-acetyl-glucuronoxylan, arabinoglucuronoxylan, arabinogalactan and xyloglucan.The cellulose composites were investigated by X-ray diffraction and sugar analysis [48]. Differactions of bacterial celluloses cultured in the presence of xylan (Xyl) and arabinogalactan (AraGal) retain the plane orientation of I_α-rich form. In the cases of glucomannan (GlcMan) and xyloglucan (xylGlc) indicating the increase in the disordered structure in these celluloses. Hemicellulosic polysaccharide contents in bacterial celluloses cultured in the presence of hemicellulosic polysaccharides were 3.0% (arabinogalactan), 28.2% (xylan), 48.6% (glucomannan) and 47.0% (xyloglucan). Based on these results, glucomannan was concluded to have the highest affinity to bacterial cellulose [48]. Interaction between bacterial cellulose and glucomannan have been studied through molecular and ultrastructural analysis of composites formed by deposition of cellulose into solutions containing either glucomannan or galactomannans of varying Man:Gal ratio. 13C NMR suggests that unsubstituted mannan segments can bind to cellulose by undergoing a conformational transition to an extended 2-fold form. The presence of cellulose fibrils promotes network formation from galactomannan solutions under conditions where this would not normally occur [49].

Carboxymethyl Cellulose (CMC): The effects of carboxymethyl cellulose (CMC) and xyloglucan (XG) added to the incubation medium were examined [28,29]. Because CMC and XG are known to interfere with the aggregation of microfibrils into normal ribbon assembly [50,51]. It was found that the mass fraction of cellulose I_α is greatly decreased with increasing concentration of CMC, XG in the incubation medium and this decrease in the mass fraction of cellulose I_α corresponds to the enhanced crystallization of cellulose I_β. Hence, the cause of the decrease in the fraction of cellulose I_β is expected to be associated with the reduction in the lateral width of microfibril induced by the addition of CMC or XG [50]. Recently, the effect of the CMC and XG additives with different degree of polymerization (DP) or substitution (DS) on the formation of the microfibril has been examined by transmission electron microscope (Hirai *et al.*, 1998). CMC with DP=80 and DS=0.57 is the most effective in producing separate, smaller-size microfibrils. By increasing the concentration of CMC in cultivation medium from 0.1 to 1.5%, the percentage of microfibrils measuring 3-7 nm wide is increased. Whereas, XG are less effective than CMC in producing microfibrils with smaller sizes and the resulting microfibrils still tend to aggregate.

Direct dyes: *Gluconacetobacter xylinus* normally produces crystalline 30Å microfibers in association with intracellular synthesizing sites in the lipopolysaccharide layer of the bacterium [52]. When actively synthesizing *Gluconacetobacter xylinus* cell were incubated in the medium or phosphate buffer containing more than 0.01% calcofluor (fluorescent brightener), ribbon assembly was disrupted, instead of twisting ribbons and a broad bands of bent fibrils were formed [53]. High-resolution micrographs reveal that, the smallest fibrils in the band product measured 15 Å (± 4 Å) and larger fibrils appeared to arise by fascination of smaller 15 Å. Fibrils frequently show pronounced curvature and undulation, which suggest low crystallinity. While the larger aggregate often bend sharply and appear more rigid. The altered cellulose synthesized is known to be high molecular weight, noncrystalline β-1,4-D-glucans. X-ray crystallography indicate that, the calcofluor induce product of cellulose I crystallinity after drying although it had no detectable crystallinity in the wet state [52]. This lack of crystallinity before drying indicate that, calcofluor effectively separate polymerization from crystallization in the wet state by preventing microfibril assembly and the calcofluor induced products profoundly different from native bacterial cellulose [23]. On drying, the hydrogen bonds of the nascent crystal lattice must be so favorable than the association between calcofluor and glucan. So that, calcofluor is displaced and the cellulose I lattice form [23]. The effect of direct dyes was due to extensive lateral precipitation of the dyes onto the growing cellulose microcrystals, which prevented them from reaching their normal size [53]. A number of direct dyes (Direct Red 28, Direct Red 79, Direct Blue 14, Direct Blue 15, Direct Blue 53 and Direct Blue 75) were selected to clarify the effect of them on the nascent structure of bacterial cellulose [54-57]. Cellulose IV_I and cellulose II are regenerated from Direct Red 28 and Direct Red 79, Direct Blue 14, Direct Blue 15, Direct Blue 53 and Direct Blue 75 complexes, respectively. According to the chemical structure of direct dyes which are used. The sulfonate group plays an important factor in the effect of direct dyes on the crystal structure of nascent bacterial cellulose. Kai and Keshk [58,23] confirmed the influence of sulfonate group by using four substituent of fluorescent brightener. All of them have the same skeletal structure (stilbene group) but the difference in number and position of sulfonate group. The fluorescent brightener substituent that has more sulfonyl group gave cellulose II after dye extraction. While fluorescent brightener substituent that has less sulfonyl group gave cellulose I after dye extraction [59].

DCB (2,6-dicholoronitrobenzonitrile): 2,6-dichlorobenzonitrile (DCB) is herbicide, which has been shown to be specific and effective inhibitor for cellulose synthesis in algae, higher plants and also bacteria. The extent of the inhibition is dependent on the concentration of DCB. Normally, the bacterial cellulose produced is in the cellulose I allomorph, which is predominant native form. Whereas in presence of 12 μM DCB, a mixture of cellulose I and II allomorphs were formed. X-ray diffraction method and Raman spectrum confirmed this result [60].

Reactivity of BC toward chemical reaction

Few researches have been discussed the chemical reactivity and either heterogeneous or homogenous reactions of BC [61,62].

Heterogeneous reaction: Bacterial cellulose differs from plant cellulose with respect to its high crystallinity and purity (free of lignin and other biogenic products), high water-absorption capacity and better mechanical strength [63]. The typical properties of BC have also inspired investigations on the reactivity and availability of the hydroxyl groups for chemical reactions as well as the subsequent characterization of the products obtained [61]. The carboxmethylation and cyanoethylation reactions were performed on BC, viscose grade pulp and cotton linter to study the reactivity of these cellulose samples [61]. Bacterial cellulose differs from plant celluloses (viscose grade pulp and cotton linter) with respect to its high crystallinity and purity from lignin. BC is more reactive to cyanoethylation and carboxymethylation than that of plant cellulose according to its high crystallinity with low degree of polymerization [61].

Homogenous reaction: Homogenous reaction of dissolved BC with acetic anhydride and phenyl isocyanate showed products with high degree of substitution value up to 3.0 compared to plant cellulose without explaining the reason of this reactivity. Since cellulose dissolves in 85% phosphoric acid with limited hydrolysis, its homogenous oxidation of the primary hydroxyl groups with halogen oxides have been achieved. Interestingly, iodination of cellulose in 85% phosphoric acid using mixture of KIO_3 and KI gave triester of hypoiodous acid. Homogenous derivatizations (selective Iodination and Oxidation) using phosphoric acid have been performed on different cellulose samples (BC, Kenaf and microcrystalline cellulose) to reveal their chemical reactivates and their physicochemical characterizations [64]. The results indicate that BC in phosphoric acid has the highest reactivity toward iodination reaction. This may be attributed to the use of phosphoric acid that degrades the crystal structure of cellulose. So the crystallinity does not play an important factor as in the case of heterogeneous reactions [61].

Bacterial cellulose and medical application

Due to its high purity, hydrophilicity, structure forming potential, chirality and biocompatibility offers a wide range of special applications, e.g. as a food matrix (nata de coco), as dietary fiber, as an acoustic or filter membrane, as ultra-strength paper and as reticulated fine fiber network with coating, binding, thickening and suspending characteristics up to now several applications of bacterial cellulose in human and veterinary medicine are known.

Skin therapy

The high mechanical strength in the wet state, substantial permeability for liquids and gases and low irritation of skin indicated that the gelatinous membrane of bacterial cellulose was usable as an artificial skin for temporary covering of wounds (Figure 4). Biofill® and Gengiflex® are products of bacterial cellulose with wide applications in surgery and dental implants and realities in the human health-care sector. Cases of second and third degree of burns, ulcers and others could be treated successfully with Biofill® as temporary substitute for human skin [65]. The authors documented the following advantages for Biofill® in more than 300 treatments: immediate pain relief, close adhesion to the wound bed, diminished post-surgery discomfort, reduced infection rate, easiness of wound inspection (transparency), faster healing, and improved exudates retention, spontaneous detachment following reepithelization, and reduced treatment time and costs.

Only one disadvantage was mentioned: limited elasticity in areas of great mobility. Gengiflex was also developed to recover periodontal

Figure 4: The never-dried bacterial cellulose membrane is a nonpyrogenic and fully biocompatible biomaterial with high mechanical strength.

tissues. Further results and applications of both Biofill and Gengiflex were published by different authors. Schmauder et al., described the application of bacterial cellulose (Cellumed) in veterinary medicine to treat recent, large surface wounds on horses and dogs [66].

Artificial blood vessels

If the coronary vessels around the heart are blocked as a result of hardening of the arteries, it may be necessary to carry out a bypass operation. The cellulose produced by bacteria could be used for artificial blood vessels as it carries a lower risk of blood clots than the synthetic materials currently used for bypass operations. BC can carry a lower risk of blood clots than the synthetic materials currently in use. This means that the cellulose works very well in contact with the blood and is a very interesting alternative for artificial blood vessels. Real blood vessels have an internal coating of cells that ensure that the blood does not clot. This tube has an inner diameter of 1mm, length of about 5 mm and wall thickness of 0.7 mm. So these parameters are sufficient for experimental microsurgical requirements. A sufficient mechanical strength of the BASYC tubes is one of the essential properties for their use in microsurgery. The material must resist both mechanical strains during microsurgical preparation and anatomizing and blood pressure of the living body. The native bacterial cellulose has mechanical properties, including shape retention and tear resistance, which are superior to many synthetic materials. In comparison with organic sheets, like polypropylene, polyethylene-terephthalate or cellophane. BC processed into a film or sheet show remarkable mechanical strength. Compliance mismatch between the synthetic graft and the surrounding native tissue has been reported as a major factor in ultimate failure of the currently used cardiovascular graft replacements. Thus, developing biomaterials that display close mechanical properties as the tissue it is replacing is an important objective in biomedical devices design. Polyvinyl alcohol (PVA) is a biocompatible hydrogel with characteristics desired for biomedical applications. It can be cross linked by a low temperature thermal cycling process. By using a novel thermal processing method under an applied strain and with the addition of a small amount of bacterial cellulose (BC) nano fibers, an anisotropic PVA-BC nano composite was created. The stress–strain tensile properties of porcine aorta were closely matched in both the circumferential and the axial directions by one type of anisotropic PVA-BC nano composite (10% PVA with 0.3% BC at 75% initial strain and cycle 2) within physiological range, with improved resistance to further stretch beyond physiological strains. The PVA-BC nano composite gives a broad range of mechanical properties, including anisotropy, by controlling material and processing parameters. PVA-BC nano composites with controlled degree of anisotropy that closely match the mechanical properties of the soft tissue it might replace, ranging from cardiovascular to other connective tissues, can be created. Where, BC has been investigated as a potential new vascular graft material by evaluating cell and blood interactions with BC. The specific aims were to evaluate if surface modifications could promote human endothelial cells and to investigate the thrombogenic properties of BC compared to conventional graft materials. Modification of BC with a novel technique, were xyloglucan is used as a carrier molecule for the adhesion-promoting peptide RGD, resulted in increased cell adhesion, metabolism and cell spreading. Luminal coating of BC-tubes with fibrin glue resulted in increased cell adhesion during static experiments and good cell retention during physiological shear stress. The evaluation of thrombogenicity in human blood plasma revealed that BC induces a slower coagulation compared with clinically available materials such as Gore-Tex® and Dacron®. In addition, BC induced the least contact activation evaluated by XIIa generation. A Chandler loop system with freshly drawn blood showed that BC consumed low amounts of platelets and generated low thrombin values compared with Dacron® and Gore-Tex®.

Potential scaffold for tissue engineering

The use of scaffolds in the tissue engineering of cartilage is essential in order to support cell proliferation and maintain their differentiated function in addition to definition of the shape of the new growing tissue. To this end, a variety of scaffold materials have been evaluated including natural polymers like collagen, alginate, hyaluronic acid, fibrin glue and chitosan and synthetic polymers including poly glycolic acid (PGA),polylactic acid (PLA), polyvinyl alcohol (PVA), polyhydroxy ethyl methacrylate (pHEMA) and polyN iso propyl acryl amide (pNIPAA). However, tissue constructs with native mechanical properties have not yet been described in the literature. *G. xylinus* was explored as a novel scaffold material due to its unusual material properties and degradability. Native and chemically modified BC materials (phosphorated and sulfonated BCs) were evaluated using bovine chondrocytes. The results indicate that unmodified BC supports chondrocyte proliferation at levels of approximately 50% of the collagen type II substrate while providing significant advantages in terms of mechanical properties. Compared to tissue culture plastic and calcium alginate, unmodified BC showed significantly higher levels of chondrocyte growth. Chemical sulfonation and phosphorylation of BC, performed to mimic the glucose amino glycans of native cartilage, did not enhance chondrocyte growth while the porosity of the material did affect chondrocyte viability. BC did not induce significant activation of pro-inflammatory cytokine production during in vitro macrophage screening. Hence, unmodified BC was further explored using human chondrocytes. TEM analysis and RNA expression of the collagen II from human chondrocytes indicated that unmodified BC supports proliferation of chondrocytes. In addition, in growth of chondrocytes into the scaffold was verified by TEM. The results suggest the potential for this biomaterial as a scaffold for tissue engineering of cartilage.

Wound care products

In the early 1980's Johnson & Johnson pioneered in exploratory investigations on the use of microbial cellulose as a liquid loaded pad for wound care (see US Patent 4,655,758; 4,588,400). Since that time, a company in Brazil, Biolfill Industries, has continued to investigate the properties of microbial cellulose and is beginning to market specific microbial cellulose products in the wound care market [65,67]. In Brazil, the purified gelatinous membrane of bacterial cellulose was developed and commercialized as an artificial skin (wound dressing) [65]. A high mechanical strength in the wet state, substantial permeability for liquids and gases and low irritation of skin, indicated that the gelatinous membrane of the bacterial cellulose was superior to conventional quaze as an artificial skin for temporary covering off wounds [67]. Bacterial cellulose composites made by blending chitosan, poly (ethylene glycol) (PEG), and gelatin for potential biomedical application of tissue-engineering scaffold and wound-dressing material [68]. The bacterial cellulose composites were successfully prepared by immersing a wet bacterial cellulose pellicle into chitosan, PEG, or gelatin solutions followed by freeze-drying. The products look like a foam structure. Scanning electron microscopy images show that chitosan molecules penetrated into bacterial cellulose forming a multilayer and a well interconnected porous network structure with a large aspect surface. The morphology of the bacterial cellulose/gelatin scaffold indicates that the gelatin molecules could penetrate well between the individual nano-fibers of the bacterial cellulose. Cell adhesion studies for these composites were carried out using 3T3 fibroblast cells [69]. They showed much better biocompatibility than pure bacterial cellulose.

Tablet modification

A new preparation method of microcrystalline cellulose from *G. xylinus* (BC) and kenaf (KF) is reported [63]. The developed cellulose (DBC and DKF) materials showed different crystalline structures. DBC exhibited cellulose I lattice with high crystallinity (85%) whereas DKF showed cellulose II lattice with high crystalinity (70%) The particle size of DKF was 5-20 μm while that of DBC was 1-5 μm. The physical properties of the DBC and DKF materials were compared with those of the commercially available microcrystalline cellulose Avicel® PH 101. DBC exhibited lower value of the loose density than those of DKF and Avicel PH 101 [70-74]. Both microcrystalline DBC and Avicel PH 101 demonstrated similar behavior during flow and binding processes. The thermal properties of DBC and DKF materials were investigated by thermo-gravimetric analysis [75]. The TGA results reveal increased thermal stability for DBC compared to DKF. Moreover, the weight loss of DBC occurred in one step degradation process from about 320°C to 380°C, which is mainly due to the decomposition of cellulose. These differences in the thermal properties of both cellulose materials are mainly due to differences in their degree of crystallinity. This agrees with XRD data where DBC exhibits 85% degree of crystallinity and DKF 77%. This revealed a relationship between crystal structure and the thermal degradation of cellulose. A greater crystalline structure required a higher degradation temperature and, therefore, DBC degraded at 320°C whereas DKF at 215°C.

Bacterial cellulose and paper industry

Microbial cellulose has been investigated as a binder in papers, and because it consists of extremely small clusters of cellulose microfibrils, this property greatly adds to strength and durability of pulp when integrated into paper. Ajinomoto Co. along with Mitsubishi Paper Mills in Japan are currently active in developing microbial cellulose for paper products (JP patent 63295793 and [68]. Moreover, when the agitated bacterial cellulose and static bacterial cellulose were added at wet-end, both tensile strength and filler retention of the hand sheets were improved [69]. Especially, the bacterial cellulose from agitated culture had higher effects on filler-retention than that from static one. Thus the bacterial cellulose may be useful as a wet-end additive for papermaking. Furthermore, Suspended solids in the nutrient medium for *A. xylinium* in a rotating disk bioreactor become incorporated into the gelatinous mat of bacterial cellulose as it forms [76]. Embedding fibers of ordinary cellulose creates composites with enhanced strength and the toughness of bacterial cellulose. Purified cellulose and elongated fibers from paper are incorporated differently than are spherical particles such as silica gel. About 90% of the final cellulose can come from scrap paper, and dried composite sheets were much stronger than plain bacterial cellulose per unit area. Basta and Elsaied [70] confirmed that incorporating 5% of BC with wood pulp during paper sheet formation was found to significantly improve kaolin retention, strength, and fire resistance properties as compared to paper sheets produced from incorporating BC [76].

Bacteial cellulose and food industry

The Philippine desert, Nata de Coco, has been a cottage industry for the past seventy years in the Philippines [77]. The import of Nata from the Philippines into Japan has had a major impact on the global outlook for expansion of microbial cellulose production. In 1992, a fad originated in Japan with the introduction of microbial cellulose into diet drinks. As a result of production of Nata de Coco through *G. xylinus*, David [78] showed that Nata de Coco has the plasma cholesterol-lowering effect. In which the coconut milk was incubated with the *G.xylinus* as a carbon source. BC has been applied as a functional food additive as a thickener and disperser, in these applications for food; the bacterial cellulose from agitated culture has a much higher emulsifying effect than that from static culture [79]. *Monascus*-bacterial cellulose complex, which combined the properties of bacterial cellulose and *Monascus* fungi, showed the potential to be a new foodstuff as vegetarian meat or seafood replacement. Growth of *Monascus* mycelium did not impart a flavor to this new product creating it a good base as a flavor-added food. The color and texture of the complex were like liver or lean meat. It also provides high fiber content, limited calories and healthy nutrients. Moreover, the waste broth from the fermentation could be further used as a source of water-soluble pigments. Moreover, Shirai et al. [80] reported that, gelatinous cellulose prepared by fermentation with *Acetobacter aceti* AJ 12368 consisted of cellulose fibrils (0.9%), bound water (0.3%) and free water (98.8%). The cellulose network absorbed water weakly in capillaries of about 0.5-1.0 microns. When stressed the gel released its water and deformed without fracture. The gel by itself was too tough to bite, but it became edible through processing either with sugar alcohol or with alginate and calcium chloride. The textures resemble fruit, such as grapes and molluscs such as squids, respectively. The mechanism is the immobilization of the water of gelatinous cellulose by viscous or gel-forming material, and accordingly the gel becomes easy to cut off with the teeth. These results show that gelatinous cellulose with its fibrous texture can be a new material for salads, low-calorie desserts, fabricated foods.

Future prospective and conclusions

Economic feasibility of bacterial cellulose is primarily dependent on its productivity. The choice of fermentor design is particularly critical because it must withstand vigorous mechanical agitation of the rapidly growing *A. xylinum* (obligate aerobe) culture, and also prevent the cellulose fibril and fibrillar matrix from mechanical disruption. Agitated culture results in highly branched, three-dimensional, reticulated structure, whereas static culture produces a normal cellulose pellicle with a lamellar structure and less significant branching. Desirable modifications in the fibrillar and/or macroscopic nature of the cellulosic product may be achieved by varying fermentor design factors such as the shape of vessels and agitating impellers. The advantage of the stirred tank reactor is its ability to prevent the heterogeneity of the culture broth by strong mechanical agitation, while its drawback is high energy cost for generating the mechanical power. On the contrary, the energy cost of an airlift reactor is one sixth of that of a stirred tank reactor. However, the agitation power of an airlift reactor is limited, resulting in low fluidity of the culture broth, especially at high cellulose concentrations. To meet different demands, the combined use of an airlift and a stirred tank reactors, or continuous cultivation, may be a possible solution. Use of some modified reactors like rotating disk reactor, rotary biofilm contactor, bioreactors equipped with a spin filter or reactor with silicone membrane can also be a useful solution. One of the major obstacles encountered in the industrial adaptation of *A. xylinum* is the wasteful and noxious accumulation of metabolic byproducts.

References

1. Kai A, Arashida T, Hatanaka K, Akaike T, Matsuzaki K, et al. (1994) Analysis of the biosynthetic process of cellulose and curdlan using ^{13}C-labeled glucose. Carbohyd Polym 23: 235-239.
2. Delmer PD, Yehudit A (1995) Cellulose biosynthesis. The Plant Cell 7: 987-1000.
3. Tonouchi N, Tsuchida T, Yoshinaga F, Beppu T, Horinouchi S (1996) Characterization of the biosynthetic pathway of cellulose from glucose and fructose in *Acetobacter xylinum*. Biosci Biotech Biochem 60: 1377-1379.
4. Yoshinaga F, Tonouchi N, Watanabe K (1997) Research progress in production of bacterial cellulose by aeration and agitation culture and its application as a new industrial material. Biosci Biotech Biochem 61: 219-224.
5. Montezinos D, Brown Jr RM (1976) Cellulose microfibrils. Proc Natl Acad Sci USA 84: 6985- 6989.
6. Son HJ, Heo MS, Kim YG, Lee SJ (2001) Optimization of fermentation conditions for the production of bacterial cellulose by a newly isolated Acetobacter. Biotechnol Appl Biochem 33: 1-5.
7. Colvin JR (1980) The biosynthesis of cellulose: Plant biochemistry, Academic Press Inc., New York: 543-570.
8. Ross P, Mayer R, Benziman M (1991) Cellulose biosynthesis and function in bacteria. Microbiol Rev 55: 35-58.
9. Tanaka M, Murakami S, Shinke R, Aoki K (2000) Genetic characteristics of cellulose-forming acetic acid bacteria identified phenotypically as Gluconacetobacter xylinus. Biosci Biotechnol Biochem 64: 757-60.
10. Matthysse AG, Thomas, DOL, White AR (1995) Mechanism of cellulose synthesis in *Agrobacterium tumefaciens*. J Bacteriology 177: 1076-1081.
11. Matthysse AG, white S, Lightfoot R (1995) Genes required for cellulose synthesis in *Agrobacterium tumefaciens*. J Bacteriology 177: 1069-1075.
12. Koyama M, Sugiyama J, Itoh T (1997) Systematic survey on crystalline features of algal celluloses. Cellulose 4: 147-160.
13. Blanton RL, Northcote DH (1990) A 1,4-β-D-glucan synthase system from Dictyostelium discoideum Planta 180: 324-332.
14. Bulone V, Glrard V, Fevre M (1990) Separation and partial purification of 1,3-β-glucan and 1,4-β-glucan synthases from Saprolegnia. Plant Physiol 94: 1748-1755.
15. Ryser U (1985) Cell wall biosynthesis in differentiating cotton fibers. Eur J Cell Biol 39: 236-265.
16. Ross P, Weinhouse H, Aloni Y, Michaeli D, Ohana P, et al. (1987) Regulation of cellulose synthesis in *Acetobacter xylinum* by cyclic diguanylic acid. Nature (London) 325: 279-281.
17. Boisset C, Chanzy H, Henrissat B, Lamed R, Shoham Y, et al. (1999) Digestion of crystalline cellulose substrates by the clostridium thermocellum cellulosome: structural and morphological aspects. Biochem J 340: 829-835.
18. Chang AL, Tuckerman JR, Gonzalez G, Mayer R, Weinhouse H, et al., (2001) Regulator of cellulose synthesis in *Acetobacter xylinum*. Biochem 40: 3420-3426.
19. Kimura S, Chen HP, Saxena IM, Brown RM, Itoh T (2001) Localization of c-di-GMP-binding protein with the linear terminal complexes of *Acetobacter xylinum*. J Bacteriol 183: 5668-5674.
20. Wang X, Cnops G, Vanderhaeghen R, Block S, Montagu M, et al., (2001) AtCSLD3, Cellulose synthase-like gene important for root hair growth in arabidopsis. Plant Physiol 126: 575-586.
21. Vergara CE, Carpita NC (2001) Beta-D-glycan synthases and the CesA gene family: lessons to be learned from the mixed-linkage (1-->3),(1-->4)beta-D-glucan synthase. Plant Mol Biol 47: 145-160.
22. Vanderhart DL, Atalla RH (1984) Studies of microstructure in native cellulose using solid-state 13C NMR. Macromolecules 17: 1465-1472.
23. Keshk SMA (1999) Nascent structure of microbial cellulose. Ph.D. Thesis, Tokyo Metropolitan University.
24. Hirai A, Horii F, Kitamaru R (1987) Transformation of native cellulose crystals from cellulose Ib to Ia through solid state chemical reactions. Macromolecules 20: 1440-1442.
25. Horii F, Yamamoto H, Hirai A (1997) Microstructural analysis of microfibrils of bacterial cellulose. Macromol Symp 120: 197-205.
26. Yamamoto H, Horii F (1993) CP/Mass Carbon- 13 NMR analysis of the crystal transformation induced for Valonia cellulose by annealing at high temperatures. Macromolecules 26: 1313-1317.

27. Hirai A, Masaki T, Yamamoto H, Horii F (1998) In situ crystallization of bacterial cellulose III. Influences of different polymeric additives on the formation of microfibrils as revealed by transmission electron microscopy. Cellulose 5: 201-213.

28. Yamamoto H, Horii F (1994) In situ crystallization of bacterial cellulose I. Influences of polymeric additives, stirring and temperature on the formation cellulose Iα and Iβ as revealed by cross polarization/magic angle spinning (CP/Mass) 13C NMR spectroscopy. Cellulose 1: 57.

29. Yamamoto H, Horii F, Hirai A (1996) In situ crystallization of bacterial cellulose II. Influences of polymeric additives with different molecular weight on the formation cellulose Iα and Iβ at the early stage of incubation. Cellulose 3: 229-242.

30. Haigler CH, Weimer PJ (1991) Biosynthesis and biodegradation of cellulose. Marcel Dekker, Inc. New York, USA.

31. Watanabe K, Tabuchi M, Morinaga Y, Yoshiinaga F (1998) Structural features and properties of bacterial cellulose produced in agitated culture. Cellulose 5: 187-200.

32. Hiroshi O, Watanabe K, Morinaga Y, Yoshinaga F (1997) Emulsion-stabilizing effect of bacterial cellulose. Biosci Biotech Biochem 61: 1541-1545.

33. Hestrin S, Schramm M (1954) Synthesis of cellulose by *Acetobacter xylinum*. 2. Preperation of freeze-dried ceels capable of polymerizing glucose to cellulose. Biochem J 58: 345-352.

34. Masaoka S, Ohe T, Sakota N (1993) Production of cellulose from glucose by *Acetobacter xylinum*. J Fermen Bioeng 75: 18-22.

35. Oikawa T, Morino T, Ameyama M (1995) Production of cellulose from D-Arbitol by *Acetobacter xylinum* KU-1. Biosci Biotech Biochem 59: 1564-1565.

36. Tajima K, Fujiwara M, Takai M (1995) Biological control of cellulose. Macromol Symp 99: 149-155.

37. Embuscado M, Marks J, Bemiller J (1994) Bacterial cellulose. I. Factors affecting the production of cellulose by *Acetobacter xylinum*. Food Hydro 8: 407-418.

38. Embuscado M, Marks J, Bemiller J (1994) Bacterial cellulose. II. Optimization of cellulose production by *Acetobacter xylinum* through response surface methodology. Food Hydro 8: 419-430.

39. Delmer PD (1999) Cellulose biosynthesis: Exciting times for a difficult field of study. Annu Rev Plant Physiol Plant Mol Biol 50: 245-76.

40. Tonouchi N, Tahara N, Tsuchida T, Yoshinaga F, Beppu T (1995) Addition of a small amount of an endoglucanase enhances cellulose production by *Acetobacter xylinum*. Biosci Biotech Biochem 59: 805-808.

41. Premjet S, Ohtani O, Someshima K (1994) The contribution of high molecular lignosulfonate to the powerful bacterial cellulose production system with *Acetobacter xylinum* ATCC 10245. Sen'i Gakkaishi 50: 458-463.

42. Premjet S, Ohtani Y, Saeshima K (1996) X-ray diffraction diagram of the bacterial cellulose membrane produced by *Acetobacter xylinum* in the medium with lignosulfonate. Sen, i Gakkaishi 52: 169-174. 42

43. Keshk SMAS (2014) Vitamin C Enhances Bacterial Cellulose Production in Gluconacetobacter Xylinus. Carbohydrate Polymer 99: 98-100.

44. Keshk SMAS, Suwinarti W, Sameshima K (2006) Physicochemical Characterization of Different Treatment Sequences on Kenaf Bast Fiber. Carbohydr Polym 65: 202-206.

45. Yamanaka S, Watanabe K, Kitamura N, Iguchi M, Mitsuhashi S, et al. (1989) The structure and mechanical properties of sheets prepared from bacterial cellulose. J Mater Sci 24: 3141-3145.

46. Hackney J, Atalla R, Vanderhart D (1994) Modification of crystallinity and crystalline structure of *Acetobacter xylinum* in the presence of water-soluble beta-1,4-linked polysaccharides: 13C-NMR evidence. Int J Biol Macromol 16: 215-218.

47. Uhlin K, Atalla R, Thompson N (1995) Influence of hemicelluloses on the aggregation patterns of bacterial cellulose. Cellulose 2: 129-144.

48. Iwata T, Indrarti L, Azuma J (1998) Affinity of hemicellulose for cellulose produced by *Acetobacter xylinum*. Cellulose 5: 215-228.

49. Whitney S, Brigham J, Darke A, Grant Reid J, Gidley M (1998) Structural aspects of the interaction of mannan-based polysaccharides with bacterial cellulose. Carbohyd Res 307: 299-309.

50. Haigler H, White R, Brown M, Cooper M (1982) Alteration of in vivo cellulose ribbon assembly by carboxymethylcellulose and other cellulose derivatives. J Cell Biol 94: 64-69.

51. Hayashi T, Marsden E, Delmer P (1987) Pea xyloglucan and cellulose: VI. Xyloglucan cellulose interactions in vitro and in vivo. Plant Physiol 83: 384-389.

52. Haigler H, Brown M, Benziman M (1980) Calcofluor white ST alters the in vivo assembly of cellulose microfibrils. Science 210: 903-906.

53. Haigler H, Chanzy H (1988) Electron diffraction analysis of the altered cellulose synthesized by *Acetobacter xylinum* in the presence of fluorescent brightening agents and direct dyes. J Ultrastruct Mol Struct Res 98: 299-311.

54. Kai A, Mondal IH (1997) Influence of substituent of direct dye having bisphenylenebis (azo) skeletal structure on structure of nascent cellulose produced by *Acetobacter xylinum* [I]: different influence of direct red 28, blue 1 and 15 on nascent structure. Int J Biol Macromolecules 20: 221-231.

55. Mondal I, Kai A (1998) Structure of nascent microbial cellulose I. Effects of methyl and methoxy groups of direct blue 1 and 53 on nascent microbial cellulose. Polym J 30: 78-83.

56. Mondal I, Kai A (1998) Structure of nascent microbial cellulose II. Effects of methyl and methoxy groups of direct blue 14 and 15 on nascent microbial cellulose. Polym J 30: 84-89.

57. Mondal I, Kai A (1999) Influence of substitution of direct dye having biphenylenebis (azo) skeletal structure on structure of nascent cellulose produced by *Acetobacter xylinum* [II]. J Appl Poly Sci 71: 1007-1015.

58. Kai A, Keshk SMA (1998) Structure of nascent microbial cellulose V. Influence of positions of sulfonate groups in fluorescent brightener on crystal structure of microbial cellulose. Polym J 30: 996-1000.

59. Keshk SMA, Kai A (2000) Influence of numbers and positions of sulfonate group in fluorescent brightener on nascent structure of bacterial cellulose. The 6th Arab international Conference on Materials Science 1: 115, Alexandria, Egypt.

60. Yu X, Atalla R (1996) Production of cellulose II by *Acetobacter xylinum* in the presence of 2,6-dichlorobenzonitrile. Int J Biol Macromol 19: 145-146.

61. Keshk SMAS, Nada A (2003) Heterogeneous Derivatization of Bacterial and Plant Celluloses. Biotech Biosci Asia 1: 39.

62. Keshk SMAS, Sameshima K (2005) Evaluation of Different Carbon Sources for the Production of Bacterial Cellulose. Afri J Biotech 4: 478-482.

63. Keshk SMAS, Haijia M (2011) A New Method for Producing Microcrystalline Cellulose from *Gluconacetobacter xylinus* and Kenaf. Carbohydr Polym 84: 1301-1305.

64. Keshk SMAS (2008) Homogenous Reaction of Cellulose from Different Natural Sources. Carbohydr Polym 74: 942-945.

65. Fontana JD, Souza A M, Fontana CK, Toriani IL, Moreschi JC, et al. (1990) Acetobacter cellulose pellicle as a temporary skin substitute. Appl Biochem Biotechnol 24-25: 253-264.

66. Kawecki M, Krystynowicz A, Wysota K, Czaja W, Sakiel S, et al. (2004) Bacterial cellulose biosynthesis, properties and applications. International Review Conference Biotechnology, Vienna, Austria, 14–18.

67. Fontana JD, Franco VC, Souza SJ, Lyra IN, Souza AM (1991) Nature of plant stimulators in the production of *Acetobacter xylinum* ("Tea Fungus") biofilm used in skin therapy. Appl Biochem Biotechnol 28-29: 341-351.

68. Ougiya H, Watanabe K, Morinaga Y, Yoshinaga F (1997) Emulsion effect of bacterial cellulose. Biosci Biotech Biochem 61: 1541-1545.

69. Hioki N, Hori Y, Watanabe K, Morinaga Y, Yoshinaga F, et al. (1995) Bacterial cellulose; as a new material for papermaking. Kami-Gikyo-Shi 49: 82-87.

70. Basta AH, El-Saied H (2009) Performance of improved bacterial cellulose application in the production of functional paper. J Appl Microbiol 107: 2098-107.

71. Lapuz MM, Gallardo EG, Palo MA (1969) The nata organism-cultural requirements, characteristics and identity. The philippine J of Science 96: 91-109.

72. Kuwana Y (1997) Production and new function of Nata de Coco. Cell Commun 4: 25-28.

73. Brown Jr RM, Willison JHM, Richardson CL (1976) Cellulose biosynthesis in *Acetobacter xylinum*. Proc Nat Acad Sci USA 73: 4565-4569.

74. Czaja W, Krystynowicz A, Bielecki S, Brown Jr RM (2006) Microbial cellulose the natural power to heal wounds. Biomater 27: 145-151.

75. Gouda M, Keshk SMAS (2010) Evaluation of Multifunctional Properties of Cotton Fabric Based on Chitosan-Metal Film. Carbohydr Polym 80: 504-512.

76. Kai A, Keshk SMA (1999) Structure of nascent microbial cellulose VI. Influence of positions of sulfonate groups in fluorescent brightener on crystal structure of microbial cellulose. Polym J 31: 61-65.

77. Keshk SMAS, Sameshima K (2006) Utilization of Sugar Cane Molasses With/without the Presence of Lignosulfonate for the Production of Bacterial Cellulose. Appl Microb Biotech 72: 291-296.

78. David NS (1996) Chemical modification of Lignocellulosic Materials: Chemical structures of cellulose, hemicelluloses and lignin, Marcel Dekker. Inc., New York, USA.

79. Ogawa R, Tokura S (1992) Preparation of bacterial cellulose containing N-acetylglucosamine residues. Carbohydr Polym 19: 171-178.

80. Shirai A, Takahashi M, Kaneko H, Nishimura S, Ogawa M, et al. (1994) Biosynthesis of novel polysaccharide by *Acetobacter xylinum*. Int J Biol Macromol 16: 297-300.

Pectin Methylesterases: A Review

Pooja Kohli, Manmohit Kalia and Reena Gupta

Department of Biotechnology, Himachal Pradesh University, Summer Hill, Shimla-171005, INDIA

Abstract

Pectin methylesterase (PME) is the first enzyme acting on pectin, a major component of plant cell wall. PME catalyzes reactions according to the double-displacement mechanism. In plants, PMEs can be classified on the basis of presence or absence of the PRO domain in pectin methylesterase into Type 1 and Type II. Type1 contains one to three PRO domains and two or three introns, type II PMEs are without PRO domain and with five or six introns. Once PMEs are secreted into the cell wall, mature PMEs exhibit three different modes of action: single chain mechanism, multiple chain mechanism and multiple attack mechanism. For bacterial and plant PMEs, single chain and multiple chain mechanism is used and for fungal PMEs only multiple chain mechanism has been proposed. PME activity is regulated by differential expression both spatially and temporally.

Keywords: Pectin methylesterase; Demethylestrification; Catalysis; Transacylation; Fruit firming

Introduction

Pectinolytic enzymes or pectinases are a heterogeneous group of enzymes that hydrolyze the pectic substances present in plants. They include polygalacturonases, pectin lyase, and PME that hydrolyze the glycosidic bonds of pectic substances. Pectin methylesterase (E.C. 3.1.1.11), a methyl ester group hydrolytic enzyme is produced by plants, pathogenic fungi, and bacteria also constantly being used in the wine, juice, and other food industries [1]. In *Arabidopsis* PMEs, it was revealed that PMEs have unique patterns of methylesterification that provide immunity to *Pseudomonas syringae pv maculicola* ES4326 (Pma ES4326) [2]. Pectin methylesterase catalyzes reactions according to the double-displacement mechanisms, de-esterification through transferring the C_6 carboxyl groups in the pectin-PME complexes to water molecules altering the degree and pattern of methyl esterification and transacylation through transferring the C_6 carboxyl groups to the hydroxyl groups of another pectin molecules and resulting in the formation of high molecular weight pectins with new non-methoxy ester linkages [3-8]. PME has been purified and characterized from several fruit sources including tomato [9], orange, papaya [10], apple [11], kiwi [12], grapefruit pulp and mandarin orange fruit [13]. PMEs having diverse and well-defined activities could be useful for making "designer pectins" for fulfilling new demands from food industry. It has been reported that PME activity is partly regulated by PMEI (Pectin methylestearse inhibitor) [14,15] For example, in flax (*Linum usitatissimum*) Lu PMEs gene family, most highly conserved residues were catalytic residues, while in Lu PMEIs gene family, cysteine's forming disulphide bridges between helices α2 and α3 were found to be conserved residues [16]. The modification of pectin by PMEs is known to modify the quality of plant-based food products. It has been recently confirmed that PME has a definite role in plant defence system by virus induced gene silencing of PME gene in *Nicotiana benthamiana* [17].

Types of Pectin Methylesterases

Several pectin methylesterase isoforms differing in molecular weight, pI and biochemical activity were detected in all higher plants examined so far as well as in a number of plant pathogenic fungi and bacteria [18]. In a study, a basic PME which is an isoform of B3a found to be encoded by LuPME3 from *Linum usitatissimum* and play great role in flax root development [19]. Pectin methylesterase belongs to class 8 (CE-8) of the carbohydrate esterase [20]. Plant pectin methylesterase belong to large multigene families. For instance, in *Arabidopsis thaliana*, 66 open reading frames (ORF) have been annotated as putative full-length PMEs. There are 89 PME ORFs in *Populus trichocarpa* [21] whereas this number is substantially lower in Oryza sativa with 35 ORFs, probably due to the low level of pectin in their cell wall. Higher plant PMEs are frequently organised in pre-pro-proteins. The PRE domain leading to the export of PMEs to the cell wall is formed by a common type signal peptide (SP) and by a transmembrane domain (TM or signal anchor). Different PMEs possess one, both, or neither of these motifs. Those with neither motif are classified as putatively soluble isoforms. The mature, active part of the protein (PME domain) is preceded by an N-terminal extension (PRO region) that shares similarities with the pectin methylesterase inhibitors. The PRO-regions can vary in length and shows a relatively low level of amino acid identity between isoforms [22]. In *Arabidopsis*, the isoelectric point of the 26 PME domains tends to be basic whereas that of the PRO region is neutral or acidic [23,24]. Similar differences in length and isoelectric point between PRO and PME domains have also been found in several other species [25], and might influence the optimal pH activities of specific isoforms.

Pectin methylesterases can be classified on the basis of presence or absence of the PRO domain into Type I and Type II. In PMEs Type I (500-900 amino acids; 52-105 kDa) there are 1-3 PRO domains and two or three introns (Figure 1) whereas PMEs Type II (250 to 400 amino acids; 27-45 kDa) contain no PRO domain with five or six introns [26,27]. The Type II sequences have a structure close to that of the PMEs identified in phytopathogenic organisms (bacteria, fungi).

Mode of Action of Mature Pectin Methylesterase

After their secretion into the cell wall, mature PMEs could exhibit three different modes of action (I) a single-chain mechanism where the enzyme converts all substrate sites on the polymeric chain, (II)

***Corresponding author:** Reena Gupta, Professor, Department of Biotechnology, Himachal Pradesh University, Summer Hill, Shimla-171005, India

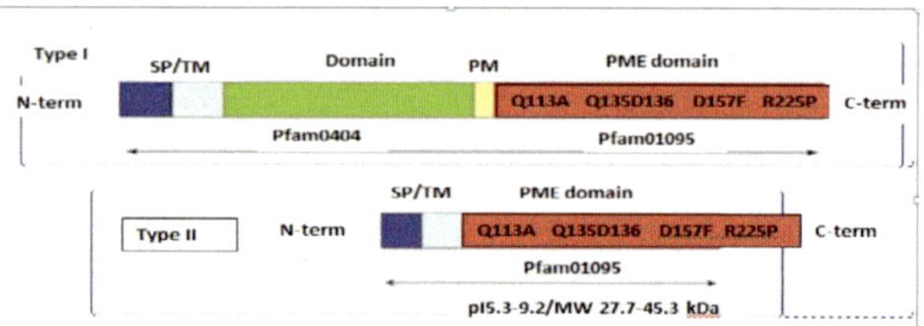

Figure 1: Pectin methylesterase(PME) structural motifs. Type I and Type II PMEs possess a conserved PME domain with characteristic, highly conserved amino acid fragments. Type I PMEs possess an N-terminal extension designated the PRO domain and a processing motif (PM) that might be a putative target for subtilisin like proteases. The targeting to the endomembrane system leading to the export of PMEs to the cell wall is mediated either by a signal peptide (SP) or a transmembrane domain (TM or signal anchor). Different PMEs possess one, both, or neither of these motifs.

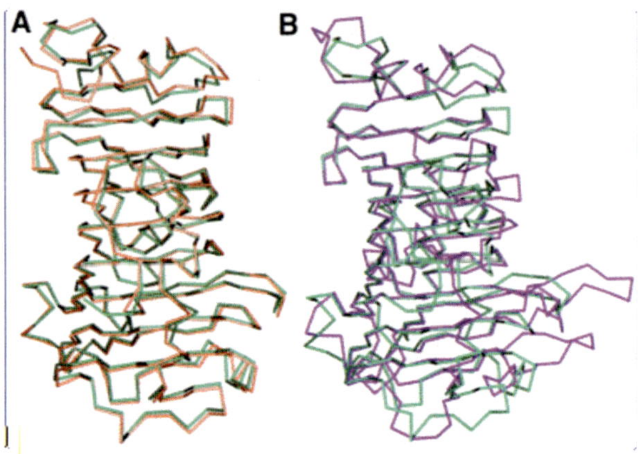

Figure 2: Comparison of the known structures of PMEs. (A) Overlay of the Cα trace of PME from tomato (green) and PME from carrot (orange). Structures are almost completely superimposable. (B) Superimposition of PME from tomato (green) and PME from *E. chrysanthemi* (violet). Although the β-helices are completely superimposable, main differences (showed by arrow) are located in the length of the turns protruding out from the β-helix in proximity of the putative active site cleft.

a multiple-chain mechanism where the enzyme catalyses only one reaction and then dissociates from the substrate and, (III) a multiple-attack mechanism where the enzyme catalyses a number of reaction cycles before the enzyme-polysaccharide complex dissociates. Plant and bacterial PMEs produce products with contiguous regions of galacturonic acid and both a single chain and multiple-attack mechanism have been proposed [28-30]. In contrast, fungal PMEs attack more randomly and a multiple-chain mechanism has been proposed for these enzymes [31-34]. When PMEs act randomly on pectic polymer, the demethylesterification releases protons that promote the action of endopolygalacturonases [33], and contribute to cell wall loosening. When PMEs act linearly on methylated pectin, PMEs give rise to blocks of free carboxyl groups that could interact with Ca^{2+}, creating the so called junction zones. Because the action of endopolygalacturonases in such a gel is limited, this action pattern of PMEs contributes to cell wall stiffening.

Regulation of Pectin Methylesterase Activity

Pectin methylesterases in the apoplasm catalyze the demethylesterification of homoglactouronic acids (HGAs), exposing carboxyl residues which can be cross-linked by calcium. These changes affect the rheological properties, porosity, and ionic status of cell wall. It follows that the tight control of PME activity, both spatially and temporally, occupies a central position in the control of cell wall growth and development. PMEs convert the methoxyl groups on the polygalacturonic acid chain into negatively charged carboxyl groups by releasing both protons and methanol. It is generally believed that most plant PMEs remove methyl esters in a block-wise fashion, creating long contiguous stretches of deesterified pectins [34]. PME activity is regulated by differential expression, multiple PME isoforms have been shown to be expressed in certain tissues at the same time. From the action pattern of the few PME isoforms studied so far, it appears that the intrinsic activity of different isoforms is same but certain substrate specificities and reaction mechanisms require different environmental conditions [34-36]. The distribution of carboxyl units along the pectin backbone and to a lesser extent the methylation degree, are important in controlling PME activity. Free carboxyl groups in the vicinity of the ester linkage increase the affinity of the enzyme for its substrate [37]. Immobilization of PMEs at these anionic sites has either increased or decreased their deesterification rate, possibly due to conformational changes induced by the microenvironment [35]. Different isoforms have different pH optima, and the action pattern of certain isoforms has been shown to be pH dependent [35,36]. Cations are also necessary for PME activity, where they affect the binding of the enzyme to its substrate [38].

Structure of Pectin Methylesterase

Three dimensional crystallographic structures of PMEs facilitated to elucidate the mechanism of action and substrate specificity of PMEs. Up to now 3 three dimensional structures have been solved: a bacterial PME, from *Erwinia chrysantemi* [39,40] and two plant PMEs, are from *Daucus carota and Solanun lycopersicon* (Figure 2A) [15]. The enzyme folds into a right handed parallel β-helix, first observed in pectate lyase C [41] and typical of pectic enzymes [39].

The three dimensional structures of the carrot and tomato PMEs showed striking similarities, and their structures are superimposable. In addition, both of these plant PMEs showed similar folding topology to PME from *Erwinia chrysanthemi*, although one major difference lies in the length of turns protruding from the β-helix in proximity of the substrate cleft (Figure 2B), in particular, turns that protrude out of the β-helix are much longer in the bacterial enzyme, making its putative active site cleft deeper and narrower than that of plant PMEs.

In tomato PME-1 the β-helix consists of seven complete coils, which have different lengths because the number of amino acids located in the loops connecting the β-strands is variable. Each coil consists of three β-strands that line up to form three extended parallel β-sheets (PB) called PB1, PB2, and PB3. The N-terminal region of PME is composed of a short α-helix followed by a β-strand that lines up with PB1. The C-terminal region has an extended conformation in which a long tail and four short and distorted α-helices protrude out of the parallel β-helix flanking PB1. The active site of PME is located on the PB3. Many aromatic residues (Phenylalanine (Phe) 80, Tyrosine (Tyr) 135, Phe 156, Tyr 218, Trp 223, and Trp 248) putatively involved in substrate binding are located in active site pocket. These residues are well conserved in plant PMEs Tyr 135, also the Phe 156 and Trp 223 residues are conserved in PME of *E. chrysanthemi*. The structures of the Michaelis complexes of PME from *Erwinia chrysanthemi* with methylated hexagalacturonates, suggest the following residues are involved in the reaction mechanism: Asp 199, Asp 178 and Gln 177 (Figure 3). Arg 267 is conserved amongst all PMEs, is not a direct participant in catalysis [40].

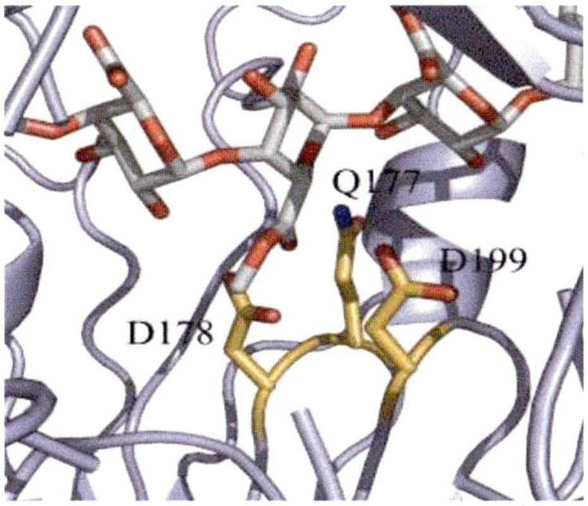

Figure 3: Stereo-view of the active site residues in the Michaelis complex.

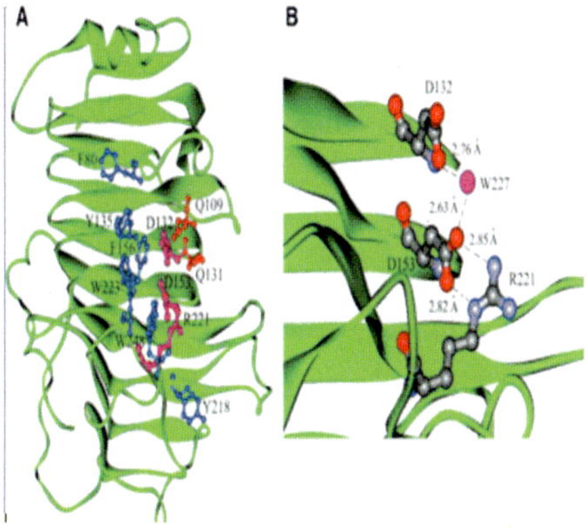

Figure 4: Close-Up View of the Tomato PME Active Site. (A) Structure of tomato PME in which residues involved in catalysis (violet), in stabilization of the catalytic intermediate (orange), and in substrate binding (blue) are shown in ball and stick representation. (B) Further close-up view representation of amino acid residues and a water molecule (blue ball) putatively involved in catalysis; H-bond pattern is highlighted.

In analogy with the proposed mechanism of action of PME from carrot and from *Erwinia Chrysantemi* [42], a mechanism of catalysis of PME-1 from tomato, has been described in which Asp 153, polarized by the proximity with Arg 221, performs a nucleophilic attack on the carboxymethyl group of the substrate. The tetrahedral anionic intermediate formed is stabilized by the interaction with two conserved Glutamine (Gln) residues (Gln109 and Gln131). Afterwards, aspargine (Asp) 132 likely acts as a proton donor in the cleavage step where methanol is released. The resulting carboxylate group of Asp 132then behaves as a base and receives a proton from an incoming water molecule (W227), thus restoring the active site of the enzyme (Figure 4).

Location of Disulfide Bonds in Pectin Methylesterase from Tomato and Comparison with Pectin Methylesterase from Other Sources

Till date, four different types of pectinesterases have been determined in primary structure, the pectinesterase from tomato [43,44] *Aspergillus niger* [45], *Erwinia chrysanthemi* [46] and *Pseudomonas solanacearum* [47,48]. These four enzymes are distantly related, exhibiting a low degree of sequence similarity, with residue identities at the 18-33% level (disregarding an elongated N-terminal overshooting end of the *Pseudomonas* enzyme). Attempts at chemical modifications have shown the presence of reactive tyrosine residues [44], and overall comparisons suggest relationships between pectinesterases, and polygalacturonases, which are also pectin degrading enzymes [46]. Analysis of tomato pectinesterase by carboxymethylation, with and without reduction, showed that the enzyme has two intrachain disulfide bridges. Analysis of fragments obtained from the native enzyme after digestion with pepsin identified bridges connecting cysteine (Cys) Cys-98 with Cys-125, and Cys-166 with Cys-200. The locations of disulfide bridges in tomato pectinesterase are not identical to those in three distantly related pectinesterases (18-33% residue identities) from microorganisms. However, one half-Cys (i.e.Cys-166) position is conserved in all four enzymes. Sequence comparisons of the overall structures suggest a special importance for three short segments of the entire protein. One segment is at the N-terminal part of the tomato pectinesterase, another in the C-terminal portion near the distal end of the second disulfide loop, and the third segment is located in the central part between the two disulfide bridges. The latter segment, encompassing only 40 residues of the entire protein, appears to highlight a functional site in a middle chain segment.

Functions of Pectin Methylesterases

Ubiquitous enzymes involved in many physiological processes

Several studies have shown a strong correlation between PME activity or PME gene expression and physiological processes such as fruit maturation microsporogenesis and pollen tube growth [47,48], cambial cell differentiation [48,49] seed germination and hypocotyl elongation [35]. Interestingly, two recent studies have shown unequivocally that a PME is a host-cell receptor for the Tobacco Mosaic Virus (TMV) movement protein [50,51]. Thus, the interaction between the virus movement protein and the PME is required for viral cell-to-cell movement through plasmodesmata. One hypothesis proposed is that binding of the TMV movement protein interferes with PME activity, altering the cell wall ion balance and consequently inducing changes in the permeability of the plasmodesmata [51].

Pectin methylesterases as a methanol source

In 1998, a close correlation was reported between PME activity and levels of methanol in fruit tissues from both wild-type tomato and a PME antisense mutant, indicating that PME is on the primary biosynthetic pathway for methanol production in tomato fruit [52]. Because methanol oxidation to CO_2 could result in the incorporation of methanol carbon into metabolites via the Calvin-Benson cycle, PMEs could play an appreciable, albeit indirect, role in the photosynthetic metabolism of the plant.

Pectin methylesterases in cellular adhesion

During separation of the border cells of the root cap of pea, PME activity increases and is correlated with an increase in the amount of acidic pectin and a decrease in cell wall pH [53]. This study was performed using an antisense transgenic plant transformed with a PME gene (*rcpme1*) obtained by screening a root cDNA library [54]. Analysis of transgenic plants showed that *rcpme1* expression is required for the maintenance of extracellular pH, elongation of the cells within the root tip and for cell wall degradation leading to border cell separation.

Pectin methylesterases in stem elongation

Transgenic potato plants were transformed with a *Petunia-inflata* derived cDNA encoding a PME in sense orientation under a constitutive promoter. This showed that the apex of the stem contained less PME activity than the wild type. Furthermore, during the early stages of development, stems of transgenic plants elongated more rapidly than those of the wild type [55].

Pectin methylesterase genes influence solid wood properties

Wood is primarily composed of lignin, cellulose, and noncellulosic polysaccharides (including neutral and pectic polysaccharides), all of which interact to form a complex structure that provides the foundation for the massive stature unique to the woody life form known as "trees". Quantitative genetic studies have revealed that wood properties are under strong additive genetic control in representative species across the Eucalyptus genus [56-58]. Associations were identified between two PME genes and a range of solid wood properties in *Erwinia pilularis.* Associations in PME7 were primarily with cellulose, pulp yield, and lignin. In contrast, the associations with PME6 were primarily with shrinkage and collapse. The associations between PME6 and PME7 with multiple wood traits are consistent with the current understanding of wood formation and the biochemical activity of PME [59-62].

Transacylation properties of pectin methyl esterase

PME from *Aspergillus niger* is found to catalyze the transacylation reaction and forms polymerized pectin as observed by an increase in the particle size of PME-catalyzed pectin solution the transacylation mechanism of plant and microbial is different, and there are at least two kind of PME-catalyzed transacylation [63].

Methyl Esterification of Pectin Plays a Role during Plant-Pathogen Interactions and Affects Plant Resistance to Diseases

Esterification status of pectin plays a critical role in plant-pathogen interactions. De-esterification of pectin affects the susceptibility of the cell wall barrier to fungal and bacterial CWDEs and influences the structure and functionality of the translocation and spreading of viruses. In addition, de-esterification of pectin influences the functionality of the feeding structures evolved by biotrophs and nematodes to sustain their parasitic growth within the plant tissue. PME activity is also required for the production of defense molecules such as OG elicitors or methanol.

Industrial Applications of Pectin Methylesterases

Pectinases have been utilized in the commercial sector for wine and fruit juice industry since 1930. Commercial pectinase preparations, account for about 25% of the global food enzymes, are produced generally from fungal sources (*Aspergillus niger*) and used in many applications in food industry without purification. According to the specific objectives of the processes, the enzyme PME is used alone or in combination with other enzymes, mostly with other pectinases. The processes that PME involved are introduced below.

Clarification of Juices

In fruit juices, pectic substances are the main substances responsible for the cloudiness. Thus, in the production of clear fruit juices the pectin in juices should be degraded by the application of pectinases. The degradation of pectin reduces also the viscosity of fruit juices and eases their filtration and concentration. To degrade pectin, PME is used in combination with other pectinases such as PG and PL. Following the degradation of pectin, the viscosity of fruit juice drops and suspended cloud particles lost their stability and precipitated. In the clarification of fruit juices, the PME enzyme may also be used alone. However, such a treatment needs the addition of $CaCl_2$ to fruit juice. This removes pectin from fruit juice as insoluble calcium pectates and causes the clarification [64-66]. Another application of pectinases in fruit juice industry involves liquefaction of fruit mashes. In such an application, the pectinases are combined with cellulases and hemicellulases for the complete disruption of the cell walls. This method can alternatively be used for the production of fruit juices from tropical fruits (e.g. banana) that cannot be processed with classical methods [67,68]. In all these enzymatic treatments, degradation of pectin facilitates pressing and increase juice yield. Also, the aromatic quality of fruit juices increases and amount of waste material reduces. Also, the presence of enzyme in the final product and alteration of organoleptic properties are inevitable [69]. The use of immobilized commercial pectinase named 'Pectinex Ultra SP-L' was studied for mash treatment of carrot puree [68]. This treatment maintains almost 93% of the enzyme after the fifth treatment. However, in practice the application of pectinases are still conducted by the classical methods. However, their use increases the volume offree-run juice (by breakdown of polysaccharides and solubilization of middle lamella), extraction yield of polyphenols, and color and aromatic compounds [70,71]. Pectin methylesterase from *Datura stramonium* in combination with polygalacturonase increased clarity of orange, apple, pomegranate and pineapple juices by 2.9, 2.6, 2.3, and 3.6 fold respectively [72].

Firming of Fruits and Vegetables before Processing

Texture, an important quality attribute of fresh and processed fruits and vegetables, is closely related with the structural integrity of the primary cell walls and middle lamella, which are mainly composed of pectic substances. In industry, most of the processes such as blanching, freezing, dehydration, pasteurization and sterilization applied to preserve fruits and vegetables cause irreversible physical damages on cellular tissues. However, the negative effects of processing on texture can be overcome by applying different processes, which involve the use of PME enzyme. For example, the application of PME and $CaCl_2$ by vacuum infusion is now used for firming fruits and vegetables [73]. The mechanism of firming in this method is based on the demethylation of naturally occurring pectin in plant tissues by the action of PME and the chelation of the added or natural calcium with the free carboxyl groups generated in pectin molecules. Such a chelation causes the formation of networks among pectin molecules (the egg box model), stabilizes pectin and increases the firmness of plant tissues. Some of the successful applications of enzymatic firming by use of PME involves, firming of strawberries before jam making and freezing [74,75] and firming of apples, strawberries, and raspberries before pasteurization [73]. Besides commercial PMEs, the in situ PME in the product may also be used for firming of fruits and vegetables. In this method, the in situ PME should be activated by low temperature blanching. The free carboxyl groups produced then were cross-linked with divalent ions in the medium or $CaCl_2$ added to form a stable network. Firming by this method is suitable to apply for whole or sliced potatoes to be processed to French fries [76].

Modification of Pectin

Pectin produced from apple and orange peels has widespread application in food industry as gelling, thickening, and stabilizing agents. The degree of esterification of pectin molecule greatly influences its functional properties. The Degree of Esterification (DE) of pectin may range between 0 and 100% and on the basis of DE, pectins are divided into two groups; high methoxylated (HM) pectins with a DE higher than 50% and low methoxylated (LM) pectins with a DE lower than 50%. Pectin can be modified by PME to obtain the required DE value [77,78]. Pectins with low DE are particularly useful to obtain gels without using sugar and acid.

Production of Low Sugar Jams and Jellies

Traditionally, HM pectin is used for the preparation of jams and jellies. The gelling mechanism of HM pectin is based on hydrophobic interactions and dehydration at low pH (<4.0). In such gels, the presence of high concentrations of sugar (>60 %) is essential for gelling. In contrast, LM pectin forms gels by ionic interactions in which calcium or other divalent cations interact with free carboxylic acid of two adjacent chains, and give rise to cross-linking of these chains. In such a gelling mechanism, the sugar concentration is not very important. Thus, LM pectins are suitable for the production of low sugar (diabetic) jams and jellies [67].

Other Applications

In a recent study on Arabidopsis PMEs, revealed that PMEs affect patterns of methylesterification in a way that impart pattern-triggered immunity to Pseudomonas syringae pv maculicola ES4326 (Pma ES4326) [79]. Oils from coconut germ, sunflower seed, palm kernel, rape seed are industrially extracted with organic solvents. The most commonly used solvent is hexane, which is a potential carcinogen. Thus, alternatively, cell wall degrading enzymes, PME and other pectinases are used in combination to extract oil in different crops by liquefying the structural components of their cell walls. Pectinase preparations (such as Olivex) are also used in olive oil industry to increase the oil extraction output and to improve certain olive oil quality indicators [71,80]. Another application of combinational use of PME, other pectinases and cellulases is the peeling of fruits. Peeling of fruits has traditionally been applied by hand or treatment with steam, boiling water, acid or alkali. But these methods sometimes cause poor product quality (losses of fruit juice or disintegrations at the fruit surface) in delicate fruits. Also, in chemical methods, the disposal of the used peeling solution is a great problem. Therefore, application of pectinases and cellulases by vacuum infusion can be used as an alternative method for peeling of delicate fruits. For example, a commercial preparate (Rohament PC) containing pectinases and cellulases could be applied to remove peels and skins of oranges and to obtain whole fruit segments [81].

References

1. Gainvors A, Frézier V, Lemaresquier H, Lequart C, Aigle M, et al. (1994) Detection of polygalacturonase, pectin-lyase and pectin-esterase activities in a *Saccharomyces cerevisiae* strain. Yeast 10: 1311-1319.
2. Bethke G, Grundman RE, Sreekanta S, Truman W, Katagiri F, et al. (2014) Arabidopsis PECTIN METHYLESTERASEs contribute to immunity against *Pseudomonas syringae*. Plant Physiol 164: 1093-1107.
3. Hou WC, Chang WH (1996) Pectinesterase-catalyzed firming effects during precooking of vegetables. J Food Biochem 20: 397-416.
4. Jiang CM, Wu MC, Chang WH, Chang HM (2001) Change in particle size of pectin reacted with pectinesterase isozymes from pea (*Pisum sativum* L.) sprout. J Agric Food Chem 49: 4383-4387.
5. Jiang CM, Lai YJ, Lee BH, Chang WH, Chang HM (2001) De-esterification and transacylation reaction of pectinesterase from jelly fig (*Ficus awkeotsang makino*) achenes. J Food Sci 66: 810-815.
6. Jiang CJ, Lai YJ, Lee BH, Chang WH, Wu MC, et al. (2002) Changes in physico-chemical properties of pectin from jelly fig (*Ficus awkeotsang makino*) seeds during extraction and gelling. Food Res Int 35: 31-35.
7. Lee CW, Wu MC, Lee BH, Jiang CM, Chang HM (2003) Changes in molecular weight of transacylated pectin catalyzed by tomato and citrus pectinesterases as determined by gel permeation chromatography. J Agric Food Chem 51: 5455-5461.
8. Wu, MC, Chen YW, Hwang JY, Lee BH, Chang HM (2004) Transacylation of citrus pectin as catalyzed by pectinesterase from tendril shoots of chayote *Sechium edule*. Food Res Int 37: 759-765.
9. Lee M, Macmillan JD (1968) Mode of action of pectic enzymes. I. Purification and certain properties of tomato pectinesterase. Biochemistry 7: 4005-4010.
10. Fayyaz A, Asbi BA, Ghazali HM, Che man YB, Jinap S (1994) Purification and molecular properties of papaya pectinesterase. Food Chem 49: 373-378.
11. Castaldo D, Quagliuolo L, Servillo L, Balestrieri C, Giovane AX (2005) Isolation and characterization of pectinmethylesterase from apple fruit. J Food Sci 54: 653-655.
12. Giovane A, Laratta B, Loiudice R, Quagliuolo L, Castaldo D, et al. (1996) Determination of residual pectin methylesterase activity in food products. Biotechnol Appl Biochem 23 : 181-184.
13. Rillo L, Castaldo D, Liovane A, Servillo L, Balestrieri, C, et al. (1992) Purification and properties of pectin methylesterase from mandarin orange fruit. J Agric Food Chem 40: 591-593.
14. Hothorn M, Wolf S, Aloy P, Reiner S, Scheffzek K (2004) Structural insights into the target specificity of plant invertase and pectin methylesterase inhibitory proteins. Plant Cell 16: 3437-3447.
15. Di Matteo A, Giovane A, Raiola A, Camardella L, Bonivento D, et al. (2005) Structural basis for the interaction between pectin methylesterase and a specific inhibitor protein. Plant Cell 17: 849-858.
16. Pinzón-Latorre D, Deyholos MK (2013) Characterization and transcript profiling of the pectin methylesterase (PME) and pectin methylesterase inhibitor (PMEI) gene families in flax (*Linum usitatissimum*). BMC Genomics 14: 742.
17. Bubici G, Carluccio AV, Cillio F, Stavolone L (2015) Virus induced gene silencing of pectin methylesterase protects *Nicotiana benthamiana* from lethal symptoms caused by tobacco mosaic virus. Eur J Plant Pathol 339-347.
18. Collmer N, Keen T (1986) The role of pectic enzymes in plant pathogenesis. Annu Rev Phytopathol 24: 383-409.
19. Mareck A, Lamour R, Schaumann A, Chan P, Driouich A, et al. (2012) Analysis of LuPME3, a pectin methylesterase from *Linum usitatissimum*, revealed a variability in PME proteolytic maturation. Plant Signal Behav 7: 59-61.
20. Henrissat B (1991) A classification of glycosyl hydrolases based on amino acid sequence similarities. Biochem J 280: 309-316.
21. Geisler-Lee J, Geisler M, Coutinho PM, Segerman B, Nishikubo N, et al. (2006) Poplar carbohydrate-active enzymes. Gene identification and expression analyses. Plant Physiol 140: 946-962.
22. Markovic O, Janecek S (2004) Pectin methylesterases: sequence-structural features and phylogenetic relationships. Carbohydr Res 339: 2281-2295.
23. Bosch M, Cheung AY, Hepler PK (2005) Pectin methylesterase, a regulator of pollen tube growth. Plant Physiol 138: 1334-1346.
24. Bosch M, Hepler PK (2005) Pectin methylesterases and pectin dynamics in pollen tubes. Plant Cell 17: 3219-3226.
25. Al-Qsous S, Carpentier E, Klein-Eude D, Burel C, Mareck A, et al. (2004) Identification and isolation of a pectin methylesterase isoform that could be involved in flax cell wall stiffening. Planta 219: 369-378.
26. Micheli F (2001) Pectin methylesterases: cell wall enzymes with important roles in plant physiology. Trends Plant Sci 6: 414-419.
27. Tian GW, Chen MH, Zaltsman A, Citovsky V (2006) Pollen-specific pectin methylesterase involved in pollen tube growth. Dev Biol 294: 83-91.
28. Dongowski G, Bock W (1984) On the influence of molecular parameters of Pektinsubstrate on the activity of pectinesterases from *Aspergillus niger* and from higher plants. Nahrung-Food 28: 507-516.
29. Kohn R, Dongowski G, Bock W (1985) The distribution of free and esterified carboxyl groups within the pectin molecule after the action of pectin esterase from *Aspergillus niger* and oranges. Nahrung 29: 75-85.
30. Christensen TM, Nielsen JE, Kreiberg JD, Rasmussen P, Mikkelsen JD (1998) Pectin methyl esterase from orange fruit: characterization and localization by in-situ hybridization and immunohistochemistry. Planta 206: 493-503.
31. van Alebeek GJ, van Scherpenzeel K, Beldman G, Schols HA, Voragen AG (2003) Partially esterified oligogalacturonides are the preferred substrates for pectin methylesterase of *Aspergillus niger*. Biochem J 372: 211-218.

32. Duvetter T, Fraeye I, Sila DN, Verlent I, Smout C, et al. (2006) Mode of de-esterification of alkaline and acidic pectin methyl esterases at different pH conditions. J Agric Food Chem 54: 7825-7831.

33. Moustacas AM, Nari J, Borel M, Noat G, Ricard J (1991) Pectin methylesterase, metal ions and plant cell-wall extension. The role of metal ions in plant cell-wall extension. Biochem J 279: 351-354.

34. Limberg G, Körner R, Buchholt HC, Christensen TM, Roepstorff P, et al. (2000) Analysis of different de-esterification mechanisms for pectin by enzymatic fingerprinting using endopectin lyase and endopolygalacturonase II from *A. niger*. Carbohydr Res 327: 293-307.

35. Bordenave M, Goldberg R (1994) Purification and characterization of pectinmethylesterases from mung bean hypocotyl cell walls. Phytochem 33: 999-1003.

36. Denès JM, Baron A, Renard CM, Péan C, Drilleau JF (2000) Different action patterns for apple pectin methylesterase at pH 7.0 and 4.5. Carbohydr Res 327: 385-393.

37. Goldberg R, Pierron M, Bordenave M, Breton C, Morvan C, et al. (2001) Control of Mung bean pectinmethylesterase isoform activities. Influence of pH and carboxyl group distribution along the pectic chains. J Biol Chem 276: 8841-8847.

38. Catoire L, Pierron M, Morvan C, du Penhoat CH, Goldberg R (1998) Investigation of the action patterns of pectinmethylesterase isoforms through kinetic analyses and NMR spectroscopy. Implications In cell wall expansion. J Biol Chem 273: 33150-33156.

39. Jenkins J, Mayans O, Smith D, Worboys D, Pickersgill RW (2001) Three-dimensional structure of *Erwinia chrysanthemi* pectin methylesterase reveals a novel esterase active site. J Mol Biol 305: 951-960.

40. Fries M, Ihrig J, Brocklehurst K, Shevchik VE, Pickersgill RW (2007) Molecular basis of the activity of the phytopathogen pectin methylesterase. EMBO J 26: 3879-3887.

41. Yoder MD, Keen NT, Jurnak F (1993) New domain motif: the structure of pectate lyase C, a secreted plant virulence factor. Science 260: 1503-1507.

42. Johansson K, El-Ahmad M, Friemann R, Jörnvall H, Markovic O, et al. (2002) Crystal structure of plant pectin methylesterase. FEBS Lett 514: 243-249.

43. Ray J, Knapp J, Grierson D, Bird C, Schuch W (1988) Identification and sequence determination of a cDNA clone for tomato pectin esterase. Eur J Biochem 174: 119-124.

44. Khanh NQ, Albrecht H, Ruttkowski E, Löffler F, Gottschalk M, et al. (1990) Nucleotide and derived amino acid sequence of a pectinesterase cDNA isolated from *Aspergillus niger* strain RH 5344. Nucleic Acids Res 18: 4262.

45. Plastow GS (1988) Molecular cloning and nucleotide sequence of the pectin methyl esterase gene of *Erwinia chrysanthemi* B374. Mol Microbiol 2: 247-254.

46. Spök A, Stubenrauch G, Schörgendorfer K, Schwab H (1991) Molecular cloning and sequencing of a pectinesterase gene from *Pseudomonas solanacearum*. J Gen Microbiol 137: 131-140.

47. Wakeley PR, Rogers HJ, Rozycka M, Greenland AJ, Hussey PJ (1998) A maize pectin methylesterase-like gene, ZmC5, specifically expressed in pollen. Plant Mol Biol 37: 187-192.

48. Futamura N, Mori H, Kouchi H, Shinohara K (2000) Male flower-specific expression of genes for polygalacturonase, pectin methylesterase and beta-3-glucanase in a dioecious willow (*Salix gilgiana Seemen*). Plant Cell Physiol 41: 16-26.

49. Micheli F, Sundberg B, Goldberg R, Richard L (2000) Radial distribution pattern of pectin methylesterases across the cambial region of hybrid aspen at activity and dormancy. Plant Physiol 124: 191-199.

50. Dorokhov YL, Mäkinen K, Frolova OY, Merits A, Saarinen J, et al. (1999) A novel function for a ubiquitous plant enzyme pectin methylesterase: the host-cell receptor for the tobacco mosaic virus movement protein. FEBS Lett 461: 223-228.

51. Chen MH, Sheng J, Hind G, Handa AK, Citovsky V (2000) Interaction between the tobacco mosaic virus movement protein and host cell pectin methylesterases is required for viral cell-to-cell movement. EMBO J 19: 913-920.

52. Frenkel C, Peters JS, Tieman DM, Tiznado ME, Handa AK (1998) Pectin methylesterase regulates methanol and ethanol accumulation in ripening tomato (*Lycopersicon esculentum*) fruit. J Biol Chem 273: 4293-4295.

53. Stephenson MB, Hawes MC (1994) Correlation of pectin methylesterase activity in root caps of pea with root border cell separation. Plant Physiol 106: 739-745.

54. Wen F, Zhu Y, Hawes MC (1999) Effect of pectin methylesterase gene expression on pea root development. Plant Cell 11: 1129-1140.

55. Pilling J, Willmitzer L, Fisahn J (2000) Expression of a *Petunia inflata* pectin methyl esterase in *Solanum tuberosum* L. enhances stem elongation and modifies cation distribution. Planta 210: 391-399.

56. Raymond CA, Henson AM, Pelletier MC, Boyton S, Joe B, et al. (2008) Improving dimensional stability in plantation grown *Eucalyptus pilularis* and *E. Dunnii*. Forest and Wood Products Australia, Melbourne, Australia, pp. 1-65.

57. Stackpole D, Vaillancourt R, Aguigar MD, Potts B (2010) Age trends in genetic parameters for growth and wood density in *Eucalyptus globulus*. Tree Genet Genomes 6: 179-193.

58. Stackpole DJ, Vaillancourt RE, Downes GM, Harwood CE, Potts BM (2010) Genetic control of kraft pulp yield in *Eucalyptus globules*. Can J For Res 40: 917-927.

59. Hertzberg M, Aspeborg H, Schrader J, Andersson A, Erlandsson R, et al. (2001) A transcriptional roadmap to wood formation. Proc Natl Acad Sci U S A 98: 14732-14737.

60. Willats WG, Orfila C, Limberg G, Buchholt HC, van Alebeek GJ, et al. (2001) Modulation of the degree and pattern of methylesterification of pectic homogalacturonan in plant cell walls: implications for pectin methyl esterase action, matrix properties, and cell adhesion. J Biol Chem 276: 19404-19413.

61. Kirst M, Myburg AA, De León JP, Kirst ME, Scott J, et al. (2004) Coordinated genetic regulation of growth and lignin revealed by quantitative trait locus analysis of cDNA microarray data in an interspecific backcross of eucalyptus. Plant Physiol 135: 2368-2378.

62. Goulao LF, Vieira-Silva S, Jackson PA (2011) Association of hemicellulose- and pectin-modifying gene expression with *Eucalyptus globulus* secondary growth. Plant Physiol Biochem 49: 873-881.

63. Huang AC, Wang YT, Yen HH, Jiang CM, Wu MC (2011) Transacylation properties of pectin methyl esterase from *Aspergillus niger*. Afr J Food Sci 13: 710-716.

64. Massiot P, Perron V, Baron A, Drilleau JF (1997) Release of methanol and depolymerization of highly methyl esterifed apple pectin with an endopolygalacturonase from *Aspergillus niger* and pectin methylesterases from *Aspergillus niger* or from orange. Food Sci and Technol 30: 697-702.

65. Alkorta I, Garbisu C, Llama J, Serra JI (1998) Industrial applications of pectic enzymes: A review. Process Biochem 33: 21-28.

66. Wicker L, Ackerley JL, Corredig M (2002) Clarification of juice by thermolabile valencia pectinmethylesterase is accelerated by cations. J Agric Food Chem 50: 4091-4095.

67. Cemeroglu B, Acar J, Meyve ve Sebze Isleme Teknoloji, (Gida Teknolojisi Dernegi, Yayin No: 6, Ankara), pp. 363.

68. Demir N, Acar, J, Sarioglu, K, Mutlu M (2001) The use of commercial pectinase in fruit juice industry. Part 3: Immobilized pectinase for mash treatment. J Food Eng 47: 275-280.

69. Pimenta-braz PN, Ricardo-da-silva JM and Laureano O (1998) Evaluation of pectolytic activities of enological interest in industrial enzyme preparations. Z Lebensm Unters Forsch A 206: 14-20.

70. Rogerson FSS, Vale E, Grande H, Silva MCM (2000) Alternative processing of port-wine using pectolytic enzyme. Cienc Technol Aliment 2: 222-227.

71. Kashyap DR, Vohra PK, Chopra S, Tewari R (2001) Applications of pectinases in the commercial sector: a review. Bioresour Technol 77: 215-227.

72. Dixit S, Upadhyay SK, Singh H, Pandey B, Chandrashekar K, et al. (2013) Pectin methylesterase of Datura species, purification, and characterization from *Datura stramonium* and its application. Plant Signal Behav 8: doi: 10.

73. Degraeve P, Saurel R, Coutel Y (2003) Vacuum impregnation pretreatment with pectinmethylesterase to improve firmness of pasteurized fruits. J Food Sci 68: 716-721.

74. Suutarinen J, Honkapaa K, Heinio L, Autio K, Mokkila M (2000) The effect of different prefreezing treatments on the structure of strawberries before and after jam making. Food Sci and Technol 33: 188-201.

75. Suutarinen J, Honkapaa K, Heinio L, Autio K, Mustranta A, et al. (2002) Effects of calcium chloride-based prefreezing treatments on the quality factors of strawberry jams. J Food Sci 67: 884-894.

76. Yemenicioglu A (2002) Control of polyphenol oxidase in whole potatoes by low temperature blanching. Eur Food Res Technol 214: 313-319.

77. Morris GA, Foster TJ, Harding SE (2000) The effect of the degree of esterification on the hydrodynamic properties of citrus pectin. Food Hydrocolloids 14: 227-235.

78. Schmelter T, Wientjes R, Vreeker R, Klaffke W (2002) Enzymatic modifications of pectins and the impact on their rheological properties. Carbohydr Poly 47: 99-108.

79. Bethke G, Grundman RE, Sreekanta S, Truman W, Katagiri F, et al. (2014) Arabidopsis pectin methylesterases contribute to immunity against *Pseudomonas syringae*. Plant Physiol 164: 1093-1107.

80. Vierhuis E, Korver M, Schols HA, Voragen AGJ (2003) Structural characteristics of pectic polysaccharides from olive fruit (*Olea europaea cv moraiolo*) in relation to processing for oil extraction. Carbohydr Poly 51: 135-148.

81. Pretel MT, Lozano P, Riquelme F, Romojaro F (1997) Pectic enzymes in fresh fruit processing: optimization of enzymic peeling of oranges. Process biochem 32: 43-49.

Heat and Mass Balance for Baking Process

El-Sayed G Khater* and Adel H Bahnasawy

Agricultural Engineering Department, Faculty of Agriculture, Benha University 13736, Egypt

Abstract

The main objective of this research was to develop a mathematical model of heat and mass balance of the bread baking process to predict the temperature and water content of bread at different heating temperatures and times. The model was able to predict the bread temperature and water content at different oven temperatures (180, 190, 200, 210 and 220°C). The results showed that the bread temperature and weight loss of bread increase with increasing oven temperature, when, the oven temperature increased from 180 to 220°C, the temperature of bread increased from 112.73 to 168.49°C and the weight loss of bread increased from 22.40 to 52.46%. The results also showed that the bread temperature and weight loss of bread increase with increasing time, this increment starts to decline after 18-20 min, until it reach equilibrium after 30 min. The weight loss of bread increases with increasing bread temperature. It indicates that when the bread temperature increased from 20.00 to 131.69°C, the weight loss of bread increased from 0.00 to 40.79% at 200°C oven temperature. The model was validated with an experimental data and showed a reasonable agreement with those measured, where; it ranged 20.00 to 131.69°C theoretically while it was from 25.00 to 119.00°C experimentally during the baking at 200°C oven temperature. The weight loss data was in a reasonable agreement with those measured, where; it ranged 0.00 to 40.79% theoretically while it was from 0.00 to 48.69% experimentally during the baking at 200°C oven temperature.

Keywords: Heat balance; Mass balance; Mathematical model; Bread temperature; Baking

Introduction

Baking is the final and most important step in bread production, and can be defined as the process which transforms dough, basically made of flour, water and leavening agents, in a food with unique sensory features by application of heat inside an oven. In particular, white or French bread is the most popular type of bread, and is distinguished for having a crunchy and golden-yellow (or brown) crust, a sponge and light crumb with soft texture and intermediate moisture, and a typical flavour. All these quality aspects are the result of a series of physical and chemical changes produced by simultaneous heat and mass transfer occurring within the product during baking [1-6].

The first bread was made around 10,000 years BC or over 12,000 years in the past, which may have been developed by deliberate experimentation with water and grain flour. Egyptians are the pioneers who make the art of bread making popular throughout the world. Control of the production and distribution of bread has been used as a means of exercising political influence over the populace for at least the last 2000 years. Even today, a shortage of bread is synonymous with hard times, while the promise of its service is used as rallying call for better life [2].

Simulation can be defined as the process of developing a model of a real system and carrying out experiments through the model, with the aim of studying, analyzing, designing or re-designing, controlling and predicting a certain real process. Besides the validation of the model, numerical simulation does not imply field experiments; it only includes the development of a mathematical model and computational effort, mostly more inexpensive than real tests. In addition, simulation gives the possibility of working under standardized operative conditions, minimizing the uncertainties of complex processes, especially those which are traditional non-automated processes. In this way, it will be very useful to have an accurate mathematical model for bread baking simulation [7].

There are several researchers have been studied the baking process, [8] studied of sponge cake batter baking process (Modeling and parameter estimation) and [9] developed a model of heat and mass transfer phenomena and quality changes during continuous biscuit baking using both deductive and inductive (neural network) modelling principles. [10,11] developed a simulation of bread making process using a direct 3D numerical method at microscale (Analysis of foaming phase during proofing, and Analysis of baking step). Also, [12] studied behavior of baking of semi-sweet short dough biscuits mathematically.

Bread is closely related to people's daily life and bread making is an important unit operation in food industry. Although people have practiced baking for a very long time, quantitative understanding of physical processes in baking is still very limited, therefore, understanding the physical processes of baking such as volume expansion, heating temperature, moisture change is the main aim of this work which could be achieved by developing a mathematical model of the heat and mass balance of the baking process to study the heating temperature and bread moisture content.

Model Development

The model includes the main distinguishing features of bread baking, i.e. the rapid heating of bread core and the development of a dry outer crust. Bread is modeled as a system containing three different regions: (1) crumb: wet inner zone, where temperature does not exceed 100°C and dehydration does not occur; (2) crust: dry outer zone, where temperature exceeds 100°C and dehydration occurs; (3) evaporation front: between the crumb and crust, where temperature is 100°C and water evaporates (liquid-vapour transition) [7,13].

***Corresponding author:** El-Sayed G Khater, Agricultural Engineering Department, Faculty of Agriculture, Benha University 13736, Egypt

Mathematically, the Moving Boundary Problem (MBP) is formulated using a physical approach, where the enthalpy jump corresponding to phase change is incorporated in the model by defining equivalent thermo physical properties [14]. Such definition states that evaporation occurs within a temperature range rather than at a fixed temperature. Other major assumptions of the model are the following:

- Bread is homogeneous and continuous; the concept of porous medium is included through effective or apparent thermo physical properties.
- Heat is transported by conduction from surface to bread core according to Fourier's law, but an effective thermal conductivity is used to incorporate the evaporation-condensation mechanism in heat transfer.
- Only liquid diffusion in the crumb and only vapour diffusion in the crust are assumed to occur [15].
- Volume change is neglected.
- Energy from oven ambient to bread surface is transferred by convection and radiation.
- Liquid water migrates from the core towards to the evaporation front under a water content gradient and liquid water flux is described by Fick's diffusion law.
- Water vapour migrates from the evaporation front to bread surface under a water vapour concentration gradient and mass flux is described by Fick's diffusion law.
- Water vapour is transferred to oven ambient through convective flux.

Governing Equations

Bread is considered as an infinite cylinder of radius r, so the problem is reduced to a single dimension via the axial symmetry assumption and it is assumed that the volume. For initial conditions, uniform temperature and water content are assumed. Figure 1 represents the heat and mass balance for bread. Heat balance include: heat conduction, heat convection and heat radiation.

Heat balance equation

$$\rho C_P \frac{\partial T}{\partial t} = \frac{1}{r}\frac{\partial}{\partial r}\left(rk\frac{\partial T}{\partial r}\right) \tag{1}$$

Boundary conditions

The heat arrives to the bread surface by convection and radiation, and is balanced by conduction inside the bread:

$$-k\frac{\partial T}{\partial r} = h\left(T_s - T_\infty\right) + \varepsilon\sigma\left(T_s^4 - T_\infty^4\right) \tag{2}$$

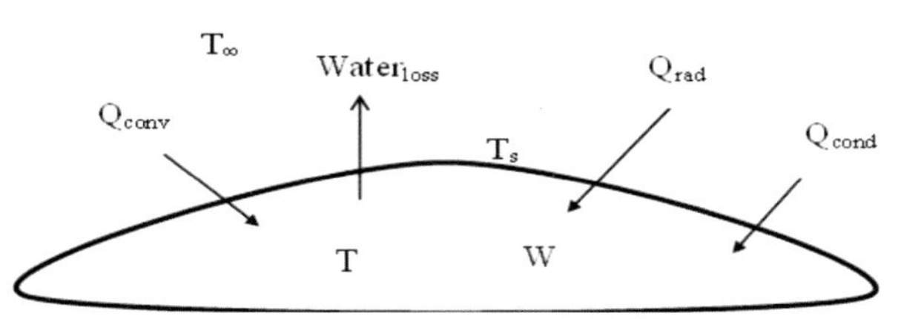

Figure 1: Heat and mass balance of bread.

Mass balance equation

$$\frac{\partial W}{\partial t} = \frac{1}{r}\frac{\partial}{\partial r}\left(rD\frac{\partial W}{\partial r}\right) \tag{3}$$

Boundary conditions

The water migrating towards the bread surface is balanced by convective flux:

$$-D\rho_s\frac{\partial W}{\partial t} = k_g\left(P_s\left(T_s\right) - P_\infty\left(T_\infty\right)\right) \tag{4}$$

$$p_s = a_w p_{sat}\left(T_s\right) \tag{5}$$

$$p_\infty = \left(\frac{RH}{100}\right)p_{sat}\left(T_\infty\right) \tag{6}$$

At the center, i.e. r = 0

$$\frac{\partial T}{\partial t} = 0 \tag{7}$$

$$\frac{\partial W}{\partial t} = 0 \tag{8}$$

Thermo physical properties

According to the MBP formulation, equivalent thermo physical properties are defined including the phase transition occurring during the process, i.e. an equivalent property is valid for dough/ crumb and crust [7,16].

Specific heat

$$C_p\left(T,W\right) = C_p^*\left(T,W\right) + \lambda_v W\delta \tag{9}$$

$$C_p^*\left(T,W\right) = C_{p,s}\left(T\right) + WC_{p,w}\left(T\right) \tag{10}$$

$$C_{p,s} = 5T + 25 \tag{11}$$

$$C_{p,w} = \left(5.207 - 73.17\times10^{-4}T + 1.35\times10^{-5}T^2\right)\times1000 \tag{12}$$

Latent heat evaporation is calculated with the following set of equations [10]:

$$\lambda_v = \left(2502535.259 - 212.56384\left(T - 273\right)\right) \tag{13}$$

Water activity

$$a_w\left(T,W\right) = \left(\left(\frac{100W}{\exp\left(-0.0056T + 5.5\right)}\right)^{-1/0.38} + 1\right)^{-1} \tag{14}$$

The heat transfer coefficient (h) is a model input for process simulation, and the mass transfer coefficient (kg) is determined by using the Chilton–Colburn (or heat–mass) analogy and a correction factor [7]:

$$\frac{h}{k_g^*} = \frac{M_{air}}{M_w}P_{atm}C_{p,air}\left(\frac{Sc}{Pr}\right)^{2/3} \tag{15}$$

$$k_g = 7.83\times10^{-3}k_g^* \tag{16}$$

$$Sc = \frac{\nu}{D} \tag{17}$$

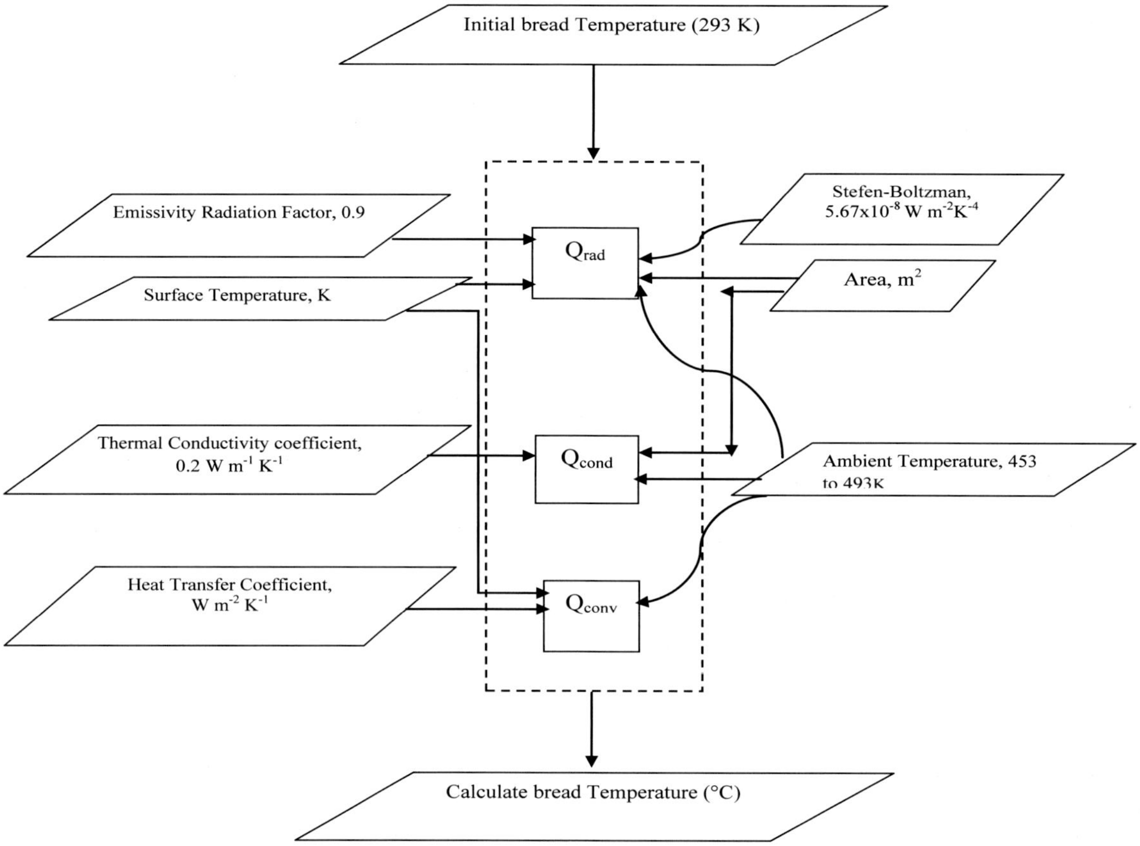

Figure 2: Flowchart of the mass balance of bread.

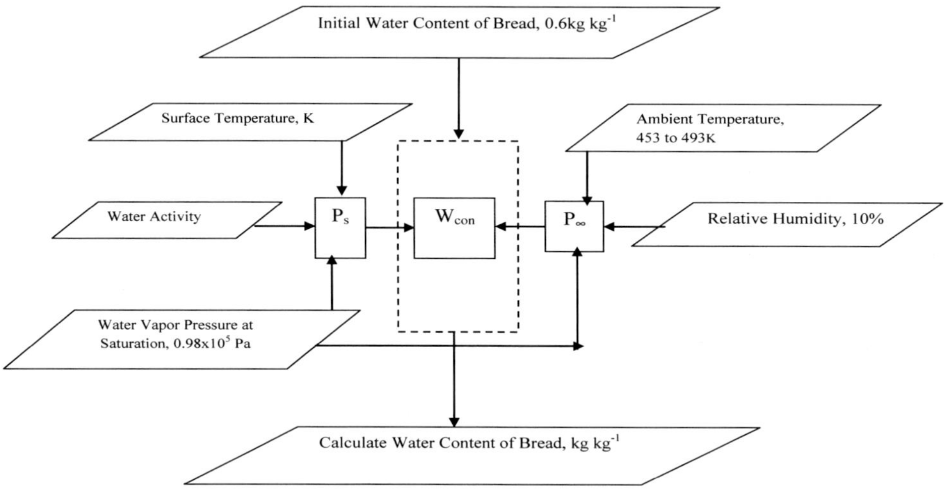

Figure 3: Flowchart of the mass balance of bread.

$$\Pr = \frac{\nu}{\alpha} \tag{18}$$

Regarding heat transfer by radiation, the emissivity of bread surface is considered equal to 0.9 [17,18].

All computational procedures of the model were carried out using Excel spreadsheet. The computer program consisted of two parts in addition to the input parts. The first part was devoted to heat balance for predicting the temperature at the middle cross-section of bread. The second part was devoted to mass balance for predicting water content of bread. Figures 2 and 3 show the flowcharts of the model steps and sequences. They are representing the heat and mass balance and the

Parameter	Units	Value	Source
Thermal conductivity, k	$W\ m^{-1}\ K^{-1}$	0.20	[6]
Bread density, ρ	$kg\ m^{-3}$	321.31	[18]
Density of solid, ρ_s	$kg\ m^{-3}$	241.76	[18]
Water diffusion coefficient, D	$m^2\ s^{-1}$	1×10^{-4}	[19]
Water vapor pressure at saturation, P_{sat}	Pa	0.98×10^{5}	[20]
Mass transfer coefficient, k_g^*	$kg\ Pa^{-1}\ m^{-2}\ s^{1}$	1.1×10^{-7}	[5]
Emissivity, ε		0.90	[18]
Kinematic viscosity, v	$m^2\ s^{-1}$	28.86×10^{-6}	[21]
Thermal diffusivity, α	$m^2\ s^{-1}$	1.17×10^{-7}	[21]
Bread radius, r	M	0.03	
Relative humidity, RH	%	10	
Delta-type function, δ		1	
Stefan-Boltzmann constant, σ	$W\ m^{-2}\ K^{-4}$	5.67×10^{-8}	
Initial temperature	K	293	
Initial water content	$kg\ kg^{-1}$	0.6	

Table 1: The parameters used in the heat and mass balance.

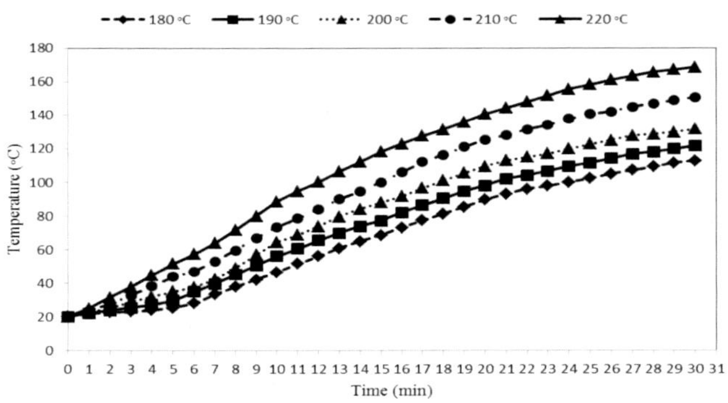

Figure 4: The predicted temperature of the bread crust at different oven temperatures.

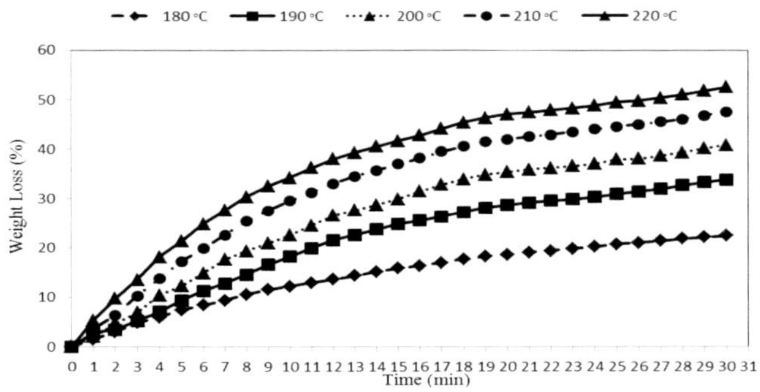

Figure 5: The predicted bread weight loss at different oven temperatures.

calculation steps of their model. Table 1 show the parameters used in the model.

Results and Discussions

Model experimentation

Temperature: Figure 4 shows the predicted temperature of the crust of bread process during the baking at different oven temperatures (180, 190, 200, 210 and 220°C). The results indicate that the bread temperature increases with increasing oven temperature. It also indicates that when the oven temperature increased from 180 to 220°C, the temperature at the crust of bread increased from 112.73 to 168.49°C. The results show that the temperature increases with increasing time, it increased from 20 to 112.73°C when the time increased from 0 to 30 min at oven temperature 180°C. On the other hand, the temperature increased from 20 to 116.18°C, 20 to 131.69°C, 20 to 144.05°C and 20 to 168.49°C at oven temperature 190, 200, 210 and 220°C, respectively. These results were in agreement with those obtained by [7,16,22-25].

It is worth to note that the bread temperature reached 62.6% of the heat balance at temperature of 180°C, while it reached 76.7% of the heat balance at 220°C oven temperature, this could lead to that using the higher temperature (200-220°C) more effective in utilization of heat energy. Multiple regression analysis was carried out to find a relationship between baking time, oven temperature and the bread temperature. The best form obtained was as follows:

$$T_P = -210.217 + 4.45t + 1.137T_o \qquad R^2 = 0.97 \quad (19)$$

Weight loss: Figure 5 shows the predicted bread weight loss during the baking process at different oven temperatures (180, 190, 200, 210 and 220°C). The results indicate that the weight loss increases with increasing oven temperature. It indicates that when the oven temperature increased from 180 to 220°C, the bread weight loss increased from 22.40 to 52.46%. The results show that the weight loss increases with increasing time; this increment starts to decline after 18-20 min, until it reached equilibrium after 30 min. It increased from 0 to 22.40% when the time increased from 0 to 30 min at oven temperature 180°C. On the other hand, the weight loss increased from 0 to 52.46% at oven temperature 220°C. These results were in agreement with those obtained by [16,25]. Weight loss of bread decrease by 22.40, 33.79, 40.49, 47.46 and 52.46% for the oven drying temperatures of 180, 190, 200, 210 and 220°C, respectively.

At the lower oven temperature (180°C), the weight loss reached 22.40% of the total weight which is lower by about 17.60% that should be lost to reach the recommended weight loss of the bread (40.00%, [26]). At the higher oven temperature (220°C), the weight loss reached 52.46% of the total weight which is higher by about 12.46% that should be lost to reach the recommended weight loss of the bread (40%). Multiple regression analysis was carried out to find a relationship between baking time, oven temperature and the bread weight loss. The best form obtained was as follows:

$$WL_P = -103.106 + 1.213t + 0.555T_o \qquad R^2 = 0.90 \quad (20)$$

Relationship between the average bread temperature and average weight loss of bread

Figure 6 shows the relationship between the predicted average

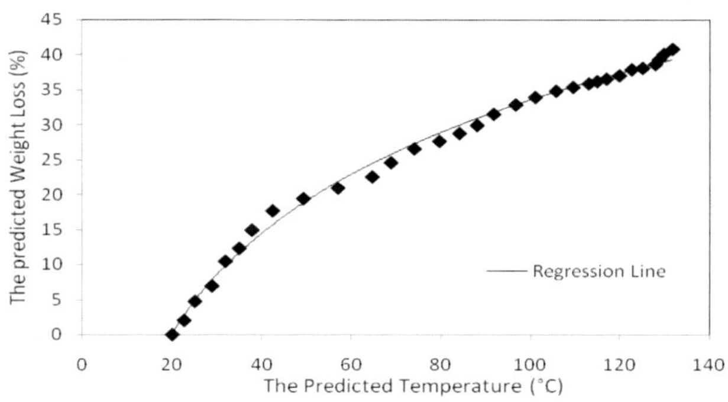

Figure 6: The relationship between the predicted average bread temperature and the predicted average weight loss of bread at the average oven temperature.

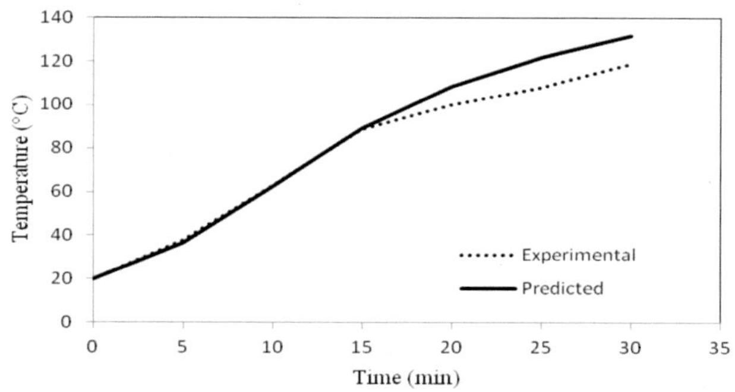

Figure 7: The predicted and the experimental temperatures of beard during the baking.

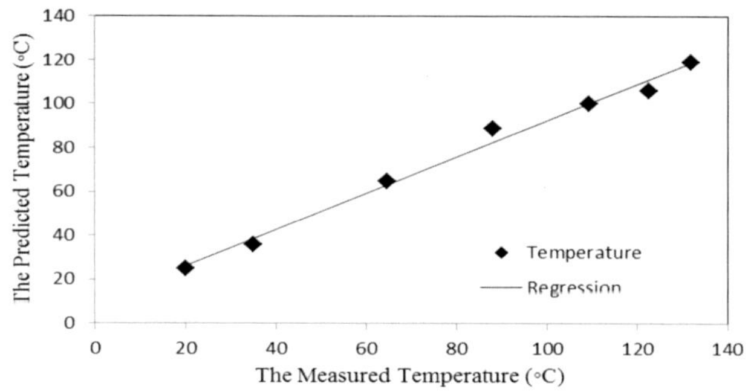

Figure 8: The comparison between the predicted and the experimental temperatures of bead during the baking.

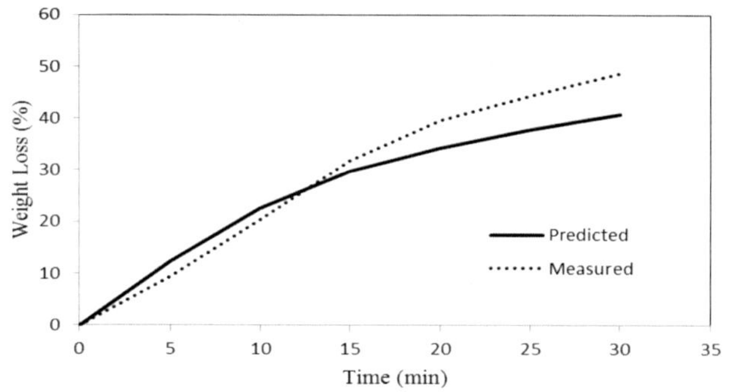

Figure 9: The predicted and the experimental weight losses of bread during the baking

bread temperature and the predicted average weight loss of bread at the average oven temperature. The average weight loss of bread increases with increasing average bread temperature, the average weight loss increased linearly during the temperature range of 20-100°C, then increment. It indicates that when the average bread temperature increased from 20.00 to 137.03°C, the average weight loss of bread increased from 0.00 to 39.38% at the average oven temperature.

The relationship between the predicted average bread temperature and the predicted average weight loss of bread at average oven temperature is shown in figure 6. This relationship is expressed by the following equation:

$$WL_P = 20.47\ln T_P - 61.3226 \qquad R^2 = 0.99 \qquad (21)$$

Model validation

The model was validated using the data published by Purlis and Salvadori [7].

Temperature

Figure 7 shows the predicted and the experimental temperatures of bread during the baking. It could be seen that, the temperature by the model was in a reasonable agreement with those measured, where; it ranged 20.00 to 131.69°C theoretically while it was from 20.00 to 119.00°C experimentally during the baking at 200°C oven temperature.

Regression analysis was carried out to find a relationship between the predicted and experimental bread temperatures and the most suitable form is shown as follows (Figure 8):

$$T_P = 1.1456T_M - 9.4539 \qquad R^2 = 0.99 \qquad (22)$$

Weight loss

Figure 9 shows the predicted and the experimental weight losses of bread during the baking. It could be seen that, the weight loss by the model was in a reasonable agreement with those measured during the period between 0-15 min, where it ranged from 0.00 to 33.16% then there were variations between the predicted and experimental data, where the model results were lower than those of the experimental, where it ranged from 33.16 to 49.79% theoretically while it was from 0.00 to 48.69% experimentally during the baking at 200°C oven temperature in the period between 15-30 min, this underestimation of bread weight loss is probably due to not considering the evaporation-condensation contribution in this process.

The variations between the predicted and the experimental are shown in figure 10. The relationship between the predicted and measured temperatures is expressed by the following equation:

$$WL_P = 1.2308WL_M - 3.5117 \qquad R^2 = 0.97 \qquad (23)$$

Conclusions

A mathematical model for the baking process was developed successively according to heat balance and mass balance to study the bread temperature and water content at different oven temperatures. Also the model was validated using an experimental data from [7]. The most important results obtained can be summarized as follows:

- The model showed that the predicted bread temperature increases with increasing oven temperature, where, the oven temperature increased from 180 to 220°C, the temperature of the crust of bread increased from 112.73 to 168.49°C.

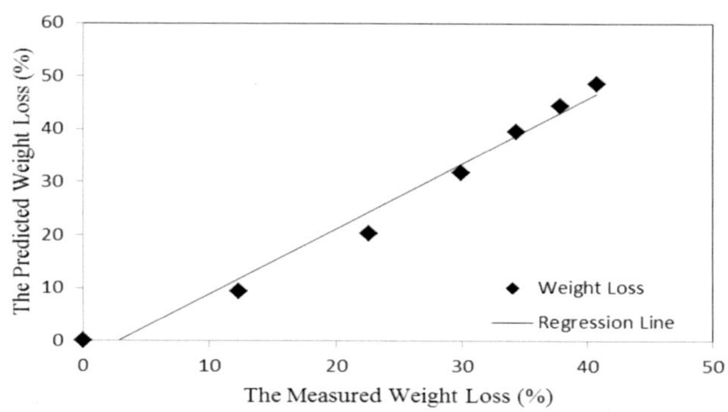

Figure 10: The comparison between the predicted and the experimental weight losses of bread during the baking.

- The model showed that the temperature increases with increasing time, where, it increased from 20 to 112.73°C when the time increased from 0 to 30 min at oven temperature 180°C. On the other hand, the temperature increased from 20 to 168.49°C at oven temperature 220°C.
- The model showed that the weight loss increases with increasing oven temperature, where, the oven temperature increased from 180 to 220°C, the weight loss of bread increased from 22.40 to 52.46%. The model showed that the weight loss increases with increasing time, this increment starts to decline after 18-20 min, until it reach equilibrium after 30 min. It increased from 0 to 22.40% when the time increased from 0 to 30 min at oven temperature 180°C. On the other hand, the weight loss increased from 0 to 52.46% at oven temperature 220°C.
- The predicted bread temperature was in a reasonable agreement with the experimental temperature with a coefficient of determination of 0.98. The bread temperature ranged from 20.00 to 131.69°C theoretically while it was from 25.00 to 119.00°C experimentally during the baking at 200°C oven temperature.
- The predicted weight loss of bread was in a reasonable agreement with the experimental weight loss of bread with a coefficient of determination of 0.97. The weight loss of bread ranged from 0.00 to 40.79% theoretically while it was from 0.00 to 48.69% experimentally during the baking at 200°C oven temperature.

Nomenclature

ρ bread density, kg m^{-3}
C_p specific heat of bread, J kg^{-1} K^{-1}
T temperature of bread at as a function of time, K
t time, s
r bread radius, m
k bread thermal conductivity, W m^{-1} K^{-1}
h heat transfer coefficient, W m^{-2} K^{-1}
T_s surface temperature, K
T_∞ ambient temperature, K
ε emissivity, dimensionless
σ Stefan-Boltzmann constant, dimensionless
W water content of bread, kg kg^{-1}
D water diffusion coefficient, m^2 s^{-1}
k_g corrected mass transfer coefficient, kg Pa^{-1} m^{-2} s^{-1}
p_s water vapour pressure of the air at the bread surface, Pa
P_∞ water vapour pressure of the air at the oven ambient, Pa
ρ_s density of solid (dough/crumb and crust), kg m^{-3}
p_{sat} water vapour pressure at saturation, Pa
a_w water activity, dimensionless
RH relative humidity, %
λ_v latent heat of evaporation, J kg^{-1}
δ delta-type function, dimensionless
k_g^* mass transfer coefficient, kg pa^{-1} m^{-2} s^{-1}
M_{air} molecular mass of air, g mol^{-1}
M_w molecular mass of water, g mol^{-1}
Sc Schmidt number, dimensionless
Pr Prandlt number, dimensionless
v kinematic viscosity, m^2 s^{-1}
α thermal diffusivity, m^2 s^{-1}
T_P predicted bread temperature, °C
T_M measured bread temperature, °C
T_o oven temperature, °C
WL_P predicted weight loss, %
WL_M measured weight loss, %

References

1. Sablani SS, Marcotte M, Baik OD, Castaigne F (1998) Modeling of simultaneous heat and water transport in the baking process. LWT - Food Science and Technology 31: 201-209.
2. Scanlon MG, Zghal MC (2001) Bread properties and crumb structure. Food Research International 34: 841-864.
3. Mondal A, Datta AK (2008) Bread baking - a review. Journal of Food Engineering 86: 465-474.
4. Vanin FM, Lucas T, Trystram G (2009) Crust formation and its role during bread baking. Trends in Food Science & Technology 20: 333-343.
5. Purlis E, Salvadori VO (2009) Modelling the browning of bread during baking. Food Research International 42: 865-870.
6. Purlis E (2010) Browning development in bakery products-A review. Journal of Food Engineering 99: 239-249.
7. Purlis E, Salvadori VO (2009b) Bread baking as a moving boundary problem. Part 2: model validation and numerical simulation. Journal of Food Engineering 91: 434-442.
8. Lostie M, Peczalski R, Andrieu J, Laurent M (2002) Study of sponge cake batter baking process.II. Modeling and parameter estimation. Journal of Food Engineering, 55: 349-357.
9. Broyart B, Trystram G (2003) Modelling of heat and mass transfer phenomena and quality changes during continuous biscuit baking using both deductive and inductive (neural network) modelling principles. Food and Bioproducts Processing 81: 316-326.
10. Bikard J, Coupez T, Valle GD, Vergnes B (2008) Simulation of bread making process using a direct 3D numerical method at microscale: Analysis of foaming phase during proofing. Journal of Food Engineering 85: 259-267.
11. Bikard J, Coupez T, Della Valle G, Vergnes B (2012) Simulation of bread making process using a direct 3D numerical method at microscale: Analysis of baking step. Int J Mater Form 5: 11-24.
12. Baldino N, Gabriele D, Lupi FR, de Cindio B, Cicerelli L (2014) Modeling of baking behavior of semi-sweet short dough biscuits. Innovative Food Science and Emerging Technologies 25: 40-52.
13. Purlis E (2012) Baking process design based on modeling and simulation: Towards optimization of bread baking. Food Control 27: 45-52.
14. Bonacina C, Comini G, Fasano A, Primicerio M (1973) Numerical solution of phase-change problems. International Journal of Heat and Mass Transfer 16: 1825-1832.
15. Luikov AV (1975) Systems of differential equations of heat and mass transfer in capillary-porous bodies (review). International Journal of Heat and Mass Transfer 18: 1-14.
16. Purlis E, Salvadori VO (2009a) Bread baking as a moving boundary problem. Part 1: mathematical modelling. Journal of Food Engineering 91: 428-433.
17. Anonymous (1992) ASAE Standards. 39th edition. American Society of Agricultural Engineers. St. Joseph, MI. 781 pages.
18. Hamdami N, Monteau JY, Le Bail A (2004) Heat and mass transfer in par-baked bread during freezing. Food Research International 37: 477-488.
19. Rogers EK, Brimel CJ (2005) Instrumentation and sensors for the food industry. 2nd edition. Woodhead Publishing Limited. Cambridge, England.
20. Eckert ER, RM Drake (1959) Heat and Mass Transfer. 2nd edition. McGraw-Hill, New York, USA.
21. Magee TRA, Branshurg T (1995) Measurement of thermal diffusivity of potato, malt bread and wheat flour. Journal of Food Engineering 25: 223-232.
22. Zhang J, Datta AK (2006) Mathematical modelling of bread baking process. Journal of Food Engineering 75: 78-89.
23. Wang R, Zhou W, Jiang X (2008) Mathematical modeling of the stability of green tea catechin epigallocatechin gallate (EGCG) during bread baking. Journal of Food Engineering 87: 505-513.
24. Mondal A, Datta AK (2010) Two-dimensional CFD model and simulation of crustless bread baking process. Journal of Food Engineering 99: 168-174.
25. Purlis E (2011) Bread baking: Technological considerations based on process modeling and simulation. Journal of Food Engineering 103: 92-102.
26. Bakers Journal (2001) The effect of hard water and soft water on baked goods.

Response Surface Optimization of Critical Extraction Parameters for Anthocyanin from *Solanum melongena*

Arunachalam Thirunavukkarasu[1]* and Rajarathinam Nithya[2]

[1]*Lecturer, Dept. of Biotechnology, PSG College of Technology, Coimbatore, India*
[2]*Asst.Professor, Faculty of Biotechnology, PRIST University, Thanjavur, India*

Abstract

The study investigated the parameter optimization for anthocyanin extraction from egg plant (*Solanum melongena*) with the approach response surface methodology. Extractions were carried out using acid–ethanol with the temperature range (60–90°C), time (30–90 min) and solid–liquid ratio (1:15–1:30). Three level three factor Box-Behnken design was followed to observe the anthocyanin yield for the studied parameters. The maximum yield of anthocyanin was observed at the temperature 76.5°C in 70 min in the ratio of 1:26.

Keywords: Anthocyanin; Egg plant; Response surface methodology; Box-Behnken design

Introduction

Anthocyanins widely exist in naturally occurring pigments of red fruits such as cherries, plums, strawberries, raspberries, blackberries, grapes, red currants and black currants [1]. Also it has been found in the roots, caudexes and leaves as well as flowers and fruits acting as a substitute for synthetic pigments. Anthocyanins are primarily used by humans for its therapeutic purpose such as antineoplastic, radiation-protective, vasotonic, vasoprotective, anti-inflammatory, chemo- and hepatoprotective [2-5]. The skin of eggplants (*Solanum melongena*) contains different kind of anthocyanins and have been extracted and identified [6,7]. However no studies were focussed on optimizing the parameters for anthocyanin extraction from eggplants peel, although other authors have already optimized extraction methods from different anthocyanins of plant sources [8-10]. Hence the present study made an attempt to utilize response surface methodology for developing optimized condtions for anthocyanin extraction from eggplant peel. The prime focus of Response surface methodology (RSM) is to use a sequence of designed experiments to obtain an optimal response. Here the study designed in such a way that differing the three independent variables temperature, time of extraction and solvent-to-solid ratio within a selected range to obtain an optimal response of anthocyanin yield. The basic theoretical and fundamental aspects of RSM have been discussed [11].

Material and Methods

Sample preparation

Solanum melongena were identified and collected locally and washed in running tap water. Then the peel layers with uniform thickness (1 mm) were simply removed using a sharp knife. The layers were finally reduced to 20 mm2 particle size and stored at 20°C as followed by Aldo Todaro et al., [12].

Methods

Peels of eggplant were set into a 50ml Erlenmeyer flask, then added with acid–ethanol (HCl, 1.5 mol/l) with different solid–solvent ratio (1:15–1:30) and placed in thermostatic water bath in the temperature range of (60–90°C) for the time duration (30-90 min), then, centrifuged at 4000 rpm for 15 min. The supernatant was collected and transferred into 50 ml volumetric flask for the determination of anthocyanins yield. About 1 g of the samples was used for each treatment.

Experimental design

RSM was employed to determine the optimum condition for anthocyanin extraction of egg plant. The experimental design and statistical analysis were performed using Stat-Ease software (Design-Expert 6.0.10 Trial, Delaware, USA Echip, 1993). A three-level three-factor Box– Behnken design was chosen to evaluate the combined effect of three independent variables extraction temperature, time and solid–liquid ratio, coded as A1, A2 and A3, respectively. The minimum and maximum values for extraction temperature were set at 60 and 90°C, extraction time between 30 and 90 min and solid–liquid ratio 1:15 and 1:30 Table 1. The response values were anthocyanin pigment yield, L*, C* and H. The complete design consisted of 17 combinations including five replicates of the center point Table 2 [13]. The second-order model is preferred for several reasons that flexible, easy to estimate the parameters [14]. The responses function (η) was partitioned into linear, quadratic and interactive components as followed by Gongjian Fan et al., [15].

$$\eta = \beta 0 + \sum_{i=1}^{n} Bi\ X1 + \sum_{i=1}^{n} Bii\ X2 + \sum_{i=1}^{n} Bij\ Xi\ Xj$$

where β0 is defined as the constant, Bi the linear coefficient, Bii the quadratic coefficient and Bij the cross-product coefficient. Xi and Xj are levels of the independent variables while n equals to the number of the tested factors (n=3). The analysis of variance (ANOVA) tables were generated and the effect and regression coefficients of individual linear, quadratic and interaction terms were determined. The significances of all terms in the polynomial were judged statistically by computing the

***Corresponding authors:** Thirunavukkarasu Arunachalam, Lecturer, Department of Biotechnology, PSG College of Technology, Coimbatore, India

F-value at a probability (P) of 0.001, 0.01 or 0.05. The regression coefficients were then used to make statistical calculations to generate contour maps from the regression models.

Quantitative analysis

Anthocyanins quantification was done by following the spectrophotometric method of Francis [16]. The molar absorptivity value for the acid-ethanol solvent is 98.2 and it refers to the absorption of a mixture of cranberry anthocyanins in acid-ethanol, measured in a 1-cm cell at 535 nm, at a concentration of 1% (w/v). The total anthocyanins (mg/100 g) were determined by the method proposed by Gongjian Fan et al., [15].

Total anthocyanins (TA) = A530 X dilution factor/98.2.

Statistics

The trials have been carried out in triplicate and all the data were reported as means±SD (Standard Deviation). Student's t-test was performed for the statistical significance at $P<0.01$ or $P<0.001$ level.

Results and Discussion

Statistical analysis

Table 3 summarizes the results of each dependent variable with their coefficients of determination (R2) [16]. The R2 values for anthocyanins yield was 0.9673 indicated that the proposed model was adequate, possessing no significant lack of fit. Since the closer the value of R2 to unity, the better the empirical models fits the actual data. On the other hand, the smaller the value of R2 the less relevance the dependent variables in the model have in explaining the behavior of variations [17]. The probability (P) values of all regression models were less than 0.000, with no lack-of fit.

Independent variable	Units	Symbols	Code levels -1	0	1
Extraction temperature	°C	A1	60	75	90
Time	min	A2	30	60	90
Solid-Liquid ratio	1:X	A3	1:15	1:25	1:35

Table 1: Independent variables and their coded and actual values used for optimization.

Treatment	Independent variables			Dependent variable
	Temp (°C)	Time (min)	Solid-Liquid ratio	Anthocyanin yield (mg/100g)
1	-1	-1	0	110
2	1	-1	0	122
3	-1	1	0	136
4	1	1	0	138
5	-1	0	-1	124
6	1	0	-1	126
7	-1	0	1	120
8	1	0	1	128
9	0	-1	-1	127
10	0	1	-1	135
11	0	-1	1	124
12	0	1	1	138
13	0	0	0	144
14	0	0	0	144
15	0	0	0	142
16	0	0	0	144
17	0	0	0	144

Table 2: Box–Behnken design and experiment data for anthocyanins extraction.

Effects of extraction Temperature, Time and Solid-Liquid Ratio

The effect of the independent variables, Temperature and time on the anthocyanins yield was presented as the contour map Figure 1. As followed by the Table 3, anthocyanins yield mainly depend on the extraction temperature since both linear and quadratic effects were significant at ($P<0.01$) giving an overall curvilinear effect. The figure 1 also shows that the Anthocyanins yield can be increased with the increase of extraction temperature till it reaches 75°C. The linear effect ($P<0.01$) was positive, whereas the quadratic effect ($P<0.01$) was negative, which resulted in a curvilinear increase in Anthocyanins yield for all the extraction times. From this, it is evident that extraction temperature is the prime variable to be considered in the efficiency of extraction. The effect of the independent variables, temperature and S/L ratio on the anthocyanins yield was presented as the contour map Figure 2. The anthocyanins yield depends on solid-liquid ratio where its linear ($P<0.01$) and quadratic ($P<0.01$) effects were significantly affected by pigments concentration. Hence, the overall effect was curvilinear in nature. As the temperature progresses, anthocyanins yield of egg plant was found to increase rapidly till it reaches 75°C. It is evident from the Table 3 that the positive linear and negative quadratic effects of solid-liquid ratio and temperature explained the observed nature of the curve shown in Figure 2. The effect of the independent variables, time and S/L ratio on the anthocyanins yield was presented as the contour map (Figure 3).

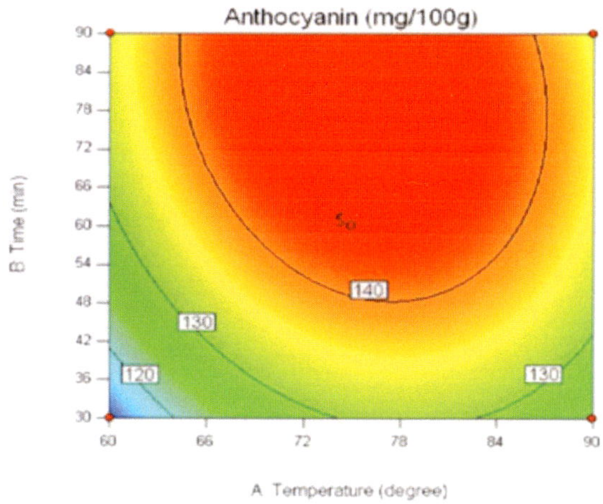

Figure 1: Effects of extraction temperature and time at a constant S/L ratio of 1:25 on the yield of Anthocyanins.

Regression Coefficient	Anthocyanins Yield
β0	143.6
b1	+3.00**
b2	+8.00
b3	+0.25**
b11	-2.50**
b12	+1.50
b22	+1.50
b13	-11.80
b23	-5.30
b33	-7.30**
R2	0.9673
P value	0.0002***

significant at 0.01 level * significant at 0.001 level

Table 3: Regression coefficients, R2, and P values for dependent variable for anthocyanins extraction.

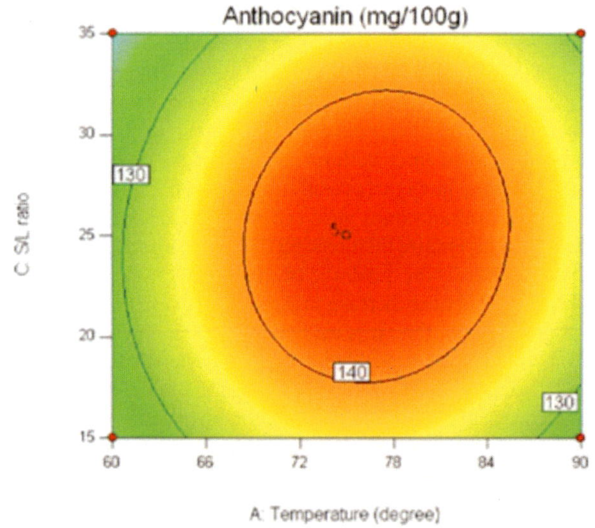

Figure 2: Effects of extraction temperature and S/L ratio at a constant time of 90min on the yield of Anthocyanins.

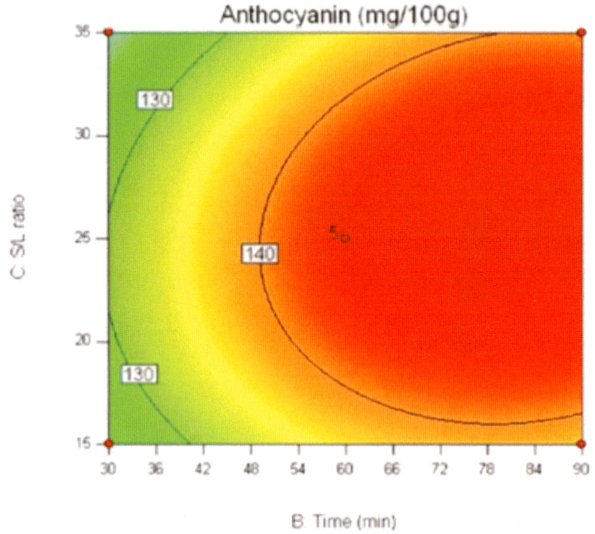

Figure 3: Effects of time and S/L ratio at a constant extraction temperature of 60°C on the yield of Anthocyanins.

The anthocyanins yield depends on solid–liquid ratio where it's linear ($P<0.01$) and quadratic ($P<0.01$) effects were significantly affected by pigments concentration. As the time proceeds, the anthocyanins yield was found to increase rapidly at the beginning but with a slower rate towards the end. It can be seen in Table 3 that the positive linear and negative quadratic effects of time and solid–liquid ratio explained the observed nature of the curve shown in Figure 3.

Optimization

The contour plots show the optimum conditions of the extraction process to anthocyanins yield, respectively. It was noted that the optimum conditions for Anthocyanins yield were slightly different. There are a number of combinations of variables that could give maximum levels of anthocyanins yield. Since the optimum response for each dependent variable did not fall exactly in the same region, the superimposition of all the contour plots obtained was done. Figure 1 shows the superimposed contour plot for optimization of anthocyanins yield. The zone of optimization depicts temperature to be over 76.5°C and solid–liquid ratio 1:26 and time to be maintained at 70 min. During the anthocyanin extraction, the extraction temperature and solid–liquid ratio are important. Therefore, the best combinations of process variables for response functions are found. The process variables for best combination of response function are extraction temperature 76.5°C, time 70 min, and solid–liquid ratio 1:26. The response anthocyanins yield from the optimized conditions was found to be 148 mg/100g.

Conclusions

The parameters extraction temperature, time and solid–liquid ratio for anthocyanins extraction were subjected to second order polynomials to determine their significance in the anthocyanin yield. The optimum conditions for the three independent variables were obtained from the contour plots and the extraction process were carried out to cross-validate their reliability and the yield exactly following at the optimized conditions. The anthocyanin yield from egg plant is estimated to be maximal if the extraction process maintaining the temperature at 76.5°C, time at 70 min, and solid–liquid ratio at 1:26.

References

1. Mazza G, Cacace JE, Kay CD (2004) Methods of analysis for anthocyanins in plants and biological fluids. Journal of AOAC International 87: 129–145.
2. Cacace JE, Mazza G (2003) Mass transfer process during extraction of phenolic compounds from milled berries. Journal of Food Engineering 59: 379–389.
3. Cacace JE, Mazza G.(2003) Optimization of extraction of anthocyanins from black currants with aqueous ethanol. Journal of Food Science 68: 240–248.
4. Mazza G, Miniati E (1993) Anthocyanins in fruits vegetables and grains Boca Raton FL: CRC Press.
5. Wang SY, Lin HS. (2000) Antioxidant activity in fruits and leaves of blackberry, raspberry, and strawberry varies with cultivar and developmental stage. Journal of Agricultural and Food Chemistry 48: 140–146.
6. Pifferi PG, Zamorani A (1969) Natural pigments VI Anthocyanins of Solanum melongena. Industrie Agrarie 7: 51–56.
7. Matsuzoe N, Yamaguchi M, Kawanobu S, Watanabe Y, Higashi H, et al. (1999) Effect of dark treatment of the eggplant on fruit skin color and its anthocyanin component. Journal of the Japanese Society for Horticultural Science 68: 138–145.
8. Cacace JE and Mazza G (2002) Extraction of anthocyanins and other phenolics from black currants with sulfured water. Journal of Agricultural and Food Chemistry 50: 5939–5946.
9. Turker N, Erdogdu F (2006) Effects of pH and temperature of extraction medium on effective diffusion coefficient of anthocynanin pigments of black carrot. Journal of Food Engineering 76: 579–583.
10. Naczk M, Shahidi F (2006) Phenolics in cereals fruits and vegetables: Occurence, extraction and analysis. Journal of Pharmaceutical and Biomedical Analysis 41:1523– 1542.
11. Chandrika LP and Fereidoon S (2005) Optimization of extraction of phenolic compounds from wheat using response surface methodology. Food Chemistry 93: 47–56.
12. Aldo Todaro, Francesco Cimino, Paolo Rapisarda, Anna E. Catalano, Riccardo N. Barbagallo, et al. (2009) Recovery of anthocyanins from eggplant peel, Food Chemistry 114: 434–439.
13. Myers RH and Montgomery DC (2002) Response surface methodology: Process and product optimization using designed experiments New York: Wiley.
14. Kathleen M Carley, Natalia Y Kamneva and Jeff Reminga (2004) Response surface methodology, CMU-ISRI, 04, 136.

15. Gongjian Fan, Yonbin Han, Zhenxin Gu, Deming Chen (2008) Optimizing conditions for anthocyanins extraction from purple sweet potato using response surface methodology (RSM) 41: 155–160.

16. Francis FJ (1989) Food colorants: Anthocyanins. Critical Reviews in Food Science and Nutrition 28: 273-314.

17. Sin HN, Yusof S, Sheikh Abdul Hamid N, Abdul Rahman R (2006) Optimization of hot water extraction for sapodilla juice using response surface methodology. Journal of Food Engineering 74: 352–358.

18. Ravikumar K, Ramalingam S, Krishnan S and Balu K (2006) Application of response surface methodology to optimize the process variables for reactive red and acid brown dye removal using a novel adsorbent. Dyes and Pigments, 70: 18–26.

Effect of Inducers and Physical Parameters on the Production Of L-Asparaginase Using *Aspergillus Terreus*

Moses Jeyakumar Rajesh[1], Leelavathy Rajesh[1], Vanapalli V Satya veni[2], G.Thirumurugan[2*] and Rajaram mohan Sivasubramanian[3]

[1]*Deparment of Biomedical Engineering, Jimma Institute of Technology, Jimma University, Jimma, Ethiopia*
[2]*Research lab, GIET School of Pharmacy, NH-5, Chaitanya nagar, Rajahmundry-533294, India*
[3]*School of Chemical and Biotechnology, SASTRA University, Tanjore, Tamilnadu, India*

Abstract

An effort has been made to find the best medium and operating conditions for the production of extracellular L-asparaginase by *Aspergillus terreus* in shake culture fermentation through stepwise optimization strategies like studies on the effect of pH and temperature on L-asparaginase, studies on the enzyme activity in various concentration of L-glutamine and study the growth kinetics for L-glutamine medium. Our study shows that L-glutamine combined with L-asparagine influence the asparaginase production level and physical parameter such as pH and temperature has more contribution in enzyme production. The pH 8 and temperature 35°C found to be good environmental condition the production of extracellular L-asparaginase by *Aspergillus terreus*.

Keywords: L-asparaginase; *Aspergillus terreus*; Shake culture fermentation; L-glutamine

Introduction

L-asparaginase (L-asparagine amidohydrolase; EC.3.5.1.1) catalyzes the deamidation of L-asparagine to L-aspartic acid and ammonia. L-asparaginase is present in many animal tissues, bacteria, plants and in the serum of certain rodents, but not in mankind. L-asparaginase is produced by a large number of micro-organism include *E. coli* [1], *Erwinia cartovora, Enterobacter aerogenes* [2], *Cornybacterium glutamicum, Candida utilis* [3], *Staphylococcus aureus* [4], *Thermus thermophilus* [5], *Pisum sativum.* Although Clementi in 1922 had reported its presence in guinea-pig serum, the anti-tumour properties of the enzyme were only recognized some time later. Tsuji first reported Deamidation of L-asparagine by extracts of *E. coli* in 1957. Broome in 1961 discovered that the regression of lymph sarcoma transplants in mice treated with guinea-pig serum was due to the nutritional dependence of the malignant cells on exogenous L- asparagines.

Several micro-organisms including *Serratia marcescens* produce L-asparaginases with antitumor activity. Although extensive studies have been carried out on the isolation and on the anti leukemia properties of this enzyme [6,7], very little information is available on the production of this enzyme by *S.marcescens* [7, 8]. A large amount of research has been conducted upon the biosynthesis of L-asparaginase since Masburn and Wriston demonstrated antitumor activity. Kinetic studies would allow the prediction of fermentation rate, product yield, and the control of the fermentation process. A continuous fermentation has advantages of productivity, ease of control, uniformity and low labor costs. Several species of *Erwinia* have been reported to provide promising sources of L-asparaginase and have proved to be effective against solid tumors [9]. Tumor cells, more specifically lymphatic tumor cells, require huge amounts of asparagines to keep up with their rapid, malignant growth. This means they use both asparagine from the diet as well as what they can make themselves to satisfy their large asparagines demand. L-asparaginase is an enzyme that destroys asparagine external to the lymphatic tumor cells. Normal cells are able to make all the asparagine they need internally whereas tumor cells become depleted rapidly and die. L-asparaginase is used as a chemotherapeutic agent for acute *lymphocytic leukemia* and less frequently for acute *myeloblastic leukemia, chronic lymphocytic leukemia*, Hodgkin's disease, *melonosarcoma* and non-Hodgkin's lymphoma. L-asparaginase produced by bacteria leads to adverse side effects in human trials. Therefore there is a search for the other sources of L-asparaginase production. It has been observed that eukaryotic microorganisms like yeast and filamentous fungi genera such as *Aspergillus, Penicillium* and *Fusarium* are commonly reported in scientific literature to produce asparaginase with less adverse effects. Asparaginase is an anti-neoplastic agent, used in the lymphoblastic leukemia chemotherapy. Neoplastic cells cannot synthetase L-asparagine due to absence of L-asparagine synthetase. For this reason the commonest therapeutic practice is to inject intravenously free enzyme in order to decrease the blood concentration of L-asparagine affecting selectively the neoplastic cells. The importance of micro-organism as L-asparaginase sources has been focused since the time it was obtained from *E. coli* and its antineoplastic activity demonstrated in guinea pig serum [9-11]. The *E.coli* and Erwinia enzymes were isolated, purified, and experimentally used as an anti-leukaemic agent in human patients. It demonstrated high potential against children's acute lymphoblastic leukemia. Several research groups have studied asparaginase production and purification in attempt to minimize impurities that produce allergic reactions. It has been observed that eukaryotic micro-organisms like yeast and filamentous fungi have a potential for asparaginase production. For example, the mitosporic fungi genera such as *Aspergillus, Penicillium,* and *Fusarium* are commonly reported in scientific literature to produce asparaginase. Objective of this research work is to find the best medium and operating conditions for the production of extracellular L-asparaginase by *Aspergillus terreus* in shake culture fermentation through stepwise optimization strategies like studies on the

***Corresponding author:** G.Thirumurugan, Research lab, GIET School of Pharmacy, NH-5, Chaitanya nagar, Rajahmundry-533294, India

effect of pH and temperature on L-asparaginase, studies on the enzyme activity in various concentration of L-glutamine and study the growth kinetics for L-glutamine medium.

Materials and Methods

Microorganism

The filamentous fungus selected to use throughout this study is *Aspergillus terreus* MTCC 1782 was obtained from Institute of Microbial Technology, Chandigarh, India.

Culture conditions

Stock culture: The spores and mycelial fragments of strain were cultivated in Czapek-Dox medium which containing the following ingredients (g/l of distilled water). Solution A: Sodium Nitrate 40.0; Potassium Chloride 10.0; $MgSO_4.7H_2O$, 0.52; $FeSO_4.7H_2O$, 0.2; dissolved in 1L of distilled water and stored in refrigerator. Solution B: K2HPO4 20.0; dissolved in 1L of distilled water and stored in refrigerator. Solution C: 1 gram of $ZnSO_4$. 7 H_2O dissolved in 100ml of distilled water. Solution D: 0.5 gram of CuSO4. H_2O dissolved in 100ml of distilled water. For 1L of Czapek-Dox medium; 50 ml of solution A, 50 ml of solution B, 1 ml of both solution C & solution D, 900ml distilled water, 30 gram of Sucrose and 20 gram of Agar. *Aspergillus terreus* 1782 strain was cultivated in Czapek agar slants at 37°C for 4 days.

Production and isolation of crude enzyme:Culture suspensions were prepared from the stock culture, using 10 ml of sterile water. 10 ml of this suspension was transferred to conical flask with 100 ml of liquid Czapek-Dox medium prepared based on statistical designs and kept in orbital shaker at 160 rpm. Then culture was filtered on Whatman 2 filter paper and the cell-free filtrate was used as crude enzyme solution.

Variation in medium pH

The pH of the Czapek Dox production media was varied with phosphate buffer having various pH. The experimental pH was selected as 6, 7, 8 & 9. Among the four pH levels, the optimum pH for the maximum enzyme activity was taken into further studies.

Variation in medium Temperature

The production medium was incubated at different temperature levels, to study the effect on production of enzyme. The selected experimental temperatures in the medium were 25°C, 30°C, 35°C and 40°C. Among the four temperature levels, the optimum temperature for the maximum enzyme activity was taken into the further studies.

Variable substrate concentration in media preparation

The spores and mycelial fragments of strain were cultivated in modified Czapek-Dox medium which containing the following ingredients (g/l of distilled water). Solution A: Sodium Nitrate 40.0; Potassium Chloride 10.0; $MgSO_4.7H_2O$, 0.52; $FeSO_4.7H_2O$, 0.2; dissolved in 1L of distilled water and stored in refrigerator. Solution B: K_2HPO_4 20.0; dissolved in 1L of distilled water and stored in refrigerator. Solution C: 1 gram of ZnSO4. 7 H_2O dissolved in 100ml of distilled water. Solution D: 0.5 gram of $CuSO_4$. H_2O dissolved in 100ml of distilled water. For 1L of Czapek-Dox medium; 50 ml of solution A, 50 ml of solution B, 1 ml of both solution C & solution D, 900ml distilled water, different L-glutamine concentration and 0.5 gram of L-asparagine. *Aspergillus terreus* 1782 strain was cultivated in modified medium (at 160 rpm)

The production medium composition modified by the addition of L-glutamine with an inducer L-asparagine; the L-glutamine concentration was varied from 0.5 % to 2.5 % but L-asparagine concentration was 0.5% for all medium. Here the production medium was maintained under optimum pH and temperature condition which were obtained from earlier studies. Enzyme activity and biomass in dry weight were determined at equal time interval upto 72 hours. L-Glu concentration 0.5%, 1.0%, 1.5% and 2.0% represented as medium M1, M2, M3 and M4 respectively. Initially the culture was inoculated at uniform concentration in each production medium.

Standard ammonia preparation

1M Ammonia stock solution was prepared by dissolving 114 g of anhydrous ammonium sulfate in 1 litre deionized water. From that working standard was prepared. Working standard was prepared by serial dilution. From 1μM working standard solution, 0.4, 0.8, 1.2, 1.6, 2.0 and 2.4 μM standards were prepared by taking 0.4, 0.8, 1.2, 1.6, 2.0 and 2.4 ml respectively. The total volume was made upto 2.5 ml using deionized water. 1ml of Tris-Hcl buffer was added to each test tube and incubates it for 30 minutes at 37°C. 0.5 ml 15% TCA was added followed by 1 ml NaOH. Immediately the solution was mixed and allowed to react for 20 minutes. The final solutions were red at 480nm in spectrophotometer. Blank was prepared by without anhydrous ammonium sulfate.

Assay of L-asparaginase activity

Enzyme activity of the culture filtrates was determined in at the end of cultivation time by quantifying ammonia formation in a spectrophotometric analysis using Nessler's Reagent. A 0.1ml sample of culture filtrate (enzyme solution), 0.9 ml of 0.1 M Tris-hyroxy methyl aminomethane buffer (pH 8.5), and 1 ml of 0.04 M L-asparagine solution were combined and incubated for 10 min at 37°C. The reaction is stopped by the addition of 0.5 ml of 15% (w/v) trichloroacetic acid. After centrifugation, a 1 ml portion of the supernatant fluid is diluted to 3 ml with distilled water and treated with 1.0 ml of Nessler's reagent and 1.0 ml of 2.0 M NaOH. The color reaction was allowed to proceed for 20 min before the OD at 480nm is determined. The OD was then compared to a standard curve prepared from solutions of ammonium sulfate as the source. Blank was prepared by without asparaginase enzyme sample (Produced in production medium).

One unit (IU) of L-asparaginase is that amount of enzyme which librates 1 μmole of ammonia in 1 min at 37°C.

$HCOOCCHNH_2CH_2COONH_2 + H_2O \rightarrow HOOCCHNH_2CH_2COOH + NH_3$

Determination of μmoles of ammonia liberated using the standard curve.

Units/ml enzyme= (μmole of NH_3 liberated) (1) (2.5) / (0.1) (20)

0.1 = Volume of filtrate
1.0 = Volume of supernatant
2.5 = Total volume of reaction mixture
20 = Time of assay in minutes

Biomass estimation

5 ml of sample was collected from the production medium, it was filtered with the Whattman 2 filter paper and dried in hot air oven at 35oC or Centrifuge the sample at 5000rpm for 5 minutes and dried in hot air-oven at 35oC. A graph is plotted of cell biomass (Dry weight g/l) against time intervals of 12, 24, 36, 48, 60 and 72 hrs for different media. Specific growth rate is found using μ= (In xt -In xo)/t and a maximum growth rate chart is drawn.

Results

Standard graph for ammonia estimation

The standard line showed the R2 = 0.9997 (Y = 0.3883 X) Figure 1.

Effect of pH on enzyme production

In the present study, a high enzyme activity was obtained at pH 8, while 48 hr productions. Enzyme production was less in other pH ranges at all the deviations. The maximum production of asparaginase was identified in 48 hr in all the pH levels (0.10, 0.15, and 0.175 IU/ml in pH 6, 7 and 9 respectively).The enzyme production was nil in all pH levels (6, 7 and 9) except pH 8 in 24 hr Figure 2.

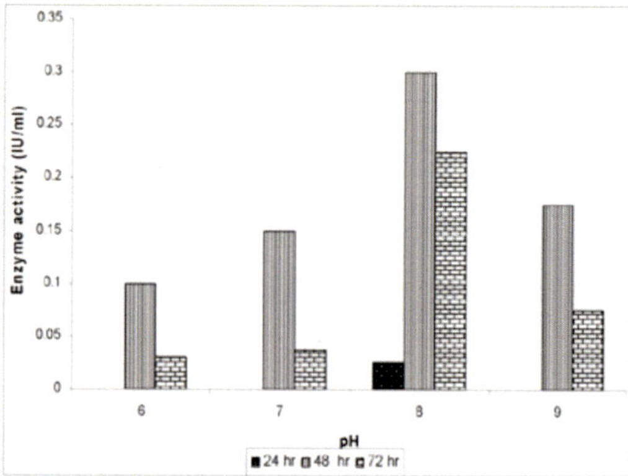

Figure 2: Enzyme asparaginase activity (IU/ml) on various pH levels (6,7,8 and 9) at different time intervals (24, 28 and 72 hrs).

Effect of temperature on asparaginase activity

The enzyme activity on the temperature at different time intervals showed in Figure 3. The maximum asparaginase activity (0.25, 0.3, 0.203 and 0.163 IU/ml at 24, 48, 72 and 96 hr, respectively) was found at 35oC throughout culture duration. The minimum enzyme production (0.075 IU/ml) was recorded in 25oC at 72 hr duration. Similarly it was maximum (0.3 IU/ml) in 35oC at 48 hr. During the study it was observed that, the enzyme production was maximum at 48 hr and recorded thereafter in all the temperatures.

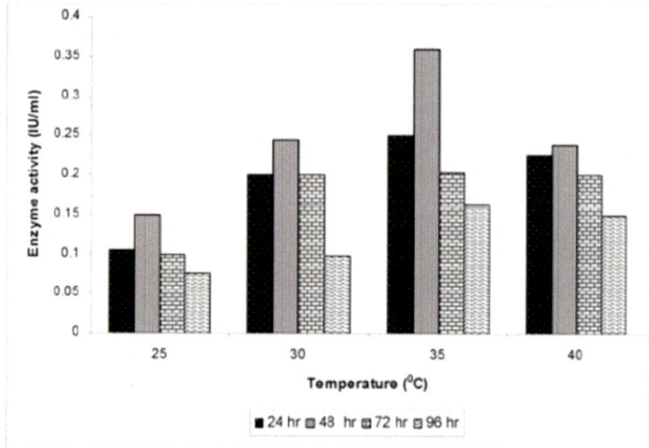

Figure 3: Asparaginase activity (IU/ml) on various temperatures (25, 30, 35 and 40oC) at different time intervals (24, 48, 72 and 96 hrs).

Asparaginase activity at different concentration of substrate

The asparaginase activity in different concentrations (M1, M2, M3 and M4) was given in Figure 4 at optimum pH and temperature. The enzyme activity was maximum in the M4 (3.40, 4.55, 4.30 and 3.75 IU/ml in 24, 48, 72 and 96 hr respectively) when compared with other medium. When compared with culture duration, it was recorded maximum at 48th hr (0.95, 2.05, 3.15 and 4.55 IU/ml in M1, M2, M3 and M4 respectively).

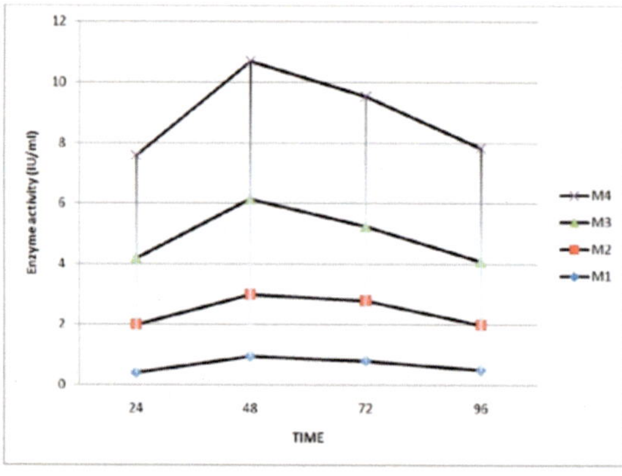

Figure 4: Asparaginase activity (IU/ml) in substrates M1, M2, M3, M4 at different time intervals (24, 48, 72 and 96 hrs).

Biomass concentration in different substrate concentration

The biomass concentration (Dry weight – g/l) was increased gradually with the culture duration from 12 to 72 hr. At the 36th hr of incubation, M2 showed the maximum (4.90 g/l) biomass production, whereas in the 96th hr, it was maximum (14.6 g/l) in medium M4, when compared with other medium Figure 5.

For each medium the specific growth rate were calculated from Figure6a, 6b, 6c, 6d. The glutamine substrate concentration 1/S (g/l)-1 and 1/Y (hr-1) was plotted and Ym (Maximum specific growth rate) and KS (Substrate constant) were calculated as 0.029 hr-1 and 2.43 (g/l) respectively Figure 7.

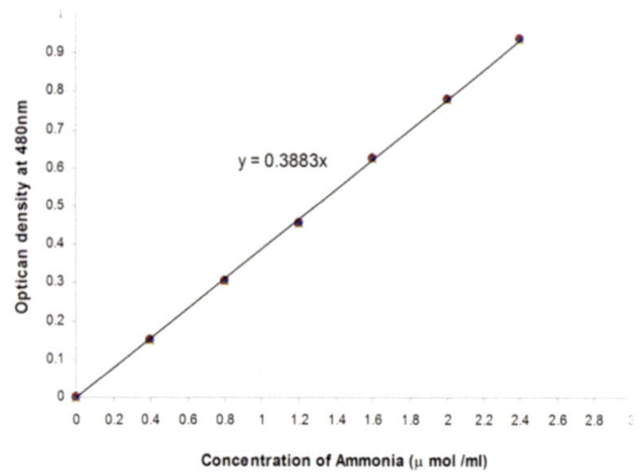

Figure 1: Standard Graph for Ammonia for the Estimation of Asparaginase Activity.

Discussion

Environmental factors are significant for biosynthesis of asparaginase with *Serratia marcescens* ATCC 60.Aeration of the fermentation medium is necessary for the growth of *Serratia marcescens*, but not for

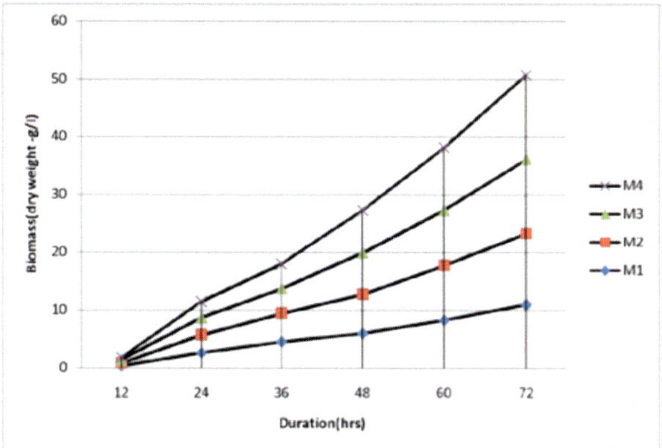

Figure 5: Biomass concentration (Dry weight g/l) in the M1, M2, M3, M4 medium at different time intervals (12, 24, 36, 48, 60 and 72 hrs).

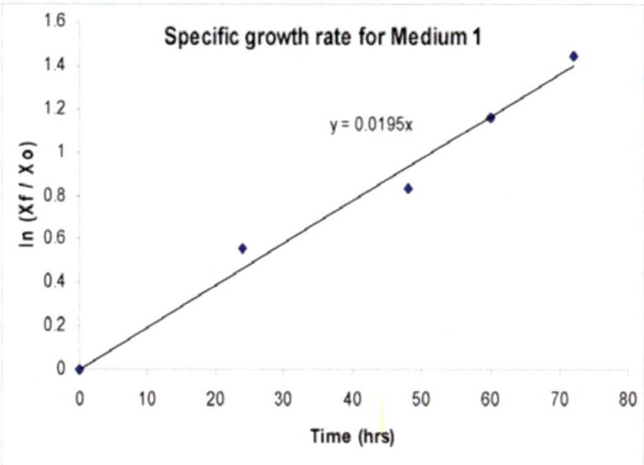

Figure 6a: Biomass specific growth rate for medium M1.

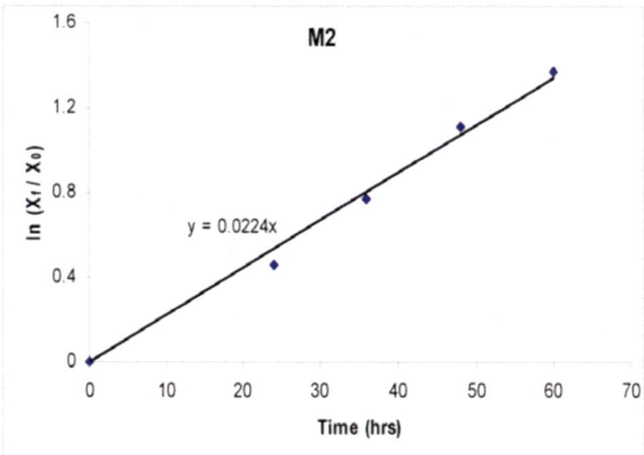

Figure 6b: Biomass specific growth rate for medium M2.

the production of asparaginase. Enhanced biosynthesis of asparaginase by *E. coli* K-12 with anaerobic growth condition has been reported by Cedar and Schwartz [12].As with number of enzymes, formation of asparaginase is inhibited by the addition of sugars, particularly glucose. In the case of asparaginase biosynthesis the depressive effect of carbohydrates may be a function of their ability to lower the pH value of the fermentation beer Heinemann and Howard [6]. The quality of L-asparaginase produced by *Streptomyces griseus* ATCC 10137 in most media did not differ significantly; yields ranged from 0.6 to 1.2 IU of L-asparaginase in cells from 100ml of broth. The glucose asparagines (GA) medium supported only poor growth and poor L-asparaginase production. Heinemann and Howard [6] reported that addition of glucose to the medium of *Serratia marcescens* resulted in substantial reduction in the quantity of L-asparaginase produced.

In *E. coli*, higher aeration rate affect the L-asparaginase production. Maximum activity occurred between pH 7 and 8 [12]; enzyme activity was essentially absent below pH 4.5, reported by Peterson and Ciegler [13]. Maximum L-asparaginase activity of 1.17 IU/100ml of broth found at pH 8.5 [14] and pH 9, *Pseudomonas aeruginosa* 10145 [11], and many other microbial asparaginase activities. Similarly, in the present study the maximum enzyme activity (0.3 IU/ml) was found at pH 8. The enzyme had maximum activity at 37oC, and more than 52% activity was attained at 50oC [15]. Similar results were recorded for asparaginases from *Pseudomonas stutzeri* MI-405, *Erwinia carotovora*, and *Staphylococcus* [16]. However in the present study, the optimum temperature for the L-asparaginase activity was recorded at 350C for *A. terreus*. Biomass production of *Aspergillus tamarii* with different nitrogen sources shows highest amount at urea 0.2% and L-glutamine 1% as nitrogen source. The biomass production of *Aspergillus terreus* with different nitrogen sources presented different growth curves when compared with A. tamarii. The biomass production of *A. terreus* pre-

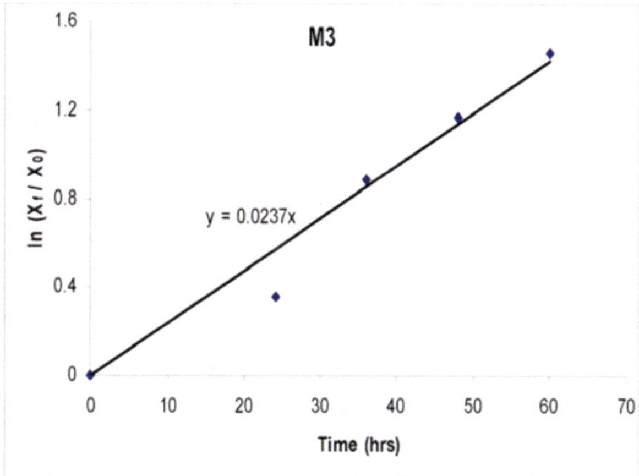

Figure 6c: Biomass specific growth rate for medium M3.

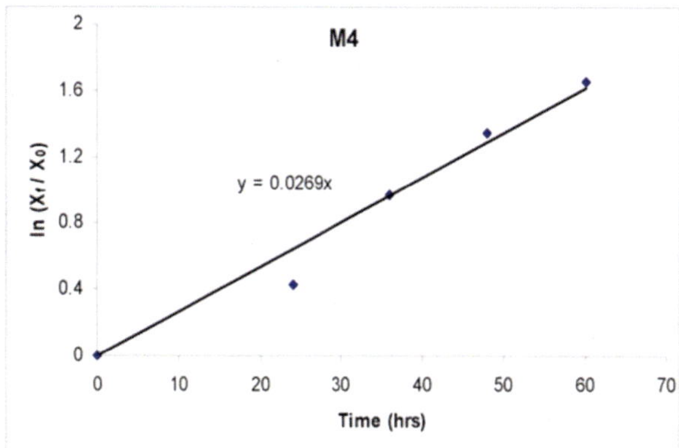

Figure 6d: Biomass specific growth rate for medium M4.

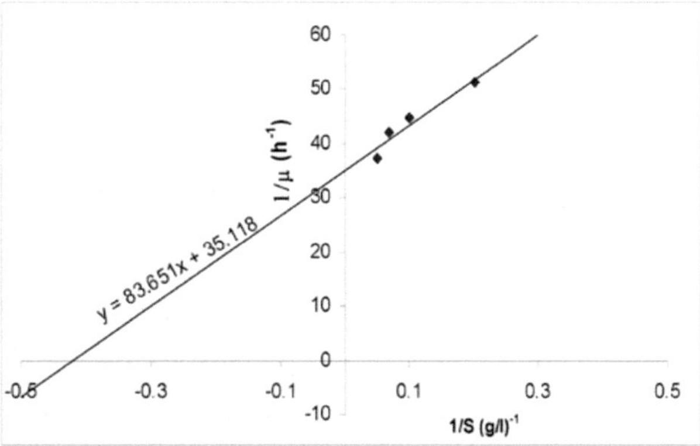

Figure 7: Maximum Specific growth chart.

sented an average of 1.5 times greater than A. tamari and L-glutamine 0.1% medium revealed itself as a good nitrogen source for the *A. terreus*. The protease activity could be responsible for decrease in L-asparaginase activity since protease levels increase when L-asparaginase levels decrease. L-asparaginase enzyme activity was 20 IU/l found at glutamine 1% reported by Maria et al. In the *A. terreus*, the biomass production was maximum in glutamine 2% medium with L asparagine 0.5% was identified in the present study. The L-asparaginase behaves like an intraperiplasmic enzyme, since whole cells possessed equivalent activity as broken cells, the enzyme not being produced extracellularly. The enzyme has been subsequently purified to near-homogeneity and characterized .

Although L-asparagine has been known to be an excellent nitrogen source for the growth of eukaryotic micro algae, the report of an algal L-asparaginase appeared only fairly recently (Paul and Cooksey, 1979). The organism, a marine *Chlamydomonas* species, could use L-asparagine, but no other amino acid for autotrophic growth (Paul and Cooksey, 1979). Nitrogen deprivation stimulated enzyme synthesis and the enzyme was repressed in the presence of combined nitrogen [17]. These are characteristics common to extra cellular [17] enzymes or enzymes found in periplasmic space were condition ionic strength, pH and substrate concentration are poorly controlled. The requirement of ammonium for good enzyme production is interesting because these ions are a product of the asparaginase reaction. It is also apparent that maximum enzyme production occurs in a medium containing ammonium ions and fumarate: asparaginase has no obvious metabolic role in such an environment. The purified enzyme has a specific activity of 200 IU/mg of protein [18]. Therefore, in crude cell extract with specific activity of 10 IU/mg of protein, asparaginase represents 5% of the total protein [19]. It is possible that asparaginase catalysis at 4oC is responsible for the loss of viability. However, the cells grown in the Wolin medium also contains significant amounts of asparaginase. The data obtained in this investigation may be compared with the published data for *E. coli* A-1. Barnas et al. showed that this strain, a good asparaginase producer when, compared with literature reports for other organism, yields a maximum of 27 IU/ml of culture. *Vibrio succinogenes*, in the best medium developed in this investigation, yielded about 2.5 IU/ml of culture. This 10 fold difference must be viewed against the 30 to 40 fold energetic advantage possessed by *E. coli* growing aerobically on glucose compared to anaerobic growth of *Vibrio succinogenes* on formate- fumarate. In *Azotobacter*, the maximal enzyme activity is reached in 30 min and is maintained at the same level during further growth of the organism in the presence of ammonium sulfate. *Erwinia aroideae* is not able to grow in the presence of ammonium nitrogen. The *Azotobactor vinelandii* asparaginase is induced in all stages of growth. Asparaginase from *A. vinelandii* had an optimum pH different from asparaginases of *Bacillus coagulans*, guinea pig serum, and *Serretia marcescens*, which have been reported to have broad pH optima [20]. External asparaginase activity of less than 1 nmol/min per mg of cells would be very difficult to detect this level of activity above the normal background on ammonia using the Nessler method. The current study also proved that, the maximum L-asparaginase production (4.55 IU/ml) found to be at glutamine (2%) and L-asparagine (0.5%) at 48th hour culture. The stationary phase was prolonged up to 72hrs showed the maximum yield of biomass.

L-asparaginase synthesis was increased by addition of L-aspartic acid, L-glutamine, L-glutamic acid and L-asparagine to the level of 20, 20.7, 19.8 and 25 IU/ 10ml respectively using *Serratia marcescens* reported by Sugumaran et al. [21].The addition of L-aspartic acid or L-glutamic acid to the medium containing sodium fumarate and corn steep liquor slightly enhanced enzyme production, but these amino acids may not be considered specific inducers for enzyme formation [22]. At 5.0% Yeast Extract concentration, the higher aeration provided yields of 5.2 IU/ml in this experiment. L-asparaginase levels up to 6.6 IU/ml were observed during these studies. With greater concentration Yeast Extract, viscosity increase could not be compensated for, and subsequent enzyme levels were greatly reduced, reported by Peterson and Ciegler [14]. L-asparaginase is produced throughout the world by sub-merged fermentation (SF). This technique has many disadvantages, such as the low concentration production, and consequent handling, reduction, and disposal of large volumes of water during the downstream processing. Therefore, the SF technique is a cost intensive, highly problematic, and poorly understood unit operation.

Conclusions

L-asparaginase production has been regulated by the physical parameters and the nitrogen sources. L-glutamine combined with L-asparagine influence the asparaginase production level. Physical parameter such as pH and temperature has more contribution in enzyme production. The pH 8 and temperature 35°C found to be good environmental condition for *Aspergillus terreus*. Our studies proves that L-asparagine induce the culture to produce more asparaginase enzyme. We can increase production level of asparaginase by adding inducer, L-asparagine in industrial level.

References

1. Mercado L, Arenas G (1999) *E.coli* L-asparaginase induced phosphorylation of endogenous polypeptide in human immune cells. Sangre (Brac) 44: 438-442.
2. Mukherjee J, Majumadar S, Scheper T (2000) Studies on nutritional and oxygen requirements for production of L-asparaginase by *Enterobacter aerogenes*. ApplMicrobial Biotechnol 53: 180-184.
3. Kil JO, Kim GN , Park I (1995) Extraction of extracellular L-asparaginase enzyme from *Candida utilis*. Biosc Biotechnol Biochem 59: 749-750.
4. Muley RG, Sarker S, Ambedkar S, Nail SR (1998) Influence of alkali treated corn steep liquor containing medium on protein production by *Staphylococcus aureus*. Folia Microbiology 43: 31-34.
5. Prista AA, Kyridio DA (2001) L-asparaginase of *Thermo thermophilus*: purification, properties and identification of essential amino acids for catalytic activity. Mol Cell Biochem 216: 93-101.
6. Heinemann B , Howard A J (1969) Production of Tumor-inhibitory L-asparaginase by submerged growth of Serratia mercescens. Appl Microbiol18: 550-554.
7. Khan AA, Pal SP, Raghavan SRV, Bhattacharyya PK (1970) Studies on *S.marcescens* L-asparaginase. Biochem BiophysRes Common 41: 525-533.

8. Netval J (1977) Follia Microbiol 22: 106.

9. Mashburn LT, Wriston JC (1964) Tumor inhibitory effect of L-asparaginase from *E.coli*. Arch Biochem Biophys 105: 451-452.

10. Broome JD (1965) Antilymphoma activity of L-asparaginase in vivo: Clearance rates of enzyme preparation from guinea pig serum and yeast in relation to the effect on tumor growth. J Nat Cancer Inst 35: 967-974.

11. Roberts J, Burson G, Hill J M (1968) New procedures for purification of L-asparaginase with high yield from *E. coli*. J Bacteriol 95: 2117-2123.

12. Cedar H, Schwartz JH (1968) Production of L-asparaginase II by Escherichia coli. JBacteriol 96: 2043-2048.

13. Peterson RE , Ciegler A (1969) L-asparaginase production by *Erwinia aroideae*. Appl Microbiol 18: 64-67.

14. Peterson RE , Ciegler A (1972) Factors influencing L-asparaginase production by *Erwinia aroideae*. Appl Microbiol 23: 671-673.

15. Bessoumy AA, Sarhan, Mansour J (2004) Production, isolation, and purification of L-asparaginase from *Pseudomonas aeruginosa* 50071 using solid state fermentation. J Biochem and Molecular Biol 37: 387-393.

16. Sobis M, Mikucki J (1991) *Staphylococal* L-asparaginase: Enzyme kinetics. Acta Microbiol Pol 40: 143-152.

17. Paul J H, Cooksey KE (1981) Regulation of Asparaginase, Glutamine synthetase, and Glutamate dehydrogenase in response to medium nitrogen concentration in a euryhaline *Chlamydomonas* Species. Plant Physiol 68: 1364-1368.

18. Distasio JA, Nredreman RA, Kafkewitz D , Goodman D (1976) Purification and characterization of L-asparaginase wiyh antilymphoma activity from *Vibrio succinogenes*. JBiolog Chem 251: 6929-6933.

19. Albanese E, Kafekewitz D (1978) Effect of medium composition on the growth and asparaginase production of *Vibrio succinogenes*. Appl And Environ Microbiol 36: 25-30.

20. Gaffar SA, Shethna YI (1977) Purification and some biological properties of asparaginase from Azotobacter vinelandii. Appl and Environ Microbiol 33: 508-514.

21. Sukumuran CP, Singh DV, Mahadevan P R (1979) Synthesis of L-asparaginase by *Serratia mercescens* (Nima). J Biosci 1: 263-269.

22. Tosa T, Sano R, Yamamoto K, Nakamura M, Ando K, et al. (1971) L-asparaginase from *Proteus vulgaris*. Appl Microbiol 22: 387-392.

Phenotypic and Genotypic Diversity of Rhizobia Nodulating Faba Bean from Various Egyptian Locations

Omar A Hewedy*, Ragaa A Eissa, Elzanaty AM, Nagaty HH and Abd Elbary MI

Genetics Department, Faculty of Agriculture, Minoufiya University, Shibin El-Kom, Egypt

Abstract

Thirteen Rhizobial isolates from root nodules of Faba beans (*Vicia faba* L.) were obtained from eleven governorates, representing different agro-ecological, agro-climatic and soil conditions in Egypt. A polyphasic approach, including phenotypic and genotypic techniques were used to study their diversity. Rhizobium isolates were examined for their ability to tolerate salt stress and antibiotic resistance. We used four NaCl concentrations ranging from 0.1%, 1%, 3% and 5% in liquid media and plates, moreover we used 24 different antibiotic disks to determine resistance or sensitive among the tested isolates on plates. The obtained results after salt tolerance and antibiotic response of the R. *leguminosarium strains* showed that the isolate No. RL9 was the superior strain for salt tolerance . The following strains for salt tolerance were the isolates (RL4) and (RL12), although the isolate No. RL11 was the superior strain for antibiotic resistant then isolates No. RL 13 and Rh3. Growth of all isolates were inhibited above 3% NaCl except five isolates RL4, RL9, RL10, RL12 and RL 13. Also the antibiotic resistance of the isolated strains showed a high level of resistance against Cefoxitin (FOX_{30}) and Cefuroxime (Cxm_{30}). The genetic diversity was studied using RAPD-PCR technique and we used specific primer (*nodC*) amplification. In this study we found that OPE_{15} and OPG_{04} primers showed the highest polymorphism level among the tested isolates, however OPJ_{20} and OPC_{06} primers showed the lowest polymorphism level.

Keywords: Faba bean; Antibiotic resistance; Salt tolerance; RAPD-PCR; nodC

Introduction

Faba bean (*Vicia faba* L.) is a major leguminous crop grown worldwide, it is most intensively cultivated in the Middle East and in North East Africa [1]. Rhizobium leguminosarium symbiovar. Vicia forms a nitrogen-fixing root nodule symbiosis with faba bean [2,3]. Since Faba bean has been grown for centuries in Egypt, it was of interest to determine the diversity of rhizobia forming nitrogen-fixing nodules with this important legume crop. The soils used in our studies distributed in geographically different locations in Egypt. Rhizobia are soil bacteria which are capable of forming nitrogen-fixing nodules with different leguminous plants and have a significant role in nutrient cycling due to biological nitrogen fixation and enhancing crop productivity [4]. The symbiotic relationships between rhizobia and leguminous plants provide soil with nitrogen. Many studies have addressed the diversity level of *V. faba rhizobia*; mainly focusing on rhizobial populations from the same location [5] or for comparison with Rhizobium leguminosarum isolates from other legume species [6]. The diversity of rhizobia provides valuable bio resource for the search of bacterial isolates in attempt to find isolates that maximize nitrogen fixation, and hence increase legume crop productivity [7]. Salinity stress is one of the most serious factors limiting the productivity of agriculture. High salt can directly impair rhizobia–legume early interactions during nodule formation [8]. In general, Egypt suffering from increasing population in both drought and arid climate. The symbiotic interaction between rhizobia and legumes is initiated by an exchange of complex molecular signals that confer host-specificity. Rhizobia respond to these by one or more of the constitutive nod D genes encoding for a soluble cytoplasmatic protein activating the other nod genes when they interact with the appropriate plant signal compounds. This leads to the biosynthetic enzymes of lipo-chitin oligosaccharides (Nod Factors) encoded by nod ABC genes. Laguerre et al., [9] used the nod C gene, a common nod gene essential for nodulation in all rhizobial species, to characterize a collection of 83 rhizobial strains which represented 23 recognized species distributed in the genera Rhizobium, Sinorhizobium, Mesorhizobium and Bradyrhizobium. Many techniques were developed and widely used to detect polymorphisms in many organisms including bacteria. Among these techniques, RAPD technique is a polymerase chain reaction (PCR) to detect the polymorphisms in genomic DNA [10]. Genomic DNA Fingerprinting using random amplification of polymorphic DNA (RAPD) has been found to be useful in differentiating between Rhizobial strains. [11] proposed this method for identification and phylogenetic grouping of Rhizobium isolates. RAPD PCR technique provides reliable information on the diversity of Rhizobium populations in soils [12].

The objectives of this study are to isolate salt tolerant Rhizobium and investigate the effects of antibiotics and NaCl on growth of isolates. Furthermore detect the diversity of these isolates by using RAPD.

Material and Methods

Isolation of Rhizobia from nodules

Thirteen Rhizobial isolates isolated from the root nodules of Faba bean (*Vicia faba* L.) isolates were obtained from eleven governorates as presented in (Table 1). All nodules were cut off in a laminar flow cabinet with small pieces of root and washed thoroughly with 2.5% NaOCl and sterile water. Nodules were surface sterilized with 70% ethanol

***Corresponding author:** Omar A Hewedy, Genetics Department, Faculty of Agriculture, Minoufiya University, Shibin El-Kom, Egypt

Isolates	Nitrogenase Activity	Nodules No.	Shoot Dry (g)	Shoot Fresh (g)
1	1.60	23.0	2.10	13.70
2	0.80	14.7	1.30	11.30
3	1.35	17.3	3.00	12.50
4	1.75	23.0	4.60	19.90
5	1.00	18.7	2.00	12.70
6	0.75	24.3	26.01	14.60
7	2.15	24.3	1.00	11.10
8	0.50	15.3	2.60	16.50
9	3.30	32.3	8.80	29.00
10	2.00	22.7	5.00	24.60
11	2.50	28.3	4.10	24.70
12	2.70	25.3	6.90	26.10
13	3.10	30.7	7.50	27.60

Table 1: Locations of isolates measured for each isolate.

alcohol for 5 min and exhaustively washed in sterile distilled water. After washing the root nodules were taken from the roots with care not to damage the surface, washed thoroughly in distilled water. Nodules were then transferred to 0.2% (W/V) solution of mercuric chloride for 3-5 minutes. Each nodule was crushed under aseptic conditions and streaked onto a Yeast-Mannitol Agar plate (YMA) using a sterile loop and incubated at 28°C. Single colonies were picked up from the original streaked plates. Pure cultures of *Rhizobium leguminosarum* were isolated according to Vincent methods and retreated on YMA containing Congo red to ensure purity before storage in 20% glycerol at -20°C. Stock cultures of the Rhizobium isolates were maintained on slants of YEM at 4°C and refreshed periodically.

Phenotypic characterization of isolates

Antibiotic resistance pattern: All isolates were evaluated for their responses as resistant or sensitive against 24 different antibiotic disks as shown in Table 2.

Rhizobium isolates were tested on YMA agar supplemented with following antibiotic disks that presented in Table 2 according to Eaglesham's technique [13]. This was done as the following each isolate plated on YEM plates and the antibiotic disks were distributed on the surface of the plate. The plates were incubated for 3 days at 28°C. After this period the presence or absence of an inhibition zone around antibiotic disks was recorded, indicating sensitivity or resistance, respectively.

Effect of Salinity on tested Rhizobium isolates:

- YEMA: The ability of the Rhizobial isolates to grow in various concentration of NaCl was tested by plating them on YEM agar Petri dishes. YEMA was prepared with different concentrations of NaCl ranging from (0.1%, 1%, 3% and 5% wt/vol) NaCl. All the plates were incubated at 28°C for three days in triplicate in addition to control.
- YEM broth: Testing of salt tolerance was measured as the ability of the bacterial cells to grow and divide under the stress of used NaCl concentration. Test tubes (10 ml) each containing 5 ml YEM medium, were inoculated with a bacterial isolates and incubated at 28°C for 72 hours. 100 µl of suspension (1×10^{-7} cfu/ml) of the grown isolate was used for the inoculation of 5 ml YEM medium containing various NaCl concentration (0.1 %, 1%, 3% and 5% wt/vol) and the test tubes were incubated at 28°C for different time ranging from zero time, 12, 24, 36, 48, 60 and 72 h of incubation and the growth rate in Rhizobial isolates tested using optical density (OD) at three replicates. The turbidity was measured using Jenway UV-VIS. spectrophotometer model UV-6305 at 600 nm against the blank (sterilized uninoculated YEM broth) (Table 4).

DNA isolation

Total genomic DNA of each isolate was extracted from bacterial cultures grown in yeast extract mannitol media (YEM) according to Ausubel et al., [14] method and kept at -20°C. The cells were lysed with sodium dodecyl sulphate (SDS), protein–lipopolysaccharides complexes were removed by using cetyl tri methyl ammonium bromide (CTAB), and DNA was precipitated from solution using isopropanol.

RAPD–PCR reaction and amplification

Oligonucleotide sequences of the random primers:

OPE20---5′AACGGTGACC 3′---OPA04---5′AATCGGGCTG 3′

OPJ20---5′AAGCGGCCTC 3′---OPE 15---5′AACGGTGACC 3′

OPG 04---5′AGCGTGTCTG 3′---OPC 06---5′GAACGGACTC 3′

Total genomic DNA was extracted from exponentially grown cultures and PCR reactions were carried out using an arbitrary primer RAPD-PCR were used in this study RAPD-PCR was performed in the total volume of 25 µl of reaction mixture containing 1 µl of DNA template, 1 µl dNTPs mix,4 µl (10 pmol) of an arbitary primer and 1 µl of Taq polymerase with 2.5 µl Taq buffer and 15.5 µl sterilized water.

PCR conditions were as follows template DNA was denatured at 94°C for 4 min then the PCR was carried out for 35 cycles (1 min at 94°C, 1 min at 34°C, 2 min at 72°C, for each cycle). Finally, a 7 min extension period at 72°C was performed. Amplified products were resolved on a 1% (w/v) Agarose-TBE gel electrophoresis and separated by running 5 µl of PCR reaction mixture at 100 V for 1 h and stained

No.	Disks	Name
1	AK30	Amikacin
2	AM 10	Ampicilin
3	ATM 10	Aztreonam
4	B 0.04	Bacitracin
5	C 30	Chloramphenicol
6	CIP5	Ciprofloxacin
7	CLR 15	Cefuroxime sodium
8	CN 10	Gentamicin
9	CRO 30	Ceftriaxone
10	Cxm30	Cefuroxime sodium
11	E 15	Erythromycin
12	FOX 30	Cefoxitin
13	IPM 10	Imipenem
14	K 30	Kanamycin
15	MA 30	Cefamandole
16	Mem 10	Meropenem
17	N 30	Neomycin
18	NOR10	Norfloxacin
19	OX 1	Dloxaciolin
20	S 10	Streptomycin
21	Sam 20	Sulbactampicillin
22	SXT25	Sulphamethoxazole
23	T 30	Tetracycline
24	VA 30	Vancomycin

Table 2: Antibiotic disks used in this study.

Antibiotic / Isolates	AK 30	AM 10	ATM 10	B 0.04	C 30	CIP 5	CLR 15	CN 10	CRO 30	CXM 30	E 15	FOX 30	IPM 10	K30	MA 30	Mem 10	N 30	Nor 10	OX 1	S 10	Sam 20	SXT 25	T 30	VA 30
t	-	-	+	-	-	-	-	-	+	+	-	+	-	-	-	-	-	-	-	+	+	-	-	-
Rh 2	-	-	-	+	-	-	-	-	-	+	-	+	+	-	-	+	-	-	+	-	+	-	-	-
Rh 3	-	+	+	+	-	-	-	-	+	+	-	+	-	-	+	+	-	-	-	-	-	+	-	+
Rh 4	-	-	-	+	-	-	-	-	-	+	-	+	+	-	-	+	-	+	+	-	+	-	-	-
Rh 5	-	-	-	-	-	-	-	-	-	+	-	+	+	-	-	+	-	+	-	-	+	-	-	-
Rh 6	-	-	-	-	-	-	-	-	+	+	-	+	-	-	+	+	-	-	+	-	+	-	-	+
Rh 7	-	-	+	-	-	-	-	-	-	+	-	+	-	-	+	-	-	+	+	-	+	-	-	-
Rh 8	-	-	+	+	-	-	-	-	-	+	-	+	+	-	-	+	-	-	-	-	+	-	-	-
Rh 9	-	-	-	-	-	-	-	-	+	+	-	+	-	-	-	-	-	-	-	-	+	-	-	-
Rh 10	-	-	+	+	-	-	-	-	+	+	-	+	-	-	+	+	-	-	-	-	+	-	-	-
Rh 11	-	+	+	+	-	-	-	-	-	+	-	+	-	+	+	+	-	+	+	+	+	-	-	-
Rh 12	-	-	-	-	-	-	-	-	-	+	-	+	-	+	-	+	-	-	+	-	+	-	-	-
Rh 13	-	+	+	+	-	-	-	-	+	+	-	+	-	-	-	+	-	+	-	-	+	-	+	-

Table 3: Response of Rhizobium isolates to 24 different types of antibiotic disks.

Isolates	Rhl	Rh2	Rh3	Rh4	Rh5	Rh6	Rh7	RhS	Rh9	Rhl 0	Rh11	Rh12	Rh13
Rhl	1												
Rh2	0.3	1											
Rh3	0.33	0.3	1										
Rh4	0.27	0.87	0.28	1									
Rh5	0.33	0.62	0.23	0.75	1								
Rh6	0.4	0.5	0.5	0.45	0.4	1							
Rh7	0.44	0.4	0.31	0.5	0.44	0.5	1						
Rh$	0.44	0.75	0.47	0.67	0.62	0.36	0.4	1					
Rh9	0.67	0.37	0.27	0.33	0.43	0.5	0.37	0.37	1				
Rh10	0.55	0.5	0.64	0.45	0.4	0.6	0.5	0.67	0.5	1			
Rh11	0.38	0.46	0.47	0.54	0.38	0.43	0.58	0.46	0.23	0.58	1		
Rh12	0.33	0.62	0.23	0.55	0.5	0.55	0.44	0.44	0.43	0.4	0.5	1	
Rh13	0.45	0.42	0.54	0.5	0.45	0.38	0.42	0.54	0.4	0.64	0.57	0.33	1

Table 4: Similaritybetween different tested isolatesbased on antibiotics resistant analyses.

with ethidium bromide (0.5 µl/ml). The sizes of the amplified fragments were determined by comparison with 100 bp.

Plus DNA Ladder (Gene ruler) as molecular weight marker. Presence of amplified bands were observed using UV-transilluminator and the gel was photographed. Variations between amplified bands were recorded as presence or absence of DNA fragments. The dendrogram was constructed based on UPGMA cluster analysis according to dice square co-efficiant method. The dendrogram tree among the studied Rhizobium isolates based on RAPD markers were constructed using NTSYSpc version 2.1 programs depending on the similarity matrix present in Table 5.

Amplification of the nodC gene

The nodC gene from the Rhizobium isolates was amplified by PCR by using forward primer nodCf (5-GCTGCCTATGCAGACGATG-3) and reverse primer nodCr (5-GGTTACTGGCTTTCATTTGGC-3), PCRs were carried out in 50 ml reaction mixtures [14] using the following conditions: initial denaturation at 94°C for 5 min, followed by 30 cycles of 94°C for 1 min, 55°C for 1 min, and 3 min at 72°C. A final extension was done at 72°C for 7 min. PCR products were separated on 2% agarose gels at 120 V for 1.5 h in TBA buffer, stained with ethidium bromide, and visualized under UV light.

Statistical analysis: Data obtained were statistically analyzed using SPSS analysis program (version 11.5). The significant differences among individual means were analyzed by Duncan's multiple range tests.

Isolate No.	Control (0.1%)	NaCl at different concentration		
		1%	3%	5%
1	130	110	50	0
2	150	130	70	0
3	110	100	80	0
4	140	120	90	50
5	120	90	40	0
6	120	110	40	0
7	90	80	50	0
8	100	80	60	0
9	150	120	90	50
10	160	120	50	40
11	100	70	40	0
12	140	110	90	60
13	170	140	100	80

Table 5: Response of tested isolates to different concentrations of NaCl on YEMA Plates ($1*10^7$cfu/ml).

Results and Discussion

Faba bean has been cultivated for over 6000 years, and is grown in 57 countries worldwide. Since faba bean has been grown for centuries in Egypt, it was of interest to determine the diversity of rhizobia forming nitrogen-fixing symbioses with this important legume crop. The soils used in our studies differed in texture and class, and were distributed in geographically different sites in Egypt.

Phenotypic characteristics

On the basis of morphological parameters, we have confirmed differences between the isolates. Table 1 represent thirteen isolates of *Rhizobium leguminosarum* symbiovar *Vicia* isolated from nodules of Egyptian Faba beans (*Vicia faba* L.) collected from different geographic areas. This ten isolates were tested by Congo red technique [15], to ensure that all isolates were rhizobia and did not contaminated. These results are in agreement with those obtained by [16] they sampled two hundred and eighty seven isolates of Rhizobium in France from geographically distant field populations and characterized by their colony morphology. Rhizobial isolates were found to be microscopically similar, isolates identified as Rhizobium *leguminosarium symbiovar. Vicia*. According to the negative reaction to gram stain. Strains were found to be motile. On YEM agar, the growth is generally moist, whitish, smooth and gummy. The culture on YEM broth showed uniform turbidity, white sediment and need 8-10 days or longer time to attain maximal growth. These results are in agreement with those obtained by [17]. Colonies on agar after 2-3 days were large, gummy, translucent, spreading about 2–4 mm in diameter, similar results were reported by [18].

Antibiotic resistance pattern

Resistance to different disks of antibiotic is one of the simple and rapid methods for rhizobial isolates characterization and identification. Resistance patterns of the isolates to various antibiotics were studied to provide phenotypic data for differentiating the rhizobial isolates from each other and to determine the diversity among the isolates. Apparently, the antibiotic resistance of the tested strains showed a high level of resistance against Cefoxitin (FOX30r) and Cefuroxime (Cxm30 r). More than 70% of the isolates showed resistance to Meropenem (Mem10r) except isolate No.1,7 and 9. Some isolates were resistant to Sulbactampicillin (Sam20r). Moreover all the isolates showed generalized sensitivity to Amikacin (AK30s), Chloramphenicol (C30s), Ciprofloxacin (CIP5s), Cefuroxime sodium (CLR15s), Gentamicin (CN10s) and Erythromycin (E15s). Furthermore all the isolates showed sensitivity to Tetracycline (T30s) except isolate No.13, The generalized sensitivity to Tetracycline in this study agrees with the results reported previously by Jordan (1984) for the genus Rhizobium and by Hagedorn [19] for *R.leguminosarum sv. trifolli*. Finally all the isolates showed Sensitivety to Streptomycin (S 10 s) except isolate No.1

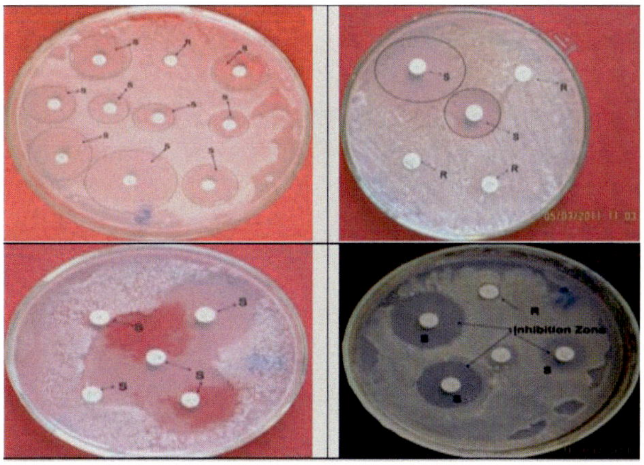

Figure 1: Response of different isolates to kinds of different antibiotic discs.

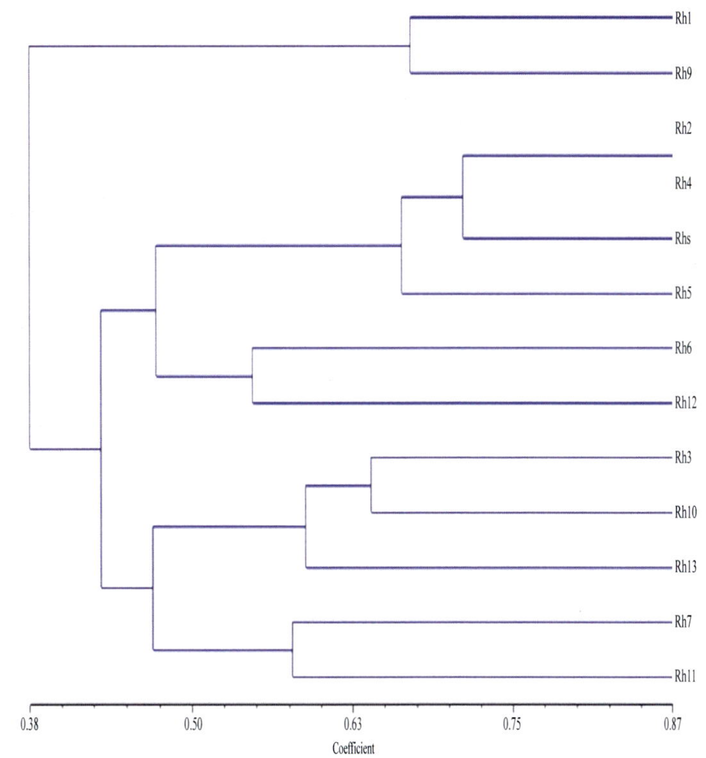

Figure 2: Phylogenetic relationship between 13 isolates using UPGMA method based on antibiotics resistant analyses.

and isolate No.11. Whereas isolates No. Rh11, Rh13 and Rh3 showed 50% resistance to the tested antibiotics as shown in Table 3, The same Table 3 and Figure 1 showed the wide range of variations among the isolates for its resistant to these discs. These results were similar to those reported by [20].

Salt tolerance: These strains were first screened to examine their resistance to environmental stresses e.g. salt tolerance. The majority of the tested strains were sensitive to salt stress . All 13 rhizobial isolates tolerated 3% NaCl. Whereas only five isolates are able to grow at 5% NaCl. only 5 isolates still grew, but 8 isolates failed to grow at the same concentration (Rh1, Rh2, Rh3, Rh5, Rh6, Rh7, Rh8 and Rh11) and showed great inhibition . In fact,Ismailia isolate showed the most salt tolerancy (Figures 2 and 3).

Results in Table 5 showed that all of the 13 isolates were able to survive in salt concentrations ranging between 0.1% till 3.0% NaCl

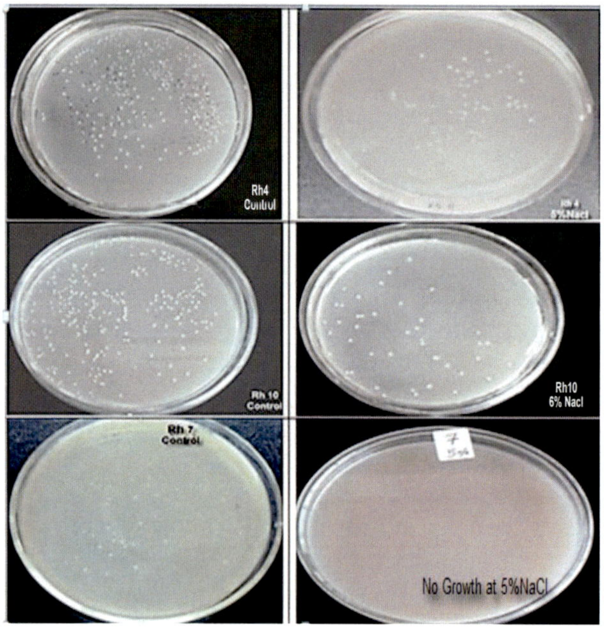

Figure 3: Some isolates growing on YEMA media supplemented with NaCl as a control and 5% NaCl concentration.

and a few of them tolerated and survived at 5% NaCl. These tolerated isolates were isolate No.4 from Zefta City, Ghrbia Governorate ,isolate No.9 from Rafah City, North Sinai Governorate, isolate No.10 from El-Menia Governorate , isolate No.12 from Cairo City, Cairo Governorate and finally isolate No.13 from Ismailia City, Ismailia Governorate. These results were similar to those reported by [21] who found that the Faba bean-nodulating strains EFBRI 35, 41, 92 and 93 that we isolated were salt tolerant, and grew well in 1% NaCl, and a large number of strains that we later identified as S. *meliloti* and isolated from Faba bean nodules were resistant to 3% NaCl. Similar results were reported by who noted that strains belonging to this species are salt tolerant.

The O.D. of all isolates was affected significantly by salt concentration (Figure 4 and Table 6). O.D. decreased with increasing salinity above 1% concentration. Most isolates had maximum O.D. At 1% NaCl; all isolates continued to grow strongly at 3% NaCl, whereas growth of all isolates was inhibited 5% NaCl except for isolate No.4, 9, 10, 12 and 13. As a result, the salt tolerance of rhizobia is critical for symbiosis in that it facilitates the survival and growth of rhizobial strains in saline conditions and, in particular, is one of the key mechanisms to enhance the symbiosis under saline conditions [22]. The Ismailia strain was able to tolerate the maximum concentration tested at 5% NaCl, while without NaCl we noticed heavy growth of Rh.7 (control), whereas plate No.7 at 5% NaCl showed no bacterial growth at all. Therefore this isolate is very sensitive to NaCl at used concentration.

Data in Table 6 showed that the isolate No. RL9 was the superior strain for salt tolerance The following strains for salt tolerance were the isolates RL12, RL4, RL10 and RL13 The higher growth rate of the Rhizobium isolates may be considered as an indicator of salt tolerance and subsequently, could be considered as an estimate parameter for high nodulation under the saline soil.

Genetic diversity of Faba bean (*Rlv*) rhizobia using different molecular approaches

RAPD PCR: Several investigators have studied the genetic diversity of Rlv isolated from several countries around the world [23-26].

RAPD profiles were used to show the genetic polymorphism among isolates of *Rhizobium leguminosarum bv. Viciae* and to discriminate their genetic differences [26]. The genetic diversity of Rhizobium isolated from several countries around the world was studied by [21]. The genetic diversity of thirteen studied strains were

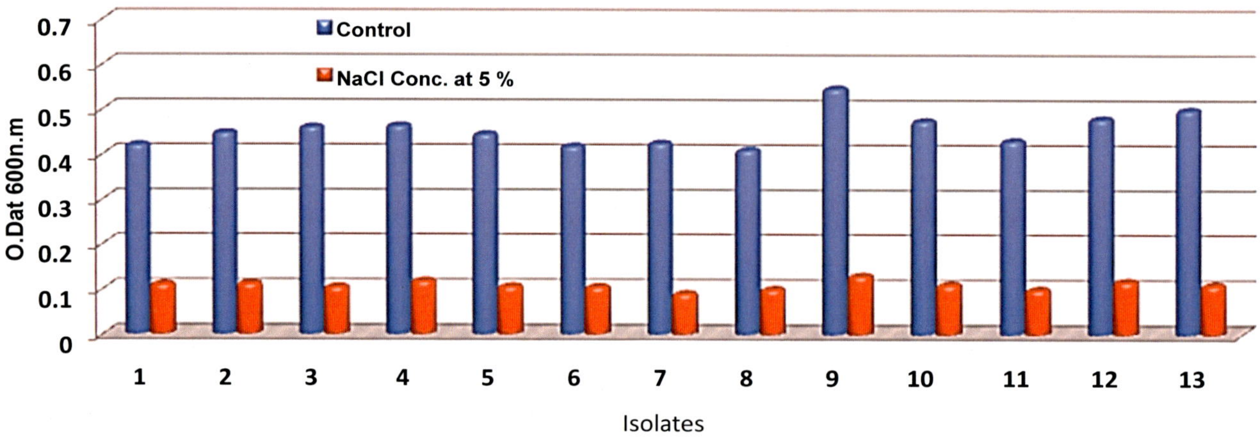

Figure 4: OD of all isolates at High Concentrations of NaCl (5%) after 72(hr) from inoculation.

	Isolates												
Conc. (%)	RL 1	RL 2	RL 3	RL 4	RL 5	RL 6	RL 7	RL 8	RL 9	RL 10	RL 11	RL 12	RL 13
0.1	0.421^{a}	0.448^{a}	0.461^{a}	0.464^{a}	0.445^{a}	0.418^{a}	0.425^{a}	0.410^{a}	0.547^{a}	0.475^{a}	0.431^{a}	0.480^{a}	0.500^{a}
1	0.256^{b}	0.361^{b}	0.256^{b}	0.253^{b}	0.321^{b}	0.258^{b}	0.321^{b}	0.321^{b}	0.265^{b}	0.248^{b}	0.361^{b}	0.321^{b}	0.266^{b}
3	0.150^{c}	0.160^{c}	0.140^{c}	0.155^{c}	0.172^{c}	0.150^{c}	0.172^{c}	0.172^{c}	0.18^{bc}	0.145^{b}	0.160^{b}	0.150^{c}	0.152^{c}
5	0.110^{c}	0.112^{c}	0.105^{c}	0.118^{c}	0.106^{c}	0.105^{c}	0.090^{c}	0.100^{c}	0.130^{c}	0.110^{c}	0.100^{b}	0.118^{c}	0.108^{c}

a,b,c=values of each trait within each column don't have the same letter are significantly different

Table 6: Effect of different concentrations of NaCl on Growthafter 72 hr. (measuredas optical density) of isolates.

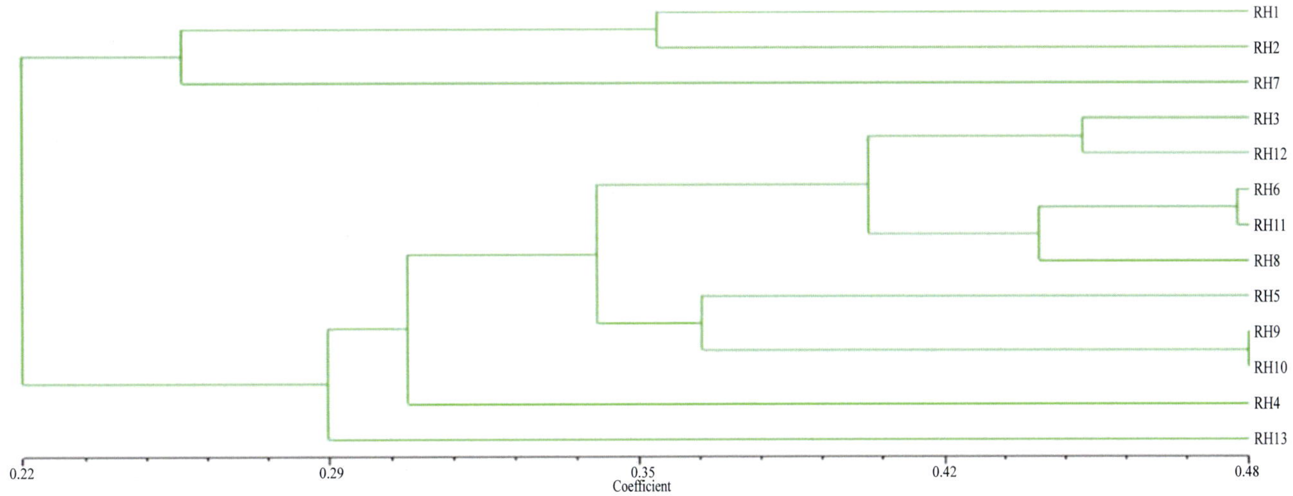

Figure 5: Phylogenetic relationship between 13 Rhizobial isolates using UPGMA method based on RAPD analyses.

examined: Rh1,Rh2, Rh3 , Rh4 , Rh5 , Rh6 , Rh7 , Rh8 , Rh9 , Rh10 , Rh11 , Rh12 and Rh13 using six random primers (OPE 15, OPA 04, OPC 06, OPE 20, OPJ 20 and OPG 04), and were used to screen for the polymorphism between the tested isolates. The results of RAPD analysis showed that all used primers were polymorphic. OPE 15 and OPG 04 primers showed the highest polymorphism level among the tested isolates, while OPJ 20 and OPC 06 primers showed the lowest polymorphism level Figure 6.

The similarity matrix values were converted into Dendrogram using UPGMA (Unweight Pair Group Method with the Arithmetic average) clustering method. Cluster analysis divided into two main clusters (Figure 5). The first cluster included isolates from Quesna, Benha, and South Sinai; whereas the second cluster included isolates from Sadat, Cairo, Sues, South Sinai , Beni Suef , Al- Dakahlia , North Sinai , El-Menia , Zefta, and Ismailia. Remarkably, the first cluster was divided into three groups for the isolates were highly related to each other. The second cluster presented ten isolates, the isolates within each group were overlapped, principal coordinates analysis showed similar pattern as cluster analysis. The analysis was based on the number of markers that were similar between any given pair of isolates Table 7.

All Rhizobia isolates genotypically characterized by RAPD_PCR. Total genomic DNA amplified with six oligonucleotide primers, amplification patterns revealed a high level of polymorphism. All primers produced multiple DNA products ranging in size from 0.2 to 1.5 Kb. Primer OPJ20 Produced multiple DNA products ranging in number from one in isolates Rh 11 and Rh 12 to three bands in Rh 1, Rh 3, Rh 4, Rh 5, Rh 6, Rh 8, Rh 10 and Rh 11. Whereas only Rh 2, Rh 7 and Rh 13 produced two bands. Primer OPE15 Produced multiple DNA products ranging in number from TWO bands in isolate Rh 5 to five bands in isolates Rh 3 and Rh 13. Whereas isolates Rh 4, Rh 8, Rh 10, Rh 11 and Rh 12 produced four bands . Also isolates Rh 1, Rh 2, Rh 6, Rh7 and Rh 9 produced three bands. OPC 06 Produced multiple DNA products ranging in number from two bands in isolates Rh 4, Rh 10, Rh 12 and Rh 13 to four bands in isolates Rh 1, Rh 3 and Rh 6 . On the other hand isolates Rh 2, Rh 5, Rh 7, Rh 8, Rh9 and Rh 11. OPG 04 Produced multiple DNA products ranging in number from three bands in isolates Rh 2, Rh 3, Rh 5, Rh 6 and Rh 8 to five bands in isolates Rh 7, Rh 11 and Rh12. Whereas isolates Rh 1 , Rh 4 ,Rh 9 , Rh10 and Rh 13 showed four bands. OPA 04 Produced multiple DNA products ranging in number from three bands in isolates Rh 2, Rh 3, Rh 7 and Rh 11 to six in isolates Rh9. Also five bands in isolates Rh 4 and Rh10 were produced, further to isolates Rh 1, Rh 5, Rh 6, Rh8, Rh 12 and Rh 13 showed four bands. Finally primer OPE 20 produced multiple DNA products ranging in number from two bands in only isolate Rh 2 to five bands in several isolates Rh 10, Rh 12 and Rh 13. Also the same primer produced four bands in isolates Rh 1, Rh 3, Rh 7 and Rh 9. Whereas isolates Rh 4, Rh 5, Rh 6, Rh 8 and Rh 11 showed only three bands. All six primers were successfully generated reproducible polymorphic bands to evaluate the degree of genetic diversity and calculating the genetic distances of Rhizobia based on the DNA nucleotide sequence using RAPD and used to identify the genetic variability and genetic relationship among the thirteen genotypes. Also we used Sequence analysis of 16S rDNA and subsequent BlastN analyses indicated that the majority of isolated strains were *Rlv.* [27].

To further confirm the taxonomic status of the Faba bean

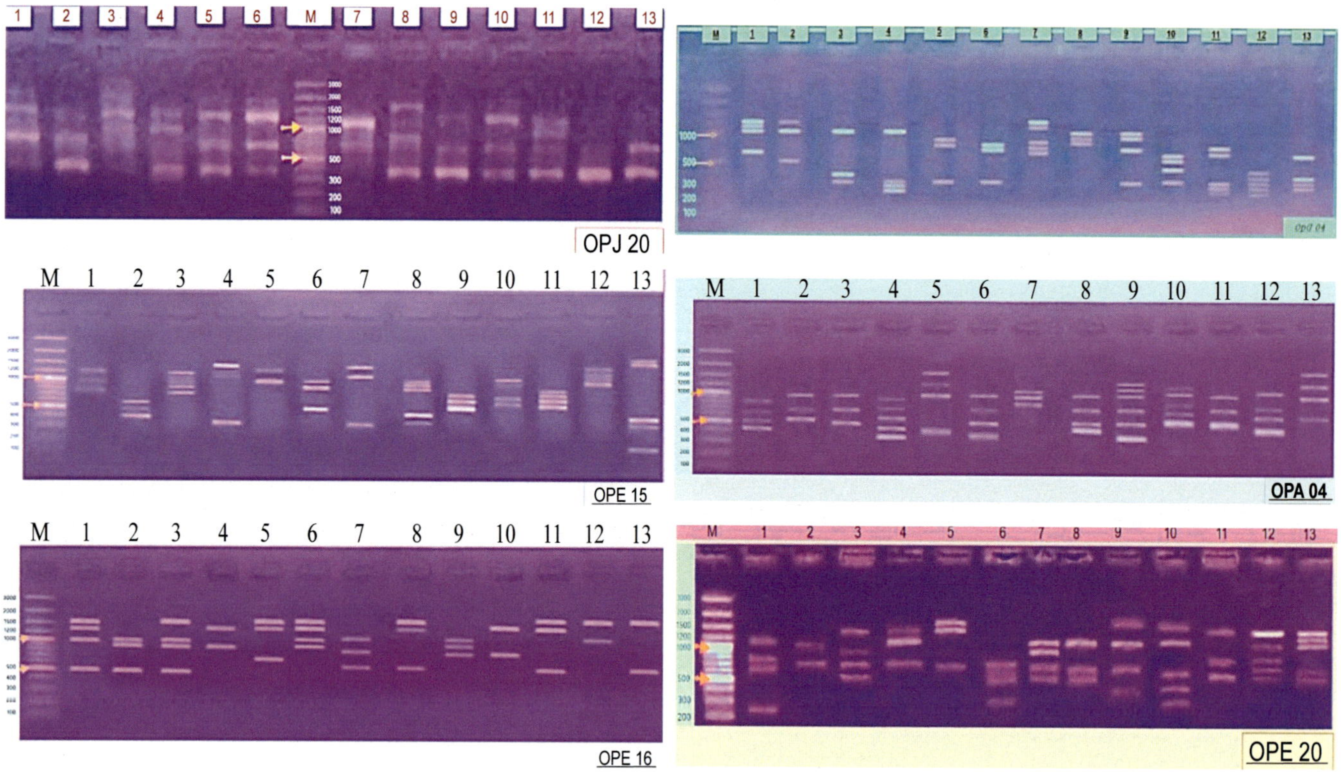

Figure 6: 1% Agarose gel electrophoresis of RAPD PCR products. (M) DNA size marker (Gene Ruler™ 100bp Plus DNA Ladder, Fermentas); 1=Rh. Quesna; 2=Rh. Banha; 3=Rh. Sadat ; 4=Rh. Zefta ; 5=Rh. Baneswaif; 6=Rh. Sues; 7=Rh. South Sina; 8=Rh. Al Arish; 9=Rh. Rafah; 10=Rh. New El-Menia; 11=Rh. Moshtohor; 12=Rh. Cairo and 13=Rh. Ismailia.

Isolates	**Rh 1**	**Rh 2**	**Rh 3**	**Rh 4**	**Rh 5**	**Rh 6**	**Rh 7**	**Rh 8**	**Rh 9**	**Rh 10**	**Rh 11**	**Rh 12**	**Rh 13**
Rh 1	1												
Rh 2	0.35	1											
Rh 3	0.34	0.27	1										
Rh 4	0.31	0.16	0.32	1									
Rh 5	0.17	0.06	0.25	0.26	1								
Rh 6	0.23	0.16	0.36	0.29	0.35	1							
Rh 7	0.31	0.2	0.28	0.25	0.22	0.21	1						
Rh 8	0.34	0.19	0.35	0.36	0.39	0.41	0.32	1					
Rh 9	0.15	0.3	0.29	0.3	0.36	0.38	0.22	0.25	1				
Rh 10	0.18	0.21	0.33	0.22	0.36	0.38	0.13	0.29	0.48	1			
Rh1 1	0.2	0.2	0.36	0.29	0.35	0.48	0.25	0.46	0.38	0.48	1		
Rh 12	0.22	0.15	0.44	0.36	0.34	0.41	0.28	0.44	0.25	0.37	0.46	1	
Rh 13	0.15	0.18	0.26	0.27	0.29	0.2	0.23	0.3	0.21	0.25	0.35	0.43	1

Table 7: Similarity between different tested isolates.

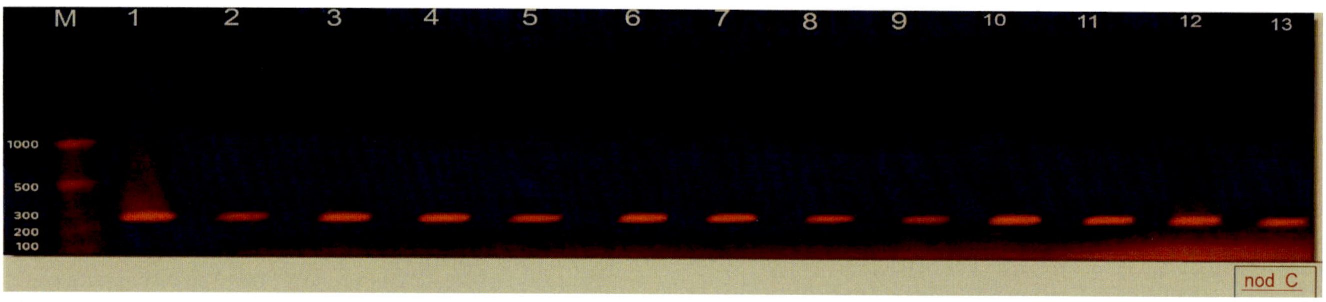

Figure 7: Identification of Rhizobial isolates by amplifying a specific region of nodC gene (220 bp).

nodulating Rlv strains, and to examine variation in a plasmid-borne, symbiotic specific gene, primers nodCf and nodCR were designed to amplify a region of the nodC specific for *Rlv.* as well as the nodC gene amplification results (Figure 7). PCR analyses done in this study showed that all of the strains that were identified as Rlv produced about a 220 bp DNA fragment using a nodC-specific primer pair.

Conclusions

Salt-stress are the major constraints to plant productivity and isolation of effective rhizobia to inoculate the leguminous crop plants could be an important strategy to improve the efficiency of rhizobium-legume symbiosis and thereby productivity. In our study we isolated of Faba plants thirteen *R. leguminosarum sv. Viciae* isolates, identified by phenotypic and genotypic characteristics.

References

1. Bond DA (1976) Evolution of Crop Plants. Longman and Hall, London, UK.
2. Allen ON, Allen AK (1981) Leguminosae: Source Book of Characteristics, Uses and Nodulation. Macmillan, London, UK.
3. Zakhia F, de Lajudie P (2001) Taxonomy of rhizobia. Agronomie 21: 569-576.
4. Stocker R, Seymour JR, Samadani A, Hunt D (2008) Rapid chemotactic response enables marine bacteria to exploit ephemeral microscale nutrient patches. Proc Natl Acad Sci 105: 4209-4214.
5. Amarger N, Bours M, Revoy F, Allard MR, Laguerre G (1994) Rhizobium tropici nodulates field-grown Phaseolus vulgaris in France. Plant Soil 161: 147-156.
6. Moreira FMS, Haukka K, Young JPW (1987) Biodiversity of rhizobia isolated from a wide range of forest legumes in Brazil. Mol Ecol 7: 889-895.
7. Binde DR, Menna P, Bangel EV, Barcellos FG (2009) Rep-PCR fingerprinting and taxonomy based on the sequencing of the 16S rRNA gene of 54 elite commercial rhizobial strains. Appl Microbiol Biotechnol 83: 897-908.
8. Singleton PW, Bohlool BB (1984) Effect of salinity on nodule formation by soybean. Plant Physiol 74: 72-76.
9. Laguerre G, Nour SM, Macheret V, Sanjuan J, Drouinb P, et al., (2001) Classification of rhizobia based on nod C and nif H gene analysis reveals a close phylogenetic relationship among Phaseolus vulgaris symbionts. Microbiol 147: 981-993.
10. Williams JGK, Kubelik AR, Livak KJ, Raifalski A, Tingey SV (1990) DNA polymorphisms amplified by arbitrary primers are useful as genetic markers. Nucleic Acids Res 18: 6531-6535.
11. Labes G, Ulrich A, Lentzsch P (1996) Influence of bovine slurry deposition on the structure of nodulating Rhizobium leguminosarum bv. Vicia soil populations in a natural habitat. Appl Environ Microbiol 62: 1717-1722.
12. Oliveira IAD, Vasconcellos MJ, Seldin L, Paiva E, Vargas MA, et al., (2000) Random Amplified Polymorphic DNA Analysis of effective Rhizobium sp. Associated with Beans cultivated in Brazilian Cerrado Soils. Braz J Microbiol 31: 39-44.
13. Eaglesham AR (1987) The use of intrinsic antibiotic resistance for Rhizobium study. In: Elkan, H. (Ed.), Symbiotic Nitrogen Fixation Technology. Marcel Dekker, NY, USA.
14. Moschetti G, Peluso A, Protopapa A, Anastasio M, Pepe Q, et al., (2005) Use of nodulation pattern, stress tolerance, nodC gene amplification, RAPD-PCR and RFLP-16S rDNA analysis to discriminate genotypes of Rhizobium leguminosarum symbiovar viciae. Sys & App Microbiol 28: 619-631.
15. Hahn NJ (1964) The Congo red reaction in bacteria and its usefulness in the identification of Rhizobia. Can J Microbiol 12: 725 - 733.
16. Laguerre G, Allard MR, Revoy F, Amarger N (1994) Rapid identification of rhizobia by restriction fragment genes. Appl Environ Microbiol 60: 56-63.
17. Navarro DN, Buendiab AM, Camachoa M, Lucasc MM, Santamaria C (2000) Characterization of Rhizobium spp. bean isolates from South-West Spain. Soil Biol Biochem 32: 1601-1613.
18. Hafeez FY, Asad S, Ahmad T, Malik KA (1995) Host specificity and characterization of fast-growing rhizobia from Macroptilium atropurpureum cv. Siratro in Pakistan. Soil Biol Biochem 27: 729-733.
19. Hagedorn C (1979) Relationship of antibiotic resistance to effectiveness in Rhizobium trifolii populations. Soil Sci Soc Am J 43: 921-925.
20. Rashid MD, Sattar A, Uddin MDI, Young JPW (2009) Molecular characterization of symbiotic root nodulating rhizobia isolated from lentil. EJEAF Che 8: 602-612.
21. Shamseldin A, El-Saadani M, Sadowsky MJ, An CS (2009) Rapid identification and discrimination among Egyptian genotypes of Rhizobium leguminosarum sv. viciae and Sinorhizobium meliloti nodulating Faba bean (Vicia faba L.) by analysis of nodC, ARDRA, and rDNA sequence analysis. Soil Biology & Biochemistry 41: 45-53.
22. Cordovilla MP, Ligero F, Lluch C (1999) Effect of salinity on growth, nodulation and nitrogen assimilation in nodules of faba bean (Vicia faba L.). Appl Soil Ecol 11: 1-7.
23. Amanuel G, Kuhne RF, Tanner DG, Viek PLG (2000) Biological nitrogen fixation in Faba bean (Vicia faba L.) in the Ethiopian highlands as affected by P fertilization and inoculation. Biol & Fertil Soils 32: 353-359.
24. Mutch LA, Tamimi SM, Young JPW (2003) Genotypic characterization of rhizobia nodulating Vicia faba from soils of Jordan: a comparison with UK isolates. Soil Biol & Bioch 35: 709-714.
25. Mutch LA, Young PW (2004) Diversity and specificity of Rhizobium leguminosarum symbiovar viciae on wild and cultivated legumes. Molecular Ecology 13: 2435-2444.
26. Ventorino V, Chiurazzi M, Aponte M, Pepe O, Moschetti G (2007) Genetic diversity of a natural population of Rhizobium leguminosarum sv. Vicia nodulating plants of Vicia faba in the Vesuvian area. Current Microbiology 55: 512-517.
27. Ismail M, El-Zanatay AM, Eissa RA, Hewedy OA (2013) Genetic Diversity of Rhizobium leguminosarum as Revealed by 16S rRNA Gene Sequence. American-Eur J Agric & Environ Sci 13: 797-801.

Production of Xylitol by the Thermotolerant *Kluyveromyces marxianus* IMB Strains

Michael Mueller[1], Mark R. Wilkins[1*] and Ibrahim M. Banat[2]

[1]*Department of Biosystems an Agricultural Engineering, Oklahoma State University, 111 Agricultural Hall, Stillwater, Oklahoma, USA 74078*
[2]*School of Biomedical Sciences, University of Ulster, Coleraine, BT52 1SA, Northern Ireland, UK*

Abstract

Xylitol is a fermentation product of xylose that can be used as a sweetener by diabetics. Five novel Kluyveromyces marxianus yeast strains, IMB1, IMB2, IMB3, IMB4 and IMB5, have shown the ability to grow on xylose and produce xylitol. In comparison to chemical synthetic xylitol production, biological production can be more cost effective since no expensive metal catalysts and high temperatures are needed. The five K. marxianus IMB strains were tested at temperatures ranging from 25°C to 45°C with xylose as the sole carbon source. IMB2 was found to be the best xylitol producer among the IMB yeast. Product yields ($Y_{P/S}$) up to 0.90 g/g at 40°C with 1.0 g/l initial cell mass and 50 g/l initial xylose concentration were observed for IMB2.

Keywords

Bioproducts; Sweeteners; Xylose; Fermentation

Introduction

Xylitol is a by-product of xylose fermentation and is a sugar alcohol [1]. Xylitol contains 40% fewer calories than sucrose on a mass basis and has similar sweetness [2]. It is odorless and has no aftertaste. Researchers have found that xylitol prevents dental cavities [3] and has an anti-ketonic [4] and an anti-infection effect [5]. Xylitol is produced by catalytic hydrogenation of xylose [6], which is costly and needs harsh reaction conditions. Biological xylitol production could be more cost effective, since no expensive catalysts are needed and the reaction takes place at ambient temperatures. Yeasts are well-known for their potential in industrial applications. *Candida mogii* was found to produce xylitol with a high yield ($Y_{P/S}$ = 0.62) [7]. The researchers observed that initial xylose concentration, co-factor regeneration and oxygen transfer were key factors for xylitol production [7]. *Debaryomyces hansenii* was observed to produce xylitol with product yields from 0.76 g/g to 0.78 g/g at 30°C and 200 rpm using Erlenmeyer flasks [8-10]. In a study using corn cob hemicellulose hydrolyzate, xylitol yields from xylose up to 0.73 g/g were observed using *Candida tropicalis* W103 at 35 °C [11].

In this study five *K. marxianus* strains, IMB1, IMB2, IMB3, IMB4, and IMB5, were investigated for their ability to produce xylitol from xylose. The yeast were isolated from an Indian distillery and showed the ability to grow at high temperatures using glucose (52°C) and ferment xylose as a carbon source at temperatures up to 45°C [12,13]. The thermotolerant nature of the IMB yeasts allows them to be used in tropical locations while reducing the need for fermenter cooling systems. The optimum growth rates of IMB strains were found at 40°C and a value of 0.93 h^{-1} with glucose as substrate [13]. Wilkins et al. carried out tests by using *K. marxianus* IMB2, IMB4, and IMB5 with different pH values and different temperatures under anaerobic conditions [14]. IMB5 had the greatest xylitol production at 40°C and an initial pH of 4.5 with a value of 1.33 g/l and a product yield of 0.25 g/g [14].

The enzymes necessary for xylose fermentation, xylose reductase (XR) and xylitol dehydrogenase (XD) must be activated by the presence of xylose. XR and XD are not available during glucose fermentation [15]. Xylose is reduced to xylitol, catalyzed by XR, and further oxidized to xylulose, catalyzed by XD. Xylulose enters the pentose phosphate cycle and is eventually converted to ethanol [16-18]. Some yeast tend to produce mainly xylitol and others ethanol. This is caused by different co-factor regeneration systems. It is known that yeast like *Pichia stipitis*, which produce ethanol, can use NADH as a co-factor for XR; whereas, xylitol producers primarily use NADPH [19]. XD uses NAD^+ as a co-factor in all xylose fermenting yeast [19]. In ethanol producing yeast, a regeneration cycle for NADH is established and xylulose production is favored. In xylitol producing yeast, NADPH is regenerated with the subsequent pentose phosphate cycle, but NADH accumulates and XD is inhibited, resulting in xylitol accumulation. This process only occurs under anaerobic or microaerobic conditions since NAD^+ is regenerated by oxygen [19].

In this study all five *K. marxianus* IMB yeast strains were screened at 25, 30, 35, 40, and 45°C at microaerobic conditions to ascertain their ability to produce xylitol using xylose as the sole carbon source. Growth rates, xylitol and ethanol concentrations, conversion efficiencies, as well as product and biomass yields were determined. Also, XR and XD activities at 40°C for all IMB strains were analyzed. In addition, the effect of initial substrate concentration on fermentation was analyzed.

Materials and Methods

Organisms

The organisms used were five *K. marxianus* yeast strains called IMB1, IMB2, IMB3, IMB4, and IMB5. These yeast were first discovered in an Indian distillery and are maintained at the University of Ulster [13]. Cultures were stored on slants of solid media in a refrigerator. The slant media contained 3 g/l yeast extract, 3 g/l malt extract, 5 g/l peptone, 20 g/l xylose, and 20 g/l agar.

***Corresponding author:** Mark R. Wilkins Department of Biosystems and Agricultural Engineering, Oklahoma State University, 111 Agricultural Hall, Stillwater, OK 74078

Preparation of seed culture

Prior to use all glassware, stoppers, and other equipment used in direct contact with the organisms were sterilized by autoclaving at 121°C for 30 min. Cell transfer and sample collection were done in a biological safety cabinet (Coy, Grasslake, MI, USA). A seed culture was prepared in media containing 10 g/l yeast extract, 20 g/l peptone and 20 g/l xylose (YPX media). Sterilization of media was achieved by using a filter assembly with a 0.45 μm filter. A loopful of cells for each *K. marxianus* IMB strain was added to 100 ml of YPX media in 250 ml baffled flasks. The flasks were closed with a stopper containing a filter to allow air exchange and prevent microbial contamination (Bug Stopper, Whatman Inc., Florham Park, NJ, USA). The seed culture was incubated for 18 h at 45°C. Aeration was achieved by shaking at 220 rpm by using an incubator with an orbital shaker (MaxQ Mini 4450, Barnstead International, Dubuque, IA, USA).

IMB yeast screening

The inoculum media contained 10 g/l yeast extract, 20 g/l peptone, and 20 g/l glucose. The media was sterilized with a filter assembly using a 0.45 μm filter. Baffled flasks (250ml) were filled with 100 ml of inoculum media, inoculated with 1ml of seed culture containing one of the *K. marxianus* strains and closed with the stopper previously described. Conditions were the same as for the seed culture (45°C, 18 hours, 220 rpm). To maintain pH during fermentation, a 50 mM sodium citrate buffer at pH 5.5 was used to dissolve the media components. The fermentation media contained 0.50 g/l yeast extract, 2.0 g/l KH_2PO_4, 1.0 g/l $(NH_4)_2SO_4$, 1.0 g/l $MgSO_4*7H_2O$, 0.10g/l $MnSO_4$, and 20.0 g/l xylose [12].

After 18 h the optical density (OD) of the inocula at a wave length of 660 nm was determined using a UV visible spectrophotometer (Varian, Palo Alto, CA, USA). An initial OD of 0.5 (representing 0.14 g dry cells/l) was set for each fermentation culture by mixing centrifuged (5 min, 4°C, and 6,100 g) (Sorvall Legend RT, Thermo Scientific, Waltham, MA, USA), DI water washed cells from the inocula with 10 ml DI water. Ten ml inoculum was added to 90 ml fermentation media [20]. The flasks were enclosed with the previously described stoppers. Growth conditions were 25, 30, 35, 40 or 45°C with shaking at 100 rpm. Constant temperature and shaking was achieved by using the incubator/shaker. According to Yablochkova et al. [19], shaking at 100 rpm on an orbital shaker like the one used in this study resulted in an oxygen transfer rate of 5 mmol/(l*h) and led to a microaerobic environment for ethanol and xylitol production from xylose. For the IMB screening experiment, samples were taken at 0, 1, 2, 3, 4, 5, 6, 24, 48, 72, 96, 120, and 144 h.

OD was measured via the UV-Vis spectrophotometer and samples were filtered through a 0.2 μm filter from Fisher Scientific (Pittsburgh, PA., USA). The permeate was used for high performance liquid chromatography (HPLC) analysis, which is described below. Every IMB strain was tested twice at each temperature.

Effect of initial cell mass concentration on xylitol production

To investigate the effect of different initial cell mass concentrations, a test with IMB2 at 40°C and high xylose concentration was performed. The inoculum was prepared the same as the inoculum for the screening experiment except the incubation temperature was 40°C and xylose was the carbohydrate source. Xylose was used since it was found that xylitol production was better for cells produced from medium containing xylose as opposed to cells produced from medium containing glucose (data not shown). The fermentation media was prepared as the fermentation media for the screening test containing a xylose concentration of 50 g/l. The fermentation media was inoculated with cells to achieve 0.5, 1.0, and 1.5 g/l of dry cell mass using the same procedure for cell harvest and wash as described above [20]. Samples were taken at 0, 6, 12, 24, 48, 72, 96, 120 and 144 hours and analyzed via HPLC as described later.

HPLC analysis

Concentrations of xylose, xylitol, ethanol, acetic acid, and glycerol were analyzed on a HPX-87H column (Bio-Rad, Sunnyvale, CA, USA). The eluent was 0.01 N H_2SO_4 with a flow rate of 0.6 ml/min at 60°C, and a refractive index detector (1100 Series, Agilent, Santa Clara, CA, USA) was used for detection [21].

Assay for XR and XD Activity

The procedure was modified from the one described by Smiley and Bolen [15]. All IMB yeast strains were grown in YPX medium containing 10 g/l yeast extract, 20 g/l peptone, and 20 g/l xylose for 18 h at 40°C and 220 rpm. The cells were centrifuged at 4°C and 6,100 g for 10 min. After the liquid was discarded, the cells were washed with DI water and centrifuged again under the same conditions. The cells were re-suspended with 1 ml 0.1 M potassium phosphate buffer per gram cell pellet. The cells were stored until usage at -18°C. After the cells were thawed, 0.3 ml of 0.1 M 2-mercaptoethanol per ml cell suspension was added. A cell disruption was achieved via sonication (Misonix Sonicator 3000, Misonix Incorporated, Farmingdale, NY, USA) for 35 min in 10 sec pulses with 10 sec intervals on level 5. The mass was centrifuged at 16,060 g for 20 min (AccuSpin Micro, Fisher Scientific, Pittsburgh, PA, USA). The following solutions were added into a 1 ml spectrophotometer cuvette: 0.05 ml of 1.0 M potassium phosphate buffer pH 7.4, 0.1 ml of 0.1 M 2-mercaptoethanol, 0.1 ml of 0.5 M D-xylose, 0.1 ml of cell free extract, and 0.55 ml DI water. At time zero 0.1 ml of 3.4 mM NADH was added. The absorption was measured in a spectrophotometer (Varian, Palo Alto, CA, USA) at 340 nm for 6 min. The same procedure was repeated with NADPH as cofactor. NADH and NADPH were purchased from Sigma-Aldrich (St. Louis, MO, USA) and EMD Chemicals (Gibbstown, NJ, USA), respectively.

Statistical analysis

Two level factorial tests were performed for both the yeast screening and effect of carbohydrate sources tests. Yeast strain and temperature were the independent variables andxylitol concentration was the dependent variable. An analysis of variance was calculated using SAS Release 9.1 (SAS, Carey, NC) and means were separated for each variable by Fisher's protected least significant difference analysis at 95% confidence level.

Results and Discussion

IMB Yeast screening

Table 1 presents the product and substrate yields, maximum xylitol concentrations, conversion efficiencies, and growth rates for each strain at each temperature. Table 2 summarizes the mean xylitol concentration for 72 to 144 h, including p-values for strain, temperature, and the interaction of strain and temperature. Values with the same letter are not significantly different ($p>0.05$). The growth rates differed among temperatures. In general the growth rate was highest at 40°C. At 40°C

Temp. [°C]	IMB	YP/S (g xylitol/g xylose)	YP/X (g xyiltol/g cells)	CE [%]	Maximum Xylitol Concentration	Growth rate [h^{-1}]
25	1	0.06	0.18	3.7	0.75	0.08
	2	0.09	0.26	5.8	1.18	0.08
	3	0.03	0.09	2.0	0.41	0.08
	4	0.08	0.20	5.0	1.01	0.07
	5	0.18	0.68	10.3	2.09	0.07
30	1	0.26	1.13	18.6	3.77	0.10
	2	0.15	0.54	10.1	2.05	0.11
	3	0.12	0.45	8.3	1.68	0.11
	4	0.19	0.60	13.3	2.70	0.11
	5	0.17	0.61	11.5	2.33	0.10
35	1	0.26	1.18	24.0	4.86	0.12
	2	0.26	1.18	23.6	4.78	0.13
	3	0.26	1.35	23.5	4.76	0.11
	4	0.23	0.96	20.9	4.24	0.14
	5	0.24	1.24	21.8	4.42	0.12
40	1	0.41	3.37	35.8	7.26	0.15
	2	0.42	3.42	38.8	7.86	0.16
	3	0.34	2.18	26.1	5.29	0.14
	4	0.40	2.06	30.2	6.12	0.20
	5	0.32	2.27	27.3	5.53	0.16
45	1	0.36	3.45	27.4	5.55	0.11
	2	0.42	4.10	30.1	7.72	0.12
	3	0.31	3.21	21.5	4.36	0.08
	4	0.41	3.42	33.3	6.75	0.08
	5	0.29	2.76	21.8	4.42	0.09

Table 1: Xylitol yields from xylose and cells (YP/S, YP/X), conversion efficiency (CE) and growth rates of K. marxianus IMB strains at 100 rpm in 250 ml baffled flasks.

Time (h)		72	96	120	144
Strain	IMB1	$2.63^{AB} \pm 0.47$	$3.74^{A} \pm 0.63$	$4.01^{A} \pm 0.60$	$3.37^{A} \pm 0.48$
	IMB2	$2.85^{A} \pm 0.59$	$3.90^{A} \pm 0.77$	$4.10^{A} \pm 0.75$	3.50A ± 0.61
	IMB3	$2.32^{C} \pm 0.47$	$2.74^{C} \pm 0.51$	$2.58^{C} \pm 0.49$	$2.07^{C} \pm 0.51$
	IMB4	$2.50B^{C} \pm 0.52$	$3.23^{B} \pm 0.57$	$3.35^{B} \pm 0.53$	$2.86^{B} \pm 0.55$
	IMB5	2.75AB ± 0.43	$3.17^{B} \pm 0.42$	$3.07^{B} \pm 0.31$	$2.63^{B} \pm 0.35$
Temp (°C)	25	$0.73^{A} \pm 0.14$	$0.83^{A} \pm 0.12$	$0.99^{A} \pm 0.14$	$1.09^{A} \pm 0.17$
	30	$1.42^{B} \pm 0.10$	$2.04^{B} \pm 0.17$	$2.40^{B} \pm 0.22$	$2.36^{B} \pm 0.23$
	35	$3.03^{C} \pm 0.14$	$4.31^{C} \pm 0.16$	$4.57^{C} \pm 0.16$	$3.74^{C} \pm 0.23$
	40	$5.71^{E} \pm 0.12$	$6.24^{E} \pm 0.40$	$4.73^{C} \pm 0.66$	$2.42^{B} \pm 0.72$
	45	$3.48^{D} \pm 0.32$	$4.80^{D} \pm 0.44$	$5.57^{D} \pm 0.45$	$5.55^{D} \pm 0.37$
P values	Strain	0.0001	<.0001	<.0001	<.0001
	Temp	<.0001	<.0001	<.0001	<.0001
	Str. +Temp.	<.0001	<.0001	<.0001	<.0001

Table 2: Mean values with standard error of xylitol concentration [g/l] during *K. marxianus* IMB strains screening. Factor strain shows the mean values over all temperatures and factor temperature shows the mean values over all strains. Values with the same letter for each column were not significantly different (α = 0.05).

the highest growth rate of 0.2 h^{-1} was achieved with IMB4 Table 1. Banat and Marchant [12], who first investigated the IMB strains, found growth rates on xylose at 45°C between 0.18 h^{-1} and 0.26 h^{-1} for IMB5 and IMB4, respectively. The higher growth rates in their study can be explained by the difference in aeration. In this study, microaerobic conditions were used; whereas, Banat and Marchant [12] determined growth rates for aerobic conditions. Microaerobic condition is defined by aeration with oxygen concentrations below the oxygen concentration of air. Yablochkova et al. [19] determined an oxygen transfer rate of 5 mmol O_2/(l*h) as microaerobic conditions leading to optimum xylose fermentation. Low growth rates are favorable, since more of the substrate can be used for product formation.

Xylose consumption varied with temperature as well. At 40°C all strains consumed almost all xylose; whereas, at all other temperatures xylose remained after 144 h. The highest xylose concentrations remained in the media at 25°C with values ranging from 8.2 g/l with IMB5 to 6.6 g/l with IMB3 (data not shown). Similar characteristics where shown in a previous study that was conducted with IMB2, IMB4, and IMB5 [14]. According to the figures shown in their work, IMB4 did not utilize xylose completely at 45°C with a value of about 10 g/l after 120 h. In that study it was concluded that IMB4 produced the highest ethanol concentration. However, that study was under anaerobic conditions; whereas, this study was conducted under microaerobic conditions [14].

The IMB yeast showed a greater capability for xylitol production in comparison to ethanol production. Ethanol production was negligible. Temperature has a strong influence on xylitol production with the highest mean concentration at 40°C (6.24 g/l after 96 h) and the lowest at 25°C (1.09 g/l after 144 h) Table 2. This is in alignment with the higher xylose utilization at 40°C. The results indicate the thermotolerant nature of the K. marxianus IMB yeast, which was, to the authors' knowledge, not found in any other publication on xylitol producing yeast. Yeast strain also affected xylitol production. IMB2 had the highest mean maximum xylitol concentration of all strains with the same value at 40 and 45°C (4.1 g/l), but it was not significantly different than IMB1. The lowest production was achieved by IMB3 with a mean value of 2.74 g/l Table 2.

The greatest xylitol yield from xylose ($Y_{P/S}$) and initial cell mass ($Y_{P/X}$) was achieved by IMB2 with 0.42 g xylitol/g xylose and 4.18 g xylitol/g cells at 45°C Table 1. The highest conversion efficiency (maximum xylitol concentration/theoretical xylitol concentration) was 38.1%. In general, the yeast with the greatest YP/S also had the greatest YP/X. The same was true for the lowest YP/S and YP/X with the exception of 40°C, where the lowest YP/S and YP/X was shown by IMB5 and IMB4, respectively. It should be noted that the xylitol concentration for all IMB strains at 40°C decreased rapidly after reaching the maximum concentration (data not shown). At all other temperatures, the maximum xylitol concentrations were observed at the end of fermentation. Within xylose metabolism xylitol is an intermediate that can be further converted to ethanol. Indeed, ethanol concentration increased while xylitol concentration decreased after 96 h at 40°C (data not shown). However, ethanol production was very low with the greatest observed concentration of 1.64 g/l by IMB1 at 40°C after 144 h. It can be assumed that the decrease in xylitol concentration is caused by the usage of xylitol for other metabolites and/or cell maintenance.

Effect of initial cell mass concentration on xylitol production

During the test with three different initial cell mass concentrations, similar cell mass concentrations were observed after 120 h, ranging between 2.22 and 2.45 g/l for 1.5 g/l and 0.5 g/l, respectively. Cultures initially containing 0.5 g/l and 1.5 g/l cell mass showed an increase in cell mass during the last 24 h (between 120 and 144 h) to 3.68 and 2.95 g/l, respectively Figure 1. The cultures initially containing 1.0 g/l cell mass did not grow in the last 24 h. It can be assumed, that during the 120 h sampling, the 0.5 g/l and 1.5 g/l cultures were introduced to higher oxygen concentration, which led to elevated growth rates.

The cultures with 0.5 and 1.5 g/l initial cell mass concentration consumed the most xylose after 144 h with a final xylose concentration of

4.90 and 6.28 g/l, respectively Figure 1A. Cultures initially containing 1.0 g/l cell mass had a final xylose concentration of 15.3 g/l.

The highest xylitol concentration for all tests was achieved after 144 h. Cultures initially containing 0.5 g/l cell mass produced the highest xylitol concentration with 34.6 g/l Figure 1B, which represents an YP/S of 0.76 g/g Tables 4 and 5. Cultures initially containing 1.0 g/l cell mass produced a lower concentration with 29.9 g/l of xylitol, but had a product yield of 0.86 g/g since less xylose was used. Cultures initially containing 1.5 g/l cell mass produced a xylitol concentration of 30.6 g/l and a product yield of 0.70 g/g. There was no significant difference between the three initial cell mass concentrations in terms of 144 h xylitol concentration (p>0.05). Comparing YP/S values at 120 h, which excludes the great increase in cell mass in the last 24 h, shows that cultures with1.0 g/l initial cell mass had the highest yield with 0.90 g/g compared to 0.80 and 0.76 g/g with cultures with 0.5 g/l and 1.5 g/l initial cell mass, respectively. Xylitol concentrations after 120 h were different among the initial cell mass concentrations Table 3. The greatest mean xylitol concentration produced was 29.76 g/l by cultures with 0.5 g/l initial cell mass.

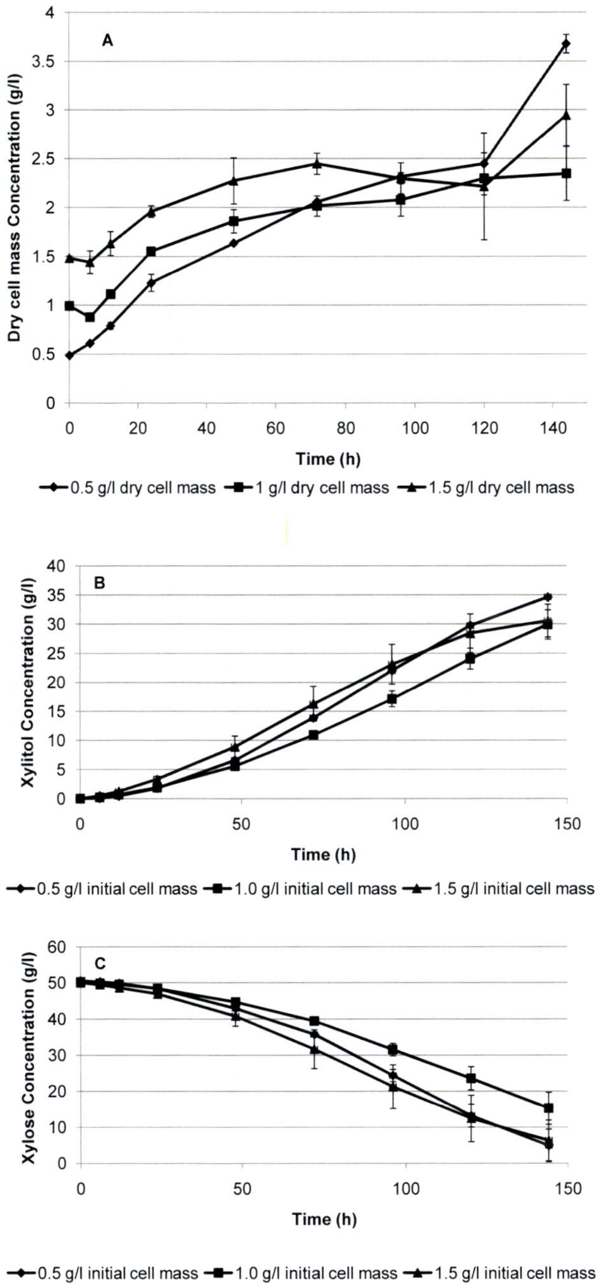

Figure 1: Dry cell mass concentration (A), xylitol production (B) and xylose cuncumption (C) by K. marxianus IMB2 at 40°C three different initial cell mass concentrations (0.5, 1.0, 1.5 g/l). Error bars represent standard error.

Cultures with 1.0 g/l initial cell mass still had the potential to produce more xylitol after 144 h, since 18 g/l of xylose remained in the media. Comparing the YP/X values, 0.5 g/l initial cell mass showed the highest yield of 71.16 g xylitol/g initial dry cell mass after 144 h Table 4. Cultures initially containing 1.0 g/l and 1.5 g/l cell mass have lower yields since less xylitol was produced with higher initial cell mass concentrations. The yields are 30.20 and 20.64 g/g for 1.0 g/l and 1.5 g/l, respectively Table 4.

Initial substrate and cell mass concentrations have an effect on the xylitol production of the IMB yeast. The yields of all three tests with elevated initial cell mass concentrations are higher than the yields achieved during the IMB screening and the inoculum carbohydrate test. Compared to the literature, the yields achieved with IMB2 at 40°C with 50 g/l initial substrate concentration and 1.0 g/l initial dry dell mass concentration (0.86 g/g) are the highest with natural occurring yeast we have found (e.g. D. hansenii with 0.78 g/g) [8], 10]. However, the maximum observed productivity of 0.24 g/L/h is much lower than observed for C. tropicalis W103, which was 1.07 g/L/h [11]. Increased productivity is necessary for utilizing *K. marxianus* IMB2 for industrial xylitol production.

Time (h)		120	144
Initial cell mass concentration [g/l]	0.5	$29.76^{A} \pm 0.19$	34.64 ± 0.20
	1.0	$24.09^{B} \pm 1.04$	29.92 ± 1.44
	1.5	$28.43^{AB} \pm 1.92$	30.59 ± 1.62
P value	Cell mass	0.0439	0.0745

Table 3: Mean values and standard errors for xylitol concentration after 120 and 144 hours with different initial cell mass concentrations (α = 0.05). Values with different letters are significant different (t-test (LSD)).

Time [h]	Cell mass [g/l]	YP/S (g xylitol/g xylose)	YP/X (g xyiltol/g initial cells)	CE [%]	Maximum Xylitol Concentration [g/l]
120	0.5	0.80	61.15	58.4	29.76
	1.0	0.90	24.32	47.5	24.09
	1.5	0.76	19.19	56.3	28.43
144	0.5	0.76	71.16	67.9	34.64
	1.0	0.86	30.20	58.9	29.92
	1.5	0.70	20.64	60.6	30.59

Table 4: Xylitol yields from xylose and cells (YP/S, YP/X), conversion efficiency (CE) and growth rates of IMB2 with different initial cell mass concentrations (Test parameters: 40°C, 100 rpm, and 50 g/l initial xylose concentration).

Yeast	XR activity [IU/ml] NADH	NADPH
IMB1	n.d.	2.85
IMB2	n.d.	2.68
IMB3	n.d.	0.51
IMB4	n.d.	0.26
IMB5	n.d.	2.91

Table 5: XR activity of IMB yeast strains at 40°C (n.d. = non-detectable). The units are in IU/ml (μmol of cofactor oxidized / (ml cell free extract * min)).

Determination of XR Activity

To understand why the IMB yeast produce high amounts of xylitol and low amounts of ethanol, the XR activity of all IMB strains was tested with NADH and NADPH as cofactors. Table 5 lists the activities of XR in IU/ml cell free extract. The resulting XR activities using NADH versus NADPH support the proposed cofactor imbalance being responsible for high xylitol and low ethanol yields. No activity was detected using NADH as cofactor for all IMB strains. By using NADPH the activity ranged from 0.26 to 2.91 IU/ml. No correlation between XR activities and maximum xylitol concentrations or product yields was established. However, IMB1 and IMB2, which were observed to produce the greatest xylitol concentrations during the yeast screening, had two of the higher activities observed.

NADPH cannot be regenerated with the second step and an imbalance of cofactors is caused by microaerobic conditions in the yeast. This confirms the study by Yablochkova et al. (2004) [20] which presented that K. marxianus has a low xylose reductase activity while using NADH as a cofactor. The researchers reported <0.01 and 0.16 μmol/(mg*min) as the activities for XR with NADH and NADPH, respectively [20]. It should be noted that their assay was conducted at 30°C, while ours was conducted at 40°C.

Conclusion

The results indicate that IMB yeast have potential for biological xylitol production. IMB1 and IMB2 were the best xylitol producers ($p<0.05$). Xylitol production until 96h was greatest at 40°C, and from 96 to 144 h it was greatest at 45°C. The greatest xylitol yield achieved from xylose was 0.90 g/g with IMB2 at 40°C using 1.0 g/l initial cell mass and 50 g/l initial xylose concentration. Based on the present study, K. marxianus IMB2 was the most promising strain in this study for biological xylitol production with the following advantages: high product yields, no genetic modification, and a thermotolerant nature; however, productivity must be improved.

Acknowledgements

The authors would like to acknowledge the Oklahoma Agricultural Experiment Station for supporting this research.

References

1. Guo C, Zhao C, He P, Lu D, Shen A, et al. (2006) Screening and characterization of yeasts for xylitol production. J Appl Microbiol 101: 1096-1104.
2. Granstrom TB, Izumori K, Leisola M (2007) A rare sugar xylitol. Part I: the biochemistry and biosynthesis of xylitol. Appl Microbiol Biot 74: 277-281.
3. Lynch H, Milgrom P (2003) Xylitol and dental caries: an overview for clinicians. Journal of California Dental Association 31: 205-209.
4. Kinami Y, Kitagawa I (1969) Fluctuation of blood sugar, urine sugar and ketone body levels in surgical stress and application of xylitol. Shujutsu 23: 1487-1491.
5. Brown CL, Graham SM, Cable BB, Ozer EA, Taft PJ, et al. (2004) Xylitol enhances bacterial killing in the rabbit maxillary sinus. Laryngoscope 114: 2021-2024.
6. Aminoff C, Vanninen E, Doty T (1978) Xylitol-occurence, manufacture and properties. Oral Health 68: 28-29.
7. Sirisansaneeyakul S, Staniszewski M, Rizzi M (1995) Screening of yeasts for production of xylitol from D-xylose. J Ferment Bioeng 80: 565-570.
8. Sampaio FC, Chaves-Alves VM, Converti A, Passos FML, Coelho JLC (2008) Influence of cultivation conditions on xylose-to-xylitol bioconversion by a new isolate of Debaryomyces hansenii. Bioresource Technol 99: 502-508.
9. Sampaio FC, De Faveri D, Mantovani HC, Passos FML, Perego P, et al. (2006) Use of response surface methodology for optimization of xylitol production by the new yeast strain Debaryomyces hansenii UFV-170. J Food Eng 76: 376-386.
10. Sampaio FC, Mantovani HC, Passos FJV, de Moraes CA, Converti A, et al. (2005) Bioconversion of D-xylose to xylitol by Debaryomyces hansenii UFV-170: Product formation versus growth. Process Biochem 40: 3600-3606.
11. Cheng KK, Zhang JA, Ling HZ, Ping WX, Huang W, et al. (2009) Optimization of pH and acetic acid concentration for bioconversion of hemicellulose from corncobs to xylitol by Candida tropicalis. Biochem Eng J 43: 203-207.
12. Banat IM, Marchant R (1995) Characterization and Potential Industrial Applications of 5 Novel, Thermotolerant, Fermentative, Yeast Strains. World J Microb Biot 11: 304-306.
13. Banat IM, Nigam P, Marchant R (1992) Isolation of Thermotolerant, Fermentative Yeasts Growing at 52-Degrees-C and Producing Ethanol at 45-Degrees-C and 50-Degrees C. World J Microb Biot 8: 259-263.
14. Wilkins MR, Mueller M, Eichling S, Banat IM (2008) Fermentation of xylose by the thermotolerant yeast strains Kluyveromyces marxianus IMB2, IMB4, and IMB5 under anaerobic conditions. Process Biochem 43: 346-350.
15. Smiley KL and Bolen PL (1982) Demonstration of D-Xylose Reductase and D-Xylitol Dehydrogenase in Pachysolen-Tannophilus. Biotechnol Lett 4: 607-610.
16. Chiang C, Knight SG (1960) Metabolism of D-Xylose by Moulds. Nature 188: 79-81.
17. Fritsche W (2002) Mikrobiologie, Spektrum Akademischer Verlag Heidelbeg, Berlin.
18. Ligthelm ME, Prior BA, Dupreez JC, Brandt V (1988) An Investigation of D-(1-C-13) Xylose Metabolism in Pichia-Stipitis under Aerobic and Anaerobic Conditions. Appl Microbiol Biot 28: 293-296.
19. Yablochkova EN, Bolotnikova OI, Mikhailova NP, Nemova NN, Ginak AI (2004) The activity of key enzymes in xylose-assimilating yeasts at different rates of oxygen transfer to the fermentation medium. Microbiology 73: 129-133.
20. Dowe N, McMillan J (2001) SSF experimental protocols: lignocellulosic biomass hydrolysis and fermentation. LAP-008 NREL Analytical Procedure. National Renewable Energy Laboratory, Golden, CO.
21. Sluiter A, Hames B, Ruiz R, Scarlata C, Sluiter J,et al. (2004) Determination of sugars, byproducts, and degradation products in liquid fraction process samples. LAP-014 NREL Analytical Procedure. National Renewable Energy Laboratory Golden, CO.

Thermal, Pasting and Gel Textural Properties of Commercial Starches from Different Botanical Sources

Shujing Li, Yingquan Zhang, Yimin Wei, Wei Zhang and Bo Zhang*

Institute of Agro-Products Processing Science & Technology, Chinese Academy of Agricultural Sciences, Key Laboratory of Agricultural Product Processing, Ministry of Agriculture, P.O. Box 5109, Beijing 100193, People's Republic of China

Abstract

The applications of starch in food systems are primarily governed by its gelatinization, pasting, solubility, swelling and digestibility properties. To select the appropriate variety starch for end-use suitability, thermal, pasting and gel textural properties of nine starches were determined. Commercial grade wheat starch, corn starch, potato starch, sweet potato starch, cassava starch, mung bean starch, pea starch, amylose from potato, and amylopectin from waxy corn were used in the research aim to the industry end-use. Wheat starch showed the lowest transition temperatures (the onset temperature, T_o and the peak temperature, T_p). Sweet potato starch and cassava starch showed the highest transition temperatures and the lowest enthalpy changes. Mung bean starch showed the slowest transition speed. Tuber and root starches showed higher pasting temperatures, peak viscosity, final viscosity, breakdown viscosity, setback viscosity, and lower adhesiveness. Legume starches showed the highest hardness values. Person correlations and a principal component analysis showed that the transition temperatures were negatively correlated with enthalpy change ($p<0.01$). Peak viscosity, final viscosity, breakdown viscosity and setback viscosity were positively correlated with each other, but negatively correlated with adhesiveness. The hardness was positively correlated with springiness ($r=0.830$, $p<0.01$), and the cohesiveness was positively correlated with enthalpy change.

Keywords: Starch; Botanical sources; Thermal; Pasting; Gel texture; Principal component analysis

Introduction

Starch is the most abundant reserve carbohydrate in plants and a valuable ingredient in the food industry because it is widely used as a thickener, a gelling agent, a bulking agent and a water retention agent [1]. The applications of starch in food systems are primarily governed by its gelatinization, pasting, solubility, swelling and digestibility properties.

The physicochemical properties of starches depend on the botanical source from which they are isolated. The major botanical and commercial sources of starches are cereals, tubers, roots, and legumes. Many comparisons of the physicochemical and functional properties of starches from different crops have been published. Starches from cultivars of wheat [2], corn [3], potato [4], sweet potato [5], cassava [6], and mung bean [7] have been shown to vary in terms of starch composition and properties. Cereal starches contain a significant quantity of phospholipids, whereas corn and rice starches generally show higher transition temperatures than wheat and potato starches, and the highest enthalpy change values are found in potato and wheat starches. The free fatty acids in rice and corn starches contribute to their higher transition temperatures [8]. The starches from cassava and potato share similarities: these produce relatively bland pastes with higher viscosity, better clarity, and lower retrogradation rates than the starches derived from cereals. Normal cassava and potato starches have a high swelling power and a dispersed volume fraction compared with starches from other tropical root and tuber crops [9]. Tuber starches have been shown to be more susceptible than legume or cereal starches to heat-moisture treatment [10,11]. Potato starch exhibits higher swelling power, solubility, paste clarity, and viscosity than wheat, rice, and corn starches. Potato starch shows a higher tendency toward retrogradation compared with cereal starches [8]. In general, legume starches are characterized by a high amylose content (24-65%) and are reportedly more viscous than cereal starches [12,13].

Starch gelatinization refers to the disruption of the molecular order within the starch granules when heated in the presence of water. Evidence of the loss of an organized structure includes irreversible granule swelling, loss of birefringence, and crystallinity. Gelatinization is an energy-absorbing process that can be monitored through Differential Scanning calorimetry (DSC) and has been used for the study of starches [14]. In the DSC thermogram, thermal stability of samples can be monitored by the onset temperature (T_o) or peak transition temperature (T_p), while the transformed proportion is reflected by the area under the endothermic peak, representing the enthalpy change (ΔH). The sharpness of the transition peak measured as width at half peak height ($\Delta T^{1/2}$), is an index of the co-operativity of the transition from native to denatured state [15]. A study of the thermal properties of starch can guide the processing and the utilization of starch and provide information about its structure.

It is generally accepted that the increased viscosity observed during the heating of starch in water occurs mainly because of the swelling of granules. Differences in breakdown are related to differences in the rigidity/fragility of the swollen granules. The viscosity parameters during pasting are cooperatively controlled by the properties of the swollen granules and the soluble materials leached from the granules [16]. The breakdown viscosity indicates the starch's propensity for disintegration. The final viscosity indicates the ability of the starch to

***Corresponding author:** Bo Zhang, Institute of Agro-Products Processing Science & Technology, And Chinese Academy of Agricultural Sciences, Key Laboratory of Agricultural Product Processing, Ministry of Agriculture, P.O. Box 5109, Beijing 100193, People 's Republic of China

form a viscous paste. The setback viscosity indicates the synaeresis of starch upon the cooling of the cooked starch pastes [17].

The gelling ability of starch is important in food manufacturing. The textural properties of starch gels are crucial criteria that are used to evaluate the performance of starch in a food system.

Although lots of studies have been conducted on starch systems, few studies have compared commercial starches originating from different starch-rich crops using the same experimental methods. Usually, the cultivars of normal starches for research are certain and with high purity. The commercial starches are usually mixed with several cultivars, and their purities are different. However, starches used for industry are mostly commercial grade. Thus, the present investigation focuses on the comparison of commercial starch samples from different origins, i.e. wheat, maize corn, potato, sweet potato, cassava, mung bean, and pea, to document the differences in their thermal, pasting, and gel textural properties. These comparisons will aid the selection of the appropriate variety for end-use suitability.

Materials and Methods

Materials

Commercial grade wheat starch, corn starch, potato starch, sweet potato starch, cassava starch, mung bean starch, pea starch, amylose from potato, and amylopectin from waxy corn were acquired in a market. The total starch contents, measured by a third-party detection institution using the Chinese standard method GB/T 5009.9-2003, are specified in Table 1.

Thermal properties of starches

The thermal characteristics of the starches were studied using a differential scanning calorimeter (DSC, Q-200, TA Instruments, USA). The starch was mixed with water in a vial to yield a 50% moisture paste. The vial was then capped and stored at 4°C overnight. After allowing the vial to reach room temperature, the moisture-equilibrated starch was conditioned in hermetic aluminum TA pans, weighed (approximately 10 mg) using a precision balance (± 0.01 mg, Analytical Plus, Mettler Toledo) and heated at a rate of 10°C min^{-1} from 20°C to 120°C in an inert atmosphere (50 mL min^{-1} of dry N_2). The reference was a void aluminum TA pan. The onset temperature (T_o), the peak temperature (T_p), the enthalpy change (ΔH) and the width at half-peak height ($\Delta T_{1/2}$) were computed from the curves using the Universal Analysis Program (Version 1.9 D; TA Instruments). The enthalpies were calculated on a starch dry-weight basis.

Pasting properties of starches

According to AACC 76-21 STD2 [18], the pasting properties of the starches were evaluated using the Micro Visco-Amylo-Graph (Brabender, Germany). The viscosity profiles of the starches were recorded using starch suspensions (9.21%, w/w; 100 g total weight). The heating and cooling cycle program to which the samples were subjected was the following: heating from 30°C to 95°C at a rate of 6°C min^{-1}, holding 95°C for 5 min, cooling from 95°C to 50°C at a rate of 6°C min^{-1}, and holding at 50°C for 2 min. The parameters recorded were the pasting temperature (P_{Temp}), peak viscosity (PV), final viscosity (FV), breakdown viscosity (BV), and setback viscosity (SV). All of the measurements were replicated twice.

Textural properties of starch gels

A starch emulsion (9.21%, w/w) was heated from 30°C to 95°C at a rate of 6°C min^{-1}, maintained at 95°C for 20 min in the Micro Visco-Amylo-Graph, poured into small beakers and stored at 4°C to cause gelation. The textural properties of the starch gels were measured as described by Sandhu and Singh [17]. The gels' textural properties were evaluated through a texture profile analysis (TPA) using the TA-XT2i texture analyzer (Stable MicroSystems, Surrey, England). The gel was compressed at a rate of 0.5 mm s^{-1} to a distance of 10 mm with a cylindrical plunger (diameter=10 mm). The compression was repeated twice to generate a force-time curve from which the hardness (the height of the first peak) and springiness (the ratio between the recovered height after the first compression and the original gel height) were determined. The negative area of the curve during the retraction of the probe was termed the adhesiveness. The cohesiveness was calculated as the ratio of the area under the second peak to the area under the first peak. Five repeated measurements were performed for each sample and their average was used for the analysis.

Statistical analysis

One-way Analysis of Variance (ANOVA), Pearson correlation coefficients (r) for the relationships between all of the properties, and a Principal Component Analysis (PCA) were performed using the PASW Statistics 18.0 software. The comparisons between treatments were evaluated using Duncan's test. The significance level was set to 95% for all of the data analyses.

Results

Thermal properties of starches

The DSC curves of the starches are shown in Figure 1. The transition

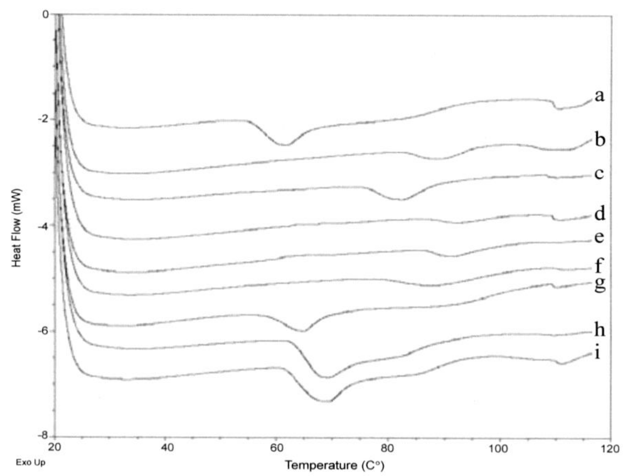

Figure 1: DSC endotherms of the gelatinization of starches: (a) wheat starch; (b) corn starch; (c) potato starch; (d) sweet potato starch; (e) cassava starch; (f) mung bean starch; (g) pea starch; (h) amylose; and (i) amylopectin. The moisture content was 50%, and the heating rate was 10°C min^{-1}.

Variety	Wheat starch	Corn starch	Potato starch	Sweet potato starch	Cassava starch	Mung bean starch	Pea starch	Amylose	Amylopectin
Starch content(%. d.b)	94.52d ± 0.00	99.51a ± 0.00	98.44a ± 0.74	99.23a ± 0.57	93.98d ± 0.02	97.37b ± 0.42	95.75c ± 0.55	98.66a ± 0.63	92.90e ± 0.31

Table 1: Total starch content of various starches.

temperatures (onset, T_o; peak, T_p), the enthalpy change (ΔH) and the width at half-peak height ($\Delta T_{1/2}$) of the gelatinization process are shown in Table 2. A significant difference was observed among the varieties of starches. Wheat starch had the lowest gelatinization temperatures. Sweet potato starch had the highest gelatinization temperatures, indicating that more energy is required to initiate the gelatinization of this starch. The ΔH for the various starches ranged from 0.49 to 5.29J/g. The $\Delta T_{1/2}$ of mung bean starch (12.14°C) was considerably higher than that of the other starches. The corn, sweet potato, and cassava starches showed the highest transition temperatures and the lowest ΔH.

Pasting properties of starches

The pasting properties of the various starches are summarized in Figure 2 and Table 3. A significant difference in the pasting properties was observed among the different starches was observed. The corn and amylopectin starches have a substantial higher pasting temperature than the other starches. The peak viscosities (PV) of the various starches ranged from 405.50 to 2148.00 m Pa•s, and the root and tuber starches exhibited had the highest PV values: 2148.00 m Pa•s for potato, 1239.00 m Pa•s for cassava, 1006.00 m Pa•s for amylose, and 983.00 m Pa•s for sweet potato. The wheat and pea starches presented a low peak viscosity. The final viscosity (FV) of the different starches ranged from 462.00 to 1647.50 m Pa•s, with the lowest values observed in wheat starch and the highest found in potato starch. Pea starch exhibited the lowest breakdown viscosity (BV) whereas potato starch presented the highest value. The setback viscosity (SV) for the various starches differed significantly. Pea starch exhibited the lowest SV (229.00 m Pa•s), whereas potato starch had the highest SV (949.50 m Pa•s). The PV, FV, BV, and SV of amylose were higher than those of amylopectin. The pasting properties of the tuber and root starches were significantly higher than those of the cereal and legume starches.

Gel texture properties of starch gels

The textural properties of starch gels, which were determined

Variety	T_o(°C)	T_p(°C)	ΔH(J/g)	$\Delta T_{1/2}$(°C)
Wheat starch	56.00g ± 0.33	61.25f ± 0.37	3.51bc ± 0.07	6.56b ± 0.30
Corn starch	83.62b ± 0.57	89.12b ± 0.02	1.08fg ± 0.34	7.20b ± 0.40
Potato starch	77.34d ± 1.46	83.16c ± 0.82	2.32de ± 0.80	6.90b ± 0.93
Sweet potato starch	87.81a ± 0.55	92.76a ± 0.19	0.49g ± 0.30	6.98b ± 1.10
Cassava starch	85.78ab ± 1.23	90.62b ± 0.80	1.74ef ± 0.42	7.20b ± 0.01
Mung bean starch	80.00c ± 1.78	89.40b ± 1.82	2.15def ± 0.15	12.14a ± 0.65
Pea starch	58.39f ± 0.75	64.60e ± 0.33	3.09cd ± 0.93	6.50b ± 0.78
Amylose	62.71e ± 0.83	67.93d ± 0.61	4.36ab ± 0.46	7.42b ± 0.19
Amylopectin	62.00e ± 0.05	68.13d ± 0.03	5.29a ± 0.16	8.04b ± 0.15

Table 2: Gelatinization properties of starches.

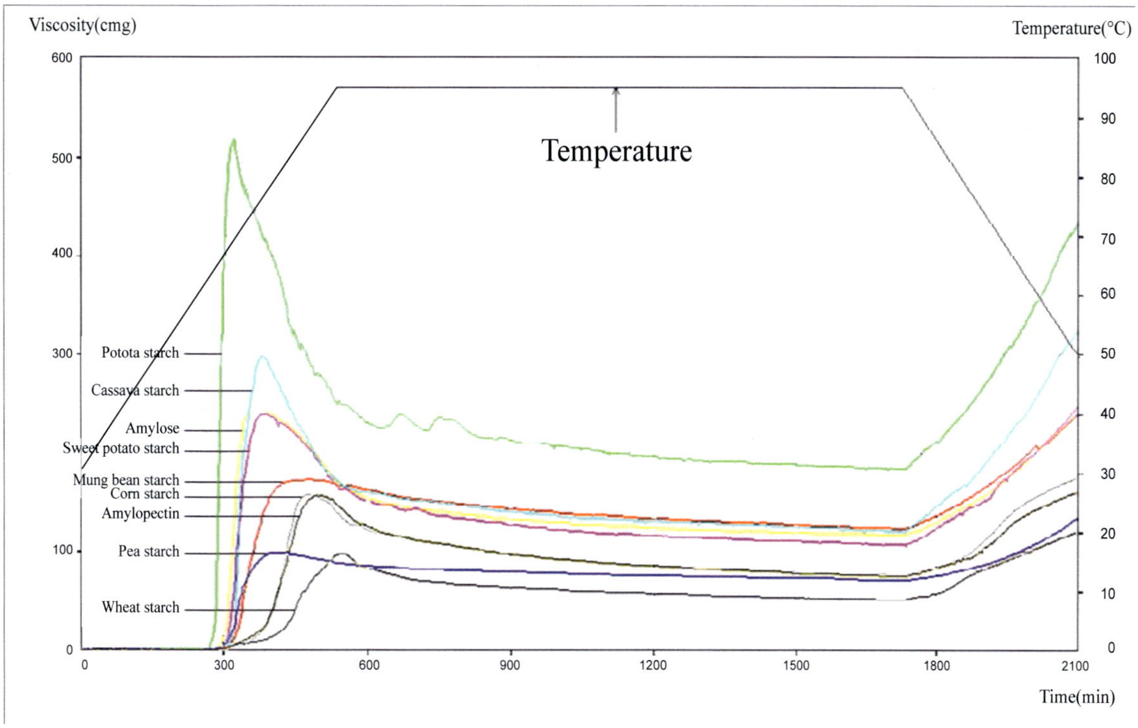

Figure 2: Pasting profiles of the various starches.

using a texture analyzer, are shown in Table 4. As shown, the textural parameters of starches varied significantly. The legume starches showed the highest hardness (701.94 g for pea starch, and 587.31 g for mung bean starch), whereas the cassava starch gel showed the lowest hardness (51.37 g). The cohesiveness of the starches varied from 0.34 to 0.49, and wheat starch, cassava starch and amylose exhibited slightly higher cohesiveness than the other starches. The springiness of the starches varied from 0.73 to 1.00, with mung bean starch having the highest springiness. The adhesiveness of cassava starch and amylose were not detected, whereas that of the other starches varied from -28.72 to -363.91 g•s, with pea starch exhibited the highest value, and tuber and root starches presenting the lowest values.

Pearson correlations among the various properties of the starches

Several significant correlations were observed between the gelatinization, and the pasting and the gel texture properties of the starches (Table 5). Positive interrelationships between the gelatinization parameters were also observed. The T_o was correlated with the T_p and ΔH (r=0.994 and -0.845, respectively, p<0.01), and the T_p was negatively correlated with the ΔH (r=-0.832, p<0.01). The adhesiveness was negatively correlated with the PV, BV, and SV (P<0.05), and the hardness was positively correlated with the springiness (r=0.830, p<0.01).

Principal component analysis of the various properties of the starches

A principal component analysis (PCA) was conducted on four of the tested attributes. The results of the PCA (Figure 3) indicated that 35.64% of the variation in the data can be explained by PC1, 23.54% can be explained by PC2, 18.85% can be explained by PC3, and 9.96% can be explained by PC4. The variables that are found close to each other in pairs or groups indicate a positive correlation. The variables describing the pasting properties, specifically PV, FV, SV, and BV, were correlated with each other (Figure 3A). The loading plot shows that the pasting properties are correlated with the total starch content. The T_o is correlated with the T_p. The hardness is correlated with the springiness (Figure 3B).

The potato starch and cassava starch presented high positive scores in PC1 and were located near the PC1 axis (Figure 3a). The sweet potato starch and mung bean starches had positive scores in the PC2 and were located near the PC2 axis. The potato, pea, and mung bean starches had positive scores in PC3 (Figure 3b).

The comparison of the score and loading plots (Figure 3A and 3a) revealed that the potato starch and cassava starch showed high values of PV, FV, BV, and SV. The sweet potato starch and mung bean starch showed high values of T_o and T_p. The legume starches (mung bean starch and pea starch) and potato starch showed high values of hardness and springiness.

Discussion

The value of T_o shows the internal structure of the granule during its disintegrationes, which results in the release of polysaccharide into the surrounding medium [19]. The heat capacity plays a major role in this process because it directly reflects the molecular structure [20].

Variety	P_{Temp}(°C)	PV(mPa•s)	FV(mPa•s)	BV(mPa•s)	SV(mPa•s)
Wheat starch	66.45d ± 0.07	405.50g ± 7.78	462.00h ± 18.38	198.50g ± 2.12	263.50h ± 12.02
Corn starch	70.85a ± 0.07	660.00f ± 4.24	683.00f ± 4.24	342.00e ± 4.24	394.00f ± 0.00
Potato starch	63.80f ± 0.57	2148.00a ± 8.49	1647.50a ± 7.78	1385.50a ± 10.61	949.50a ± 20.51
Sweet potato starch	66.90cd ± 0.28	983.00d ± 14.14	913.50e ± 14.85	547.00c ± 4.24	531.50c ± 2.12
Cassava starch	68.45b ± 0.07	1239.00b ± 8.49	1165.50b ± 2.12	732.50b ± 3.54	789.00b ± 5.66
Mung bean starch	67.55bc ± 0.21	716.50e ± 2.12	1012.00c ± 31.11	217.50f ± 16.26	434.50e ± 12.02
Pea starch	67.40c ± 0.00	411.00g ± 0.00	687.50f ± 14.85	117.50h ± 2.12	229.00i ± 8.49
Amylose	65.05e ± 0.07	1006.00c ± 5.66	962.50d ± 26.16	508.00d ± 9.90	492.00d ± 32.53
Amylopectin	70.20a ± 0.99	661.50f ± 10.61	629.50g ± 6.36	347.50e ± 3.54	342.00g ± 4.24

P_{Temp}, pasting temperature; PV, peak viscosity; FV, final viscosity; BV, breakdown viscosity; SV, setback viscosity

Table 3: Pasting properties of the different starches.

Variety	Hardness(g)	Cohesiveness	Springiness	adhesiveness (g•s)
Wheat starch	95.30e ± 4.50	0.46a ± 0.01	0.86cd ± 0.02	-304.97ab ± 29.89
Corn starch	152.88cd ± 17.00	0.35b ± 0.04	0.88e ± 0.03	-212.90bc ± 88.50
Potato starch	546.84b ± 88.28	0.38b ± 0.05	0.99ab ± 0.02	-28.72d ± 19.64
Sweet potato starch	103.35de ± 9.51	0.34b ± 0.03	0.73f ± 0.06	-45.30d ± 15.89
Cassava starch	51.37e ± 5.71	0.49a ± 0.03	0.79ef ± 0.03	nd[a]
Mung bean starch	587.31b ± 55.25	0.36b ± 0.01	1.00a ± 0.05	-230.79abc ± 187.18
Pea starch	701.94a ± 33.10	0.38b ± 0.03	0.92bc ± 0.12	-363.91a ± 182.99
Amylose	61.94e ± 9.30	0.49a ± 0.02	0.79ef ± 0.02	nd
Amylopectin	168.28c ± 34.78	0.38b ± 0.09	0.81de ± 0.03	-93.52cd ± 43.86

[a]nd stands for not detected

Table 4: Textural properties of starch gels.

	Starch content	T_o	T_p	ΔH	$\Delta T_{1/2}$	P_{Temp}	PV	FV	BV	SV	hardness	cohesiveness	springiness
T_o	0.466												
T_p	0.464	0.994**											
ΔH	-0.564	-0.845**	-0.832**										
$\Delta T_{1/2}$	0.044	0.232	0.329	-0.011									
P_{Temp}	-0.309	0.165	0.170	-0.058	0.139								
PV	0.311	0.443	0.421	-0.228	-0.117	-0.565							
FV	0.340	0.506	0.509	-0.299	0.104	-0.571	0.948**						
BV	0.261	0.409	0.374	-0.215	-0.238	-0.532	0.988**	0.894**					
SV	0.231	0.589	0.563	-0.349	-0.067	-0.443	0.959**	0.936**	0.945**				
Hardness	0.095	-0.120	-0.047	-0.029	0.327	-0.242	0.093	0.281	0.021	-0.020			
cohesiveness	-0.387	-0.339	-0.380	0.400	-0.256	-0.285	0.054	0.019	0.072	0.149	-0.468		
springiness	0.126	-0.069	0.007	-0.022	0.437	-0.216	0.179	0.319	0.119	0.086	0.830**	-0.321	
adhesiveness	-0.166	-0.453	-0.408	0.051	-0.084	0.244	-0.709*	-0.620	-0.702*	-0.749*	0.479	-0.276	0.458

*Correlation is significant (P<0.05)
**Correlation is significant (p<0.01)

Table 5: Pearson correlation coefficients between the various properties of the starches.

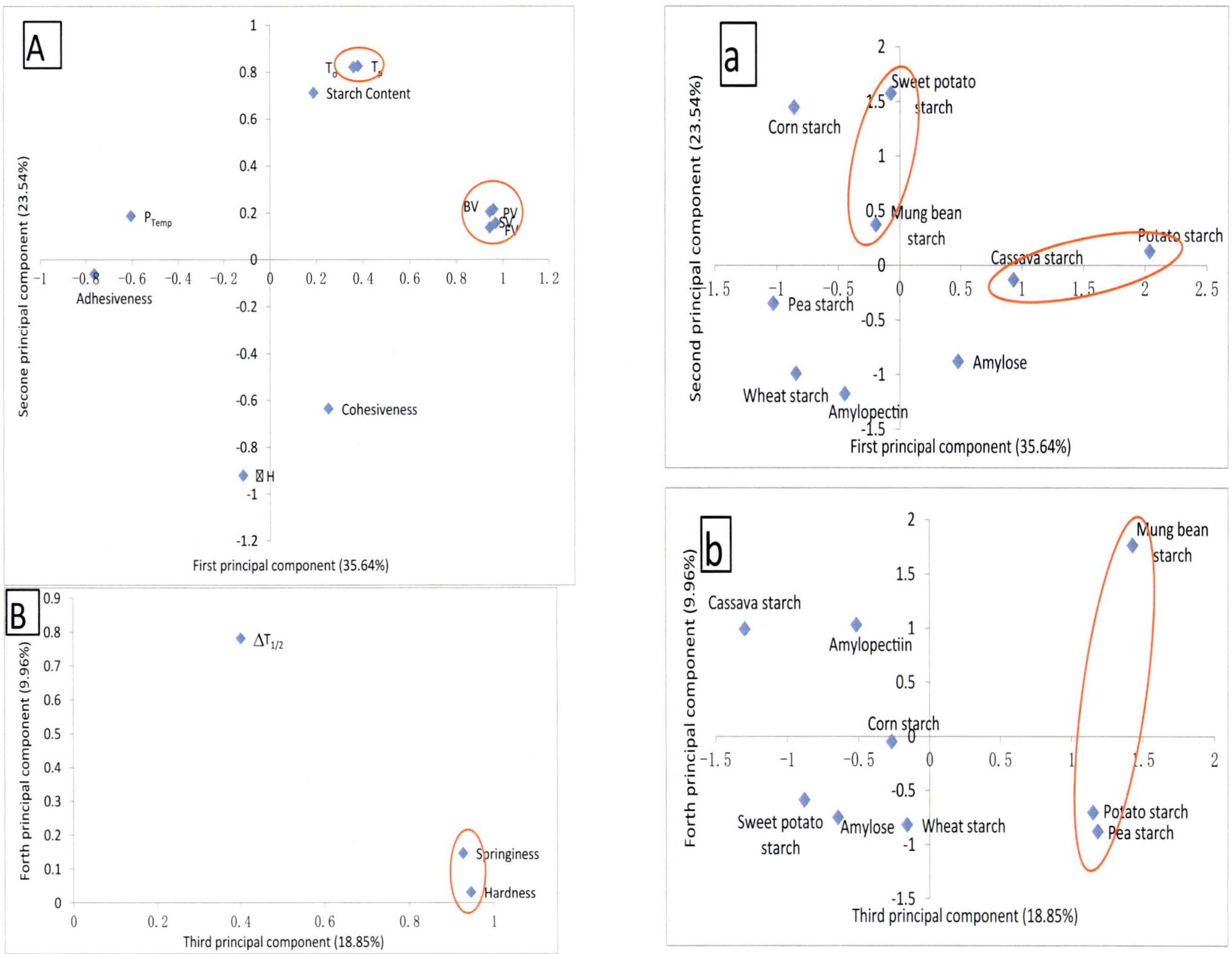

Figure 3: (A) Loading plot and (a) score plot of the principal components 1 and 2. (B) Loading plot and (b) score plot of the principal components 3 and 4, describing the variation in the physicochemical properties of the nine starch samples.

Higher enthalpies and temperatures indicate a higher degree of granule crystallinity [21]. According to our results, T_o and T_p are negatively correlated with the ΔH (n=9, r=-0.845 and -0.832, p<0.01).

Singh et al. [13] studied the thermal and pasting properties of 13 different black gram cultivars. The T_o varied from 66.1 to 71.3 °C (the moisture content was 70%); the PV varied from 422 to 514 RVU; and the BV varied from 134 to 212 RVU. The PV, and BV were negatively correlated with the T_o (n=13, r=-0.794 and -0.835, p<0.01). The P_{Temp} was positively correlated with the T_o (n=13, r=0.961, p<0.01). During the pasting process, the starch granules first swell; and this swelling is followed by the melting of crystals and then the paste becoming viscous.

The variations in ΔH may represent differences in the bonding forces between the double helices that form the amylopectin crystallites, which alter the alignment of hydrogen bonds within the starch molecules [17]. According to previous results, the PV, FV, BV, and SV of starch emulsions are slightly negatively correlated with the ΔH (n=9; r=-0.743, -0.623, -0.733, and -0.611, respectively; P<0.05), and the P_{Temp} and the hardness of the starch gel are positively correlated with ΔH (r=0.620, p<0.05; r=0.44, p<0.01) [22]. Bao et al. [23] observed a significant positive correlation between ΔH and cohesiveness (n=127, r=0.234, P<0.05) in rice starch.

Conclusion

In the present study, variations in the thermal, pasting and gel texture properties were observed between different varieties of starches. Wheat starch showed the lowest transition temperatures (the onset temperature, T_o and the peak temperature, T_p). The root starches (sweet potato starch, cassava starch) showed the highest transition temperatures and the lowest enthalpy change values. Mung bean starch showed the slowest transition speed. The tuber and root starches showed higher viscosities (peak viscosity, final viscosity, breakdown viscosity and setback viscosity) and lower adhesiveness. The legume starches showed the highest hardness values. The properties of amylose and amylopectin largely depended on their source.

The peak viscosity, final viscosity, breakdown viscosity and setback viscosity were positively correlated with each other and were positively correlated with the starch content. The onset temperature was negatively correlated with the enthalpy change. The pasting temperature was positively correlated with onset temperature and the final viscosity, setback viscosity, and pasting temperature were negatively correlated with the enthalpy change.

Acknowledgment

The authors acknowledge the Special Fund for Agro-scientific Research in the Public Interest (Grant No. 201303071) and the National Natural Science Foundation of China (31101377).

References

1. Sandhu KS, Singh N, Kaur M (2004) Characteristics of the different corn types and their grain fractions: physicochemical, thermal, morphological, and rheological properties of starches. J Food Eng 64: 119-127.
2. Wootton M, Mahdar D (1993) Properties of starches from Australian wheats. Part 2: Some Physico-chemical Properties. Starch 45: 295-299.
3. Yun SH, Matheson NK (1993) Structures of the amylopectins of waxy, normal, amylose-extender, and wx:ae genotypes and of the phytoglycogen of maize. Carbohyd Res 243: 307-321.
4. Alvani K, Qi X, Tester RF, Snape CE (2011) Physico-chemical properties of potato starches. Food Chem 125: 958-965.
5. Collado LS (1997) Physical properties and utilization of sweet potato starch and flour. PhD thesis. University of Hong Kong.
6. Asaoka M, Blanshard JMV, Rickard JE (1991) Effects on the Physico□chemical Properties of Starch from Four Cultivars of Cassava. Starch 43: 455-459.
7. Hoover R, Li YX, Hynes G, Senanayake N (1997) Physicochemical characterization of mung bean starch. Food Hydrocolloid 11: 401-408.
8. Singh N, Singh J, Kaur L, Sodhi NS, Gill BS (2003) Morphological, thermal and rheological properties of starches from different botanical sources. Food Chem 81: 219-231.
9. Sánchez T, Dufour D, Moreno IX, Ceballos H (2010) Comparison of pasting and gel stabilities of waxy and normal starches from potato, maize, and rice with those of a novel waxy cassava starch under thermal, chemical, and mechanical stress. J Agr Food Chem 58:5093-5099.
10. Hoover R, Vasanthan T (1994) Effect of heat-moisture treatment on the structure and physicochemical properties of cereal, legume, and tuber starches. Carbohyd Res 25: 33-53.
11. Jacobs H, Delcour JA (1998) Hydrothermal modifications of granular starch, with retention of the granular structure: A review. J Agr Food Chem 46: 2895-2905.
12. Ratnayake WS, Hoover R, Warkentin T (2002) Pea starch: Composition, structure and properties: A review. Starch 54: 217-234.
13. Singh N, Kaur M, Sandhu KS, Guraya HS (2004) Physicochemical, thermal, morphological and pasting properties of starches from some Indian black gram (Phaseolus mungo L.) cultivars. Starch 56: 535-544.
14. Freitas RA, Paula RC, FeitosaJPA, Rocha S, Sierakowski MR (2004) Amylose contents, rheological properties and gelatinization kinetics of yam (Dioscorea alata) and cassava (Manihot utilissima) starches. Carbohyd Polym 55: 3-8.
15. Li-Chan ECY, Ma CY (2002) Thermal analysis of flaxseed (Linum usitatissimum) proteins by differential scanning calorimetry. Food Chem 77: 495-502.
16. Doublier JL, Llamas G, Le Meur M (1987) A rheological investigation of cereal starch pastes and gels. Effect of pasting procedures. Carbohyd Polym 7: 251-275.
17. Sandhu KS, Singh N (2007) Some properties of corn starches II: Physicochemical, gelatinization, retrogradation, pasting and gel textural properties. Food Chem 10: 1499-1507.
18. AACCI (1995) General pasting method for wheat or rye flour using the rapid visco analyser. 11th Edtn, AACC International Approved Methods.
19. Singh N, Kaur L, Sandhu KS, Kaur J, Nishinari K (2006) Relationships between physicochemical, morphological, thermal, rheological properties of rice starches. Food Hydrocolloid 20: 532-542.
20. Louaer W,Meniai AH, Grolier JPE (2008) Thermal analysis of the influence of water content on glass transitions. J Therm Anal Calorim 93: 605-610.
21. Svegmark K, Helmersson K, Nilsson G, Nilsson PO, Andersson R, et al. (2002) Comparison of potato amylopectin starches and potato starches—influence of year and variety. Carbohyd Polym 47: 331-340.
22. Singh SK, Johnson LA, White PJ, JaneJL, Pollak LM (2001) Thermal properties and paste and gel behaviors of starches recovered from accessions used in the germplasm enhancement of maize project. Cereal Chem 78: 315-321.
23. Bao J, Sun M,Zhu L, Corke HJ (2004) Analysis of quantitative trait loci for some starch properties of rice (Oryza sativa L.): thermal properties, gel texture and swelling volume. Cereal Sci 39: 379-385.

Solar Light Induced Photo Catalytic Disinfection of Gram Positive and Negative Microorganisms from Water with Highly Efficient $AuTiO_2$ Nanoparticle

Sandhya S[1*], Sadhana Rayalu[2] and Benedict Bruno[1]

[1]*CSIR-National Environmental Engineering Research Institute, CSIR-Complex, Chennai, India*

[2]*CSIR-National Environmental Engineering Research Institute, Nagpur, India*

Abstract

$AuTiO_2$ nanomaterial was used for solar photocatalytic disinfection of gram negative *Escherichia coli* and gram positive *Staphylococci aureus*. In this study, time of disinfection, light intensity, concentration of nanomaterial and bacterial concentration effects has been investigated. Prepared nanomaterial characterized using Scanning Electron Microscopy (SEM), X-Ray Diffractometry (XRD). $AuTiO_2$ samples have shown antibacterial effect when irradiated in solar light. The bactericidal effect may be due to Au ions and photocatalysis of TiO_2. The photocatalytic inactivation rate constant, k for each dose were obtained from their respective inactivation curve over a 2 hours incubation period.

Keywords: $AuTiO_2$; Disinfection; *E. coli*; *S. aureus*; Solar; Photocatalyst; Antibacterial

Introduction

Contaminated water causes an estimated 6 to 60 billion cases of gastrointestinal illness annually and the majority of them are from rural areas of developing countries [1]. This is due to intermittent water supply, and water is polluted with variety of microorganisms. There is an urgent need of low cost, low maintenance and effective mode of disinfection for portable water. A safe affordable, robust and sustainable method for water purification is in great demand throughout the world [2]. Water disinfection means the removal, deactivation or killing of pathogenic microorganism including gram positive and negative bacteria, virus, fungi etc. Microorganisms are destroyed or deactivated, resulting in termination of growth and reproduction. Conventional method of water disinfection is by strong oxidant such as free chlorine [3] or UV/chlorine or ozone/chlorine combination [4]. However, virus is resistant to both UV and combined chlorine disinfection. The recent approach based on active photocatalysis by using $AuTiO_2$ suspension and targeting bacteria and virus from water.

The use of solar irradiation for treatment of contaminated water is an age old technology [5]. The bacterial inactivation rate in contaminated water is proportional to intensity of sunlight, temperature and inversely proportional to water depth [6]. The rapid growth of nanotechnology has provided many types of material for water decontamination [6]. The antibacterial mechanism of Ag, TiO_2, ZnO [7,8] and metals like platinum [9], gold [10], silver [7], Zn [8] in nano form are known for their resonance under illumination. Metal doping has been used to enhance TiO_2, photocatalysis by tapping excited electrons to prevent charge recombination as electron tapping can occur it the doping has a lower level than excited electrons [11]. Matsunga et al. first time reported photocatalytic inactivation of microorganisms. They inactivated *Lactobacillus acidophilus*, *Saccharomyces servisiae* and *E. coli* using TiO_2 particle suspension [12]. Wei et al. used P25 (TiO_2 Degussa) to inactivate *E. coli* in water [13]. Gold nanoparticle may become useful in the development of antimicrobial strategies because of their nontoxicity, versatility in surface modification polyvalent effect and photo thermal effect [14].

The comprehensive studies on gold nanomaterial have rarely been carried out in bacteria. To understand the interaction of doped $AuTiO_2$ in presence of sunlight, investigation was carried out for antibacterial effects on model gram negative *Escherichia coli* and gram positive *Streptococcus aureus.*

Materials and Methods

Preparation and characterization of nanoparticles

Pre weighed quantity of Au and TiO_2 (60 mg Au and 2960 mg of TiO_2) were loaded in a crucible and placed at the bottom of reactor. The reactor was evacuated in presence of liquid nitrogen and heated resistively to achieve condensation of Au atom with solvent molecules. After completion of reaction the reactor was brought to the room temperature under Ar atmosphere. On cooling Au solvent colloid comes in contact with TiO_2, this mixture was stirred for 2 hrs in Ar atmosphere, recovered and vacuum dried [15].

Characterization of $AuTiO_2$ nanoparticles

The crystal structure of the nanoparticle was analyzed by X-ray diffraction (XRD) using a D/MAX-2004 X-ray powder diffractometer (Rigaku Corporation, Tokyo, Japan) with Cu target (0.1540598 nm) radiation at 40 kV and 181 mA. Scanning range was varied between 10° and 90° at a scanning rate of 2°/min. The morphology of these nanoparticles was examined by both Field Emission Scanning Electron Microscopy (FESEM) and Energy Dispersive X-ray spectroscopy (EDAX). FESEM/EDAX (JEOL Ltd., Tokyo, Japan) samples were prepared by dispersing in a thin film of the photocatalyst on a grid that was operated at 15.0 KV. EDAX was performed to the ratio of the nanoparticles present.

***Corresponding author:** Sandhya S, CSIR-National Environmental Engineering Research Institute, CSIR-Complex, Chennai, India

Scanning electron microscopy

SEM observation was made with a SEM LEO SUPRA 35 (Carl ZEISS Inc., Oberkochen, Germany) at an acceleration voltage of 20 kV. Before imaging, the sample was sputtered with gold for 90 s (Cressington Sputter Coater 208HR, Cressington Scientific Instrument Ltd., Watford, U.K.). The pellet was fixed in 2.5% glutaraldehyde for 2 h in a refrigerator and was washed with phosphate buffer and dehydrated with ethanol. Then these cells were placed on the grids which was sputtered with gold for 110s (Cressington Sputter Coater 208 HR, Cressington Scientific Instrument Ltd.) and the image was captured using the SEM LEO SUPRA 35.

Determination of bactericidal activity

E. coli and *Staphylococcus aureus* were used as a model bacterium to evaluate antibacterial activity of $AuTiO_2$ nanoparticles prepared in the study. To evaluate the bactericidal effect of $AuTiO_2$ nanoparticles, *E. coli* and *S. aureus* were first precultured in flasks containing nutrient broth at 37°C to a concentration of 10^8 CFU/ml. 1 ml of the cell suspension was added in 99 ml of sterile Phosphate Buffer Saline (PBS) to get 10^5 CFU/ml. Prepared $AuTiO_2$ nanoparticles (10 ug/ml and 20 ug/ml) were added and exposed to direct sun light. The samples were collected at equal intervals and were plated on Mac Conkey agar plates to know the count of surviving *E. coli* and Nutrient agar for *S. aureus*. The dead cells were further subjected to SEM to know their morphological changes.

Results

Characterization of $AuTiO_2$

XRD pattern of $AuTiO_2$ nanocomposite synthesized is shown in figure 1. The calcined samples were examined by powder XRD analysis and the spectra suggest high degree of crystallinity in each case. The Au peak can be indexed at 2© value 38.2, 48, 64.7 peaks. This clearly indicates that the structural integrity of titanium has been retained.

The SEM micrograph of the prepared catalytic material was obtained to investigate the powder morphology and microstructures of the $AuTiO_2$ (Figure 2). The particle size of the developed material in the photograph shows a homogenous distribution of the crystalline material although not fully resolved. The image highly dispersed nanoparticles in the system. EDAX spectra Figures 2 B and 2C confirms the presence of Ti with kα line at 4.5 keV, V with Lα line at 0.5 keV and Au with Lα line 2.1 kev.

Effects of cell and $AuTiO_2$ concentrations on disinfection

The nanoparticle was successfully tested for artificially contaminated water with *E. coli* and *S. aureus*. The average irradiation intensity ranged from 500-888 Wm^{-2} with exposure up to 120 min. The final temperature during the exposure was always more than 36°C. Bacterial concentrations in the treated samples were determined by plating. *E. coli* and *S. aureus* cells were exposed to $AuTiO_2$ for 2 hrs in solar light to assess any bactericidal effect of the photocatalyst. At higher initial cell concentrations, the colony count decreased with time as shown in figure 3A (10 ug/ml $AuTiO_2$) and B (20 ug/ml $AuTiO_2$) for *E. coli* and figure 4 for *S. aureus* respectively. A typical death curve was observed for bacteria in solar light over time, thus it appeared that dyeing effect of the light irradiation was causing decrease in cell viability. Within the first 15 mins of irradiation in solar light, 91.7%-95.7% of *E. coli* and 87.7%-96.9% of *S. aureus* were viable, while the number of viable bacteria decrease progressively on further exposure.

In order to study the disinfection mechanism, a high concentration of cells is required to expose the nanomaterial resulting from $AuTiO_2$

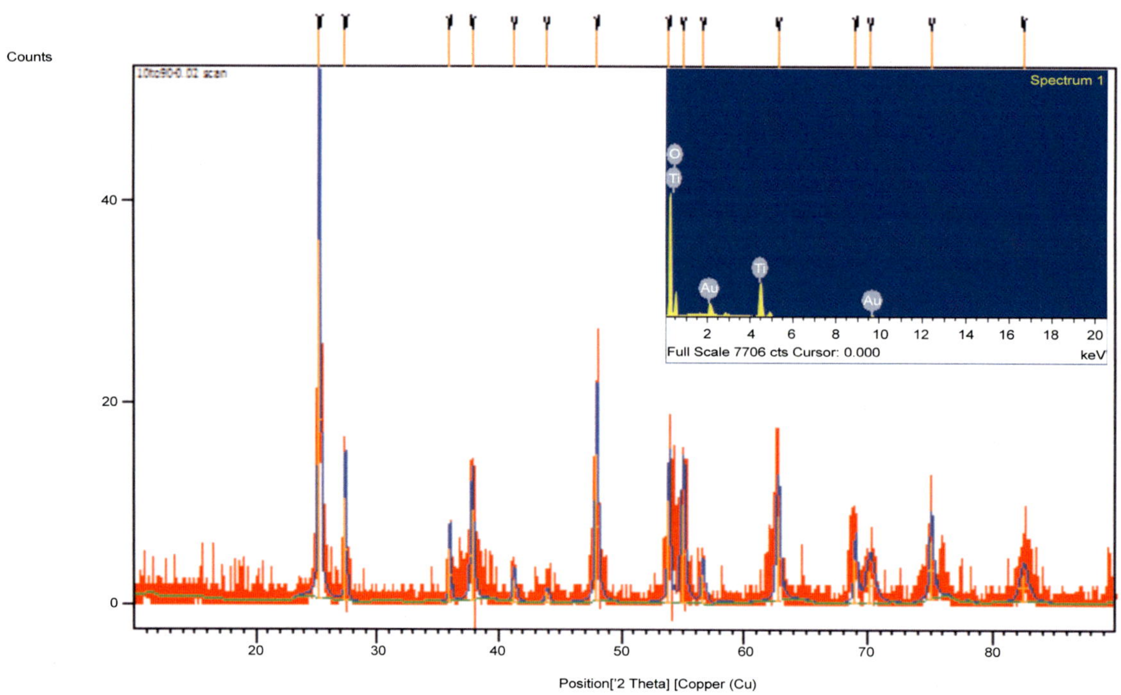

Figure 1: XRD Pattern of the $AuTiO_2$ nanoparticle.

Figure 2: EDAX and FEMS image explaining the morphology of the $AuTiO_2$ nanoparticle.

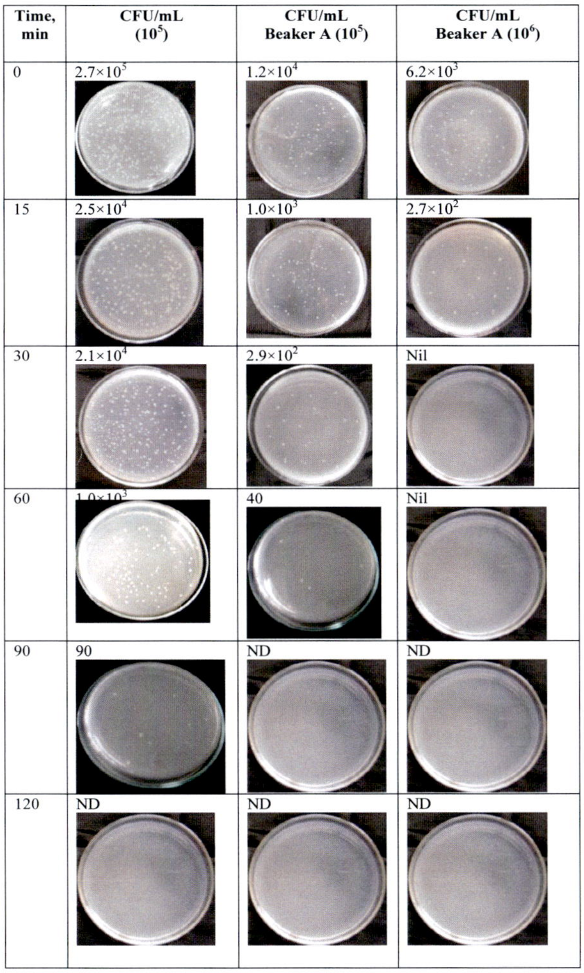

Time, min	CFU/mL (10^5)	CFU/mL Beaker A (10^5)	CFU/mL Beaker A (10^6)
0	2.7×10^5	1.2×10^4	6.2×10^3
15	2.5×10^4	1.0×10^3	2.7×10^2
30	2.1×10^4	2.9×10^2	Nil
60	1.0×10^3	40	Nil
90	90	ND	ND
120	ND	ND	ND

Figure 3A: *E.coli* disinfection in presence of $AuTiO_2$ (10 ug/ml).

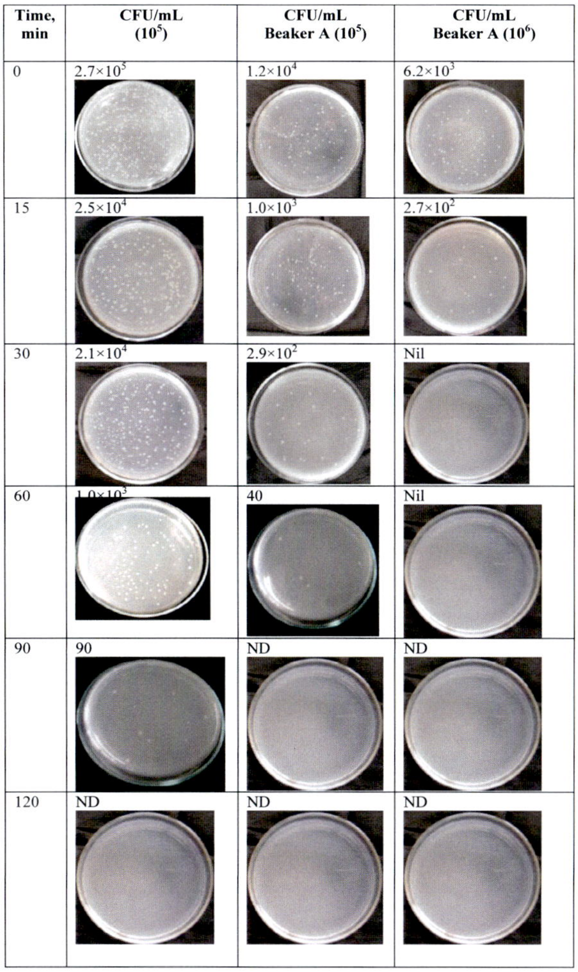

Time, min	CFU/mL (10^5)	CFU/mL Beaker A (10^5)	CFU/mL Beaker A (10^6)
0	2.7×10^5	1.2×10^4	6.2×10^3
15	2.5×10^4	1.0×10^3	2.7×10^2
30	2.1×10^4	2.9×10^2	Nil
60	1.0×10^3	40	Nil
90	90	ND	ND
120	ND	ND	ND

Figure 3B: *E.coli* disinfection in presence of $AuTiO_2$ (20 ug/ml).

biocidal action. To determine the optimal dose of $AuTiO_2$ for a certain cell concentration, a solar photocatalytic reaction were carried out with the cell concentration raging from 6.6×10^3 to 2.8×10^5 CFU/ml and $AuTiO_2$ concentrations ranging from 10 to 20 ug/ml (Table 1 and Figure 5). After 60 min of irradiation in solar light in presence of $AuTiO_2$ 10 ug/ml, 97.8 to 99.2 % of *E. coli* was killed, when the initial cell concentration was less than 10^5 CFU/ml. When the $AuTiO_2$ dose was increased to 20 ug/ml, the disinfection or killing rate was 99.6 to 99.99%. This shows significant improvement in the disinfection efficiency at higher dose. Therefore the most effective $AuTiO_2$ concentration for disinfection of *E. coli* was 20 ug/ml at initial cell concentration ranging from 10^3 to 10^5 CFU/ml. The same concentration of $AuTiO_2$ (20 ug/ml) when tested against gram positive *S.aureus* showed 99.5 to 99.99 disinfection efficiency.

Results of photocatalytic inactivation of *E. coli* and *S. aureus* using $AuTiO_2$ are shown in figures 6 and 7 respectively. The figures suggest that bacterial growth were less inhibited in presence of 0 and 10 ug/ml of $AuTiO_2$. It was observed that bacteria were inhibited by only TiO_2 particles as it generates OH radical (Data not shown). On the other hand, gold has low toxicity to biological systems or animals or human, due to elemental properties [14]. In both the cases, the concentration of bacteria seems to exponentially decrease with respect to inactivation time. This indicates that inactivation of both the bacteria follows Chick's Law, which says that the inactivation rate of bacteria can be expressed as a first ordered reaction with respect to the concentration of viable cell. According to Chick's Law the inactivation can be given as follows,

Time, min	CFU/mL (10^5)	CFU/mL (10^4)	CFU/mL (10^3)
0	2.5×10^5 	1.3×10^4 	5.9×10^3 
15	2.7×10^4 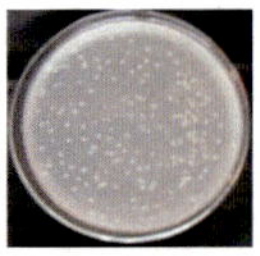	1.6×10^3 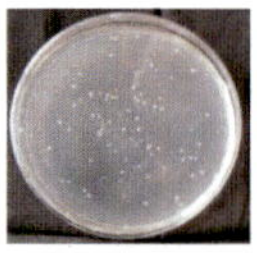	2.3×10^2 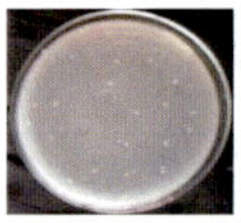
30	2.2×10^4 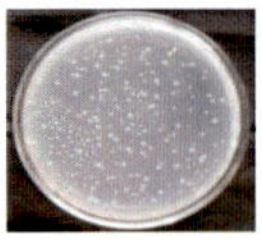	3.2×10^2 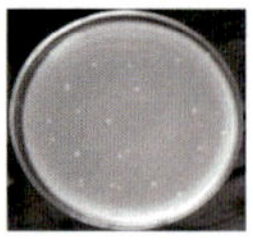	Nil 
60	1.3×10^3 	40 	Nil 
90	78 	Nil 	Nil 
120	Nil 	Nil 	Nil 

Figure 4: *Staphylococci aureus* disinfection in presence of $AuTiO_2$ (20 ug/ml).

Microorganisms	Cell conc., CFU/ml	$AuTiO_2$ 10 ug/l	Survival ratio, %*	$AuTiO_2$ 20 ug/ml	Survival ratio, %*
	2.8×10^5	2.3×10^3	0.82	-	-
	1.0×10^4	2.2×10^2	2.2	-	-
	6.6×10^3	45	0.68	-	-
E.coli	2.7×10^5	-	-	1.1×10^3	0.40
	1.2×10^4	-	-	90	0.75
	6.2×10^3	-	-	ND	0.016
	2.5×10^5	-	-	1.3×10^3	0.52
S.aureus	1.3×10^4	-	-	40	0.31
	5.9×10^3	-	-	ND	0.017

ND: Not Detected

*Ratio of cell concentration after 60 min in solar light to the corresponding initial concentration

Table 1: Effect of Escherichia coli, Staphylococcus aureus and $AuTiO_2$ concentrations on its disinfection.

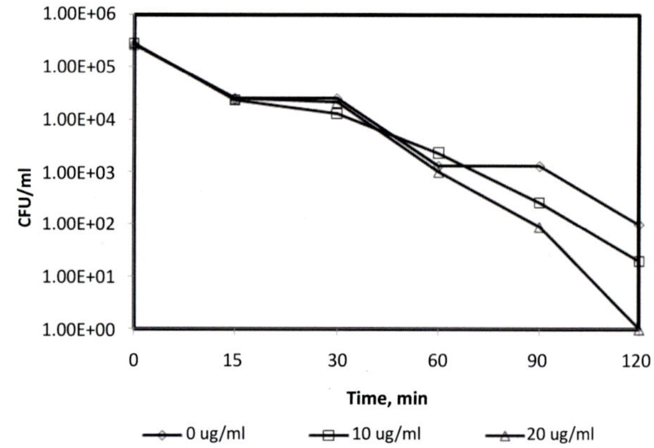

Figure 5: Survival ratio of *E.coli* with $AuTiO_2$ photocatlyst at different concentrations samples exposed to solar light.

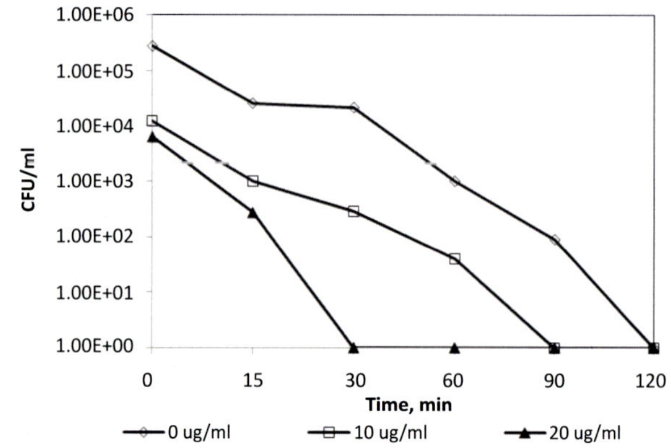

Figure 6: Survival ratio of *E.coli* at various concentration of $AuTiO_2$.

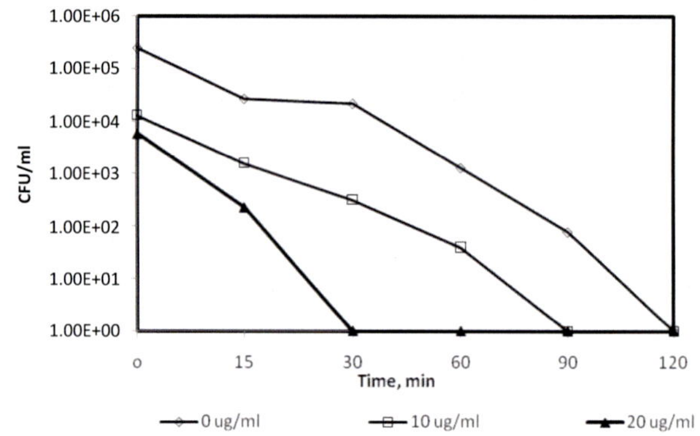

Figure 7: Survival ratio of *S. aureus* at various concentration $AuTiO_2$.

$$-kct = \frac{dct}{dt}$$

Where, ct is the active cell concentration at time t, From the equation the inactivation rate constant can be obtained as

$$k = \frac{\ln\frac{Ct}{Co}}{T}$$

Since the initial concentration was fixed as 10^5 cells/ml, the inactivation rate constant can represent the inactivation efficiency of photocatalyst. The k value for each concentration was obtained liner slop of a log scale plot of photocatalyst. The activation rate constant k for each case is given in table 2. Values of k show comparable disinfection with that of Herrera Melian et al. They required 120 min for two log inactivation of total coliforms [16]. In every experiment sample temperature was always monitored and values are higher than 36°C and therefore disinfection caused by temperature was minimum. Evidence of SEM image (Figure 8) showed that the $AuTiO_2$ nanoparticles could adhere to the outer cell of bacteria in a dark media. Though the outer membrane of the cell was not damaged, the bacteria cannot nourish from aqueous media.

Proposed mechanism of photocatalytic disinfection of microorganisms using AuTiO2 nanoparticles

The role of complex photo oxidants, such as hydroxyl radical (OH) the superoxide radical (O_2) and hydrogen peroxide (H_2O_2) etc still not fully understood. Figure 9 reports the probable disinfection mechanism using nanomaterial $AuTiO_2$. It is frequently assumed that the hydroxyl radical is a major factor responsible for the antimicrobial activity observed in $AuTiO_2$ photocatalytical reaction [17]. On the other hands Maners et al. reported that other Reactive Oxygen Species (ROS) (H_2O_2 and O_2^- etc) as well as the hydroxyl radical play significant role in microorganism's inactivation [17]. The gold nanoparticles exert their antibacterial action mainly inhibiting ATP synthesis with decline in metabolism or inhibit subunit of ribosome from the rRNA binding [10].

Discussion

The result on viability study confirmed the previous finding of Cui et al. [10] and Zhou et al.[14] that $AuTiO_2$ in solar light exhibits bactericidal activity and that the disinfection is positively correlated with the $AuTiO_2$ dose (20 ug/ml). The survival ration in table 1 compared the levels of viability at different dose of $AuTiO_2$ in solar light. Experiments were conducted using *E. coil* DH5α and *S. aureus* and selective media was used for bacterial detection during photo treatment. This point is of great importance when the purpose of treatment is to get portable water.

Bacteria	$AuTiO_2$ k/min		
	0 ug/ml	10 ug/ml	20 ug/ml
Escherichia coli	0.082	0.049	0.042
Staphylococcus aureus	0.054	0.051	0.041

Table 2: Inactivation rate constant of photocatalytic inactivation of Escherichia coli, Staphylococcus aureus in solar irradiation.

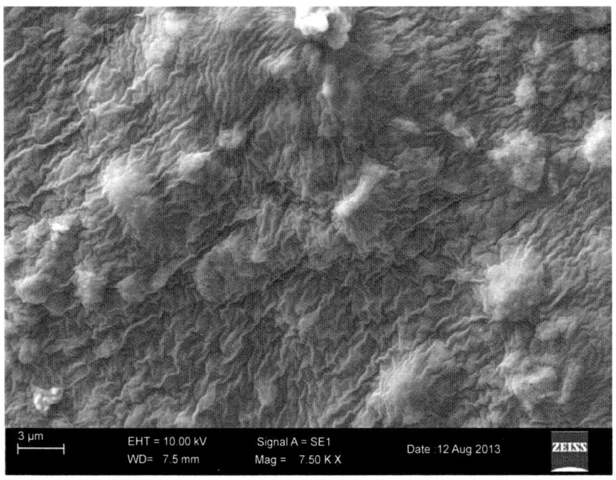

Figure 8: SEM image of the bacterial cells after treatment with $AuTiO_2$ nanoparticle in solar light.

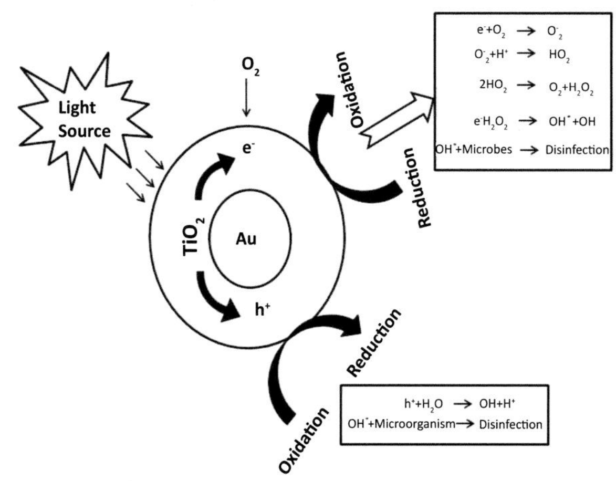

Figure 9: Proposed mechanism of photocatalytic disinfection of microorganisms using $AuTiO_2$ nanoparticles.

The inactivation effect of light in presence of photocatalyst depends on the original bacterial growth state. Always sixteen hours grown cells were used in the study. The deactivation rate increases with decrease in initial concentrations of bacteria. The deactivation rate constant was lower for *E. coli* and *S. aureus* in presence of $AuTiO_2$ (20 ug/ml) than in absence of the same.

Based on photocatalytic disinfection mechanism of nanomaterial ($AuTiO_2$) the ROS such as OH^-, O_2^- and H_2O_2 generated on the irradiated $AuTiO_2$ surface, attack either ATP synthesis or the rRNA binding in bacteria. The disinfection mechanism reported here may be outside the cell is very likely the initial mode of killing. However, we cannot rule out the possibility of photocatalytic attack inside the bacterial cells.

Acknowledgement

Authors are thankful to Ms Sumana Biswas for some study. Authors are thankful to Director, NEERI, Nagpur, India and the CSIR- SIP for funding the project.

References

1. Caslake LF, Connolly DJ, Menon V, Duncanson CM, Rojas R, et al. (2004) Disinfection of contaminated water by using solar irradiation. Appl Environ Microbiol 70: 1145-1150.

2. Weber WJ Jr (2002) Distributed optimal technology networks: a concept and strategy for potable water sustainability. Water Sci Technol 46: 241-246.

3. Szewzyk U, Szewzyk R, Manz W, Schleifer KH (2000) Microbiological safety of drinking water. Annu Rev Microbiol 54: 81-127.

4. Zhang D, Li G, Yu JC (2010) Inorganic material for photocatalytic water disinfection.J Mater Chem 20: 4529-4536.

5. Acra A, Jurdi M, Muállem H, Karahagopian Y, Raffoul Z (1996) Water disinfection by solar irradiation:Assessment& Application.Intemational Development Research Centre, Canada.

6. Yacoby I, Benhar I (2008) Antibacterial nanomedicine. Nanomedicine (Lond) 3: 329-341.

7. Li Q, Mahendra S, Lyon DY, Brunet L, Liga MV, et al. (2008) Antimicrobial

nanomaterials for water disinfection and microbial control: potential applications and implications. Water Res 42: 4591-4602.

8. Huang Z, Zheng X, Yan D, Yin G, Liao X, et al. (2008) Toxicological effect of ZnO nanoparticles based on bacteria. Langmuir 24: 4140-4144.

9. Ahmadi TS, Wang ZL, Green TC, Henglein A, El-Sayed MA (1996) Shape-Controlled Synthesis of Colloidal Platinum Nanoparticles Science 272: 1924-1926.

10. Cui Y, Zhao Y, Tian Y, Zhang W, Lü X, et al. (2012) The molecular mechanism of action of bactericidal gold nanoparticles on Escherichia coli. Biomaterials 33: 2327-2333.

11. Iliev V, TomovaD, BilyarskaL, Eliyas A, Petrov L(2006) Photocatalytic properties of TiO_2 modified with platinum and silver nanoparticles in the degradation of oxalic acid in aqueous solution.Applied Catalysis B: Environmental 63: 266-271.

12. Matsunga T, Tomoda R, Nakajima T, Wake H (1985) Photoelectrochemical sterilization of microbial cells by semiconductor powders.FEMS Microbiology Letters 29: 211-214.

13. Wei C, Lin WY, Zainal Z, Williams NE, Zhu K, et al. (1994) Bactericidal Activity of TiO_2Photocatalyst in Aqueous Media: Toward a Solar-Assisted Water Disinfection System. Environ Sci Technol 28: 934-938.

14. Zhou Y, Kong Y, Kundu S, Cirillo JD, Liang H (2012) Antibacterial activities of gold and silver nanoparticles against *Escherichia coli* and bacillus Calmette-Guérin. J Nanobiotechnology 10: 19.

15. Rayalu SS, Jose D, Joshi MV, Mangrulkar PA, Shrestha K, et al. (2013) Photocatalytic water splitting on Au/TiO_2nanocomposites synthesized through various routes: Enhancement in photocatalytic activity due to SPR effect. Applied Catalysis B: Environmental 142-143: 684-693.

16. Herrera Melián JA, Doña Rodríguez JM, VieraSuárez A, TelloRendón E, Valdés do Campo C, et al. (2000) The photocatalytic disinfection of urban waste waters. Chemosphere 41: 323-327.

17. Maness PC, Smolinski S, Blake DM, Huang Z, Wolfrum EJ, et al. (1999) Bactericidal activity of photocatalytic TiO(2) reaction: toward an understanding of its killing mechanism. Appl Environ Microbiol 65: 4094-4098.

Genetic Modifications of *Saccharomyces cerevisiae* for Ethanol Production from Starch Fermentation: A Review

Esra Aydemir, Selami Demirci, Ayşegül Doğan, Ali Özhan Aytekin and Fikrettin Sahin*

Department of Genetics and Bioengineering, Faculty of Engineering and Architecture, Yeditepe University, Turkey

Abstract

There is a huge demand for developing new technologies for alternative energy sources due to the elevated costs of petroleum and its by-products, depletion of nonrenewable fuel sources, and to eliminate the disadvantages of geopolitical location and environmental pollution caused by high levels of carbon dioxide release. Science is striving to meet this demand and as molecular biology techniques have progressed, genetic engineering tools have been presented as promising future solutions in the form of optimizing the fermentation process to increase the ethanol yield from different carbon sources such as starch. As *Saccharomyces cerevisiae* is not naturally able to ferment starch, it can be genetically manipulated and modulated to improve the fuel production from starchy materials and the amount of cost that is required to produce ethanol would be decreased with these manipulations. General modifications in *S. cerevisiae* include specific gene expressions to gain new properties or improve existing pathways. This review aims to elicit the current status of ethanol production thorough alternative techniques from starch using current genetic engineering applications and to give further directions for high-throughput fermentations using genetically modified *S. cerevisiae* strains.

Keywords: *Saccharomyces cerevisiae;* Ethanol; Fermentation; Genetic engineering; Starch

Introduction

Global energy consumption is rapidly increasing and it causes an elevation of energy cost and contributes to global warming because of the excessive use of petroleum based energy sources; hence the demand for renewable biomass-derived fuels has increased in recent years. Production of such alternative sources, which could also help establish a sustainable and renewable energy supply, has been popularized. Several countries have already used first generation biofuels, including ethanol, as a primary energy sources such that worldwide biofuel production reached 106 billion liters in 2011 and it is estimated to be 155 billion liters by 2020 [1]. Feedstocks rich in sugars are generally used for ethanol production via fermentation processes. Main sugar sources for ethanol production are presented in Figure 1. Although all of fuel ethanol is obtained from corn glucose in US [2] and sucrose in Brazil [3], starchy materials have been accepted as the major renewable biomass resources for ethanol production due to their low cost and abundance [4]. In this line, corn wheat and tubers from starchy crops are used for ethanol production in North America, Europe, and tropical countries [5].

Manipulation and utilization of starchy resources starts with enzymatic hydrolysis followed by fermentation of sugar molecules and subsequently the elimination of ethanol from culture media. Simultaneous Saccharification and Fermentation (SSF), is an alternative method used for bioethanol production from feedstock and it decreases fermentation costs by reducing equipment requirements since both processes occur in one reactor [6]. Another bonus in the SSF system is that the ethanol production rate is higher than the conventional method [7]. SSF has been applied for ethanol production from starch fermentation and remarkable ethanol yields (0.41 liter ethanol per kg of corn) have been obtained [8]. There have been attempts to increase the fermentation efficiency with immobilization of microorganisms in SSF systems [9-11]. Although SSF has advanced in the field, several points should be optimized to reduce the total cost, provide highly efficient utilization of starch, and maximize ethanol yield. These steps include: maintaining the optimal pH for the growth of fermentative microorganism, sterility, continuous substrate supplementation, and the establishment of cooling systems for high temperature fermentation systems [5]. Furthermore, several ethanol producing microorganisms used in industry, such as *Saccharomyces cerevisiae,* have strains that are not naturally able to utilize starch and they require high amounts of amylolytic enzymes, which is associated with high cost and impractical ethanol production [12]. Therefore, while ethanol production from starchy materials is racing ahead, the conventional processes used today are not favorable at the economic level; hence, improved methods are desired with microbial strains that enable efficiency and lead to high yield ethanol production in a cost-effective way [4].

A vast amount of microbial species to obtain ethanol from starch in the fermentation process have been presented in the literature. Regardless of the species, parental microorganisms remain insufficient in conventional ethanol production due to the lack of availability of sugar rich input and low ethanol yields. Over 150 amylolytic yeast strains have been reported to be impractical in industrial use because of limited characteristics [13]. Although *Clostridium* spp. and *Zymomonas mobilis* are popular bacteria for ethanol production [14,15], *S. cerevisiae,* well-known and widely used yeast in alcoholic beverages and bakery industries, is traditionally preferred because it sustains steady-state production, can consume various monomeric sugar molecules, has a high fermentation capacity, ethanol productivity and ethanol tolerance, along with having "generally regarded as safe" (GRAS) status [16]. As

***Corresponding author:** Fikrettin Sahin, Department of Genetics and Bioengineering, Faculty of Engineering and Architecture, Yeditepe University, Kayisdagi, Atasehir, Istanbul, Turkey 34755

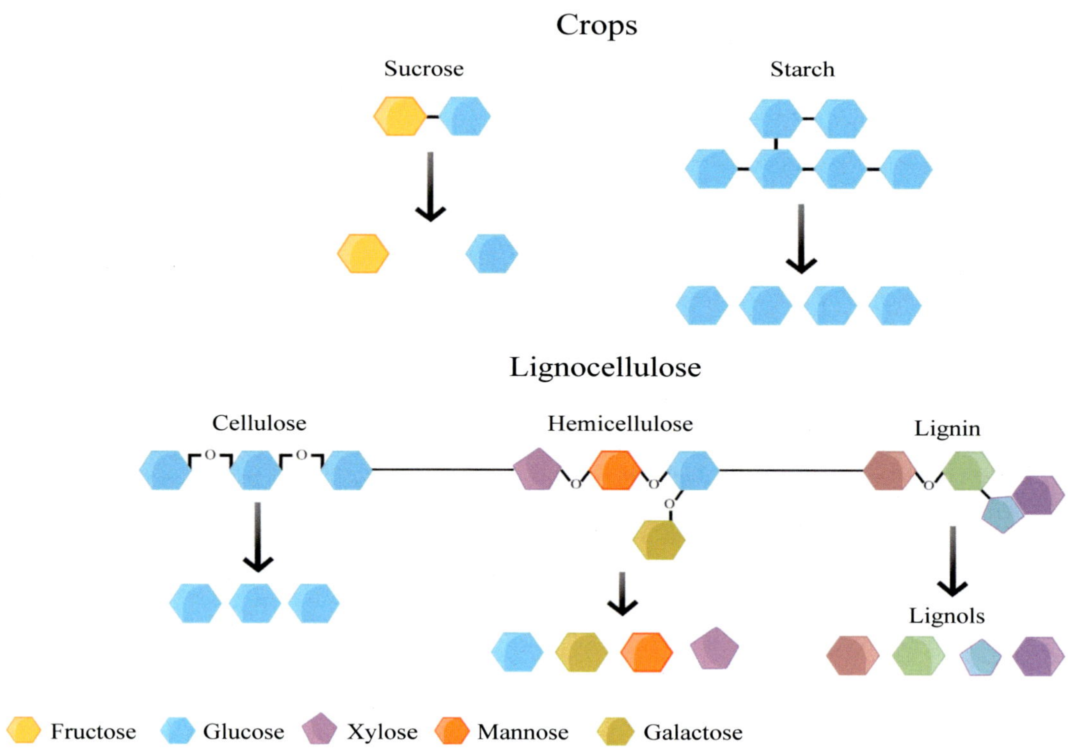

Figure 1: Main sugar sources for ethanol fermentation.

S. cerevisiae is not naturally able to ferment starch, the development of genetically modified *S. cerevisiae* strains displaying starch utilizing enzymes (α-amylase and glucoamylase), leading to saccharification and fermentation of starch at the same time has been an area of interest. In this review the genetic engineering of *S. cerevisiae* to allow it to utilize starchy sources resulting in a decreased total cost of fermentation and the realization of high ethanol levels will be discussed in detail.

Starch Fermentation

Starch is a cheap, renewable, and fermentable carbon source [17] found in all green plants in various amounts. There is approximately a starch content of 70% in corn, triticale and cassava [18]. Starch molecules are generally associated with lipids, proteins and fatty acids, and the type of association determines the pre-treatment process to be applied before fermentation [19]. Starch consists of glucose monomers joined by glycosidic bonds (α-1,4 glucan-linked D-glucopyranose chains) [18]. It is mainly composed of the linear glucose polymer amylose (20-30%) and highly branched amylopectin (70-80%) (Figure 2) [20]. Amylose has mostly α-1,4-linked D-glucopyranose and sparse α-1,6-linked D-glucopyranose chains (about 0.3% to 0.5%), and forms a flexible molecule leading to interaction with fatty acids, alcohols, and iodine [21]. Amylopectin is a highly branched helix like structured molecule consisting of α-1,4-D-glucopyranose chains [22]. Amylose and amylopectin, found in the starch structure at different rates, form an interconnected structure which directly determines the chemical characteristic of biomass and fermentation yield [23].

Fermentation of starch commonly involves two stages; i) starch hydrolysis by amylolytic enzymes (liquefaction) and saccharification, ii) fermentation of glucose into ethanol [24]. For an efficient starch fermentation, both α-1,4 and α-1,6-debranching hydrolases, with amylases and glucoamylases, and α-glucosidases displaying both α-1,4 and α-1,6-debranching activities are required [25]. As starch itself is not readily utilizable by *S. cerevisiae,* it has to be hydrolyzed by acid treatment and saccharificated by enzyme (amylase and/or glucoamylase) treatment before the main fermentation process [7]. This is a relatively expensive application as 30-40% of the total cost for the fermentation process is spent for the liquefaction and saccharification necessary for this popular microbe to be used [26]. Moreover, adding caustic soda, lime and sulphuric acid to maintain optimum pH levels for the enzymes, also increases the total cost [27]. Cold starch hydrolysis is an alternative method to decrease the total energy input. However, total enzyme requirement is much higher in cold hydrolysis than starch hydrolysis at high temperatures [28,29]. Although elevated temperatures in the initial step is advantageous to prevent bacterial contamination, high cooling cost result and become a major problem for fermentation because the temperature for industrially available yeasts should be stabilized to 30-37°C. Co-culture systems (amylolytic microorganisms and yeast) have emerged in the SSF process to avoid pure enzyme requirements [30-32]. However, the process is still expensive and the ethanol yield is not satisfactory because of the high starch content consumed for the growth of amylolytic microorganisms. Consolidated Bioprocessing (CBP), hydrolysis and fermentation of sugar molecules in a single step using microorganisms with fermentation capability, is the most effective method for ethanol production from starch fermentation. In this concept, if a microorganism does not naturally express amylolytic enzymes, genetic engineering tools offer a solution for starch fermentation by non-amylolytic microorganisms, e.g. *S. cerevisiae* as explained in this review [18]. The optimization of temperature requirements for saccharification and fermentation processes and the development of genetically modified *S. cerevisiae* strains that do not need the addition of exogenous enzymes to achieve the full potential for starch fermentation in a single step, are highly

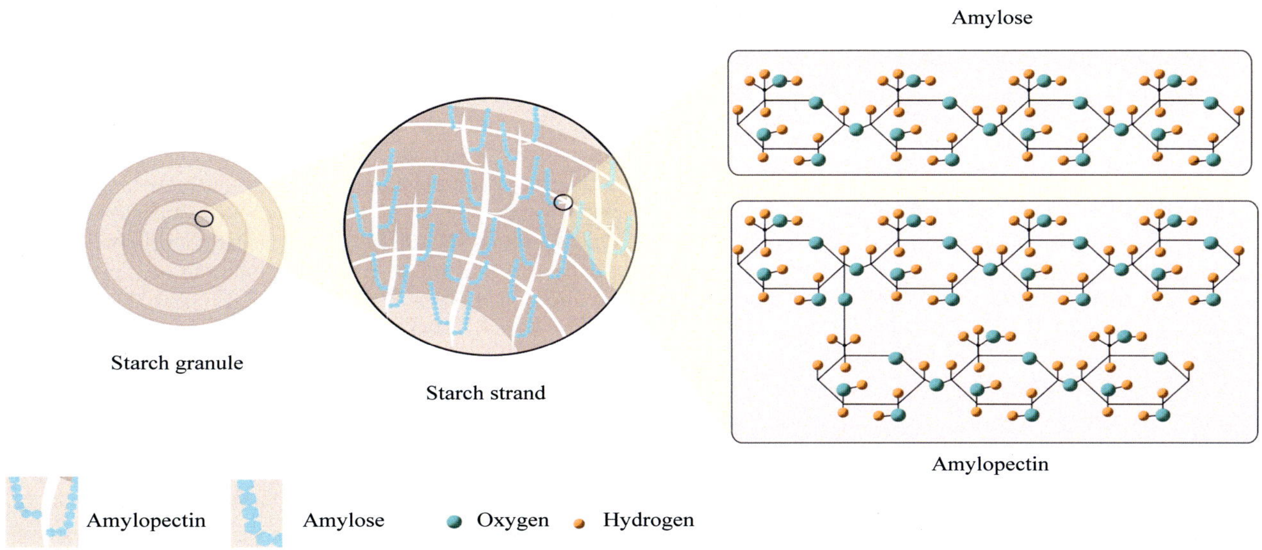

Figure 2: Components and structure of starch.

warranted to enhance ethanol production [16,33]. Therefore, genetic manipulations in *S. cerevisiae* for a direct conversion of starch to ethanol is a promising process and it would save the feedstock spent for the growth of amylolytic microorganisms and reduce the total cost along with providing high ethanol yield.

Glucoamylase expressing *S. cerevisiae*

Construction of a yeast strain that can utilize raw starch has been studied since the 1980s [34,35]. In the beginning of genetically engineered *S. cerevisiae,* fashioned for starch fermentation, 13% (v/v) ethanol yield was obtained from direct hydrolysis of starch via using glucoamylase expressing yeast [36]. The first step of the process began by constructing recombinant strains that express amylolytic enzymes (α-amylase or glucoamylase) for the liquefaction and saccharification of the starchy biomass. Glucoamylase (1,4-α-D-glucan glucohydrolase; EC 3.2.1.3) is recognized as the most important enzyme which is responsible for the progressive hydrolysis of starch from non-reducing ends to release β-D-glucose units and saccharification of the polymers [37]. The glucoamylase cDNA gene (*glu*) from *Aspergillus awamori* has been successfully incorporated into *S. cerevisiae* genome for the utilization of starch and recombinant strains have been found to be stable for 50 generations without applying any selective pressure [38]. In another study, the glucoamylase enzyme coding sequence was transferred to *S. cerevisiae* genome and hydrolyzation of unprocessed and soluble starch was attained at high performance levels (0.23 $g.l^{-1}.h^{-1}$) [39]. The same research group also increased the enzyme activity of the *A. awamori* glucoamylase by codon-optimizing, compared to native glucoamylase (791 nkat and 591 nkat per gram dry cell weight, respectively) and transformed the recombinant gene to industrial *S. cerevisiae* (27P) strain for a direct starch fermentation [27]. Although co-culture of *S. cerevisiae* with an amylolytic microorganism is a conventional option, Nakamura et al. have reported that a recombinant *S. cerevisiae,* SR93, modified to express glucoamylase has produced more ethanol (24.9 $g.l^{-1}$) as compared to a co-culture of *A. awamori* and wild type *S. cerevisiae* system (22.0 $g.l^{-1}$), because SR93 has saved the internal starch amount consumed for the growth of the amylolytic microorganism, *A. awamori* [7]. However, SR93 could not degrade all the starch content efficiently because recombinant glucoamylase enzyme originated from *Saccharomyces cerevisiae* var. *diastaticus* was not able to degrade α-1,6 glycoside bond of amylopectin units. It has been recognized that this glucoamylase enzyme, coded by the STA1 gene, lacked a starch-binding domain which made fermentation and ethanol production unsatisfactory [40,41]. Therefore, the starch-binding domain of the *Aspergillus niger* glucoamylase gene has been fused with *STA1* gene resulting in a remarkable hydrolysis and utilization of insoluble starch [42]. Instead of using a starch binding domain, the gene for glucoamylase of *Rhizopus oryzae*, capable of breaking down both α-1,4- and α-1,6-glycosidic bond efficiently, has been transferred to *S. cerevisiae* and approximately 80% of starch content was utilized in a 100h fermentation period [43]. As a long fermentation period (~150h) is required for sufficient starch fermentation by glucoamylase secreting *S. cerevisiae,* in anaerobic or minimal aerobic conditions [7,34,35,44], a recombinant strain of *S. cerevisiae* (YF207/pGA11) expressing cell surface anchored *R. oryzae* glucoamylase has been tested for ethanol production under aerobic conditions (dissolved oxygen was 2.0 ppm), using soluble starch [4]. A high ethanol production rate (0.71 $g.h^{-1}.l^{-1}$) was achieved and the fermentation process has been completed in seven repeated fermentations over 300 h without losing modified gene stability. As the same enzyme from various sources would provide different outcomes due to their divergent kinetic properties and activities in different experimental conditions, determining optimal enzyme source to be transferred is one of the most critical issues. For instance, *R. arrhizus* glucoamylase gene has been transferred to *S. cerevisiae* and up to 5% ethanol and 2400 $U.l^{-1}$ enzyme activity (one of the highest level reported to date) has been obtained in a flask ferment experiment [45]. *S. cerevisiae* has been modified by *Aspergillus oryzae* glucoamylases, encoded by *glaA* and *glaB*, and *R. oryzae* glucoamylase separately on the cell surface and compared with each other for their starch fermentation ability [46]. The highest ethanol yield has been obtained from *glaA* glucoamylase expressing yeast (15 $g.l^{-1}$ in 24 h); although *R. oryzae* glucoamylase exhibited the highest glucoamylase activity (9×10^{-9} U/cell). From a different point of view, as raw and naive starch is not favorable from a commercial perspective, Kosugi et al. have designed an experiment to produce ethanol from cassava pulp rich in starch (up to 60%) [47]. A high production rate of ethanol from

5% cassava pulp (91% of theoretical value) has been achieved using *S. cerevisiae* displaying *R. oryzae* glucoamylase enzyme.

α-amylase expressing *S. cerevisiae*

Apart from transforming glucoamylase into *S. cerevisiae* for starch hydrolysis and saccharification, α-amylase modification is another important option in recombinant technology for starch fermentation. The α-amylases (EC 3.2.1.1) breakdown starch molecules from the internal α-1,4-bonds of amylose and amylopectin and release dextrins (10-20 glucose units in length), free glucose and maltose units [37]. Wheat α-amylase has been cloned into *S. cerevisiae* and found secreted into medium successfully as early as 1987 [48]. Since then, amylase enzymes from different sources such as bacterial [49], yeast [50,51], mold [52], barley [53] and rice [54] have been cloned into *S. cerevisiae*. From the beginning of these studies, optimization of cell growth and enzyme activity has been determined as the key factors for an efficient ethanol production. The selection of the source for the enzyme is a critical issue as the nature and activity of the enzyme should be compatible with the fermentation conditions. It has been shown that decrease in pH, from 5.5 to 4.5 due to yeast growth, in fermentation resulted in a dramatic reduction in *Bacillus subtilis* α-amylase activity, whereas there was no effect, even a slight increase, on barley α-amylase activity [55]. *S. cerevisiae* strains expressing recombinant α-amylase genes (*LKA1* and *LKA2*) obtained from *Lipomyces kononenkoae* have been proven to convert starch directly into ethanol [56]. However, the rate of ethanol production has been found to be low (17.2 g.l^{-1} in 200 h fermentation period) since the capacity of starch hydrolysis by recombinant α-amylase was inadequate. Similarly, Ramachandran et al. have reported *LKA1* expressing flocculent and non- flocculent *S. cerevisiae* strains for ethanol production from raw starch [57]. Genetically modified flocculent strains produced higher ethanol levels compared to non-flocculent counterparts (4.61 kg.l^{-1} and 5.1 g.l^{-1}, respectively) in a 90 h fermentation period. However, the general accepted concept is that *S. cerevisiae* should express a high amount of α-amylase under aerobic conditions for effective starch fermentation [56,58]. As a different strategy, apart from genetically engineered *S. cerevisiae* that expresses recombinant α-amylase, ethanol production rate sometimes increased by addition of exogenous glucoamylase enzyme [53,59]. Although this is a preferable option for a high starch fermentation rate, it is not economically favorable from an industrial perspective due to the high price of pure enzymes. In this line, scientists have focused on both α-amylase and glucoamylase expressing *S. cerevisiae* to increase ethanol yield and decrease total cost.

Co-expression of α-amylase and Glucoamylase

In order to increase the rate of ethanol production from starch fermentation by *S. cerevisiae,* additional genetic manipulations such as co-expressing of α-amylase and glucoamylase have emerged as a latter strategy. Glucoamylase and α-amylase enzymes synergistically enhance the rate of corn and wheat starch hydrolysis with respect to their individual performances [51,60-62]. Hence, scientists have been working on constructing *S. cerevisiae* strains that express both enzymes. The glucoamylase gene of *A. awamori* (*GA1*), *Debaryomyces occidentalis* glucoamylase (*GAM1*), and α-amylase (*AMY*) encoding plasmids have been transformed into industrial *S. cerevisiae* strain for a direct conversion of starch to ethanol [63]. Yeast containing *GAM1*, *GA1* and *AMY* genes have exhibited the highest glucoamylase (required for debranching of starch molecules) activity (1020 U.l^{-1}) compared to only *GAM1* or *GA1* transformed strains (790 U.l^{-1} and 560 U.l^{-1}, respectively); indicating synergistic activity. Altıntaş et al. have also transformed *S. cerevisiae* with a bifunctional fusion protein that contained both the *B. subtilis* α-amylase and the *A. awamori* glucoamylase, but they have found the biomass (3.86-6.24 g.l^{-1}) and ethanol production (18.4-23.2 g.l^{-1}) insufficient in the experimental model of intermittent starch feeding system [64]. In a different study, three different recombinant strains of *S. cerevisiae* have been used for the comparison of their ethanol production capabilities [44]. YPG/AB strain expresses *B. subtilis* α-amylase and the *A. awamori* glucoamylase separately, but YPB-G strain expresses both enzymes as a fusion protein. One last strain YPG/MM expresses mouse α-amylase and *A. awamori* glucoamylase. YPG/AB strains were found to be superior in ethanol production (43.8 g.l^{-1}) than YPB-G (35.2 g.l^{-1}) and YPG/MM (24.3 g.l^{-1}) strains. Although the glucoamylase activity in YPG/AB and YPB-G strains were similar (1053 U.l^{-1} and 1100 U.l^{-1}, respectively), α-amylase of YPG/AB showed 2.2 fold higher activity than the fusion protein group which was the possible explanation for the higher rate of ethanol production. In addition, low levels of ethanol produced by YPG/MM strain have been attributed to negligible glucoamylase activity. From a different point of view, expressing amylase and glucoamylase genes together with bacterial pullulanase in *S. cerevisiae* resulted in complete (99%) utilization of the initial starch [65]. However, as it was mentioned in the previous section, using laboratory strains and raw starch resources are not economically favorable and do not always reflect real conditions. In this sense, scientists have tried to optimize starch fermentation using unprocessed biomass and industrial yeast strains. In a high-yielding brown rice fermentation, yeast strain expressing α-amylase and glucoamylase exhibited acceptable ethanol production rate (1.1 g.l^{-1}.h^{-1}) [66]. Industrial yeast strains have also been genetically modified for starch hydrolysis due to their high ethanol production rate and tolerance to harsh conditions. Viktor et al. have tested ethanol production capacities of *Aspergillus tubingensis* T8.4 α-amylase and glucoamylase expressing laboratory strain, *S. cerevisiae* Y294, and the semi-industrial strain, *S. cerevisiae* Mnuα1. Y294 and Mnuα1 strains have produced 9.03 and 6.67 g.l^{-1} ethanol, respectively, from a substrate load of 200 g.l^{-1} raw corn starch after 10 days fermentation period without any heat treatment [37]. Industrial strains of *S. cerevisiae* have been transformed with amylase and glucoamylase genes separately and co-cultured for an efficient one-step starch utilization [67]. Activities for glucoamylase and α-amylase have been determined as 920 U.l^{-1} and 7960 U.l^{-1}, respectively.

Cell surface anchored enzyme expressing *S. cerevisiae* strains

Co-expressing of starch utilizing enzymes has also been studied by anchoring enzymes on the cell wall of yeast to provide long term durability. Expressing on the cell surface and secreting to the medium has various advantages and disadvantages. Secreting of the enzymes to the fermentation environment increases the rate and amount of ethanol by providing high possibility of interaction between starch molecules and enzymes [68]. On the other hand, secreting to the environment is not favorable because of the loss of stability in the early stage of fermentation. An additional consideration for secreted enzymes is including extra ingredients such as metal ions or surfactants in the fermentation medium to provide enzyme stability for repeated large-scale productions. For example, calcium ions have been claimed to be protective for α-amylase during repeated 10-cycles of raw starch fermentation [69]. However, cell surface engineering of yeast for starch utilizing enzymes is suitable for long term stability and repeated large-scale production without the necessity of adding additional reagents. In a recent study, *S. cerevisiae* strain co-expressing glucoamylase and α-amylase on the cell surface has been used in 23 continuous cycles of ethanol fermentation without losing enzyme activity [70]. Recombinant yeast strains, glucoamylase and α-amylase anchored on the cell wall (or secreted to the medium), have been developed for a direct starch utilization [71]. Cell surface glucoamylase and α-amylase expressing *S. cerevisiae* strain has produced more ethanol (60 $g.l^{-1}$ in a 100 h fermentation period) and a higher starch degradation rate than the only cell surface glucoamylase expressing strain (50 $g.l^{-1}$ in 120 h fermentation). In another study, *S. cerevisiae* co-displaying glucoamylase of *R. oryzae* and α-amylase of *Streptococcus bovis* anchored to cell membrane via C-terminal-half region of α-agglutinin and the flocculation functional domain of Flo1p, respectively, has been effective (61.8 $g.l^{-1}$ ethanol in a 72 h fermentation period) in raw corn starch hydrolysis [12]. A recent study has brought a different point of view to starch conversion into ethanol by *S. cerevisiae*. A recombinant strain of yeast which expressed cell surface engineered aspartic protease has been constructed for the consuming of complex nitrogenous materials, other than starchy molecules found in crops, that resulted in high ethanol yield and shorter fermentation duration (3.8 ± 0.15% improvement in the final ethanol concentration compared to parental strain in 72 h fermentation period) [72].

Stable Starch Fermenting *S. cerevisiae* Strains

Although genetic manipulation is a relatively easy process by using episomal vectors which do not require integrative sites, the stability of the episomal plasmid is a serious problem for long-term and repeated fermentation such that plasmid leakage is generally observed in long-term incubation [7,73]. Gene integration into the microbial genome has been undertaken in many studies to provide long-term enzyme activity [38,74,75]. The δ-sequences of the Ty retrotransposon, or rDNA sequence, of *S. cerevisiae* are generally used elements for chromosomal integration of a recombinant gene. Enzyme activities could be increased 20-fold by means of δ-integration sequence with respect to the conventional transformation [76]. In another study, it has been shown that 90% of the initial starch content was fermented by recombinant *S. cerevisiae* that co-expressed glucoamylase and α-amylase transformed via δ-integration [77]. Although, there has been a 2-fold decrease in ethanol production and cell mass in *S. cerevisiae* strain transformed with episomal vector after 7-repeated fermentation process, the yeast strain transformed with δ-integration sequence containing plasmid has exhibited long term stability of enzyme activity up to 10 cycles [69,78] and a high ethanol production rate up to 23 cycles [70]. Ribosomal DNA sequence of yeast is another effective option for chromosomal integration via homologous recombination. An α-amylase gene has been integrated into yeast chromosome and exhibited a 2-fold increase in starch consumption compared to the episomal vector cloned strain [79]. Targeting ribosomal DNA sequence (150-200 copy in yeast genome) via homologous recombination has resulted in stable high copy number of recombinant gene [59,80]. Multiple copy integrated genes via rDNA homologous recombination displayed a higher quantity of starch consumption and ethanol production (19.2 $g.l^{-1}$) [81]. As a novel strategy, rDNA and δ-integration combination could be used by targeting two separate DNA sites for efficient cloning of two or more genes [63]. Cell fusion technique is also used to make diploid and tetraploid *S. cerevisiae* strains to obtain high level of biomass and ethanol production. Diploid and tetraploid strains have proliferated and grown faster, fermented starch more efficiently in comparison with parental strains. At the end of 72h fermentation process, haploid, diploid and tetraploid strains have produced 0.55, 0.72 and 0.93 $g.l^{-1}.h^{-1}$ ethanol, respectively [76].

Conclusion and Future Perspectives

Energy prices are growing fast, mainly because of depletion of petroleum sources. Therefore, alternative energy sources obtained from sun, wind and biomass have become attractive recently. Ethanol, produced by fermentation from biomass, as a promising alternative energy is a transportable and economically favorable source. In addition, ethanol production from various feedstocks such as starch maintains a considerable potential due to their availability, accessibility and relatively low cost in comparison to sucrose and glucose based feedstocks. Recombinant DNA technology offers a valuable opportunity for consolidated bioprocessing processes of the biomass fermentation. To date however, transforming new pathways and overexpression of a single or group of enzymes have

Engineered Enzyme	Gene Source	Cultivation Time (h)	Ethanol Yield ($g.l^{-1}.h^{-1}$)	Enzyme Activity	Starch Source	Reference
α-amylase	*Lipomyces kononenkoae*	90	0.05	87 $U.l^{-1}$	Raw corn starch	[57]
Glucoamylase	*Aspergillus awamori*	NA*	ND#	162 $U.l^{-1}$	Soluble starch	[38]
Glucoamylase	*Aspergillus awamori*	50	0.23	624 $U.l^{-1}$	Soluble starch	[39]
Glucoamylase	*Saccharomyces diastaticus* (var. *diastaticus*)	200	0.66	ND	Soluble starch	[7]
Glucoamylase	*Rhizopus oryzae*	168	0.77	60.2 U/g of wet cells	Cassava pulp rich in starch (60%)	[47]
Glucoamylase	*Aspergillus awamori*	80	0.175	ND	Soluble starch	[58]
Glucoamylase/ α-amylase	*Rhizopus oryzae/ Streptococcus bovis*	72	0.85	57/114 U/g of wet cells	Raw corn starch	[12]
Glucoamylase/ α-amylase	*Saccharomycopsis fibuligera/ Lipomyces kononenkoae*	120	0.178	1340 /<30 $U.l^{-1}$	Soluble starch	[56]
Glucoamylase/ α-amylase	*Rhizopus oryzae/ Streptococcus bovis*	120	0.74	790/1306 $U.l^{-1}$	Raw corn starch	[70]
Glucoamylase/ α-amylase	*Rhizopus oryzae/ Streptococcus bovis*	24	1.2	4700/1800 $U.l^{-1}$	High-yielding rice	[66]
α-amylase/ glucoamylase/ glucoamylase with debranching activity	*Debaryomyces occidentalis/ Aspergillus awamori/ Debaryomyces occidentalis*	168	0.45	5940/1020 $U.l^{-1}$	Soluble starch	[63]
Glucoamylase/ isoamylase	*Aspergillus awamori/ Pseudomonas amyloderamosa*	140	0.137	ND	Soluble starch	[81]

***NA:** Not available,
#ND: Not determined

Table 1: Genetic modifications on *S. cerevisiae* for high efficient ethanol fermentation.

not taken biomass fermentation via *S. cerevisiae* to a satisfactory level for industrial arena. Although there are hundreds of studies claiming higher ethanol production from starchy materials, most of these have been tested at laboratory scale only and not at large-scale (Table 1). Several laboratory yeast strains have been proven to exhibit remarkable starch fermenting capacity and produce high amounts of ethanol, but transferring engineering technology to industrial *S. cerevisiae* strains has remained insufficient. They are not convenient for repeated large scale applications because of low stability of the modification, high cost of the process, low yield of ethanol production and fermentation rate. Although using exogenous amylase and glucoamylase seems to increase total yield, developing yeast strains sufficient alone for the whole fermentation process is desirable. New amylase and glucoamylase from different sources should be presented and investigated for their activity (alone or in combination) and compatibility with industrial yeast stains. Researchers should continue to conduct studies on two main areas; efficient hydrolysis and fermentation of biomass, which result in high ethanol production rate and combination of improved bioprocess applications with genetic engineering tools in the view of productivity and large scale processes.

Acknowledgment

This review was supported by Yeditepe University, Istanbul, Turkey. The authors would like to thank to Lynne Emily Ozgur for her numerous suggestions on language.

References

1. (2012) OECD-FAO Agricultural Outlook 2011-2020.
2. (2001) US Ethanol Industry:Production capacity outlook. Results of a Survey Conducted by the California Energy Commission. California Energy Commission.
3. Rosillo-Calle F, Cortez LAB (1998) Towards proalcohol II: A review of the Brazilian bioethanol programme. Biomass Bioenergy 14: 115-124.
4. Kondo A, Shigechi H, Abe M, Uyama K, Matsumoto T, et al. (2002) High-level ethanol production from starch by a flocculent *Saccharomyces cerevisiae* strain displaying cell-surface glucoamylase. Appl Microbiol Biotechnol 58: 291-296.
5. Cardona CA, Sánchez OJ (2007) Fuel ethanol production: process design trends and integration opportunities. Bioresour Technol 98: 2415-2457.
6. Wyman CE (1994) Ethanol from lignocellulosic biomass: Technology, economics and opportunities. Bioresour Technol 50: 3-16.
7. Nakamura Y, Kobayashi F, Ohnaga M, Sawada T (1997) Alcohol fermentation of starch by a genetic recombinant yeast having glucoamylase activity. Biotechnol Bioeng 53: 21-25.
8. Madson PW, Monceaux DA (1995) Fuel ethanol production. Nottingham University Press, 1-12.
9. Chithra N, Baradarajan A (1992) Direct conversion of starch hydrolysate to ethanol using a coimmobilizate of amyloglucosidase and *Saccharomyces cerevisiae* in batch stirred tank reactor. Bioprocess Eng 7: 265-267.
10. Farid MA, El-Enshasy HA, Noor El-DeenAM (2002) Alcohol production from starch by mixed cultures of Aspergillusawamori and immobilized *Saccharomyces cerevisiae* at different agitation speeds. J Basic Microbiol 42: 162-171.
11. Kurosawa H, Nomura N, Tanaka H (1989) Ethanol production from starch by a coimmobilized mixed culture system of *Aspergillusawamori* and *Saccharomyces cerevisiae*. Biotechnol Bioeng 33: 716-723.
12. Shigechi H, Koh J, Fujita Y, Matsumoto T, Bito Y, et al. (2004) Direct production of ethanol from raw corn starch via fermentation by use of a novel surface-engineered yeast strain codisplayingglucoamylase and alpha-amylase. Appl Environ Microbiol 70: 5037-5040.
13. ToksoyOner E, Oliver SG, Kirdar B (2005) Production of ethanol from starch by respiration-deficient recombinant *Saccharomyces cerevisiae*. Appl Environ Microbiol 71: 6443-6445.
14. Gunasekaran P, Raj KC (1999) Ethanol fermentation technology-*Zymomonasmobilis*. Curr Sci 77: 56-68.
15. Pfromm PH, Amanor-Boadu V, Nelson R, Vadlani P, Madl R (2010) Bio-butanol vs. bio-ethanol: A technical and economic assessment for corn and switchgrass fermented by yeast or *Clostridium acetobutylicum*. Biomass Bioenergy 34: 515-524.
16. Görgens JF, Bressler DC, van Rensburg E (2014) Engineering *Saccharomyces cerevisiae* for direct conversion of raw, uncooked or granular starch to ethanol. Crit Rev Biotechnol .
17. Ülgen KÖ, Saygili B, Önsan ZI, Kirdar B (2002) Bioconversion of starch into ethanol by a recombinant *Saccharomyces cerevisiae* strain YPG-AB. Process Biochem 37: 1157-1168.
18. vanZyl WH, Bloom M, Viktor MJ (2012) Engineering yeasts for raw starch conversion. Appl Microbiol Biotechnol 95: 1377-1388.
19. BeMiller JN, Whistler RL (2009) Structural features of starch granules II. Starch: Chemistry and technology. (3rd edition), Academic Press, New York and London.
20. Yamada R, Bito Y, Adachi T, Tanaka T, Ogino C, et al. (2009) Efficient production of ethanol from raw starch by a mated diploid *Saccharomyces cerevisiae* with integrated a-amylase and glucoamylase genes. Enzyme Microb Technol 44: 344-349.
21. Whistler RL, BeMiller JN (1997) Carbohydrate chemistry for food scientists. Eagan Press, St. Paul, Minnesota, USA.
22. Stevnebø A, Sahlström S, Svihus B (2006) Starch structure and degree of starch hydrolysis of small and large starch granules from barley varieties with varying amylose content. Anim Feed Sci Technol 130: 23-38.
23. Klucinec JD, Thompson DB (2002) Amylopectin nature and amylose-to-amylopectin ratio as influences on the behavior of gels of dispersed starch. Cereal Chem 79: 24-35.
24. Lee WS, Chen IC, Chang CH, Yang SS (2012) Bioethanol production from sweet potato by co-immobilization of saccharolytic molds and *Saccharomyces cerevisiae*. Renew Energ 39: 216-222.
25. van der Maarel MJ, van der Veen B, Uitdehaag JC, Leemhuis H, Dijkhuizen L (2002) Properties and applications of starch-converting enzymes of the alpha-amylase family. J Biotechnol 94: 137-155.
26. Wu Q, Miao Y (2008) Mechanochemical effects of micronization on enzymatic hydrolysis of corn flour. Carbohydr Polym 72: 398-402.
27. Favaro L, Jooste T, Basaglia M, Rose SH, Saayman M, et al. (2012) Codon-optimized glucoamylasesGAI of *Aspergillus awamori* improves starch utilization in an industrial yeast. Appl Microbiol Biotechnol 95: 957-968.
28. Lang X, Macdonald D, Hill G (2001) Recycle bioreactor for bioethanol production from wheat starch II. Fermentation and economics. Energ Source 23: 427-436.
29. Robertson GH, Wong DW, Lee CC, Wagschal K, Smith MR, et al. (2006) Native or raw starch digestion: a key step in energy efficient biorefining of grain. J Agric Food Chem 54: 353-365.
30. Manikandan K, Viruthagiri T (2009) Simultaneous saccharification and fermentation of wheat bran flour into ethanol using coculture of amylotic*Aspergillusniger* and thermotolerant*Kluyveromycesmarxianus*. Front Chem Eng China 3: 240-249.
31. Suresh K, Kiran Sree NRao LV (1999) Utilization of damaged sorghum and rice grains for ethanol production by simultaneous saccharification and fermentation. Bioresour Technol 68: 301-304.
32. Verma G, Nigam P, Singh D, Chaudhary K (2000) Bioconversion of starch to ethanol in a single-step process by coculture of amylolytic yeasts and *Saccharomyces cerevisiae* 21. Bioresour Technol 72: 261-266.
33. Dogan A, Demirci S, Aytekin AÖ, Sahin F (2014) Improvements of Tolerance to Stress Conditions by Genetic Engineering in *Saccharomyces cerevisiae* during Ethanol Production. Appl Biochem Biotechnol 174: 28-42.
34. Cole GE, McCabe PC, Inlow D, Gelfand DH, Ben-Bassat A, et al. (1988) Stable Expression of *Aspercillus awamori* Glucoamylase in Distiller's Yeast. Nat Biotechnol 6: 417-421.
35. Innis MA, Holland MJ, McCabe PC, Cole GE, Wittman VP, et al. (1985) Expression, Glycosylation, and Secretion of an *Aspergillus* Glucoamylase by *Saccharomyces cerevisiae*. Science 228: 21-26.

36. Ashikari T, Kunikasi S, Matsumoto N, Amachi T, Yoshizumi H (1989) Direct fermentation of raw corn to ethanol by yeast transformants containing a modified *Rhizopus*glucoamylase gene. Appl Microbiol Biotechnol 32: 129-133.

37. Viktor MJ, Rose SH, Van Zyl WH, Viljoen-Bloom M (2013) Raw starch conversion by *Saccharomyces cerevisiae* expressing *Aspergillus tubingensis* amylases. Biotechnol Biofuels 6: 167-177.

38. Lin LL, Ma YJ, Chien HR, Hsu WH (1998) Construction of an amylolytic yeast by multiple integration of the *Aspergillus awamori* glucoamylase gene into a *Saccharomyces cerevisiae* chromosome. Enzyme Microb Technol 23: 360-365.

39. Favaro L, Jooste T, Basaglia M, Rose SH, Saayman M, et al. (2013) Designing industrial yeasts for the consolidated bioprocessing of starchy biomass to ethanol. Bioengineered 4: 97-102.

40. Adam AC, Latorre-García L, Polaina J (2004) Structural analysis of glucoamylase encoded by the STA1 gene of *Saccharomyces cerevisiae* (var. diastaticus). Yeast 21: 379-388.

41. Sevcík J, Solovicová A, Hostinová E, Gasperík J, Wilson KS, et al. (1998) Structure of glucoamylase from *Saccharomycopsis fibuligera* at 1.7 A resolution. Acta Crystallogr D Biol Crystallogr 54: 854-866.

42. Latorre-García L, Adam AC, Manzanares P, Polaina J (2005) Improving the amylolytic activity of *Saccharomyces cerevisiae* glucoamylase by the addition of a starch binding domain. J Biotechnol 118: 167-176.

43. Murai T, Ueda M, Yamamura M, Atomi H, Shibasaki Y, et al. (1997) Construction of a starch-utilizing yeast by cell surface engineering. Appl Environ Microbiol 63: 1362-1366.

44. Birol G, Önsan ZI, Kirdar B, Oliver SG (1998) Ethanol production and fermentation characteristics of recombinant *saccharomyces cerevisiae* strains grown on starch. Enzyme Microb Technol 22: 672-677.

45. Yang S, Jia N, Li M, Wang J (2011) Heterologous expression and efficient ethanol production of a *Rhizopus* glucoamylase gene in *Saccharomyces cerevisiae*. Mol Biol Rep 38: 59-64.

46. Kotaka A, Sahara H, Hata Y, Abe Y, Kondo A, et al. (2008) Efficient and direct fermentation of starch to ethanol by sake yeast strains displaying fungal glucoamylases. Biosci Biotechnol Biochem 72: 1376-1379.

47. Kosugi A, Kondo A, Ueda M, Murata Y, Vaithanomsat P, et al. (2009) Production of ethanol from cassava pulp via fermentation with a surface-engineered yeast strain displaying glucoamylase. Renew Energ 34: 1354-1358.

48. Rothstein SJ, Lahners KN, Lazarus CM, Baulcombe DC, Gatenby AA (1987) Synthesis and secretion of wheat alpha-amylase in *Saccharomyces cerevisiae*. Gene 55: 353-356.

49. Kovaleva IE, Novikova LA, Luzikov VN (1989) Synthesis and secretion of bacterial alpha-amylase by the yeast *Saccharomyces cerevisiae*. FEBS Lett 251: 183-186.

50. Kang NY, Park JN, Chin JE, Lee HB, Im SY, et al. (2003) Construction of an amylolytic industrial strain of *Saccharomyces cerevisiae* containing the *Schwanniomy cesoccidentalis* alpha-amylase gene. Biotechnol Lett 25: 1847-1851.

51. Steyn AJ, Pretorius IS (1995) Characterization of a novel alpha-amylase from *Lipomyces kononenkoae* and expression of its gene (LKA1) in *Saccharomyces cerevisiae*. Curr Genet 28: 526-533.

52. Randez-Gil F, Sanz P (1993) Expression of *Aspergillus oryzae* alpha-amylase gene in *Saccharomyces cerevisiae*. FEMS Microbiol Lett 112: 119-123.

53. Liao B, Hill GA, Roesler WJ (2010) Amylolytic activity and fermentative ability of *Saccharomyces cerevisiae* strains that express barley α-amylase. Biochem Eng J 53: 63-70.

54. Kumagai MH, Shah M, Terashima M, Vrkljan Z, Whitaker JR, et al. (1990) Expression and secretion of rice alpha-amylase by *Saccharomyces cerevisiae*. Gene 94: 209-216.

55. Textor SD, Hill GA, MacDonald DG, Denis ES (1998) Cold enzyme hydrolysis of wheat starch granules. Can J Chem Eng 76: 87-93.

56. Knox AM, du Preez JC, Kilian SG (2004) Starch fermentation characteristics of *Saccharomyces cerevisiae* strains transformed with amylase genes from *Lipomyces kononenkoae* and *Saccharomycopsis fibuligera*. Enzyme Microb Technol 34: 453-460.

57. Ramachandran N, Joubert L, Gundlapalli SB, Otero RRC, Pretorius IS (2008) The effect of flocculation on the efficiency of raw-starch fermentation by *Saccharomyces cerevisiae* producing the *Lipomyces kononenkoae* LKA1-encoded a-amylase. Ann Microbiol 58: 99-108.

58. Nakamura Y, Sawada T, Komatsu A (2002) Ethanol production from raw starch by a recombinant yeast having saccharification and fermentation activities. J Chem Technol Biotechnol 77: 1101-1106.

59. Liao B, Hill G, Roesler W (2012) Stable expression of barley α-amylase in *S. cerevisiae* for conversion of starch into bioethanol. Biochem Eng J 64: 8-16.

60. Eksteen JM, Van Rensburg P, Cordero Otero RR, Pretorius IS (2003) Starch fermentation by recombinant *Saccharomyces cerevisiae* strains expressing the alpha-amylase and glucoamylase genes from *Lipomyceskononenkoae*and *Saccharomycopsis fibuligera*. Biotechnol Bioeng 84: 639-646.

61. Soni SK, Kaur A, Gupta JK (2003) A solid state fermentation based bacterial α-amylase and fungal glucoamylase system and its suitability for the hydrolysis of wheat starch. Process Biochem 39: 185-192.

62. Wong DW, Robertson GH, Lee CC, Wagschal K (2007) Synergistic action of recombinant alpha-amylase and glucoamylase on the hydrolysis of starch granules. Protein J 26: 159-164.

63. Kim JH, Kim HR, Lim MH, Ko HM, Chin JE, et al. (2010) Construction of a direct starch-fermenting industrial strain of *Saccharomyces cerevisiae* producing glucoamylase, alpha-amylase and debranching enzyme. Biotechnol Lett 32: 713-719.

64. Altintas MM, Ülgen KÖ, Kirdar B, Önsan ZI, Oliver SG (2003) Optimal substrate feeding policy for fed-batch cultures of *S. cerevisiae* expressing bifunctional fusion protein displaying amylolytic activities. Enzyme Microb Technol 33: 262-269.

65. Janse BJ, Pretorius IS (1995) One-step enzymatic hydrolysis of starch using a recombinant strain of *Saccharomyces cerevisiae* producing alpha-amylase, glucoamylase and pullulanase. ApplMicrobiol Biotechnol 42: 878-883.

66. Yamada R, Yamakawa S, Tanaka T, Ogino C, Fukuda H, et al. (2011) Direct and efficient ethanol production from high-yielding rice using a *Saccharomyces cerevisiae* strain that express amylases. Enzyme Microb Technol 48: 393-396.

67. Ghang DM, Yu L, Lim MH, Ko HM, Im SY, et al. (2007) Efficient one-step starch utilization by industrial strains of *Saccharomyces cerevisiae* expressing the glucoamylase and alpha-amylase genes from *Debaryomyces occidentalis*. Biotechnol Lett 29: 1203-1208.

68. Khaw TS, Katakura Y, Koh J, Kondo A, Ueda M, et al. (2006) Evaluation of performance of different surface-engineered yeast strains for direct ethanol production from raw starch. Appl Microbiol Biotechnol 70: 573-579.

69. Yamakawa S, Yamada R, Tanaka T, Ogino C, Kondo A (2010) Repeated batch fermentation from raw starch using a maltose transporter and amylase expressing diploid yeast strain. Appl Microbiol Biotechnol 87: 109-115.

70. Yamakawa S, Yamada R, Tanaka T, Ogino C, Kondo A (2012) Repeated fermentation from raw starch using *Saccharomyces cerevisiae* displaying both glucoamylase and α-amylase. Enzyme Microb Technol 50: 343-347.

71. Shigechi H, Uyama K, Fujita Y, Matsumoto T, Ueda M, et al. (2002) Efficient ethanol production from starch through development of novel flocculent yeast strains displaying glucoamylase and co-displaying or secreting α-amylase. J Mol Catal B Enzym 17: 179-187.

72. Guo ZP, Qiu CY, Zhang L, Ding ZY, Wang ZX, et al. (2011) Expression of aspartic protease from *Neurosporacrassa* in industrial ethanol-producing yeast and its application in ethanol production. Enzyme Microb Technol 48: 148-154.

73. Romanos MA, Scorer CA, Clare JJ (1992) Foreign gene expression in yeast: a review. Yeast 8: 423-488.

74. Kim HR, Im YK, Ko HM, Chin JE, Kim IC, et al. (2011) Raw starch fermentation to ethanol by an industrial distiller's yeast strain of *Saccharomyces cerevisiae* expressing glucoamylase and α-amylase genes. Biotechnol Lett 33: 1643-1648.

75. Okano K, Zhang Q, Shinkawa S, Yoshida S, Tanaka T, et al. (2009) Efficient production of optically pure D-lactic acid from raw corn starch by using a genetically modified L-lactate dehydrogenase gene-deficient and alpha-amylase-secreting *Lactobacillus plantarum* strain. Appl Environ Microbiol 75: 462-467.

76. Yamada R, Tanaka T, Ogino C, Fukuda H, Kondo A (2010) Novel strategy for yeast construction using d-integration and cell fusion to efficiently produce ethanol from raw starch. Appl Microbiol Biotechnol 85: 1491-1498.

77. Wong D, Batt S, Robertson G, Lee C, Wagschal K (2010) Chromosomal integration of both an α-amylase and a glucoamylase gene in *Saccharomyces cerevisiae* for starch conversion. Ind Biotechnol 6: 112-118.

78. Chen JP, Wu KW, Fukuda H (2008) Bioethanol production from uncooked raw starch by immobilized surface-engineered yeast cells. Appl Biochem Biotechnol 145: 59-67.

79. Wang JJ, Wang ZY, He XP, Zhang BR (2012) Integrated expressionof the α-amylase, dextranase and glutathione gene in an industrial brewer's yeast strain. World J Microbiol Biotechnol 28: 223-231.

80. Nieto A, Prieto JA, Sanz P (1999) Stable high-copy-number integration of *Aspergillus oryzae* alpha-AMYLASE cDNA in an industrial baker's yeast strain. Biotechnol Prog 15: 459-466.

81. Ma YJ, Lin LL, Chien HR, Hsu WH (2000) Efficient utilization of starch by a recombinant strain of *Saccharomyces cerevisiae* producing glucoamylase and isoamylase. Biotechnol Appl Biochem 31: 55-59.

Demonstrating β-glucan Clearance in CHO- and Yeast-Produced Monoclonal Antibodies during Downstream Purification Processes

Fengqiang Wang[1*], Hong Li[2], Zhi Chen[1], John P Welsh[2], Douglas Richardson[3], Jennifer M Pollard[2], Daisy Richardson[1] and Mohammed Shameem[1]

[1]*Sterile Product and Analytical Development, Merck Research Laboratories, Kenilworth, NJ, USA*
[2]*Process Development and Engineering, Merck Research Laboratories, Kenilworth, NJ, USA*
[3]*Bioprocess Technology and Expression, Merck Research Laboratories, Kenilworth, NJ, USA*

Abstract

Biologics production using yeast or CHO with Yeastolate as cell culture additives often introduces β-glucan, which could potentially pose immunogenicity risk, if not adequately removed. Although a previous study has shown the effective clearance of Yeastolate-derived β-glucan by Protein A chromatography, the clearance pattern of yeast cell derived β-glucan remains unknown. In this study, we characterized the β-glucan clearance patterns during downstream processing of three monoclonal antibody (mAb) products, one mAb fragment from *Pichia pastoris* (mAb A) and two full mAbs from CHO expression system (mAb B and mAb C), by Glucatell assay. We demonstrated effective β-glucan clearance in both small (100 L) and large scale (5000 L) batches of mAb A as well as in one batch of mAb B. Protein A purification step removed an average of 97.74% (1.7 $\log_{10}$ reduction) of β-glucan detected in the two batches of mAb A microfiltration permeates (MFP) and 99.99% (3.9 $\log_{10}$ reduction) of β-glucan detected in mAb B clarified culture fluid harvest (HCCF). Residual β-glucans post Protein A purification in the two batches of mAb A were further removed by the two polishing chromatography steps (94.76% reduction on average). Residual β-glucan measured in the mAb A and mAb B drug substance ranges from 7.8 to 19 pg/mg, which is unlikely to alter physiological concentrations significantly in healthy adults when administered with typical intravenous doses. However, in mAb C, after almost complete removal (99.99%) by Protein A purification step, β-glucan level increased more than 20 fold in Viral Filtration (VF) product sample, indicating that it can be introduced from materials used in downstream process, such as cellulose-based filters and membranes. Our study results suggest that although β-glucan can be cleared by Protein A and other chromatographic steps such as AEX, monitoring β-glucan clearance during downstream process development remains very important to identify and avoid potential contaminations to the drug substance.

Keywords: β-glucan; Yeastolate; Chromatography; Process clearance; Biotherapeutics; Monoclonal antibody

Introduction

Monoclonal antibodies (mAbs) account for a significant portion of biotherapeutics because of their target specificity and their ability to engage a wide variety of targets [1,2]. mAbs are commonly produced by genetically-engineered Chinese Hamster Ovary (CHO) cells, with other host cells of non-human origin, such as *E. coli* and yeast, also being used [3-5]. During mAbs production, the cell culture media and subsequent feed media play important roles in maintaining cell growth and promoting protein production [6]. Traditionally, cell culture media are supplemented with animal serum to promote cell growth; however, safety concerns associated with the use of serum in biotherapeutics production have triggered the use of serum-free or animal origin-free media [7]. Although chemically-defined media has gained popularity, plant- or yeast-derived hydrolysates are still commonly used in cell culture media due to their vital roles in promoting cell growth [8-10]. However, the use of yeast-derived hydrolysates or yeastolate in cell culture media as well as the use of yeast cell for mAbs production has sparked potential safety concerns due to the introduction of β-glucan into the production process.

Beta-glucans are polysaccharides of D-glucose monomers linked by β-glycosidic bonds, with the most common forms being β-1, 4 glucans (also known as cellulose) and β-1, 3 glucans. Beta-1, 3 glucans often have side-chains of D-glucose attached at the (1, 4) or (1, 6) position, known as β-1, 3/1, 4 glucans or β-1, 3/1, 6 glucans. Beta-1, 3/1, 4 glucans are usually water soluble fiber commonly found in oats (*Avena sativa*), which are often used as dietary fiber supplements [11-13]. The most active forms of β-glucans are β-1, 3/1, 6 glucans, which are known as "biological response modifiers" because of their ability to activate the immune system [14-19]. These β-glucans are commonly found in yeast cell wall and yeast extract, and are considered process-related impurities with potential immunogenicity risk in biotherapeutics. Although previous studies have shown that β-glucans introduced in raw materials can be easily cleared by Protein A chromatography [20], potential contamination of β-glucans during downstream purification processes has also been observed, with sources of contamination including raw materials used in chromatography mobile phases, formulation buffers, and cellulose-based filters [21]. Therefore, it is important to characterize the process clearance of β-glucans during downstream process development to ensure product safety according to the Q6B guidelines from the International Conference on Harmonization of Technical Requirements for Registration of Pharmaceuticals for Human Use (ICH).

CHO-produced monoclonal antibody purification often follows

***Corresponding author:** Fengqiang Wang, Associate Principal Scientist, Sterile Product and Analytical Development, Merck Research Laboratories, Kenilworth, NJ, Mailstop K15-4-E214, 2000 Galloping Hill Rd., Kenilworth, NJ 07033, USA

similar downstream steps such as harvest by centrifugation and filtration followed by capture using Protein A chromatography, low pH for viral inactivation, polishing steps by AEX or CEX chromatography, Viral Filtration (VF), and Ultrafiltration/Diafiltration (UFDF) [22-24]. For biotherapeutics produced from *Pichia pastoris* or *E. coli*, both viral inactivation and viral filtration steps are not needed since these hosts are not suitable for amplification of viruses that are infectious for humans. To characterize the β-glucan clearance patterns during downstream processing, process-intermediates retained from a monoclonal antibody fragment (mAb A) produced from *Pichia pastoris* and two full monoclonal antibodies (mAb B and mAb C) produced from CHO expression system were analyzed by Glucatell assay. The production of both mAb B and mAb C uses cell culture feed media containing Yeastolate, which has been considered as a source of β-glucan [20]. The downstream process steps used for purifying mAb A, mAb B, and mAb C are illustrated in Figure 1. Since β-glucans normally do not bind to Protein A resin, it is expected that β-glucans released into cell culture from the use of yeast cell or yeast extract can be removed during Protein A chromatography step. However, the risk of introducing additional β-glucans from raw materials or cellulose-based filters used in downstream process steps persists [21]. Therefore, in this study, we analyzed the β-glucan levels in all process-intermediates retained from downstream processing of both yeast-produced mAb A and CHO-produced mAb B and mAb C.

Materials and Methods

Materials

Microfiltration Permeate (MFP), Quenched ProA Pool (QPAP), Mix-Mode Chromatography product (MMC), AEX product (AEX), and UF retentate (UF) were collected from mAb A downstream processing at Fujifilm Diosynth Biotechnologies; Clarified Culture Fluid Harvest (HCCF), Neutralized Viral Inactivated Product (NVIP), AEX product (AEX), Viral Filtration product (VF), and UFDF product retains were collected from mAb B and mAb C downstream purification steps at Merck & Co. Inc. For mAb A, process-intermediates from both a small scale (100 L) and a large scale (5000 L) production were retained. These samples were used for residual β-glucan testing by Glucatell assay. The downstream processes used for mAb A, B, and C purification and the steps where retains were collected are shown in Figure 1. All samples were shipped frozen from Kenilworth, NJ to Associates of Cape Cod in East Falmouth, MA and stored at -70°C until analysis. The concentrations of process-intermediates described above are listed in Tables 1 and 2.

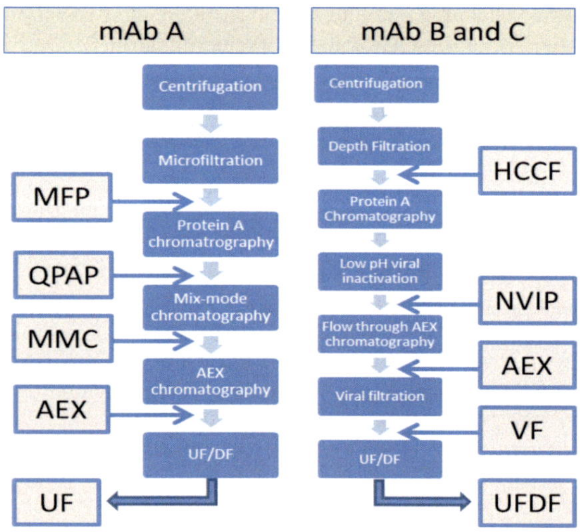

Figure 1: Schematic illustration of the downstream purification processes and the process-intermediates collected for mAb A, B, and C. Left: downstream purification process and process-intermediates collected for mAb A; Right: downstream purification process and process-intermediates collected for mAb B and mAb C.

Beta-glucans analysis

The Glucatell-kinetic chromogenic method was used to determine (1, 3)-β-D-glucan concentration in three mAbs' process intermediates and their corresponding buffer matrices. The test was performed at Associates of Cape Cod Inc (East Falmouth, MA) using the Glucatell kit following similar procedure as previously reported [20,21]. Briefly, five β-glucan calibration standards in a series of 2-fold dilution (from 50 pg/mL to 3.125 pg/mL) were added to a 96-well plate in duplicate. Limulus Amebocyte Lysate (LAL) Reagent Water (LRW) was used as negative control, while samples were diluted with LRW in serial dilutions ranging from 1:2 to 1:2,000,000 and tested in duplicate to find minimum, non-interfering dilutions. A parallel series of sample dilutions described above were spiked with 12.5 pg/mL β-glucan standard as positive controls. The plate was placed in the plate reader preheated to 37°C (±1°C), shaken for 10 seconds and the absorbance's at 405 nm were measured every 15 seconds for 1 h. The chromogenic onset time (seconds) was then plotted against standard concentration to draw a linear curve. For system suitability, the recovery % of the positive control must be > 50% and < 200%; the relative differences of the duplicated standards must be < 25%; and the correlation coefficient (R) of the linear curve must be ≥ 0.980); the negative control values must be lower than the values observed in the lowest standard concentration (3.125 pg/mL).

Results

Beta-glucan clearance patterns in mAb A

A monoclonal antibody fragment (mAb A) was produced from a glyco-engineered yeast cell line, *Pichia pastoris*. The downstream purification steps of mAb A are illustrated in Figure 1. Beta-glucan concentrations in process-intermediates collected from one small scale (100 L, Lot 1) and one large scale production batch (5000 L, Lot 2) were tested by the Glucatell assay. Testing results were expressed as both pg/ml in the sample solution and pg/mg relative concentration to the protein content in the testing samples. MFP samples from both small scale and large scale production have relatively high amount of β-glucans, with 8,847 pg/mL detected in the small scale MFP sample and 15,181 pg/ml detected in the large scale MFP sample (Table 1). The relative concentrations of β-glucans in these two samples are at 11,580 and 14,188 pg/mg of protein content, respectively, which have a relative difference of 20.2% (Table 1). After ProA purification, β-glucans were removed to lower than 1 ng/mg (1 ppm) in both cases, with 1.9 $\log_{10}$ reduction observed in Lot 1 and 1.5 $\log_{10}$ reduction observed in Lot 2 (Table 1). The scaling up of production may have contributed to the slightly increased amount (18.4%) of β-glucan in MFP and relatively lower reduction factor by Protein A chromatography in Lot 2. Additionally, the polishing chromatography steps further removed β-glucans by a total of 0.8 and 1.3 $\log_{10}$ in Lot 1 and Lot 2, respectively, as calculated by the pg/mg concentration (Table 1). The final β-glucan concentrations in mAb A Drug Substance (DS) were 10 pg/mg (Lot 1) and 19 pg/mg (Lot 2), respectively, which are not considered

Samples	Protein (g/L)		β-glucan (pg/mL)		β-glucan (pg/mg)		Log_{10} reduction	
	Lot 1	Lot 2	Lot 1	Lot 2	Lot 1	Lot 2	Lot 1	Lot 2
MFP	0.76	1.07	8,847	15,181	11,580	14,188	-	-
QPAP	8.30	10.98	1,335	4,878	161	444	1.9	1.5
MMC	5.36	7.78	330	172	62	22	0.4	1.3
AEX	4.84	7.72	128	167	26	22	0.4	0.0
UF	85.1	81.8	856	1,588	10	19	0.4	0.1

Table 1: β-glucan concentrations measured in *P. pastoris* produced mAb A process-intermediates and the log_{10} reduction from each purification step.

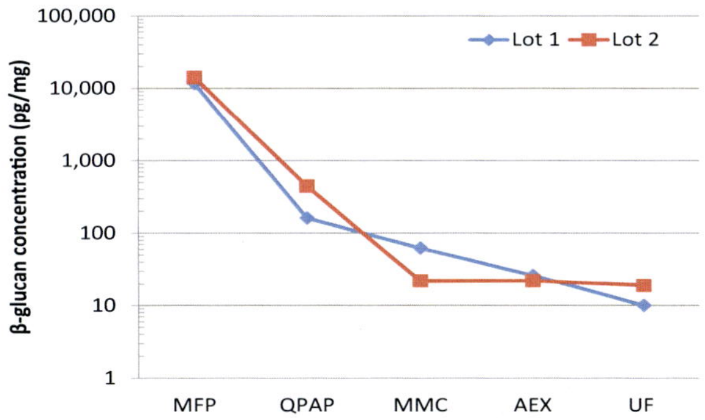

Figure 2: β-glucan clearance patterns during mAb A purification process. The majority of β-glucans found in the MF permeate of mAb A was removed by ProA affinity purification and mix-mode chromatography purification steps.

Samples	Protein (g/L)		β-glucan (pg/mL)		β-glucan (pg/mg)		Log_{10} reduction	
	mAb B	mAb C	mAb B	mAb C	mAb B	mAb C	mAb B	mAb C
HCCF	2.3	2.7	73,928	152,059	32,143	56,318	-	-
NVIP	8.9	17.5	38.2	79.6	4.3	4.6	3.9	4.1
AEX	8.2	8.9	70.7	44.6	8.6	5.0	- 0.3	- 0.0
VF	7.8	5.4	80.7	626	10.3	116.0	- 0.1	- 1.4
UFDF	66.4	50.2	517	3,974	7.8	79.2	0.1	0.2

Table 2: β-glucan concentrations measured in CHO-produced mAb B and mAb C process-intermediates and the log_{10} reduction from each purification step.

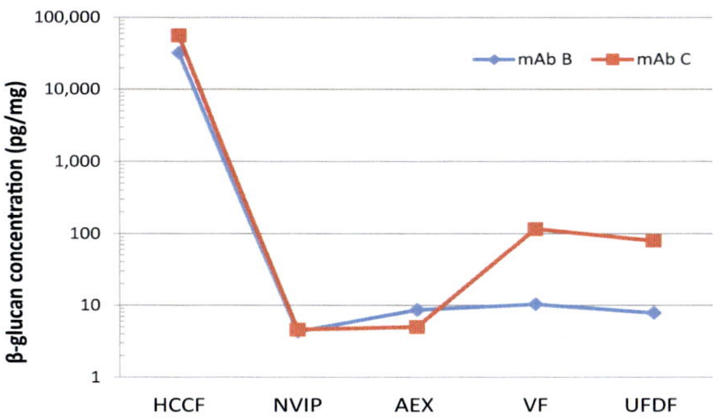

Figure 3: β-glucan clearance patterns during mAb B and mAb C purification process. The majority of β-glucans found in the HCCF of mAb B and mAb C was removed by ProA purification, however, further introduction of β-glucans in mAb C VF sample was observed.

significantly different given the accuracy (50%-200%) of the Glucatell method. The pattern of β-glucan clearance in mAb A downstream process is also shown in Figure 2.

Beta-glucan clearance patterns in mAb B

Monoclonal antibody B was produced in genetically engineered CHO cells. Similar to the downstream purification processes of mAb A, mAb B was first harvested and centrifuged to remove cell debris, filtered and then captured on Protein A chromatography, followed by low pH viral inactivation, AEX chromatography, viral filtration and UFDF steps (Figure 1). Process-intermediates testing results show that β-glucans observed in the HCCF sample of mAb B (32143 pg/mg), were almost completely removed by Protein A chromatography step, with only 4.3 pg/mg of β-glucan detected in the NVIP sample (3.9 log_{10} reduction, Table 2). Beta-glucan concentrations measured in AEX, VF, and UFDF process-intermediates (Figure 1) range from 7.8 pg/mg to 10.3 pg/mg, which are approximately twice of that in NVIP, but within the method variation of the Glucatell assay. It appears that Protein A and viral inactivation steps of mAb B purification almost completely removed β-glucans detected in HCCF, with no further removal of residual β-glucans observed during flow through AEX, viral filtration, and UFDF steps. The pattern of β-glucan process clearance during mAb B downstream process is shown in Figure 3.

Beta-glucan clearance patterns in mAb C

Monoclonal antibody C was also produced in CHO cells and purified following a similar downstream process to mAb B (Figure 1). Similarly, β-glucans detected in the HCCF sample of mAb C were almost completely removed by Protein A chromatography, with a 4.1 log_{10} reduction of β-glucans observed in the NVIP sample as compared to the HCCF sample (Table 2). However, β-glucan levels increased more than 22 times from 5 pg/mg to 116 pg/mg after viral filtration step (Table 2). A previous study by Gefroh, E. and colleagues has shown that β-glucans can be introduced into the purification process through raw materials containing β-glucan or leachables from filters made from

Samples	Test 1 (pg/mg)	Test 2 (pg/mg)	Mean value (pg/mg)	Relative difference (%)
VF	116	98.5	107.3	16.3
UFDF	79.2	70.7	75.0	11.3

Table 3: Repeated β-glucan testing results for VF and UFDF process-intermediates from mAb C.

cellulose [21]. The observed increase of β-glucan concentrations after viral filtration in mAb C suggests a contamination or leak of β-glucans from the viral filtration filter used. To rule out the possibility of testing error, VF and UFDF samples from mAb C were re-tested. Less than a 20% difference in amount of β-glucans was found between the results from the two tests (Table 3). The β-glucan concentrations remained at high levels after UFDF, which confirms the lack of β-glucan removal capability at the UFDF step (Figure 3). Since low levels of beta-glucans have also been found in raw materials such as sucrose and cellulose filters used [21], the Millipore V pro virus filter used for mAb C viral filtration could be the potential source of contamination, pending confirmation from future studies.

Discussions and Conclusions

This study reports the β-glucan testing results in process-intermediates retained from both yeast-produced mAb A and CHO-produced mAb B and mAb C where yeast hydrolysate was used as cell culture additive. Protein A chromatography step removed the majority of β-glucans from all three monoclonal antibody products. However, the extent of β glucans removal by Protein A chromatography step differs significantly between *P. p*astoris-produced product (mAb A) and CHO-produced products (mAb B and mAb C). Both our results (Table 2) and data presented from a previous study [20] show close to 4 $\log_{10}$ reduction of β-glucans by Protein A chromatography step in monoclonal antibody products produced in engineered CHO cells, where Yeastolate or Yeast Hydrolysate has been used as culture media additive. In contrast, for mAb A, Protein A purification step only reduced the amount of β-glucans by ~ 2 $\log_{10}$ (Table 1). Significant amount of β-glucans remained after Protein A purification in mAb A, which was further cleared by polishing chromatography steps. Although, the exact cause of β-glucan clearance difference observed above remains unclear, both the differences in host cells and downstream processing may have contributed to this difference. First of all, the structure and linkage differences between β-glucans leaked from yeast cell wall and those in Yeastolate may contribute to the observed β-glucan clearance differences by Protein A purification. A previous report shows that the proteins with internal repeats (PIR proteins) are covalently linked to the *β* 1,3-glucan of the yeast cell wall via alkali-sensitive ester linkage [25]. Other yeast cell wall constituents such as chitin, manno proteins also play important roles in cell wall integrity and have interactions with β-glucans [26]. Therefore, the interaction of β-glucans with other cell wall components may help them non-specifically bind to the Protein A resin and thus only free β-glucans were removed. Secondly, the downstream processing of mAb B and C includes a viral inactivation step after Protein A elution, which is not part of mAb A purification process. It is unclear but also unlikely the low pH viral inactivation step removes β-glucan. However, as indicated in the β-glucan clearance pattern of mAb A, β-glucans not removed by Protein A purification can be further removed through polishing chromatography steps. Residual β-glucan levels in the UF/DF samples of mAb A and mAb B were both at very low levels after downstream purification processes, ranging from 7.8 pg/mg to 19 pg/mg.

As previously reported [20,27], the average serum level of β-glucan in healthy adults was found to be 17 pg/mL ± 34 pg/mL. Given the average circulating blood volume of 77 mL/Kg in healthy adults, a dose of up to 207 mg/Kg (69 mg/Kg ± 138 mg/Kg) drug substance containing 19 pg/mg residual β-glucan can be injected into healthy adults to have β-glucans remained at physiological levels. Therefore, the downstream process removal of β-glucans in mAb A and mAb B is sufficient to assure the residual β-glucan levels will not affect product safety. Due to the effective removal of β-glucans by Protein A chromatography, risks related to β-glucan contamination in cell culture raw materials are not a significant concern. However, contaminants introduced during the polishing chromatography steps or leached from the filtration steps are of more concern since later steps have limited capability in β-glucan removal. The sources for these β-glucan contaminants can in general be divided into two categories: β-glucan contaminants introduced through chemicals used in the chromatography and formulation buffers, and β-glucan contaminants introduced directly into the production solution through interactions with product-contacting filters that leach β-glucan [20,21]. Additional assessment of the above mentioned β-glucan sources and hence their control will have to be performed unless the downstream processes show robust clearance of β-glucan by Protein A chromatography and no further contaminations are introduced in downstream purification steps.

Both mAb B and mAb C downstream processing follow a similar platform approach as shown in Figure 1. However, the filter used in viral filtration step of mAb C is different from what is used for mAb B. Planova 20 N filter was used for mAb B viral filtration, while Millipore VPro was used at mAb C viral filtration step. The potential introduction of β-glucan by Millipore VPro filter is currently under investigation. Additional testing on raw materials may be necessary if the filter investigation results show no leachable β-glucans.

In summary, β-glucan testing results in the process-intermediates of three mAb products provide insights on its clearance patterns during downstream processing. This study also indicates the importance of β-glucan testing and control during the process characterization of biotherapeutics.

Acknowledgement

We thank Dr. Barbara J Markley, Martha Blick, Lori Keleher, and Pamela Soque at Associates of Cape Cod Inc. for performing the Glucatell test on our samples and providing administrative assistance on data reporting. We would also like to thank our colleagues Kudrat Goswami and Dr. David T. Chin for their critical review of this manuscript.

References

1. Brekke OH, Sandlie I (2003) Therapeutic antibodies for human diseases at the dawn of the twenty-first century. Nat Rev Drug Discov 2: 52-62.
2. Leader B, Baca QJ, Golan DE (2008) Protein therapeutics: a summary and pharmacological classification. Nat Rev Drug Discov 7: 21-39.
3. Hudson PJ, Souriau C (2003) Engineered antibodies. Nat Med 9: 129-134.
4. Wurm FM (2004) Production of recombinant protein therapeutics in cultivated mammalian cells. Nat Biotechnol 22: 1393-1398.
5. Nelson AL, Dhimolea E, Reichert JM (2010) Development trends for human monoclonal antibody therapeutics. Nat Rev Drug Discov 9: 767-774.
6. Dick LW Jr, Kakaley JA, Mahon D, Qiu D, Cheng KC (2009) Investigation of proteins and peptides from yeastolate and subsequent impurity testing of drug product. Biotechnol Prog 25: 570-577.
7. Keen MJ, Rapson NT (1995) Development of a serum-free culture medium for the large scale production of recombinant protein from a Chinese hamster ovary cell line. Cytotechnology 17: 153-163.

8. Stoll TS, Muhlethaler K, von Stockar U, Marison IW (1996) Systematic improvement of a chemically-defined protein-free medium for hybridoma growth and monoclonal antibody production. J Biotechnol 45: 111-123.

9. Lee GM, Kim EJ, Kim NS, Yoon SK, Ahn YH, et al. (1999) Development of a serum-free medium for the production of erythropoietin by suspension culture of recombinant Chinese hamster ovary cells using a statistical design. J Biotechnol 69: 85-93.

10. Franek F, Hohenwarter O, Katinger H (2000) Plant protein hydrolysates: preparation of defined peptide fractions promoting growth and production in animal cells cultures. Biotechnol Prog 16: 688-692.

11. Ahmad A, Anjum FM, Zahoor T, Nawaz H, Dilshad SM (2012) Beta glucan: a valuable functional ingredient in foods. Crit Rev Food Sci Nutr 52: 201-212.

12. Cloetens L, Ulmius M, Johansson-Persson A, Akesson B, Onning G (2012) Role of dietary beta-glucans in the prevention of the metabolic syndrome. Nutr Rev 70: 444-458.

13. Majtan J (2012) Pleuran (β-glucan from *Pleurotus ostreatus*): an effective nutritional supplement against upper respiratory tract infections? Med Sport Sci 59: 57-61.

14. Chan GC, Chan WK, Sze DM (2009) The effects of beta-glucan on human immune and cancer cells. J Hematol Oncol 2: 25.

15. Lehtovaara BC, Gu FX (2011) Pharmacological, structural, and drug delivery properties and applications of 1,3-β-glucans. J Agric Food Chem 59: 6813-6828.

16. Meena DK, Das P, Kumar S, Mandal SC, Prusty AK, et al. (2013) Beta-glucan: an ideal immunostimulant in aquaculture (a review). Fish Physiol Biochem 39: 431-457.

17. Soltanian S, Stuyven E, Cox E, Sorgeloos P, Bossier P (2009) Beta-glucans as immunostimulant in vertebrates and invertebrates. Crit Rev Microbiol 35: 109-138.

18. Thompson IJ, Oyston PC, Williamson DE (2010) Potential of the beta-glucans to enhance innate resistance to biological agents. Expert Rev Anti Infect Ther 8: 339-352.

19. Xiang D, Sharma VR, Freter CE, Yan J (2012) Anti-tumor monoclonal antibodies in conjunction with β-glucans: a novel anti-cancer immunotherapy. Curr Med Chem 19: 4298-4305.

20. Jiang C, Scherfner S, Dick LW Jr, Mahon D, Qiu D, et al. (2011) Demonstrating β-glucan and yeast peptide clearance in biopharmaceutical downstream processes. Biotechnol Prog 27: 442-450.

21. Gefroh E, Hewig A, Vedantham G, McClure M, Krivosheyeva A, et al. (2013) Multipronged approach to managing beta-glucan contaminants in the downstream process: control of raw materials and filtration with charge-modified nylon 6,6 membrane filters. Biotechnol Prog 29: 672-680.

22. Chon JH, Zarbis-Papastoitsis G (2011) Advances in the production and downstream processing of antibodies. N Biotechnol 28: 458-463.

23. Liu HF, Ma J, Winter C, Bayer R (2010) Recovery and purification process development for monoclonal antibody production. MAbs 2: 480-499.

24. Shukla AA, Thömmes J (2010) Recent advances in large-scale production of monoclonal antibodies and related proteins. Trends Biotechnol 28: 253-261.

25. Khasa YP, Conrad S, Sengul M, Plautz S, Meagher MM, et al. (2011) Isolation of *Pichia pastoris* PIR genes and their utilization for cell surface display and recombinant protein secretion. Yeast 28: 213-226.

26. Chagas B, Farinha I, Galinha CF, Freitas F, Reis MA (2014) Chitin-glucan complex production by *Komagataella (Pichia) pastoris*: impact of cultivation pH and temperature on polymer content and composition. N Biotechnol 31: 468-474.

27. Odabasi Z, Mattiuzzi G, Estey E, Kantarjian H, Saeki F, et al. (2004) Beta-D-glucan as a diagnostic adjunct for invasive fungal infections: validation, cutoff development, and performance in patients with acute myelogenous leukemia and myelodysplastic syndrome. Clin Infect Dis 39: 199-205.

Insilico Screening and Comparative Study on the Effectiveness of Textile Dye Decolourization by Crude Laccase Immobilised Alginate Encapsulated Beads from *Pleurotus Ostreatus*

Chinnathambi Velu[1]*, Ezhilarasan Veeramani[1], Sridhar Suntharam[2], Kasinathan Kalimuthu[1]

[1]*Research Associate, ARMATS BIOTEK Research Institute, Kottur, Chennai, India-600085*
[2]*Assistant Professor, Jeppiaar Engineering College, Jeppiaar nagar, Chennai, India-600119*

Abstract

The necessity to safeguard the environment has increased the potential of enzyme usage in textile processing to ensure eco-friendly production. Laccase enzyme formulation has been used in textile processing such as bio-bleaching, dyeing, rove scouring, finishing, neps removal, printing, wash-off treatment, dye synthesis and effluent treatment. However, a high cost associated with biocatalyst production is still a hindrance to their use. *Pleurotus ostreatus* is a white-rot fungus that produces a ligninolytic enzyme complex rich in several laccase iso-enzymes. The main objective of this study is optimize influence of pH and stability of divalent metal ion-immobilzed crude laccase enzyme towards decolourization of prototype textile dyes such as Reactive red 80 (Red F3B) and Reactive blue 21(T Blue G). Wheat bran is used as a lead candidature for production of lignolytic enzyme using *Pleurotus osteratus* by solid state fermentation. Two divalent metal ions such as $Zn2^+$ and $Ca2^+$ were selected to study the influence of metal ions towards the dye decolourization. Results revealed that $Ca2^+$ ion was better compared to $Zn2^+$ ion towards enzyme immobilization and its influence on dye decolourization in the optimal pH 5.5. Finally, interactions between laccase and dyes were studied exclusively using Insilco structure based molecular docking methods.

Keywords: Dye degradation; *Pleurotus ostreatus*;Agricultural wastes; Laccase; Molecular docking

Introduction

Nearly ten thousand types of synthetic dyes are widely used in the textile, paper, printing and leather tanning industries, corresponding to 8×10^5 tonnes per year [1]. The chemical classes of synthetic dyes mostly used in industrial processes are azo, anthraquinone, sulfur, indigoid, triphenylmethyl (trityl) and phthalocyanine derivatives, although other types have many applications such as triazyne[2].Besides the color effect, the majority of these compounds are toxic, carcinogenic and highly persistent in the environment. Conventional biological treatment of wastewater is not effective for degradation of dyes, and so a number of chemical and physical techniques have been used to remove them, including adsorption to inorganic or organic matrices, decolorization by photo catalysis and oxidation [3].

However, some ligninolytic microorganisms are capable of degrading a wide variety of pollutants resembling lignin or its derivative, and are an attractive option due to low cost, specificity and the possibility of total mineralization of these compounds [4]. Basidiomycete's fungi are the most efficient lignin-degrading organisms that produce mainly laccase (EC 1.10.3.2), lignin peroxidase (EC 1.11.10.14) and manganese peroxidase (EC 1.11.1.13). Laccases are copper-containing oxidases biocatalyzes the oxidation of electron-rich natural and synthetic organic substrates, in a non-specific manner and degrade them into non-toxic molecular species without forming any recalcitrant metabolites. So only laccase are been used for bioremediation process due to their ability to degrade azo, heterocyclic, reactive and polymeric dyes [5].

A number of oxidative enzymes from bacteria, fungi and plants have been reported to play an important role in numerous waste treatment applications [6], Laccase is one of these oxidative enzymes, which appears to have a great potential [7] and it alone has a limited effect on bioremediation due to its specificity for phenolic subunits in lignin.

Value added bioproducts are at receiving end with greater demand in the global market. Potentially high titres of enzyme products can be produced by solid substrate fermentation [8]. Low cost alternative like agro-wastes are bio-prospected for enhanced enzyme production. Solid-state fermentation (SSF) is generally defined as a culture for the production of ligninolytic enzymes because it mimics the natural environment of the white rot fungi [9,10]. The inductive capability of rice bran to laccase production may be related with its phenolic compounds such as ferulic acid, vanillic acid which were reported to be an inducer for laccase production by white rot fungi [11,12]. Ferulic acid is found approximately 0.1% (w/v) and easily prepared in large quantity from rice bran [13]. Wheat bran provides a conducive, natural habitat for high secretion of lingo-cellulolytic enzymes without incorporation any initial amount of carbon and nitrogen supplements, thereby reducing the process economics. Natural phenolic mediators such as ferulic acid, coumaric acid and syringic acid are present as abundant source in wheat bran, which stimulates in enhanced laccase enzyme production in white rot fungi [14].

Biological processes are efficient at low contaminant concentrations, can be sensitive to shock loads, require long hydraulic retention times and form large amounts of solid residues [15]. Some of these disadvantages can be overcome in an enzyme-based reactor, since enzymes are able to operate over a broad concentration range [16,17]. The potential advantages of this enzymatic treatment, as compared to

***Corresponding author:** Chinnathambi Velu, Research Associate, ARMATS BIOTEK Research Institute, Kottur, Chennai, India-600085

conventional treatments, include application to recalcitrant materials, operation at high and low contaminant concentrations over a wide pH, temperature and salinity range. Recent research in this area has focused on enzymatic process for the treatment of wastewater and soil [18, 19].

In general, it is accepted that in order to increase the potential use of enzymes in wastewater treatment processes, their immobilization is absolutely essential for biochemical stability and reuse [20-22]. Consequently, several studies support the immobilization of laccase. By using these immobilized forms, adequate characteristics including high resistance to thermal denaturation, significant improvement of the enzymatic activity, and its preservation for long periods, have been frequently reported.

In the present study, an attempt has been made to explore the potential decolourization of synthetic textile dyes by using immobilized laccase enzymes isolated from white rot fungus *Pleurotus ostreatus* and also in remediation of the dye. The results obtained from the decolourization of two reactive dyes by the fungus have been described in this article.

Materials and Methods

Organism and inoculum

The fungal culture, *Pleurotus ostreatus*, was obtained from the Microbiology Department, Thiyagarayar college of Arts and Science, Madurai, and was maintained in Potato Dextrose Agar medium (PDA), This was used as inoculums for the experiments.

Chemicals

The azo dye Reactive Red 80 (Red F3B) and Reactive Blue 21(T Blue G) Figure 1 and 2 were obtained from local dye manufacturing unit, Tirupur, Tamil Nadu, India. The chemical structure of reactive dyes are shown in Figure 1, 2 were selected on the basis of their structural diversity and frequent use in local textile industries. The media components and chemicals were purchased from SD Fine Labs, Himedia Labs, and Bombay, India. All chemicals used were of analytical grade.

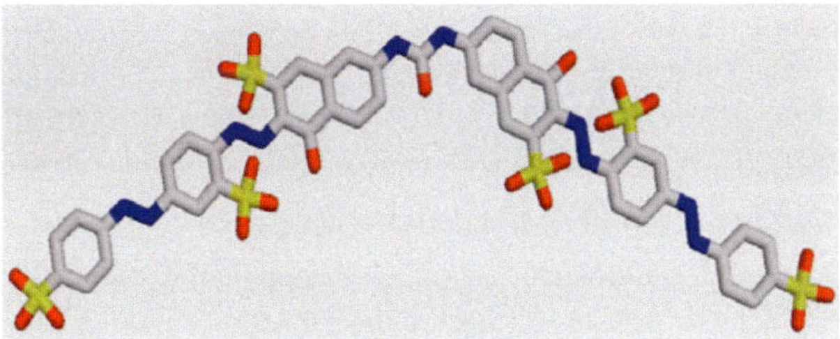

Figure 1: Reactive Red 80.

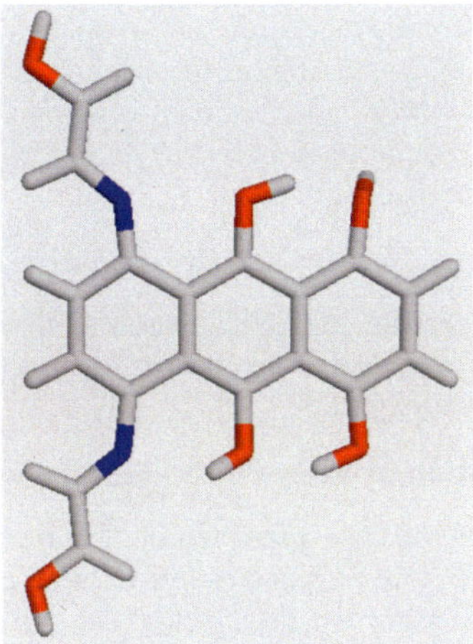

Figure 2: Reactive Blue 21.

Enzyme production

Laccase enzyme was produced using solid state fermentation method. Totally, three different agro-wastes were used as substrates, namely wheat bran, rice bran and green gram. 20 grams of each agro waste was moistened (70%) with double distilled water and transferred into the conical flasks (250ml) and sterilised. 6mm 5 discs of *Pleurotus osteratus* (5 day old culture) were inoculated and the flask incubated at room temperature for 12 days. The contents were extracted with 50 ml sodium acetate buffer (0.2 M) at pH 5.4 and kept overnight in the shaker. The filtrate was centrifuged at 5000 rpm for 20 minutes, the supernatant was collected and subjected to dialysis and stored at 4°C. Later, the partially purified enzyme was used in enzyme estimation and dye degradation studies.

Enzyme activity assay

Extra cellular laccase activity was measured spectrophotometrically following Wolfenden and Wilson method [23] by utilizing syringaldazine as substrate for partially purified enzyme. The reaction mixture contains 0.1 ml of 1 mM syringaldazine in 1.8 ml of 0.2 M sodium acetate buffer (pH 5.4) and 0.1 ml of enzyme. The reaction was monitored by measuring the change in A_{530} for 3 minutes. One unit enzyme activity was defined as the amount of enzyme that oxidizes 1 m mole of syringaldazine per minute at 25°C. The experiment was carried out three times finally the mean value has taken as enzyme activity.

$$\text{Unit / ml} = \frac{(\text{Final OD} - \text{Initial OD}) \times 1000 \times \text{total volume} \times \text{dilution factor}}{\text{Molecular enzyme coefficient} \times \text{enzyme volume} \times \text{time}}$$

pH stability of partially purified Laccase [10]

For determination of pH stability of partially purified laccase enzyme from *Pleurotus ostreatus*, using wheat bran, the enzyme was incubated with different buffers of pH ranging from 4.0, 4.6, 5.0, 5.6, 6.0, 6.6, 7.0, 7.6, and 8.0 after 60 mins incubation the samples were assayed for enzyme activity under standard condition as described before. The experiment was carried out three times; the mean value has been taken as pH stability.

Temperature stability of partially purified Laccase [11]

To determine the temperature stability of partially purified laccase enzyme from *Pleurotus ostreatus* using wheat bran, the enzyme was incubated with sodium acetate buffer of pH 5.8 was incubated with different temperature ranging from 40°C, 50°C, 60°C, 700C, 80°C, 90°C and 100°C after 15mins incubation the samples were assayed for enzyme activity under standard condition as described before. The experiment was carried out three times; the mean value has been taken as pH stability.

Enzyme activity on gel

Gel electrophoresis was carried out on glass plates of 10.5 x 10.5 cm (*Biorad*, USA) in 10% (w/v) polyacrylamide gel with Tris/glycine buffer, pH 8.3. To detect laccase activity after NATIVE- PAGE, this gel was incubated at 35°C for 30 min. It was then rinsed with sodium acetate buffer 0.2 M, pH 5.4 for 10 min and subsequently submerged in 0.1ml guaiacol. The laccase active band was highlighted in dark brown.

Dye decolourization by enzyme immobilization technique

Immobilization of laccase enzyme was carried out using two different metal ions (calcium and zinc). 3 grams of sodium alginate was added slowly to warm water which was continuously stirred. Later, 5-10 ml of the enzyme was added and mixed well. The enzyme beads were prepared at three different pH values of 4.5, 5.5 and 6.5 by adjusting the pH of the sodium alginate solution accordingly before adding the enzyme.

Calcium and zinc alginate beads formation:The dissolved sodium alginate solution was dropped from a height of 15- 20 centimetres into beak ers containing 0.2M calcium chloride solution. Each drop, falling into the beaker, settled in the form of beads and the enzyme was entrapped in the core region of each bead. The beads formed were kept in the calcium chloride solution for 3-4 hours at room temperature in order to harden. After 3 hours, the beads were washed with double distilled water and stored in 50mM calcium chloride solution, like wise prepared Zinc alginate beads by using zinc chloride solution instead of calcium chloride.

Effective pH and metal ions optimization in dye degradation

Two Reactive dyes were collected from textile units located in Tirupur, Tamil Nadu, India ie, Reactive Red 180 and Reactive Blue 21. The dyes were dissolved in double distilled water in the range of 100ppm. Later, the prepared dyes were treated with the above mentioned enzyme beads (calcium alginate and zinc alginate).

The degradation of dyes was analyzed using a spectrophotometer. Reactive Red 180 was analyzed at 540nm and Reactive Blue 21 at 630nm. The absorbance was measured every 24 hr for eleven days. The degree of decolourization was calculated using the following formula.

decolourization percentage= (Initial absorbance - final absorbance/ initial absorbance) x 100

Optimization of pH in dye decolourization with Czapek`s dox broth medium

The Czapek`s dox [24] broth medium was prepared in 250ml conical flasks for each 100ml at 3 different pH values of 4.5, 5.5 and 6.5. The dyes (Reactive Red 180 and reactive Blue 21) were amended individually for each pH value. The medium was autoclaved and inoculated in 5 discs of *Pleurotus ostreatus* fungus. The absorbance of the dye in all the flasks was measured every 24 hr for eleven days. The degree of decolourization was calculated using the following formula.

decolourization percentage= (Initial absorbance - final absorbance/ initial absorbance) x 100

In silico analysis

Homology modelling and structural analysis: The protein sequences used in this project were isolated from Universal Protein Resource [25] The templates used for homology modelling were obtained by running BLAST against the Protein Data Bank (PDB)[26]. On the basis of these hits given by BLAST, the required template PDB structures were downloaded from the protein data bank [27] and global alignment was then performed between the COX-2 sequence and the selected template. The identities between templates were retrieved in terms of the score provided by ClustalW [28]. The structure was modelled using effective and comparative molecular modelling software named MODELLER 27. Modelled structures were then validated with the help of DOPE scores [29] defined by MODELLER. Later, these structures were analyzed with the help of PROCHECK [30] all the macromolecules and ligands were viewed and analyzed with the help of two molecular viewers namely Chimera [31] and VMD [32].

Molecular docking studies: Molecular Docking is a computational technique used in Cheminformatics research. It involves the rapid Computational assessment of most favourable interacting regions between two different molecules. PyRx is virtual screening software for Computational Drug Discovery (CDD). It uses a large body of already established open source software such as Auto Dock 4 [33] and Auto Dock Vina. These two are used as docking software. Python was used as a programming/scripting language. Open Babel was used for importing SDF files, removing salts and energy minimization. Initially, dyes were energy minimized using the steepest decent method [34] with MMFF94 force field [35] later subjected to molecular docking analysis.

Results and Discussion

The main objective of this study was the application of laccase systems in dye decolourization. The activity of the laccase enzyme excreted by *Pleurotus ostreatus* was investigated through SSF on agro-industrial waste including wheat bran (WB), rice bran (RB) and green gram (GG). These wastes acted as the sole carbon source.

Laccase enzyme assay using syringaldazine

Wheat bran substrate showed the highest enzyme activity (2.565U/ml) after dialysis Chart 1, various types of oxidative enzymes are produced by white rot fungi in order to make use of lignocellulosic substrates for nutrition [36] SSF has been considered an efficient method for enzyme production in biotechnology due to its high yield.

The high level of laccase production could be attributed to the presence of ferulic acid in wheat bran. In this study, we selected SSF using various agro wastes such as wheat bran, rice bran and green gram for optimizing the best substrate that exhibits maximum laccase activity. The results indicated that after harvesting and dialysis, the maximum laccase activity was reported in the 12th day sample of wheat bran, whereas other samples showed declining activity.

pH stability of partially purified Laccase

Among the different pH values such as 4.0, 4.6, 5.0, 5.6, 6.0, 6.6, 7.0, 7.6, and 8.0 after 60 mins incubation the pH 6.0 showed high activity Chart 2.

Temperature stability of partially purified laccase

Among the different temperature such as 400C, 500C, 600C, 700C, 800C, 900C and 1000C after 60mins incubation the samples were assayed for enzyme activity under standard condition as described before. Laccase showed high stability at the temperature 700C so laccase is thermo stability in nature Chart 3.

Zymogram analysis: laccase activity on native PAGE

The activity of partially purified Laccase was determined on native PAGE using guaiacol as a substrate. The oxidized region appeared as a clear dark brown band Figure 3.

Dye decolourization by Enzyme immobilization technique

Effective pH and effective metal ion optimization in dye decolorizing process by enzyme immobilization technique:Laccase enzyme beads were prepared using two metal ions ie., calcium chloride and zinc chloride at three different pH levels (4.5, 5.5, and 6.5). Reactive Red 80 and Reactive Blue 21 were used for the analysis of degradation bye

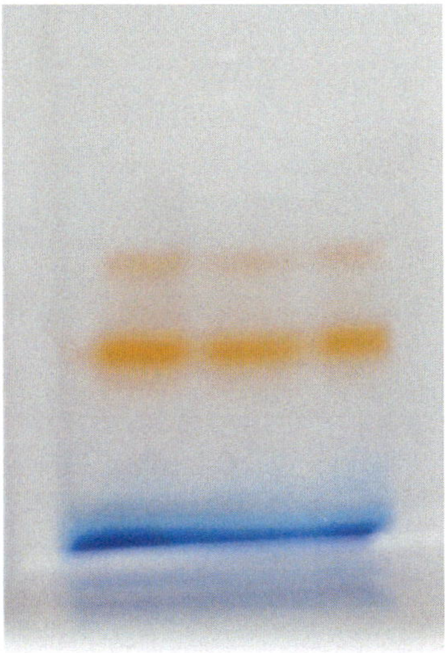

Figure 3: Laccase activity on Native page.

pH	Reactive Red 80 at 540nm		Reactive Blue 21 at 630nm	
	Calcium alginate Beads	Zinc Alginate Beads	Calcium alginate Beads	Zinc Alginate Beads
4.5	29.26	13.04	33.21	20.84
5.5	40.71	19.99	67.42	24.71
6.5	28.70	15.57	33.02	24.09

Table 1: Effective pH and effective metal ion optimization in dye decolorizing process by enzyme immobilization technique.

Reactive Red 80 at 540nm		Dye decolourization percentage for 11 days					
S.no	pH	On day 1st	On day 3rd	On day 5th	On day 7th	On day 9th	On day 11th
1	4.5	6.69	11.87	19.08	23.02	33.34	45.65
2	5.5	15.45	27.78	39.09	48.67	56.90	67.20
3	6.5	11.23	19.78	24.67	31.56	40.89	49.61

Table 2: pH optimization in dye decolorization with Czapek's dox broth medium for Reactive Red 80 at 540nm.

Reactive Blue 21 at 630nm		Dye decolourization percentage for 11 days					
S.no	pH	On day 1st	On day 3rd	On day 5th	On day 7th	On day 9th	On day 11th
1	4.5	18.03	23.35	29.23	35.45	41.66	57.83
2	5.5	15.09	28.07	37.45	49.01	71.90	88.05
3	6.5	11.21	17.56	24.65	29.37	30.48	51.65

Table 3: pH optimization in dye decolorization with Czapek's dox broth medium for Reactive Blue 21 at 630nm.

S.No	Dye	Binding Energy	Ligand Efficiency	Ref RMS	Hydrogen Bond	Molecular Weight (daltons)
1	Reactive Red 80	-1.76	-0.02	23.67	VAL 218	381.37
2	Reactive Blue 21	-4.74	-0.18	25.16	VAL 218	293.31

Table 4: Docking analysis.

immobilised laccase. From the detailed analysis, the optimum metal ion for dye decolourization using immobilized system was found to be calcium at a pH of 5.5. The reactive dye Red 80 showed maximum dye degradation percentage at pH 5.5 was 40.71 on 11th day while other pH values showed dye degradation percentage for pH 4.5 was 29.26 % and 6.5 was 28.70% by using calcium alginate beads. And by using Zinc alginate beads, the dye degradation percentage was at pH 4.5=13.04%, 5.5=19.99% and 6.5= 15.57% on 11th day. Whereas the reactive Blue 21 showed maximum dye degradation percentage at pH 5.5 was 67.42 % on 11th day while other pH values showed dye degradation percentage for pH 4.5 was 33.21 % and 6.5 was 33.02% by using calcium alginate beads. And by using Zinc alginate beads, the dye degradation percentage was at pH 4.5=20.84 %, 5.5=24.71% and 6.5= 24.09% on 11th day Table 1.

pH optimization in dye decolourization with Czapek`s dox broth medium

The optimum pH for the maximum dye degradation was found to be 5.5 through Czapek`s dox broth medium analysis Table 2 and 3.

Molecular modelling of laccase

The sequence of Laccase was retrieved from Universal Protein Resource (UniProt) and its corresponding sequence id was P51589. It consists of 529 amino acids. This sequence was subjected to similarity search against Protein Data Bank, using the BLAST tool offered by NCBI. Later, the templates were selected on the basis of structural hits and its alignment pattern against the query sequence. The selected templates were as follows: chain A of 1GYC, chain A of 3KW7 and chain A of to 1V10.

The advanced modelling tutorial package offered in MODELLER was utilized for comparative molecular modelling. The DOPE score belonging to the best modeled structure was -60304.7734. The stereo-

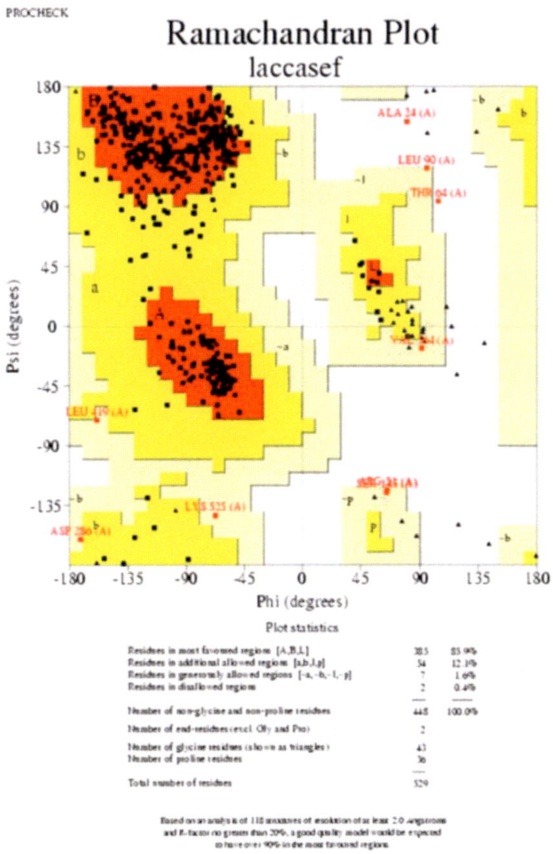

Figure 4: PROCHECK structure validation-plot.

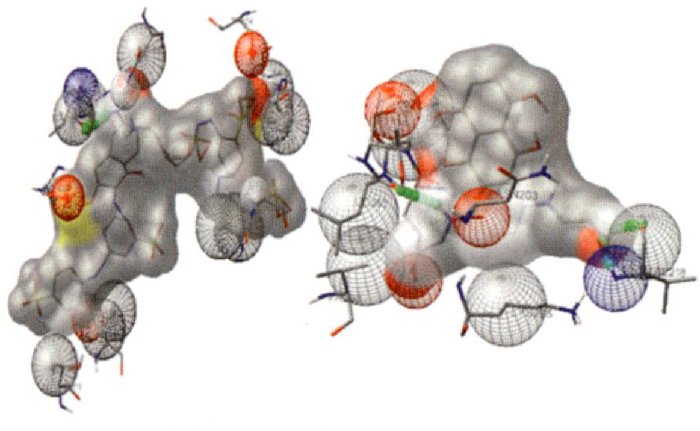

Figure 5: Molecular docking shows the interaction between dye and laccase.

chemistry qualities of the structures were validated with PROCHECK [37] structural validation tool. PROCHECK results clearly indicated the higher fidelity of modeled Laccase structure Figure 4.

Molecular docking studies

The enzyme Laccase and dyes Reactive Red 80 and reactive Blue 21 were subject to molecular docking analysis using AutoDock Module, which is available in PYRX-0.8 software. In the AutoDock Module, molecular docking was performed using Genetic algorithm parameters with a maximum of 25, 00, 000 energy evaluations Table 4. Later, results were analyzed with the help of AutoDock [38] tools 1.4.5. The interactions between the ligand and the target are given in Figure 5. The amino acids interact with dyes are exhibiting remarkably enhanced binding affinities. The residue Valine 218 was interacting with two dyes and plays a potential role in degradation. The higher affinity of this small molecule was presumably attributed to the formation of hydrogen bonds and eventually leads to the degradation of dyes. The hydrogen bond between drugs and Laccase are highlighted as green colour beads Figure 5.

Conclusion

The utilization of bioremediation techniques in the degradation of industrial pollutants is a research field of high focus. Particularly, laccase enzymes play a vital role in biodegradation of toxic substances. From this study, it is evident that immobilization of laccase enzyme using calcium metal ion would enhance its activity when compared to zinc metal ion. Also, the optimum pH for a maximum laccase enzyme activity in dye degradation is 5.5. The interaction between the dyes and enzyme were discussed effectively with the help of in silico analysis. Hence, employing immobilized laccase enzyme for the degradation of dyes Reactive Red 80 and reactive Blue 21 could be effective bioremediation method to prevent industrial pollution.

Acknowledgement

Cordially, we would like to thank our director Dr.P.Arumugam, ARMATS-BIOTEK and all my colleagues for their moral support in this study.

References

1. Revankar MS, Lele SS (2007) Synthetic dye decolorization by white rot fungus, Ganoderma sp *WR-1*. Biores. Technol 98: 775-780.
2. Forgacs E, Serháti TC, Oros G (2004) Removal of synthetic dyes from wastewaters: a review. Environ Intern 30: 953-971.
3. Behnajady MA, Modirshahla N, Fathi H (2006) Kinetics of decolorization of an azo dye in UV alone and UV/H2O2 processes. J Haz.Mat 136: 816-821.
4. Van der Zee FP, Villa Verde S (2005) Combined anaerobic-aerobic treatment of azo dyes- A short review of reactor studies. Water Res 39: 1425-1440.
5. Baldrian P, Snajdr J (2006) Production of ligninolytic enzymes by litter-decomposing fungi and their ability to decolorize synthetic dyes. Enzyme Microbiol Technol 39: 1023-1029.
6. Durán N, (1997) The impact of biotechnology in the pulp and paper industry(Eds.), Progress in Microbial Ecology.
7. Gianfreda LF, Xu JM, Bollag (1999) Laccases: a useful group of oxidoreductive enzymes. Bioremediation J 3:1-26.
8. Raimbault M (1998) General and microbiological aspects of solid substrate fermentation. J Biotechnol 1:174–188.
9. Couto SR, Sanromán MA(2005) Application of solidstate fermentation to ligninolytic enzyme production. Biochem Eng J 22: 211-219.
10. Gao JM, Weng HB, Zhu DH, Yuan MX, Guan FX, et al.(2008) Production and characterization of cellulolytic enzymes from the thermoacidophilic fungal Aspergillus terreus M11 under solid-state cultivation of corn stover. Bioresource Technol 99: 7623-7629.
11. Bollag J.M , Leonowicz A (1984) Comparative studies of extracellular fungal laccases. Applied and Environmental Microbiology 48: 849-854.
12. Munoz C, Guillen F, Martinez AT , Martinez MJ (1997) Induction and characterization of laccase in the lininolytic fungus *Pleurotus eryngii*. Current Microbiology 34: 1-5.
13. Taniguchi H, Hosoda A , Tsuno T, Maruta Y, Nomura E (1999) Preparation of ferulic acid and its application for the synthesis of cancer chemopreventive agents. Anticancer Research 19: 3757-3761.
14. Revankar MS, Lele SS (2006) Enhanced production of laccase using a new isolate of white rot fungus WR-1. Process Biochem 41:581–588.
15. Durán N , Esposito E (1997) Lignin biodegradation and effluent treatment by ligninolytic fungi, in: Microbiologia Ambiental 12: 269.
16. Edwards W, Bownes R, Leukes WD, Jacobs EP, Sanderson R, et al. (1999) Capillary membrane bioreactor using immobilized polyphenol oxidase for the removal of phenols from industrial effluents. Enzyme Microb Technol 24: 209-217.
17. Sisak C, Nagy E, Burfeind J, Schugerl K (2000) Technical aspects of separation and simultaneously enzymatic reaction in multiphase enzyme membrane reactors. Bioprocess Eng 23: 503-512.
18. Karam J, Nicell JA (1997) Potential application of enzymes in waste treatment. J Chem Technol Biotechnol 69:141.
19. Duran N, Esposito E (2000) Potential applications of oxidative enzymes and phenoloxidase-like compounds in wastewater and soil treatment: a review. Appl Catal B: Environ 28:83-99.
20. Peralta-Zamorav P, Pereira CM, Tibursio EL, Moraes SG, Minussi RC,et al.(2003) Decolorization of reactive dyes by immobilized laccase. Applied Catalysis B: Environmental 42: 131-144.
21. Cordi L, Freire RS, Kubota LT , Duran N (2000) Laccase immobilization on pyrolytic carbon in Kraft E1 effluent treatment. Proceedings of the Seventh National Meeting on Environmental Microbiology.
22. Freire RS, Durán N , Kubota LT (2001) Effect of fungal laccase immobilization procedures for the development of a biosensor for phenol compounds. Talanta 54: 681-686.
23. Benny Cheefetz, Yona Chen , Yitzhak Hadar (1998) Purification and Characterization of Laccase from *Chaetomium thermophilium* and Its Role in Humification. Applied and environmental biology 64: 3175–3179.
24. Anna jarosz-wilkolazka, Janina Kochmanska-Rdest, Elzbieta malarczyk, Wladyslaw Wardas (2002) Fungi and their ability to decolourize azo and anthraquinonic dyes. Enzyme and microbial technology 30: 566-572.
25. The Universal Protein Resource (UniProt) (2009) UniProt Consortium, Nucleic Acids Res. 37.
26. Altschul SF, Gish W, Miller W, Myers EW, Lipman DJ (1990) Basic local alignment search tool. J Mol Biol 215 : 403–410.
27. Berman HM, John Westbrook, Zukang Feng, Gary Gilliland, TN Bhat, et al. (2000) The Protein Data Bank, Nucleic Acids Res 28: 235–242.

28. Thompson JD, Higgins DG, Gibson TJ (1994) CLUSTAL W: improving the sensitivity of progressive multiple sequence alignment through sequence weighting, position-specific gap penalties and weight matrix choice. Nucleic acids research 22: 4673-80.

29. Eswar M, Marti-Renom MA, Webb B, Madhusudhan MS, Eramian D, et al. (2006) Comparative Protein Structure Modeling With MODELLER.Current Protocols in Bioinformatics 15: 5.6.1-5.6.30.

30. Eramian D, Shen MY, Devos D, Melo F, Sali A ,et al. (2006) A composite score for predicting errors in protein structure models. Protein Science 15: 1653–1666.

31. Pettersen EF, Goddard TD, Huang CC, Couch GS, Greenblatt DM,et al. (2004) UCSF Chimera visualization system for exploratory research and analysis. J Comput Chem 25: 1605-1612.

32. William H, Andrew D, Klaus S (1996) VMD - Visual Molecular Dynamics. Journal of Molecular Graphics 14: 33-38.

33. Trott O , Olson AJ (2010) AutoDock Vina: improving the speed and accuracy of docking with a new scoring function, efficient optimization and multithreading. Journal of Computational Chemistry 31: 455-461.

34. Fedoryuk MV (2001) Method of steepest descent in Hazewinkel, Michiel, Encyclopaedia of Mathematics. Springer.

35. Halgren T.A (1996) Merck Molecular Force Field. V. Extension of MMFF94 using experimental data, additional computational data and empirical rules. J Comp Chem 17: 616-641.

36. Hegde S, Kavitha S, Varadaraj MC, Muralikrishna G (2006) Degradation of cereal bran polysaccharide-phenolic acid complexes by *Aspergillus niger* CFR1105. Food Chem 96: 14–9.

37. Laskowski RA, Rullmannn JA, MacArthur MW, Kaptein R, Thornton JM (1996) AQUA and PROCHECK-NMR: programs for checking the quality of protein structures solved by NMR, J Biomol NMR 8: 477-486.

38. Morris GM, Goodsell DS, Halliday RS, Huey R, Hart WE, et al. (1998) Automated Docking Using a Lamarckian Genetic Algorithm and and Empirical Binding Free Energy Function. J Computational Chemistry 19: 1639-1662.

Determination of Pesticide Residues in Honey using the GC×GC-TOFMS Technique

Żaneta Bargańska*, Ewa Olkowska, Tomasz Dymerski and Jacek Namieśnik

Gdansk University of Technology, Faculty of Chemistry, Department of Analytical Chemistry, Narutowicza 11/12, 80-233 Gdańsk, Poland

Abstract

In this study, a QuEChERS method and gas chromatography coupled with time-of-flight mass spectrometry (GC×GC-TOFMS) was developed for rapid extraction and simultaneous determination of 12 pesticide residues in honey. The GC×GC-TOFMS method was validated according to the SANCO guidance in terms of linearity, selectivity, reproducibility and recovery. Regarding the results, recovery rates ranged between 70-120% with relative standard deviations <20% in most cases. The method Limits of Quantification (LOQ) ranged between 6-26 ng/g. According to the estimated LOQ values the analytical procedure can be applied to analysis of real honey samples.

Keywords: Environmental monitoring; GC×GC-TOF-MS; Honey; Pesticide residues

Introduction

Honey produced by honey bees from pollen, plant nectars, and/or honeydew is composed of over 300 chemical substances which belong to different chemical compound groups. These are mainly carbohydrates, water, polysaccharides, fatty acids, proteins, minerals, dyes, fragrances, enzymes, hormones and vitamins in amounts depending on the plant from which the honey was made [1].

Honey bees can bring many pollutants deposited on plants into the hive. Therefore, plant protection products used in agriculture can not only cause mass poisoning of bees, but may also be transferred to bee products, especially honey affecting its quality, properties and posing a particular threat to human health [2,3]. Pesticides are a significant group of xenobiotics affecting the biota. Regulation 396/20 of the European Parliament and of the Council established values of the Maximum Residue Levels (MRLs) of pesticides in products of plant and animal origin Regulation (EC), 2005. Since September 1st 2008, the European Commission set new MRLs of some pesticides in honey, which are within the range of 10 and 50 ng/g [4]. Honey can be used as an indicator of environmental pollution with radioactive elements [5], heavy metals [6] or pesticides [7].

Sample preparation and isolation/enrichment of the target compounds are very important analysis steps because the pollutants are present in honey at very low concentration levels [8]. There are many extraction techniques, which are designed to disallow the determination of very low pesticide residues in honey. Solid Phase Extraction (SPE) [9,10] and Liquid-Liquid Extraction (LLE) [11,12] are the most common extraction and purification techniques used in the determination of pesticide residue in honey. Other extraction techniques, such as Supercritical Fluid Extraction (SFE) [13], Matrix Solid Phase Dispersion (MSPD) [14], Solid Phase Micro Extraction (SPME) [15,16] and Stir Bar Sorptive Extraction (SBSE) [13,17] have been developed to reduce the quantities of reagents and time spent on sample preparation.

In recent years, the QuEChERS (quick, easy, cheap, effective, rugged and safe) method developed in 2003 [18] has become the most frequently employed technique for determination of pesticide residues in food (especially in fruits and vegetables). The QuEChERS method has many advantages. It is less time-consuming, does not require high financial costs and it has wide applicability. In this method, pesticides are extracted with acetonitrile; water is removed by salting out. In most stage the acetonitrile extract is cleaned up by mixing with an SPE sorbent rather than passing it through an SPE column [19].

In this study, the QuEChERS method was applied to honey samples for the extraction of antibiotics in honey and bee products [20] and also for determination of neonicotinoid and other pesticides [21,22]. So far gas or Liquid Chromatography coupled with a tandem Mass Spectrometer (LC-MS/MS and GC-MS/MS) has been mainly used for determining pesticide residues in honey or other bee products [23,24].

We have described the first evaluation and adaptation of the QuEChERS approach in combination with two-dimensional gas chromatography coupled with time-of-flight mass spectrometry (GC×GC-TOFMS) for determining pesticides in honey. The method provides good analytical results for the targeted pesticides in the method validation according to the SANCO guidance [25].

Materials and Methods

Chemicals and reagents

The solution of hexachlorobenzene (HCB) in acetonitryle, 1000 mg/mL, analytical grade, used as an internal standard was purchased from Sigma-Aldrich (Schnelldorf, Germany). Certified Reference Materials (solutions in acetonitrile, 100 mg/mL) of bifenthrin, diazinon, pyriproxyfen were purchased from LGC Standards (Łomianki, Poland). The CRM solutions (in acetonitrile, 100 mg/mL) of alachlor (in methanol, 100 mg/mL), vinclozoline and quinalphos were obtained from Ultra Scientific (North Kingston, RI, USA), and the CRM solutions of haloxyfrop-R-methyl (in acetonitrile, 10 mg/mL) was obtained from Dr. Ehrenstorfer GmBH (Germany). The CRM solutions of 4,4'-DDE (in methanol, 500 µg/mL), 4,4'-DDD and endosulfan

***Corresponding author:** Żaneta Bargańska, Gdansk University of Technology, Faculty of Chemistry, Department of Analytical Chemistry, Narutowicza 11/12, 80-233 Gdańsk, Poland

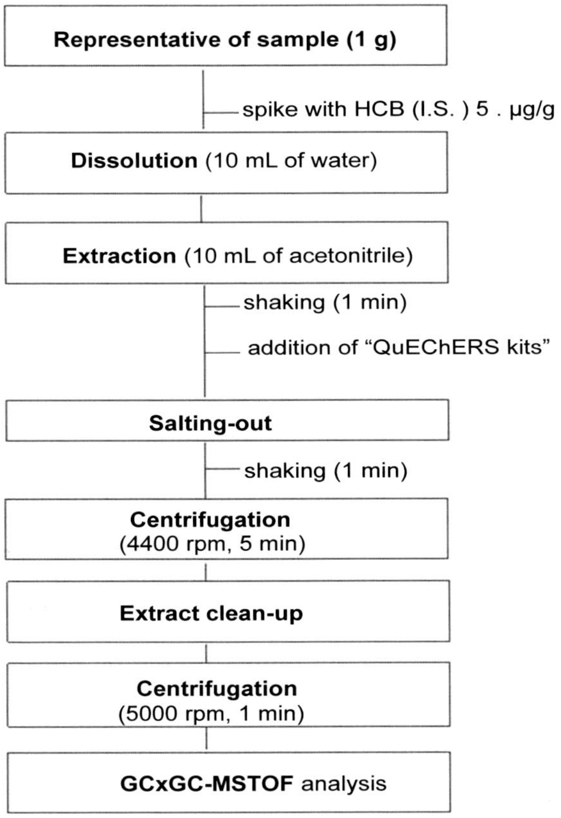

Figure 1: Analytical procedure work-up flow chart.

(analytical standard, 250 mg), dieldrin (in acetonitrile, 100 mg/mL) were obtained from Sigma-Aldrich (Schnelldorf, Germany). The stock standard solutions were stored at −18°C. The calibration standards and working standards were prepared by dilution with acetonitrile on the day of analysis.

Acetonitrile and methanol (LC-MS Chromosolv® 99.9%) were obtained from Fluka (Sigma-Aldrich, Germany). Water was purified with a Milli-Q water system (Millipore Corporation, USA). The QuEChERS kits (part no. 5682-5650) with salt packets containing 4 g anhydrous magnesium sulfate, 1 g sodium chloride, 1 g of sodium citrate and 0.5 g sodium hydrogencitrate sesquihydrate, and two-milliliter centrifuge tubes with 150 mg anhydrous magnesium sulfate and 25 mg primary-secondary amine (PSA) for dispersive solid phase extraction (dSPE, part no. 5982-5021) were purchased from Agilent Technologies (USA).

Instrumentation

GC×GC–TOFMS instrument from LECO, Co. (USA) consisted of Agilent 7890A gas chromatograph with cooled injection system of Gerstel (CIS4), 7683 Series autosampler and time-of-flight mass spectrometer Pegasus IV LECO Co. The column set consisted of a 30 m × 0.25 mm × 1 μm primary column (1D) with Equity 1 stationary phase (Supelco, Germany) and a 1.0 m × 0.25 mm × 0.25 μm secondary column (2D) with SG-Wax stationary phase (SGE, Austin, TX, USA).

Inside the GC oven, a dual-stage jet modulator and the secondary oven were utilized. Resistively heated air was used as a medium for hot jets, while cold jets were supplied by a gaseous nitrogen, previously cooled by liquid nitrogen. Instrumental parameters of the GC×GC–TOF-MS method were as mentioned below.

GC×GC–TOFMS analysis

Injection mode: Programmed Temperature Vaporization (PTV); injection volume: 25 μL; modulation time: 4 s (hot pulse 0.8 s); modulation temperature offset: 30°C and cool time between stages: 1.20 s. The PTV temperature was programmed as follows: 105°C (0.2 min) →280°C (at 10°C/s) (2 min).

Gas chromatography: The temperature program of the first column (main GC oven) was as follows: 100°C (2 min), 100°C→120°C (at 40°C/min), 120°C→280°C (at 5°C/min) (13 min). The temperature of the second oven was programmed from 130°C (2 min), 130°C→150°C (at 40°C/min), 150°C→280°C (at 5°C/min) with a final hold time of 13 min.

Mass spectrometric detection: Acquisition rate: 125 Hz; acquisition delay: 520 s; mass range: 40-550 amu; ion source temperature: 250°C; transfer line temperature: 250°C; detector voltage: 1600V, Electron Impact ionization (EI); electron energy −70 eV.

Data analysis: Data processing was done automatically using the algorithm for peak deconvolution implemented in the Chroma TOF software (LECO Corp., version 4.44). Analytes were tentatively identified through MS library search using the NIST 2011 mass spectral library. Tentative identification was confirmed for some compounds by injecting authentic standards.

Sample preparation

The honey samples were thoroughly homogenized and approximately 1 g of the homogenate was transferred into a polypropylene centrifuge tube (50 mL) containing 10 mL of water, 10 mL of Acetonitrile (ACN) and 50 mL of the internal standard solution (HCB at 100 mg/mL). The content of the tube was shaken manually and the QuEChERS salt kit was added. The sample was immediately hand-shaken for 1 min and subsequently centrifuged at 4400 rpm for 3 min.

Thereafter, 1 mL of the acetonitrile fraction was transferred into a 2 mL dSPE polypropylene tube. The tube was hand-shaken for 1 min and centrifuged for 1 min at 5000 rpm. Finally, 0.5 mL of the supernatant was taken into a glass auto sampler vial. Figure 1 illustrates the scheme of the entire analytical protocol used during the present study.

Results and Discussion

Figure 2A shows a contour plot obtained by the analysis of the honey blank sample. By mass spectral comparison to a library, major co-extracts were identified and also confirmed by a comparison with authentic standards. Figure 2B shows the contour plot of the pesticides standard at a concentration of 20 ng/g. For the identification of individual pesticides in this figure 2, data presented in Table 1 were used data mentioned in Table 1.

Method performance

For evaluation of the method performance calibration sets were prepared at concentrations of 10, 20, 50, 100, 200, and 500 ng/mL, both in solvent, thereby enabling evaluation of linearity. Validation of the overall method was performed according to an EU guideline [25], which involved analysis of two control samples five samples spiked with 12 pesticides and contaminants at 20 ng/g and five samples spiked at 50 ng/g.

All standards and extracts samples were analyzed in one sequence, starting with the solvent standards, then the standards in matrix, and

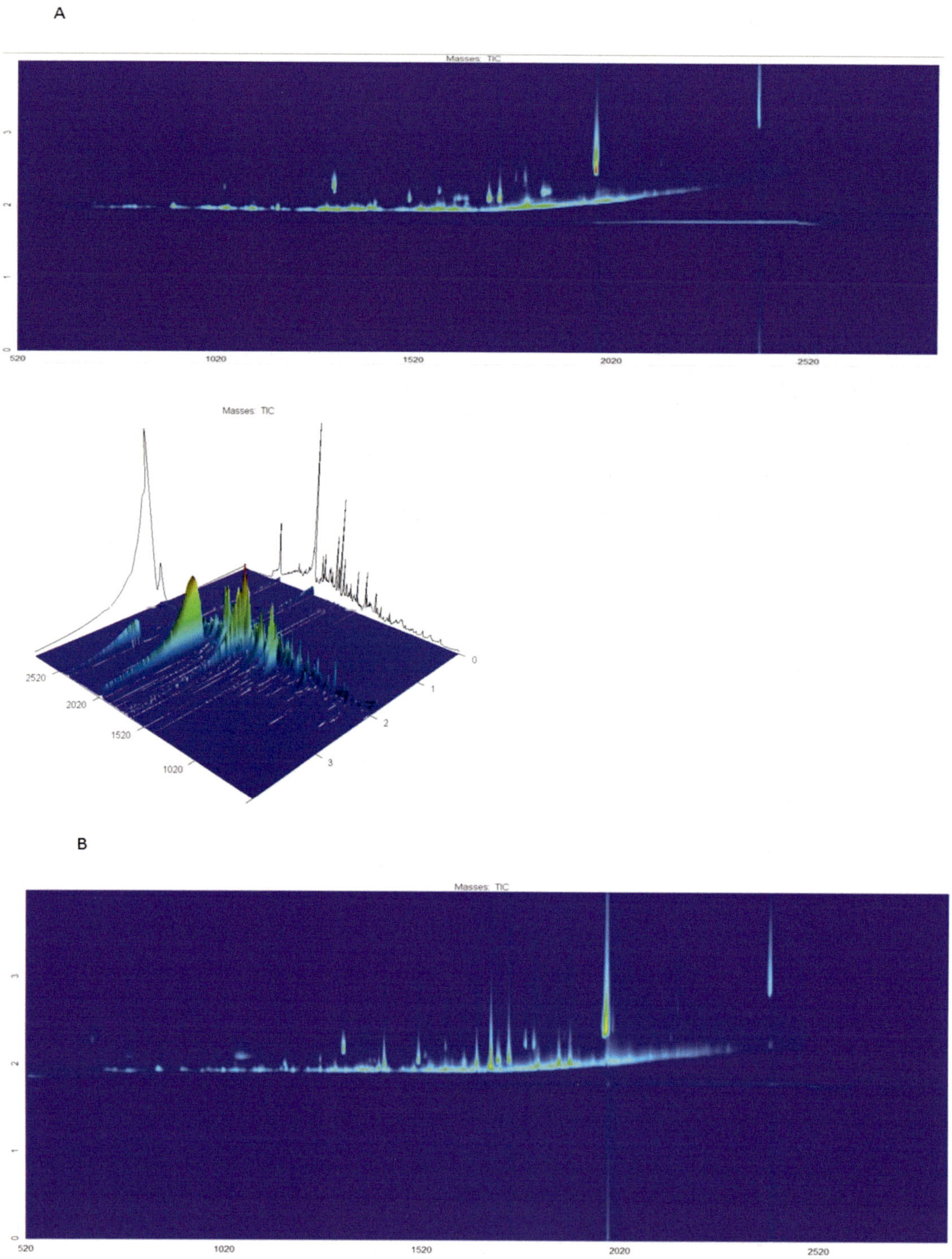

Figure 2: GC×GC chromatograms (TIC plots) of pesticides isolated by QuEChERS from (A) blank honey sample and (B) spiked honey sample at 20 ng/g.

finally the validation extracts. The validation extracts, in sets of four, were bracketed by additional calibration standards (HCB) in matrix at the level of 5 µg/mL (corresponding to 5 µg/g). The repetitive injections of the 50 ng/mL standards in matrix were used to assess the stability of the system performance.

The Limit of Detection (LOD) is the lowest concentration of analytes detectable by an analytical method and the Limit of Quantification (LOQ) is the lowest solute concentration that can be determined with acceptable precision and accuracy, under the stated experimental conditions. In this study, LOD and LOQ values were determined regarding the LOD as 3 times the baseline noise and the LOQ as the concentration that produced a relation signal to baseline noise of 10 [26].

The equations of calibration curves in the solvent are included

Pesticides	Quantify mass *m/z*	Retention time in 1 dimension [s]	Retention time 2 dimension [s]
Alachlor	270 238 162	1520	2.208
Bifenthrin	181 165 115	2080	2.520
4,4'-DDE	246	1824	2.312
4,4'-DDD	235	1832	2.432
Diazinon	304 137	1388	2.112
Dieldrin	261 263 265	1836	2.336
Endosulfan	337 339 341	1784	2.296
Haloxyfrop-R-methyl	315 91	1736	2.176
Hexachlorobenzene (HCB)	284	1320	2.216
Pyriproxyfen	321 136	2156	3.352
Quinalphos	298 146	1700	2.312
Vinclozoline	285 287	1500	2.296

Table 1: Target compounds, qualifier ions and retention times.

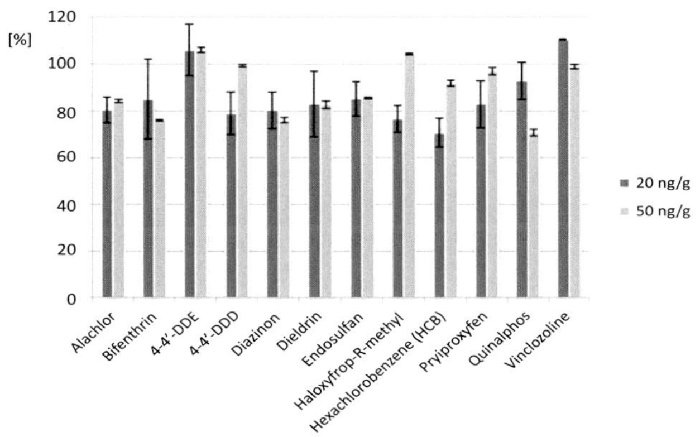

Figure 3: Pesticide recoveries (%) with error bars representing the residue standard deviation (n=5) obtained by QuEChERS in honey spiked at 20 ng/g and 50 ng/g.

in Table 2. Based on correlation coefficients of calibration curves, the proposed analytical procedure is characterized by linearity in the investigated range. The other method performance parameters (limit of detection, limit of quantification) also are included in Table 2. Figure 3 shows the mean recovery of analytes with residue standard deviation (RSD) for n=5. The obtained recoveries ranged between % (RSD 2.4-19.5). Recovery values are in accordance with the SANCO guidance [25].

Conclusions

This study describes a new, rapid, easy, efficient and robust analytical procedure based on the QuEChERS and GC×GC-TOFMS technique for the simultaneous determination of 12 pesticide residues in honey samples. The developed methodology is characterized by low values of limits of detection and quantification and relatively good recovery rates

Pesticides	Calibration curves solvent	Correlation coefficient r	LOD [ng/g]	LOQ [ng/g]
Alachlor	y=0.07154x+0.4588	0.9998	10	14
Bifenthrin	y=0.8312x+0.3235	0.9988	12	15
4,4'-DDE	y=0.7574x+0.3041	0.9997	1.5	7.5
4,4'-DDD	y=0.8148x−0.002	0.9999	1.4	9
Diazinon	y=0.7388x+0.2243	0.9999	10	19
Dieldrin	y=0.9555x+0.459	0.9996	4.5	16
Endosulfan	y=0.8112x+0.3134	0.9988	2.5	6
Haloxyfrop-R-methyl	y=0.7029x−0.3007	0.9989	10	17
Hexachlorobenzene (HCB)	y=0.9676x+0.4666	0.9997	15	26
Pyriproxyfen	y=0.6112x+0.4134	0.9996	15	19
Quinalphos	y=0.2120x+0.2132	0.9998	14	18
Vinclozoline	y=0.6212x+0.2114	0.9999	16	19

Table 2: Method performance data.

of analytes. According to the aforementioned aspects, it can be used for the determination of pesticide residues in real honey samples.

In summary, the obtained results show that high-resolution 2-D chromatography coupled with mass spectrometry can be used for identification and quantification of pesticide residues in the aforementioned samples Moreover, those analytical techniques can be applied to identify various ranges of contaminants present in honey and other bee products.

Acknowledgment

This work has been supported financially by the National Science Centre in Poland based on Decision No. DEC-2012/05/N/ST4/00211 and by the system project "InnoDoktorant – Scholarships for PhD students, 5th edition". The project is co-financed by the European Union in the framework of the European Social Fund.

References

1. Erdogrul Ö (2007) Levels of selected pesticides in honey samples from Kahramanmaras, Turkey. Food Control 18: 866-871.
2. Kujawski MW, Namiesnik J (2008) Challenges in preparing honey samples for chromatographic determination of contaminants and trace residues. TrAC 27: 785-793.
3. Barganska Z, Slebioda M, Namiesnik J (2013) Pesticide residues levels in honey from apiaries located of Northern Poland. Food Control 31: 196-201.
4. (2005) Regulation (EC) No 396/2005 of the European Parliament and of the Council of on maximum residue levels of pesticides in or on food and feed of plant and animal origin and amending Council Directive 91/414/EEC. FAOLEX.
5. Tonelli D, Gattavecchia E, Ghini S, Porrini C, Celli G, et al. (1990) Honey bees and their product as indicators of environmental radioactive pollution. J Radioanal Nucl Chem 141: 427-436.
6. Conti ME, Botrè F (2001) Honeybees and their products as potential bioindicators of heavy metals contamination. Environ Monit Assess 69: 267-282.
7. Celli G, Porrini C, Radehhieri P, Sabatini AG, Marcazzan L, et al. (1996) Honeybees (*Apismellifera* L.) as bioindicators for the presence of pesticide in the agro ecosystem. Insect Social Life 1: 207-212.
8. Barganska Z, Namiesnik J (2010) Pesticide Analysis of Bee and Bee Product Samples. Crit Rev Anal Chem 40: 159-171.
9. Amendola G, Pelosi P, Dommarco R (2011) Solid-phase extraction for multi-residue analysis of pesticides in honey. J Environ Sci Health B 46: 24-34.
10. Pang GF, Fan CL, Liu Y, Cao YZ, Zhang JJ, et al. (2006) Multiresidue method for the determination of 450 pesticide residues in honey, fruit juice and wine by double-cartridge solid-phase extraction/gas chromatography-mass spectrometry and liquid chromatography-tandem mass spectrometry. Food AdditContam 23: 777-810.
11. Kujawski MW, Namiesnik J (2011) Levels of 13 multi-class pesticide residues in Polish honeys determined by LC-ESI-MS/MS. Food Control 22: 914-919.

12. Rissato SR, Galhiane MS, de Almeida MV, Gerenutti M, Apon BM (2007) Multiresidue determination of pesticides in honey samples by gas chromatography-mass spectrometry and application in environmental contamination. Food Chem 101: 1719-1726.

13. Rissato SR, Galhiane MS, Knoll FR, Apon BM (2004) Supercritical fluid extraction for pesticide multiresidue analysis in honey: determination by gas chromatography with electron-capture and mass spectrometry detection. J Chromatogr A 1048: 153-159.

14. Sánchez-Brunete C, Albero B, Miguel E, Tadeo JL (2002) Determination of insecticides in honey by matrix solid-phase dispersion and gas chromatography with nitrogen-phosphorus detection and mass spectrometric confirmation. J AOAC Int 85: 128-133.

15. Blasco C, Font G, Picó Y (2008) Solid-phase microextraction-liquid chromatography-mass spectrometry applied to the analysis of insecticides in honey. Food Addit Contam Part A Chem Anal Control Expo Risk Assess 25: 59-69.

16. Das YK, Kaya S (2009) Organophosphorus insecticide residues in honey produced in Turkey. Bull Environ Contam Toxicol 83: 378-383.

17. Yu C, Hu B (2009) Sol-gel polydimethylsiloxane/poly(vinylalcohol)-coated stir bar sorptive extraction of organophosphorus pesticides in honey and their determination by large volume injection GC. J Sep Sci 32: 147-153.

18. Anastassiades M, Lehotay SJ, Stajnbaher D, Schenck FJ (2003) Residues and trace elements - fast and easy multiresidue method employing acetonitrile extraction partitioning and "dispersive solid-phase extraction. J AOAC Int 86: 412-431.

19. Omar N, Bakar J, Muhammad K (2013) Determination of organochlorine pesticides in shrimp by gas chromatography-mass spectrometry using a modified QuEChERS approach. Food Control 34: 318-322.

20. Lombardo-Agüí M, García-Campaña AM, Gámiz-Gracia L, Cruces-Blanco C (2012) Determination of quinolones of veterinary use in bee products by ultra-high performance liquid chromatography-tandem mass spectrometry using a QuEChERS extraction procedure. Talanta 93: 193-199.

21. Kamel A (2010) Refined methodology for the determination of neonicotinoid pesticides and their metabolites in honey bees and bee products by liquid chromatography-tandem mass spectrometry (LC-MS/MS). J Agric Food Chem 58: 5926-5931.

22. Barakat AA, Badawy HMA, Salama E, Attallah E, Maatook G (2007) Simple and rapid method of analysis for determination of pesticide residues in honey using dispersive solid phase extraction and GC determination. J Food Agric Environ 5: 97-100.

23. Walorczyk S, Gnusowski B (2009) Development and validation of a multi-residue method for the determination of pesticides in honeybees using acetonitrile-based extraction and gas chromatography-tandem quadrupole mass spectrometry. J Chromatogr A 1216: 6522-6531.

24. Tomasini D, Sampaio MRF, Caldas SS, Buffon JG, Durate FA, et al. (2012) Simultaneous determination of pesticides and 5-hydroxymethylfurfural in honey by the modified QuEChERS method and liquid chromatography coupled to tandem mass spectrometry. Talanta 99: 380-386.

25. (2009) Method validation and quality control procedures for pesticide residues analysis in food and feed.

26. Ribani M, Collins CH, Bottoli CB (2007) Validation of chromatographic methods: evaluation of detection and quantification limits in the determination of impurities in omeprazole. J Chromatogr A 1156: 201-205.

28

Biosorption of Co(II) by *Schizosaccharomyces pombe*: Kinetic and Thermodynamic Studies for Process Design

Nihat Alpagu Sayar, Selcen Durmaz-Sam, Gizay Odabasi and Ahmet Alp Sayar*

Department of Bioengineering, Marmara University, Kadikoy 34722, Istanbul, Turkey

Abstract

The characterisation of biosorption by *Schizosaccharomyces pombe* of Co(II) was done under varying experimental conditions including pH, temperature, initial metal concentration and biosorbent dose. Optimum pH was found to be 5.0 with optimal biosorbent dose for highest adsorption capacity at 1.0 g L^{-1}. Higher initial metal concentrations helped enhance q_e slightly with the highest value, 2.7 mg g^{-1}, being achieved at 25°C and 50 mg L^{-1} initial Co(II). Temkin isotherm model showed the best statistical fit for equilibrium data while Freundlich isotherm model exhibited thermodynamic consistency. Pseudo-first order kinetics was found to represent Co(II) biosorption. Increasing ΔG^0 values calculated with respect to higher operational temperatures were in agreement with kinetic profiling simulations whereby higher temperatures were found to be less favourable. Overall efficiency of *S. pombe* as a biosorbent for Co(II) was found to be significantly less than Ni(II), for this reason it can be regarded as a potential bioseparation agent for aqueous mixtures containing both Co(II) and Ni(II).

Keywords: Biosorption; *Schizosaccharomyces pombe*; Adsorption isotherms; Process thermodynamics, Biokinetics; Separation

Nomenclature

A_0 Frequency factor

A_T Temkin isotherm equilibrium constant, L g^{-1}

b_K Khan isotherm constant

b_T Temkin isotherm constant

C_e Residual aqueous metal concentration at equilibrium, mg L^{-1}

C_i Initial aqueous metal concentration, mg L^{-1}

E_a Activation energy, kJ mol^{-1}

k Kinetic rate constant

K_{eq} Selected isotherm equilibrium constant

K_F Freundlich isotherm equilibrium constant, L mg^{-1}

K_L Langmuir isotherm equilibrium constant, L mg^{-1}

k_{p1} Pseudo-first order rate constant, min^{-1}

k_{p2} Pseudo-second order rate constant, g mg-1 min^{-1}

K_{RP} Redlich-Peterson isotherm equilibrium constant, L g^{-1}

K_S Sips isotherm equilibrium constant, (mg g^{-1}) (mg L^{-1})$^{-\beta}$

m Amount of biosorbent, g

n_F Freundlich isotherm exponent

n_K Khan isotherm exponent

q_0 Maximum metal uptake, mg g^{-1}

q_e Equilibrium metal uptake, mg g^{-1}

q_i Initial metal uptake rate, mg g^{-1}

q_t Metal uptake at t, mg g^{-1}

R Universal gas constant, 8.314 J mol^{-1} K^{-1}

R^2 Correlation coefficient

RMSE Root mean square error

T Time, s or hr

T Temperature, K

V Volume of aqueous phase, L

α_{RP} Redlich-Peterson isotherm constant

α_S Sips isotherm constant, (L mg^{-1})$^{\beta}$

β_{RP} Redlich-Peterson isotherm exponent

β_S Sips isotherm exponent

ΔG^0 Gibbs free energy change, J mol^{-1}

ΔH^0 Enthalpy change, J mol^{-1}

ΔS^0 Entropy change, J mol^{-1} K^{-1}

Introduction

Biosorption of heavy metal ions from aqueous solutions is an area where there are growing interest and success [1]. Previous studies have shown that wild-type *Schizosaccharomyces pombe* is a promising agent for the biosorption of nickel(II) from aqueous solutions [2]. Considering the fact that in most naturally and technologically relevant conditions cobalt(II) is found along with nickel(II), it is a logical undertaking to investigate the biosorption properties of wild-type *S. pombe* for cobalt(II) for both biohydrometallurgical separation and environmental control purposes.

***Corresponding author:** Ahmet Alp Sayar, Department of Bioengineering, Marmara University, Kadikoy 34722, Istanbul, Turkey

A variety of biological materials have been examined as biosorption agents for Co(II). Among these are marine algae *Ascophyllum nodosum* [3], crab shell particles [4], fungi of genera *Aspergillus, Rhizopus* [5] and *Saccharomyces* [1]. This study focuses on the biosorptive properties of *S. pombe* on Co(II). Advantages of using this organism are its rapid growth rate, high biomass yield and non-pathogenic character. These advantages also ensure cheaper and easier production at large scales.

It also has to be noted that there has been mounting interest in using membrane processes for the separation of cobalt and nickel in aqueous solutions. The reported separation factors achieved using some of these processes are very high, with examples like 80 [6] or 250 [7]. Although such values are seemingly remarkable, one has to keep in mind that membrane technology displays some serious disadvantages in terms of industrial application such as scale-up and staging difficulties and membrane fouling. Therefore, adsorption and solvent extraction methods for the separation of heavy metals still hold primary importance in scientific research and industrial applications.

Characterisation of biosorption properties is an essential prerequisite for further design studies towards industrial application of such processes. Characterisation of the biosorbent properties of *S. pombe* for Co(II) comprises the main aim of the current paper. Fundamental aspects for such a characterisation include equilibrium, kinetic and thermodynamic analyses of the biosorption process. Effects of pH, initial aqueous metal concentration and biosorbent dose also constitute the key aspects of such a study. On top of these conventional and generally performed investigations, the scope of this study also covers the kinetic profiling of and preliminary design calculations for a single stage batch biosorber.

The biosorption of Co(II) by *S. pombe* is presumed to happen via a passive ion exchange process between the aqueous metal and the cellular surface of the biosorbent. Process parameters such as the pH and temperature and initial biomass and metal concentrations in the aqueous media affect the binding between the organism and the metal. Accordingly, these parameters are to be examined in order to assess the feasibility of an industrial application [8]. Industrial feasibility assessment and comparison with alternative systems also require an understanding of the performance of the system for which kinetic information and preliminary design calculations are informative [9].

The biosorption characterisation procedure follows the single sorbate-simple closed batch system methodology. The details of this approach are available in a previous study [2]. The equilibrium values of Co(II) uptake by the biomass, q_e, mg (g dry weight)$^{-1}$ can be calculated using:

$$q_e = \frac{(C_i - C_e)V}{m} \tag{1}$$

where C_i, C_e, V and m denote, initial Co(II) concentration of the aqueous phase (mg L^{-1}), equilibrium Co(II) concentration of the aqueous phase (mg L^{-1}), volume (L) of the solution contacted with sorbent and the amount of the sorbent expressed as dry mass (g), respectively.

Adsorption isotherms were used for the assessment and evaluation of biosorption of Co(II) onto *S. pombe*. These isotherms can be further employed during the preliminary equipment design for a relevant industrial process. Six different adsorption isotherm models were considered for this study. Langmuir, Freundlich, Langmuir-Freundlich (Sips) and Redlich-Peterson isotherms were chosen since they are the most studied and best understood models. Khan and Temkin isotherms [10] were also included. Khan isotherm shows characteristic similarities to the above-mentioned ones, thus offers a possibility of a good fit. Although it is argued elsewhere that Temkin isotherm might not be appropriate for complex adsorption systems including liquid-phase adsorption [11]; Temkin isotherm features a linear dependency to biosorption heat which is in agreement with the observations made of the present system. Mathematical descriptions of these isotherm models can be found in Table 1. Once an isotherm is found to be satisfactorily descriptive of the studies it will be used in an elementary design of a single-stage batch biosorber.

Another important aspect of the characterisation of biosorption is an adequate description of the kinetics of the interaction between the biosorbent and the adsorbate. Pseudo-first or pseudo-second order kinetics is usually found to be adequately descriptive of the biosorption phenomenon [1]. Relevant rate equations are given in Table 2. The appropriate rate equations that describe the currently studied system will be used in a preliminary modelling scheme to simulate the kinetic biosorption profiles.

The third essential part of biosorbent characterisation is the thermodynamic analysis of the process, in terms of feasibility and energy consumption studies as a precedent for industrial design approaches. Thermodynamic functions such as activation energy, Gibbs free energy change, enthalpy and entropy changes are crucial for process design calculations. The activation energy, E_a in kJ mol^{-1}, can be calculated using:

$$\ln k = \frac{-E_a}{RT} + \ln A_0 \tag{2}$$

The calculated value for E_a is an index to be used to distinguish if the biosorption in question is either of a physical or a chemical nature. ΔG^0 can be calculated by:

$$\Delta G^0 = -RT \ln K_{eq} \tag{3}$$

Biosorption isotherm	Formulae
Langmuir	$q_e = \frac{q_0 K_L C_e}{1 + K_L C_e}$
Freundlich	$q_e = K_F C_e^{1/n_F}$
Langmuir-Freundlich (Sips)	$q_e = \frac{K_S C_e^{\beta_S}}{1 + \alpha_S C_e^{\beta_S}}$
Redlich-Peterson	$q_e = \frac{K_{RP} C_e}{1 + \alpha_{RP} C_e^{\beta_{RP}}} c$
Khan	$q_e = \frac{q_0 b_K C_e}{(1 + b_K C_e)^{n_K}}$
Temkin	$q_e = \frac{RT}{b_T} \ln(A_T C_e)$

Table 1: Isotherm models applied to the biosorption data.

Kinetic model	Differential form	Integrated form
Pseudo-first order	$\frac{dq}{dt} = k_{p1}(q_e - q_t)$	$\log(q_e - q_t) = \log q_e - \frac{-k_{p1}}{2.303} t$
Pseudo-second order	$\frac{dq}{dt} = k_{p2}(q_e - q_t)^2$	$q_t = \frac{t}{\frac{1}{k_{p2} q_e^2} + \frac{t}{q_e}}$

Table 2: Kinetic models applied to the biosorption data.

where R is the gas constant (8.314 J mol^{-1} K^{-1}), T is the temperature in Kelvin; K_{eq} (selected equilibrium constant) used in L mol^{-1}. By employing the selected equilibrium constant, the following linear thermodynamic relation can be obtained:

$$\ln K_{eq} = \frac{-\Delta H^0}{RT} + \frac{\Delta S^0}{R} \quad (4)$$

Materials and Methods

Preparation of the biosorbent

The wild-type strain of *S. pombe (972 h)* obtained from University of Istanbul, Deparment of Molecular Biology and Genetics, Turkey, was first cultivated, then harvested and finally prepared to be used as biosorbent in this study. For the cultivation of yeast cells, cultures on agar slants were used to inoculate 100 ml of sterile preculturing medium consisting of 30 g L^{-1} glucose and 5 g L^{-1} yeast extract. Preculturing medium was incubated at 30°C by shaking at a rate of 180 rpm until the early stationary growth phase which was determined by the intermittent spectrophotometric measurements of cell culture's optical density at 600 nm. Then, 200 ml of sterile media containing 30 g L^{-1} glucose and 5 g L^{-1} yeast extract in 500 ml Erlenmeyer flasks were inoculated with previously prepared preculture at an inoculum ratio of 1/100 (v/v). The flasks were agitated at 180 rpm at 30°C for 40 h when the optical density of cultures at 600 nm showed negligible changes. The yeast cells were separated from the media by centrifugation at 9,000 rpm for 3 min at 20°C. Liquid part was discarded and cell pellet was washed twice with ultra-pure grade water. In order to dry yeast cells the pellet was kept at 50°C for 72 h. Dried cell chunks were grinded with the help of a blender. Dry biomass was preserved in a desiccator until it was used in biosorption experiments.

Sterilization of all growth media and flasks was accomplished in an autoclave (OT 032 Nüve, Turkey) at 121°C and 1.0 atm for 15 min. Co(II) stock solution with a concentration of 1 g L^{-1} was prepared by dissolving a proper amount of $CoCl_2.6H_2O$ (Analytical Grade, Carlo Erba, France) in ultra-pure water.

Biosorption experiments

In order to determine the optimum pH for the current biosorption system, experiments were conducted at varying pH values. Five Erlenmeyer flasks (250 mL), each containing 100 mL of 10 mg L^{-1} Co(II) solution, were prepared and adjusted to pH 2.0, 3.0, 4.0, 5.0 and 6.0 (±0.1) respectively, by the addition of either 0.1 (or 1.0) N HNO_3 or 0.1 (or 1.0) N NaOH. Dry biomass of *S. pombe* cells was added to each flask to obtain a concentration of 1 g L^{-1}. The flasks were shaken for duration of 2 h at 150 rpm and 25°C. pH 5.0 ± 0.1 was found to be the optimal operating value used throughout the rest of the experiments.

Further experimentation was conducted using the abovementioned setup but with different initial metal concentration values (2.7 ± 0.1, 5.2 ± 0.2, 7.3 ± 0.2, 11.1 ± 1.1, and 53.8 ± 2.8 mg L^{-1}) and at various operational temperatures (25, 30, 35 and 50 ± 0.2°C). Samples were taken at predetermined time intervals and centrifuged for 5 min at 10,000 rpm and 20°C. Initial and residual Co(II) concentrations of the samples were measured using a Thermo Scientific Atomic Absorption Spectrophotometer (Model: iCE 3500) at 240.7 nm. All experiments were conducted in triplicate.

Results and Discussion

Effect of pH

The pH of the solution within which biosorption takes place influences both the metal and cell surface chemistry and thus is an important operational parameter for a biosorption process. Determination of the optimum pH value for the current Co(II) biosorption system was done by experimentation at 25°C whereby 10 mg L^{-1} Co(II) solutions with a biomass concentration of 1 g L^{-1} each and pH values ranging from 2.0 to 6.0 (by increments of 1.0) were used. pH values over 6.0 weren't considered due to the fact that no improvement on metal uptake was observed over pH 5.0. Figure 1 displays that optimum Co(II) uptake is attained at pH 5.0 where q_e equals 2.55 mg g^{-1}. Almost a linearly correlated increase in metal uptake accompanies the increasing pH of the medium from pH 2.0 to 5.0 where a peak is reached and a slow decline can be observed afterwards. An impressive ten-fold increase in metal uptake is achieved when working at pH 5.0 compared to pH 2.0. An increase in the pH might affect the ionic configuration of some of the functional groups esp. at the cellular surface and improve the interactions with the metal ions.

Effect of sorbent dose

Four experiments with different biosorbent concentrations were conducted to observe the effect of sorbent dose on metal uptake. 50 mg L^{-1} Co(II) solutions with 0.5, 1.0, 2.0 and 3.0 g L^{-1} *S. pombe* concentrations were used at 25°C and pH 5. Although the total amount of metal adsorbed increased with higher biosorbent concentrations (from 1.1 mg L^{-1} at the lowest biosorbent concentration to 6.2 mg L^{-1} at the highest); the adsorption capacity peaked at 1.0 g L^{-1} biosorbent concentration with a value of 2.7 mg g^{-1}. Figure 2 exhibits the trend with which metal uptake at equilibrium relates to biosorbent concentration. The reason why equilibrium metal uptake falls with very high biosorbent concentrations, contrary to expectation, may be due to increased electrostatic interactions between the cells which obstruct the contact between metal cations and cellular binding sites; since it cannot be due to the insufficiency of solute to cover all available binding sites considering the significant amounts of residual metal in solution at equilibrium, observed even at optimal uptake conditions. Over 50% of metal remains in solution under all conditions tested.

A techno-economic trade-off between the cost of biosorbent and the amount of metal adsorbed per unit of the biosorbent exists which a thorough cost-benefit analysis of these parameters may resolve but it

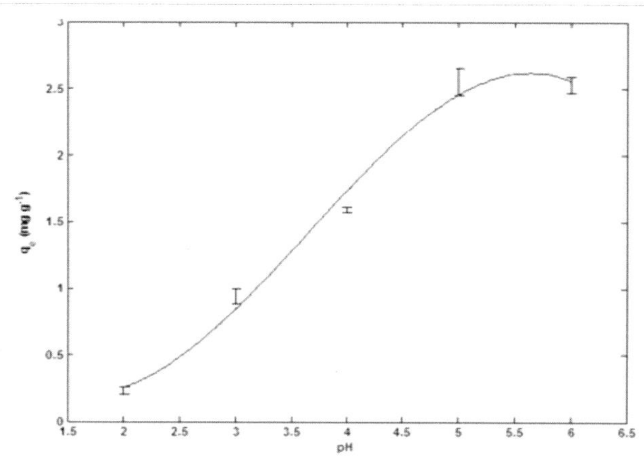

Figure 1: Effect of pH on the equilibrium uptake of Co(II) by *S. pombe* (Temperature: 25°C, initial Co(II) conc.: 10 mg L^{-1}, biosorbent dose: 1.0 g L^{-1}, agitation rate: 150 rpm. The *bars* represent standard deviation of the mean).

falls out of the scope of this study.

Equilibrium studies

Isotherm models: Elucidation of an appropriate biosorption equilibrium correlation is an essential component for reliable prediction of biosorption parameters and quantitative comparison and effective design of different biosorption systems [11]. Experimental equilibrium data of the Co(II) biosorption system was analysed using established isotherm models Langmuir, Freundlich, combined Langmuir-Freundlich (Sips), Redlich-Peterson, Khan and Temkin isotherms. Langmuir, Freundlich and Temkin are two-parameter models whereas the others have three parameters. The three-parameter Sips, Redlich-Peterson and Khan isotherms are modified versions of the Langmuir isotherm combining its features with that of Freundlich. Temkin isotherm is a two-parameter model including a term for temperature.

The correlation between the observed variables C_e and q_e has been analysed using Matlab R2012b's NonLinearModel fit function to determine the optimised values of the model parameters. The use of adjusted-R2 (square of the correlation factor) as statistical performance index is necessitated by the multivariate and nonlinear characteristics of the isotherm model equations. Additionally, root mean square error (RMSE) values for each fitted model are produced as a way to aid the statistical assessment. The results are tabulated in Table 3.

Across all experimentally observed conditions Temkin isotherm model exhibits the best fit statistically.

Effect of initial metal concentration on equilibrium metal uptake: Biosorption processes with varying initial metal concentrations between 2.5 and 50 mg L^{-1} were tested at four different operating temperatures. The experimental results are compared with Temkin isotherm models optimised with appropriate data in Figure 3. The plots show the effect of initial metal concentration, C_i, on the isothermal relationship between equilibrium values of C_e and q_e. An increase of initial metal concentration from 2.5 to 50 mg L^{-1} shows a positive effect, albeit very minor for initial metal concentrations above 10 mg L^{-1}, on the equilibrium values for all temperatures tested. It can be said that, above that level, either saturation of the binding sites on cell surfaces occurs or adsorptive (e.g. affinity between metal ions and cell surface) and desorptive (e.g. electrostatic repulsion between metal ions) forces balance. It has also to be noted that at these metal concentrations and pH value (5.0) precipitation of metal ions doesn't take place.

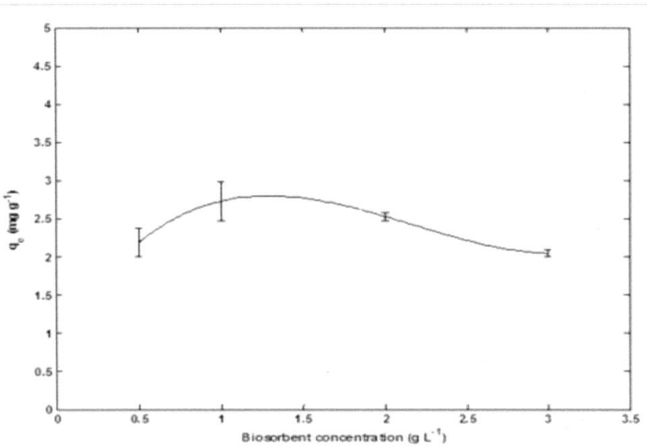

Figure 2: Effect of sorbent dose on equilibrium uptake of Co(II) by *S. pombe* (Temperature: 25°C, initial Co(II) conc.: 50 mg L-*1*, biosorbent dose: 1.0 g L^{-1}, agitation rate: 150 rpm. The *bars* represent standard deviation of the mean).

Combined effect of temperature and initial metal concentration: Figure 4 displays the combined effect of two independent process parameters, namely temperature and initial Co(II) concentration in solution on equilibrium metal uptake. Lower temperatures and higher initial concentrations seem to be favoured by the biosorption process. The kind of information inferred from such plots is valuable for guiding the overall design of the whole industrial application. It can be said that one would favour operations upstream of the biosorption process to carry lower temperature outputs and that there would be a trade-off between maintaining lower upstream temperatures and cooling efforts that might be required to increase the efficiency of the biosorption process. Quantified representation of such information aids the preliminary design calculations for the whole process and guides early design decisions.

Adsorption model	T (°C)	K_L	q_0	Adj-R^2	RMSE
Langmuir	25	0.526	2.690	0.725	0.286
	30	0.717	2.184	0.751	0.198
	35	0.924	1.796	0.830	0.114
	50	1.497	1.548	0.931	0.046

Adsorption model	T (°C)	K_F	n_F	Adj-R^2	RMSE
Freundlich	25	1.346	5.387	0.871	0.196
	30	1.234	6.328	0.849	0.154
	35	1.121	7.615	0.975	0.044
	50	1.128	11.063	0.888	0.058

Adsorption model	T (°C)	K_S	a_S	β_S	Adj-R^2	RMSE
Sips	25	1.752	0.406	0.376	0.827	0.227
	30	1.829	0.634	0.441	0.820	0.169
	35	1.804	0.721	0.355	0.992	0.024
	50	2.550	1.522	0.589	0.975	0.028

Adsorption model	T (°C)	K_{RP}	α_{RP}	β_{RP}	Adj-R^2	RMSE
Redlich-Peterson	25	9.713	6.637	0.836	0.813	0.236
	30	5.001	3.480	0.882	0.803	0.176
	35	5.689	4.525	0.898	0.992	0.025
	50	3.749	2.783	0.958	0.981	0.024

Adsorption model	T (°C)	b_K	q_0	n_K	Adj-R^2	RMSE
Khan	25	7.343	1.013	0.829	0.811	0.237
	30	3.411	1.201	0.876	0.801	0.177
	35	4.284	1.064	0.895	0.992	0.025
	50	2.717	1.277	0.956	0.982	0.023

Adsorption model	T (°C)	A_T	b_T	Adj-R^2	RMSE
Temkin	25	23.567	6386.6	0.878	0.191
	30	56.442	8814.2	0.872	0.142
	35	186.64	12536.0	0.992	0.024
	50	5066.8	20808.0	0.920	0.049

Table 3: Parameters and statistics of the fitted isotherm models.

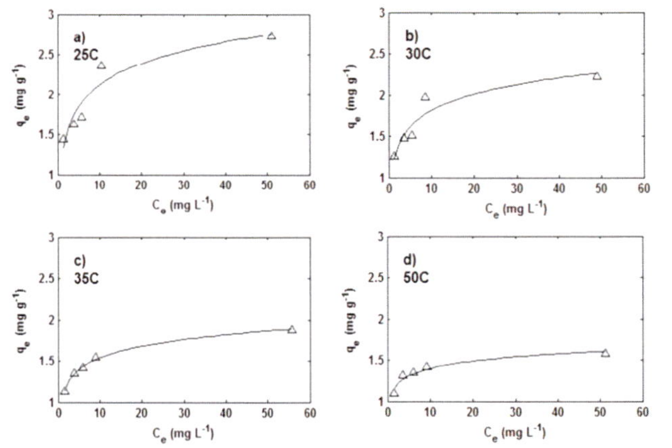

Figure 3: Effect of initial Co(II) concentration on the isothermal relationship between metal uptake and residual metal concentration at equilibrium with fitted Temkin isotherms (pH 5.0, biosorbent dose: 1.0 g L^{-1}, agitation rate: 150 rpm).

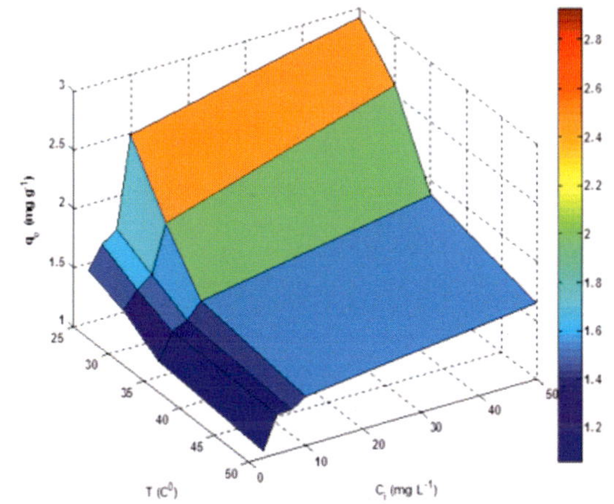

Figure 4: Combined effect of temperature and initial Co(II) concentration on biosorption by inactive *S. pombe*.

Biosorption kinetics

Determining the kinetics of a biosorption process is a key step towards reactor sizing and process optimisation. The rate of adsorption can be calculated by considering the biosorption process as a reaction system and assuming biosorbent concentration to be in great excess with respect to sorbent concentration. Following these assumptions biosorbate concentration can be grouped with the basic rate constant yielding a pseudo-constant which can be used in either a first or second order kinetic model. Both the pseudo-first and pseudo-second order approaches have been tested to correlate the present experimental data. Calculations were done using initial metal concentration values of 2.5, 5.0, 7.5, 10.0 and 50.0 mg L^{-1} at 25, 30, 35 and 50°C temperatures. It was found that pseudo-first order kinetics showed a better fit statistically. Optimised pseudo-first order rate constants with observed and calculated values for equilibrium metal uptake are reported in Table 4 along with statistical parameters.

Statistical analysis of the fit of kinetic data to pseudo-first and pseudo-second order models reveals marginally Better Square of correlation factors, R^2, for the former. A better fitting pseudo-first order kinetics might imply the absence of a chemical sorption mechanism between Co(II) and the cellular surface of *S. pombe* or its insignificance in comparison to physical interactions. It is also clear from equilibrium metal uptake values that the overall biosorption of Co(II) by *S. pombe* is not a very efficient process. These observations should be considered in an industrial application context, in terms of the reversibility of this operation and its potential use in a multi-component biosorption process.

Preliminary design calculations for a single stage biosorber

An important step towards the industrial design of a biosorption process is the utilisation of the characterisation data (biosorption isotherms) for the calculation of some relevant metrics [12]. The operating volume of a single stage operation is an important metric in terms of equipment design and operational costs. Determination of the biosorbent mass required achieving a target percentage removal for a given operating volume at a given pH and temperature is a design objective. The mass balance for Co(II) over a single stage of biosorption, depicted in Figure 5, at equilibrium is as follows:

$$V(C_i + [Spombe]_i q_i) = V(C_e + [Spombe]_e q_e) \quad (5)$$

Assuming biosorbent amount doesn't change (i.e. $[Spombe]_i = [Spombe]_e$) throughout the process and substituting in the Temkin isotherm, the above equation yields: at equilibrium.

$$[Spombe]_i = \frac{C_i - C_e}{\frac{RT}{b_T} \ln A_T C_e} \quad (6)$$

Figure 6 displays the amount of biosorbent needed to achieve predetermined percentage removal of metal in terms of varying operational volume. Although optimal biosorbent dose has been found to be 1.0 g L^{-1}, this optimality is in terms of the adsorption capacity of the biosorbent. If the cost of biosorbent is a critical factor for process design then this optimality is an important measure. However, if there is a trade-off between the cost of biosorbent and the cost of downstream operations then the kind of information gained from Figure 6 becomes important in guiding process design decisions.

T (°C)	C_i (mg L^{-1})	$q_{e\text{-}exp}$ (mg g^{-1})	Pseudo-first order model		
			k_{p1}	$q_{e\text{-}cal}$ (mg g^{-1})	R^2
25	2.790	1.437	2.501	1.413	0.9998
	5.511	1.630	2.167	1.633	0.9999
	7.440	1.724	1.712	1.725	0.9983
	12.813	2.367	0.563	2.542	0.9996
	53.716	2.728	0.187	2.997	0.9988
30	2.594	1.255	2.359	1.337	0.9999
	5.182	1.473	1.863	1.492	0.9998
	7.039	1.515	1.491	1.562	0.9933
	10.553	1.967	0.397	1.934	0.9925
	51.068	2.226	0.110	2.589	0.9942
35	2.603	1.131	1.891	1.141	0.9986
	5.088	1.355	1.557	1.361	0.9979
	7.407	1.416	1.219	1.381	0.9974
	10.528	1.544	0.283	1.564	0.9989
	57.600	1.878	0.046	1.852	0.9997
50	2.630	1.100	1.765	1.066	0.9988
	5.011	1.316	1.396	1.335	0.9934
	7.448	1.346	0.618	1.358	0.9968
	10.630	1.416	0.183	1.419	0.9975
	52.787	1.578	0.023	1.656	0.9981

Table 4: Pseudo-first order kinetic model.

Determination of thermodynamic properties

The pseudo-first order kinetic equilibrium constant can be utilised to calculate the activation energies for the predetermined experimental conditions using equation 2. Figure 7 shows the linearized ln k_{p1} versus reciprocal temperature plots for various initial sorbent concentration values. E_a values have been found to be -11.42, -13.53, -33.45, -34.88 and -66.48 kJ mol^{-1} for initial metal concentrations of 2.5, 5.0, 7.5, 10.0 and 50.0 mg L^{-1}, respectively. Negative nature of the activation energy values suggest that biosorption of Co(II) onto *S. pombe* is barrierless in these concentration ranges but no further physical interpretation is possible expect that the process is exothermic. This is an advantageous feature as it implies a low-cost operation with lower energy requirements, quick recovery and minimal loss of efficiency when compared to endothermic operations.

Further thermodynamic properties of the biosorption system were determined using equations 3 and 4. The calculated values for ΔS° and ΔH° are reported in Figure 8. The equilibrium constant from the Freundlich isotherm model was used for the thermodynamic calculations although statistically the Temkin isotherm was a better fit, albeit marginally. The reason for this is the thermodynamic consistency the calculations with the Freundlich equilibrium constants provide which is lacking in the Temkin calculation.

It can be observed from Figure 9 that increasing the operating temperature of the biosorption process has detrimental effects on two critical process metrics, namely the time required to reach equilibrium and metal uptake at equilibrium. Accordingly, it can be argued that, in terms of recovery of the metal from solution, temperature increase makes this process less favourable. Thermodynamic interpretation of such a biosorption process with above-mentioned features suggests that the ΔG° of the process should increase with increasing operating temperature. Thermodynamic properties calculated using the equilibrium constants acquired from the fitting of experimental data to biosorption isotherms produce trends that suggest otherwise with the exception of Freundlich and Redlich-Peterson models which exhibit acceptable thermodynamic consistency. Thermodynamically inconsistent predictions are expectable since these isotherm models are highly empirical and lack physical meaning; even though they offer statistically very accurate representation of equilibrium data [10,13]. Following these considerations, for the calculation of thermodynamic parameters, the equilibrium constant from the thermodynamically consistent isotherm model rather than the statistically most successful one will be used; as long as the former carries an acceptable correlation factor. Different trends depending on the isotherm model use can be observed in the calculated values of ΔG^0 as reported in Table 5.

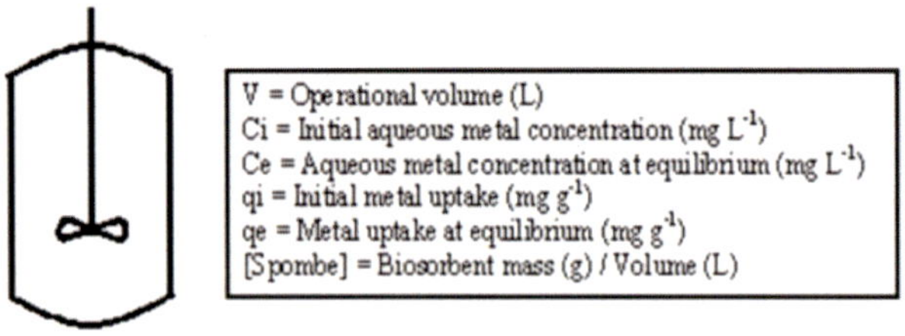

Figure 5: Batch biosorber.

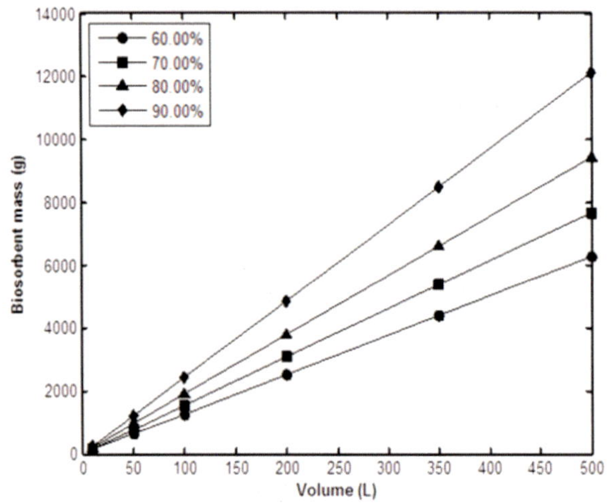

Figure 6: Biosorbent mass required for target percentage removal of metal at various operating volumes @ 25oC.

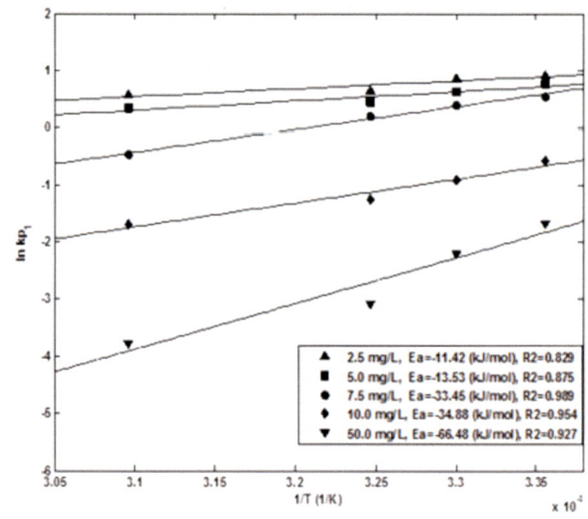

Figure 7: ln k_{p1} versus 1/T plots at different initial Co(II) concentrations.

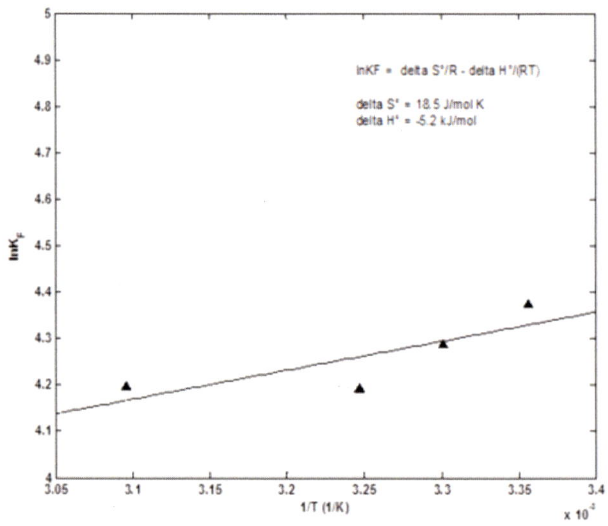

Figure 8: ln K_F versus 1/T plot.

Conclusion

Equilibrium, kinetic and thermodynamic analyses of a biosorption process are crucial for predicting its performance and comparison with alternatives. Among the various studied isotherm models, Temkin isotherm model has shown the best fit for the present data in terms of statistical success. The inclusion of a term for temperature in its mathematical structure and the experimentally observed fact that temperature has an inversely correlated effect on biosorption performance support this outcome. However, thermodynamic consistency with experimental observations in the calculation of thermodynamic properties of the system could only be achieved using equilibrium constants from the Freundlich isotherm model. This proves to be a situation where one has to choose the most appropriate model for the type of calculation needed. Indeed, equilibrium studies were conducted using the Temkin isotherm model whereas thermodynamic calculations were done incorporating the Freundlich equilibrium constant.

The ΔG^0 values calculated for operating temperatures of 25, 30, 35 and 50°C were -11.7, -11.5, -11.3 and -11.3 kJ mol^{-1}, respectively. The slight increase in the ΔG^0 values with respect to increasing operating temperatures implies thermodynamic consistency with the observation from kinetic profiling simulations (Figure 9) that higher temperatures are less favourable for this particular system.

Among the isotherm models used only Temkin isotherm takes the natural logarithm of aqueous metal concentration whereas all the others incorporate this term either linearly or exponentially. Temkin isotherm seems to minimise the effect of aqueous metal concentration

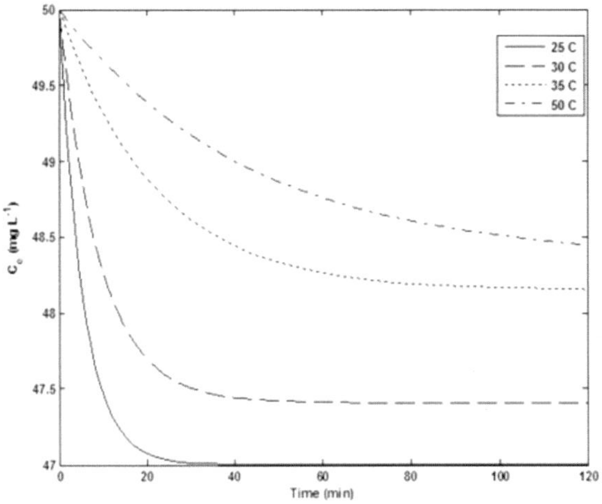

Figure 9: Kinetic profiling simulation of four biosorption processes taking place at different operating temperatures all with C_i=50 mg L^{-1} and biosorbent dose: 1 g L^{-1}.

	ΔG^0@ 25°C (kJ mol^{-1})	ΔG^0@ 30°C (kJ mol^{-1})	ΔG^0@ 35°C (kJ mol^{-1})	ΔG^0 @ 50°C (kJ mol^{-1})
Langmuir	-8.5	-9.4	-10.2	-12.0
Freundlich	-11.7	-11.5	-11.3	-11.3
Sips	-11.0	-11.3	-12.0	-13.5
Khan	-16.3	-14.2	-14.9	-13.6
Temkin	-35.0	-37.8	-41.5	-52.4
Redlich-Peterson	-16.0	-14.3	-15.0	-14.5

Table 5: ΔG^0 values calculated using equilibrium constants from different isotherm models.

on equilibrium compared to all the other models. It has to be noted that this is in parallel with experimental observation. Although increased initial metal concentrations have a positive effect on the equilibrium values of q_e and C_e, this is a marginal effect and insignificant with that of other factors such as temperature. The success of the Temkin isotherm model may be due to its reflecting this fact.

The above observation is supported by the outcome of the biosorption kinetics study as well. Pseudo-first order kinetics provides a statistically superior correlation of experimental data compared to the pseudo-second order kinetics. The mathematical difference between the two is the taking of the square of the concentration term in the pseudo-second order kinetics increasing its influence on the overall kinetic behaviour of the biosorption system. The fact that pseudo-first order kinetics better describes the observed values is in agreement with the earlier suggestion. The fact that this biosorption system follows pseudo-first order kinetics hints that a costly destructive desorption step might not be necessary in order to free the biosorbent for reuse.

It is important to integrate biosorbent characterisation information with preliminary process design calculations from an early point onwards. The use of isotherm models to simulate the single stage biosorber performance plots (Figure 6) and the subsequent identification of possible techno-economic trade-off points can be considered a tool for early stage process design.

Overall, the biosorption of Co(II) by wild-type *S. pombe* seems to be a slightly exothermic process, judging by the negative nature of the calculated E_a values and thermodynamically feasible, considering that the calculated ΔG^0 values are negative for all the conditions tested. However, this particular biosorption process is not a very efficient one since a maximum of about 50% percentage removal of metal is possible even under optimal conditions. This value is obtained when lower initial metal concentrations are used. For higher initial metal concentrations the percentage removal falls dramatically to under 20% for 10 mg L^{-1} initial Co(II) concentration and ~5% for 50 mg L^{-1} initial Co(II) operations. This means that while *S. pombe* is not an ideal biosorbent for the removal of Co(II) from aqueous solutions, it can be utilised as a bioseparation agent for aqueous mixtures containing both Ni(II) and Co(II) since its affinity towards Ni(II) is significantly higher (Durmaz-Sam et al., 2011).

Future work could focus on the characterisation of *S. pombe* in multi-component biosorption processes involving both Co(II) and Ni(II). Additionally, alternative design considerations for better utilisation of *S. pombe*, such as the use of column systems instead of a batch stirred tank reactor or the immobilisation of *S. pombe* could be studied. Inclusion of mass transfer and fluid-flow calculations would improve the preliminary design approaches. Finally, quantified techno-economic assessment of the design trade-offs using cost-benefit analyses would enhance the industrial feasibility studies.

Acknowledgements

Funding by Marmara University Scientific Research Committee (Project FEN-C-YLP-010710-0231) is gratefully acknowledged.

References

1. Wang J, Chen C (2009) Biosorbents for heavy metal removal and their future. Biotechnol Adv 27: 195-226.
2. Durmaz-Sam S, Sayar NA, Topal-Sarikaya A, Sayar AA (2011) Biosorption of Ni(II) by *Schizosaccharomyces pombe:* kinetic and thermodynamic studies. Bioprocess Biosys Eng 35: 997-1005.
3. Kuyucak N, Volesky B (1989) Accumulation of cobalt by marine algae. Biotechnol Bioeng 33: 809-814.

4. Vijaraghavan K, Palanivelu K, Velan M (2006) Biosorption of copper(II) and cobalt(II) from aqueous solutions by crab shell particles. J Hazard Mater 97: 1411-1419.

5. Pal A, Ghosh S, Paul A (2006) Biosorption of cobalt by fungi from serpentine soil of Andaman. Bioresource Technol 97: 1253-1258.

6. Li B, Liu F, Wang J, Ling C, Li L, et al. (2012) Efficient separation and high selectivity for nickel from cobalt-solution by a novel chelatin resin: batch, column and competition investigation. Chem Eng J 195-196: 31-39.

7. Kumbasar RA, Sahin I (2008) Separation and concentration of cobalt from ammoniacal solutions containing cobalt and nickel by emulsion liquid membranes using 5,6-dibromo-8-hydroxyquinoline (DBHQ). J Membrane Sci 325: 712-718.

8. Ozer A, Ozer D (2003) Comparative study of biosorption of Pb(II), Ni(II) and Cr(IV) ions onto *S. ceravisiae*: determination of biosorption heats. J Hazard Mater 100: 219-229.

9. Volesky B (2003) Bioprocess process simulation tools. Hydrometallurgy 71: 179-190.

10. Liu Y, Liu YJ (2008) Biosorption isotherms, kinetics and thermodynamics. Sep Purif Technol 61: 229-242.

11. Foo KY, Hameed BK (2010) Insight into the modelling of adsorption isotherm systems. Chem Eng J 156: 2-10.

12. Hameed BH, Mahmoud DK, Ahmad AL (2008) Equilibrium modelling and kinetic studies on the adsorption of basic dye by a low-cost adsorbent: Coconut (Cocos nucifera) bunch waste. J Hazard Mater 158: 65-72.

13. Levan MD, Vermeulen T (1981) Binary Langmuir and Freundlich isotherms for ideal adsorbed solutions. J Phys Chem-Us 85: 3247-3250.

Column Biosorption of Lead, Cadmium, Copper, and Arsenic ions onto Algae

Abbas H. Sulaymon[1], Ahmed A. Mohammed[2] and Tariq J. Al-Musawi[3]*

[1]Professor and Head of Energy Engineering Department, Baghdad University, Iraq
[2]Assistant Professor, Environmental Engineering Department, Baghdad University, Iraq
[3]Environmental Engineering Department, Baghdad University, Iraq

Abstract

A mixture of green and blue-green algae was used as an adsorbant material for biosorption of lead, cadmium, copper, and arsenic ions in fluidized bed reactor. Batch experiments showed that the algal biomass was successfully used for the removal of these metal ions from wastewater. The maximum percentage removal for 1 g dose was 89, 82, 79, and 70 for Pb^{2+}, Cu^{2+}, Cd^{2+} and As^{3+}, respectively. The experimental data fit well to an ion exchange equilibrium model. Affinity constants were calculated for each metal. A higher affinity of the biomass towards lead (Pb^{2+}) was observed due to the high electronegativity of this metal. FTIR analyses showed that hydroxyl and carboxyl groups could be very effective for capturing these metals. An ideal plug flow model was adopted to characterize the fluidized bed reactor and solved numerically using MATLAB version (R2009b), which fit well to the experimental breakthrough data. The effects of different operating conditions such as: static bed height, superficial velocity and particle diameter on the removal process were investigated. Lead showed the largest operating time compared with others.

Keywords: Algae; Metals; Ion exchange; Fluidized bed; Breakthrough curve

Introduction

One of the most challenging environmental problems today is the removal of heavy metals and other toxic contaminants from industrial wastewater. Many aquatic environments face metal concentrations that exceed water quality limits designed to protect the environment, animals, and humans [1]. Metals hazardous to humans include lead, cadmium, mercury, arsenic, copper, zinc, and chromium. Arsenic is carcinogenic metal. Cadmium can cause bone and kidney damage. Copper and lead can cause brain and bone damage [2].

Biosorption is an innovative technology that employs inactive and dead biomass for the removal and recovery of metals from aqueous solutions [3-5]. Biomass from various sources such as bacteria, yeast, algae, fungi and plants have been used to adsorb metal ions from the environment [6-9]. Among the most promising types of biosorbents studied is algal biomass [10,11]. Algal biomass has been reported to have a high metal binding capacity due to the presence of polysaccharides, proteins or lipids on the cell wall structure which contain functional groups such as amino, carboxyl, hydroxyl, sulfate, and others [12]. These groups have the ability to bind heavy metals by donation of an electron pair from these groups to form complexes with the metal ions in solution [13].

Recently, biosorption in columns and its modeling have been receiving more attention. Fixed and fluidized bed reactors have been used widely by the chemical industry, pharmaceutical industry, food industry, wastewater treatment and for recovery of different substances [14]. Fluidized bed systems are common and important reactors in process engineering because of the good mass and heat transfer rate between the fluid and the particles, and between the particles and the side wall of the column [15].

The previous literature does not include much information on biosorption of metal ions by the algal biomass with fluidized bed reactors. Therefore, the main objectives of this work are: (i) to characterize the physiochemical parameters of algal biomass such as specific surface area, particle porosity and active groups; (ii) to evaluate the application of an ion exchange model in a batch reactor; (iii) to evaluate and model the breakthrough curves of lead, cadmium, copper, and arsenic using fluidized bed of algal biomass; (iv) to evaluate the effects of some experimental conditions on the removal of these metal ions such as superficial velocity, bed height, and particle diameter.

Theoretical models

Equilibrium model: Several studies concluded that ion exchange is a principal mechanism of biosorption by dead algae [14,15]. The ion exchange isotherm model is a good representation of the biosorption process on algae, since it reflects the fact that most algal biomass contains light metal ions such as K^+, Na^+, and Mg^{2+} ions, which are released upon the binding of a heavy metal cation [11]. Therefore, it has been demonstrated that the binding of metals by algal biomass from aqueous solution can be described by the following ion exchange reaction [16]:

$$M^{2+} + (L\text{-Biomass})\ (M\text{-Biomass}) + L^{2+} \quad (1)$$

Where, M^{2+} and L^{2+} represent the divalent metal cations sorbed and released from the biomass.

The total normality, which represents the sum of the equivalent concentrations of all competing cations that can be exchanged during the reaction, remains the same when equilibrium is achieved; hence, the total normality is expressed by:

$$C_T = C_M + C_L \quad (2)$$

Where, c_M is the total normality of heavy metals that remain in the

***Corresponding author:** Tariq J. Al-Musawi, Environmental Engineering Department, Baghdad University, Iraq

liquid phase, c_L is the total normality of light metals released into the liquid phase, and c_T represents the total normality of the solution.

The total number of exchangeable binding sites (Q) is the sum of the amount of adsorbed and released metals:

$$Q = q_M + q_L \tag{3}$$

Where: q_M is the amount of adsorbed heavy metal (meq/g); q_L is the amount of released light metal into the solution due to biosorption of heavy metal (meq/g).

The equivalent fraction of one component in the liquid phase (x_M, x_L) is the ratio between its own concentration in the liquid phase and the total normality of the solution (c_T), whereas the equivalent fraction in the solid phase (y_M, y_L) is the relation between its own concentration in solid phase divided by the total number of exchangeable binding sites (Q):

$$x_M = \frac{c_M}{c^0}, \quad x_L = \frac{c_L}{c^0} \tag{4}$$

$$y_M = \frac{q_M}{Q}, \quad y_L = \frac{q_L}{Q} \tag{5}$$

Based on sorption reaction, the affinity constant ($K_{M,L}$) represents the relative power exchange of the heavy metal to the light ions and it is defined by [17]:

$$K_{M,L} = \frac{y_M . x_L}{x_M . y_L} \tag{6}$$

Where, the subscript M and L refer to the heavy metal and light metal in the solution.

Rearranging the above equation by substituting the light metal equivalent fraction, the model equation for the equilibrium uptake of a heavy metal ion in the presence of light metals can be written in the form of the following equation:

$$y_M = \frac{K_{M,L} . x_M}{1 + \left(K_{M,L} - 1\right) . x_M} \tag{7}$$

The evaluation of fitness of the equilibrium model equation with experimental data requires an error function with optimization [18]. The experimental equilibrium data for the biosorption of four metals were fitted with the ion exchange equilibrium model (Eq.7). The equilibrium model parameter (affinity constant) was evaluated and optimized by non-linear regression using STATISTICA program version 6. The model's fitness was significant as indicated by the coefficient of determination (R^2) where a large R^2 value indicates a better fit.

Breakthrough curve models: A model of fluidized bed is shown in Figure 1. Three assumptions were made to constrain the model for fluidized bed: (i) the concentration is uniform in the radial direction, (ii) there is no material product in the reactor, (iii) the fluid stream is an ideal plug flow [19]. With these assumptions, the following equations can be obtained:

$$\begin{aligned} &\textit{Rate of pollutant conc. at any depth} \\ &= (\textit{Rate of input conc. by conective flow}) + (\textit{Rate of input conc. by dispersion}) - (\textit{Rate of output conc. by convective flow}) \\ &- (\textit{Rate of output conc. by dispersion}) - (\textit{Solute lost by sorption}) \end{aligned} \tag{8}$$

$$\frac{\partial C}{\partial t}.dz = \frac{U}{\varepsilon}.C + E_z.\frac{\partial C}{\partial z} - \frac{U}{\varepsilon}\left[C + \frac{\partial C}{\partial z}dz\right] - E_z.\left[\frac{\partial C}{\partial z} - \frac{\partial}{\partial z}\left(\frac{\partial C}{\partial z}\right)dz\right] - \frac{(1-\varepsilon)}{\varepsilon}\rho_P \frac{\partial q}{\partial t}dz \tag{9}$$

where U is the superficial velocity (m/s), ε is the bed void fraction, C is the concentration of metal at any time (mg/l), E_Z is the axial dispersion coefficient (m^2/s), z is the bed height (m), t is the time, and q is the amount of adsorbed metal ion (mg/g).

Equation (9) was simplified to the following:

$$\frac{\partial C}{\partial t} = E_z \frac{\partial^2 C}{\partial z^2} - \frac{U}{\varepsilon}\frac{\partial C}{\partial z} - \frac{(1-\varepsilon)}{\varepsilon}\rho_P \frac{\partial q}{\partial t} \tag{10}$$

The mass balance for the solid phase is expressed as:

$$(1-\varepsilon)\rho_P \frac{\partial q}{\partial t} = K_{La}(C - C^*) \tag{11}$$

where C* is the equilibrium heavy metal concentration (mg/l), K_L is the mass transfer coefficient, and a is the specific surface area (m^2/m^3). Substituting Eq. (11) into Eq. (10), the following equation can be obtained:

$$\frac{\partial C}{\partial t} = E_z \frac{\partial^2 C}{\partial z^2} - \frac{U}{\varepsilon}\frac{\partial C}{\partial z} - \frac{K_L.a}{\varepsilon}(C - C^*) \tag{12}$$

The initial and boundary conditions are:

$$0<z<H,\ t=0;\ C = C^* = 0 \tag{13}$$

$$z=0,\ t\geq 0;\ C = C_i \tag{14}$$

$$z=H,\ t>0;\ \frac{\partial C}{\partial z} = 0 \tag{15}$$

Where: H is the bed height (m), C_i is the initial metal concentration (mg/l).

The partial differential equation (Eq. 12) with its initial and boundary conditions was solved numerically using MATLAB software Version (R2009b). The experimental breakthrough data for biosorption of the four metals were fitted to a theoretical model data obtained from the solution of equation (12). The model's fitness was determined to be significant by the coefficient of determination (R^2).

Materials and Methods

Materials

Adsorbent preparation: Mixture of green (Chlorophyta) algae and blue-green (Cyanobacteria) algae was used as a biosorbent material. The algae were collected in April and September 2011 from the irrigation canal of Baghdad University/Iraq. The foreign matter was removed from the algae and washed several times with tap water and distilled water to remove impurities and salts. Sample of 0.5 kg of collected algae were analyzed for species composition and percentage weight using a microscope according to the standard methods [20]. The species and percentage weight were Oscillatoria princeps (92%), Spirogyra aequinoctialis (3%), Oscillatoria subbrevis (2%), Oscillatoria Formosa (1%), and others (1%). The washed algae were sun dried and then dried in oven at 50°C for 48 h. The dried algal biomass was shredded, ground in a mortar, sieved and stored in an opaque polyethylene container at room temperature. Two mesh sizes of 0.4-0.6 and 0.6-1 mm particle diameter were used for biosorption experiments. The biomass particle size distribution was determined using a set of standard sieves. Since the algal biomass could swell in water, the biomass was initially soaked in water and then wet sieved.

A sample of powdered algal biomass was analyzed for physical properties such as particle density, surface area, particle porosity and bed void fraction. The surface area of algal biomass was determined from N_2 adsorption isotherm using a Micrometrics Nano Porosity System. In order to identify functional groups (carbonyl, carboxylic, hydroxyl and others), Fourier Transform Infrared spectra (FTIR) techniques were used. The FTIR offers excellent information on the nature of the bands present on the surface of the algae before and after the biosorption process.

Adsorbate preparation: All chemicals used in this work were analytical reagent grade and were used without further purification. Lead, cadmium, copper, and arsenic ion solutions were prepared by dissolving the corresponding analytical salts of $Pb(NO_3)_2.2H_2O$, $Cd(NO_3)_2$, $Cu(NO_3)_2.3H_2O$ and As_2O_3 , respectively in distilled water. The diluted solutions were kept in a glass container at room temperature. A solution with a concentration of 50 mg/l of each metal was used as adsorbate material. The concentration of each metal was measured using a flame Atomic Absorption Spectrophotometer (AAS, type: SHIMADZU, Model: 7200, JAPAN). Before the biosorption process was initiated, the pH of each solution was adjusted to the required value (2, 3, 4, 5, and 6) by adding 0.1 M HNO_3 or 0.1 M NaOH solutions. All the glassware used for dilution, storage and experimentation were cleaned with detergent, thoroughly rinsed with tap water, soaked overnight in a 20% HNO_3 solution and finally rinsed with distilled water before use.

Methods

Effect of experimental parameters: The effects of pH, contact time, temperature, and initial heavy metal concentration on the removal efficiency were studied in a batch reactor. All the study experiments were carried out in 250 ml flasks containing 100 ml of metal solution and a 1 g dose of algal biomass. Effect of pH was measured using a 50 mg/l metal concentration at room temperature, and the pH range was from 2-6. Effect of contact time was measured using 50 mg/l heavy metal concentration at room temperature and the pH was fixed at 4 for all metal solution. The temperature effect experiment was conducted using 50 mg/l heavy metal concentration, and the solution temperatures were 10, 20, 30, 40, and 50°C. Effect of initial heavy metal concentrations were conducted using 10, 50 and 100 mg/l heavy metal solutions at room temperature.

Different operating conditions were used in the fluid bed experiments, i.e., changing static bed height, fluid superficial velocity and particle diameter. The effect of bed height of the algal biomass on the biosorption process was investigated at heights of 2.5, 5.0, and 7.5 cm. The influence of superficial flow velocity upon adsorption performance was measured by establishing the breakthrough curves at 10% and 50% above the minimum fluidization velocity. Two particle diameter ranges were used in the column experiments of 0.4-0.6 and 0.6-1.0 mm.

Equilibrium experiments: Equilibrium experiments were carried out at different weights of algal biomass (0.05, 0.1, 0.3, 0.5, 0.8, 1, 2, and 3 g). Each weight was thoroughly mixed with 100 ml of heavy metal solution at 50 mg/l concentration. Eight flasks of 250 ml capacity were placed in a rotary shaker type (Edmund Buhler, 7400 Tubingen Shaker-SM 25, Germany) at 200 rpm for 3 h to obtain equilibrium. It was found from the preliminary experiments that these conditions lead to equilibrium conditions. The mixtures were then filtered using filter paper (type: WATMAN No.42, ash less, diameter 7 cm), then the residual concentration of lead, cadmium, copper and arsenic in solution was determined using AAS. The uptake value at equilibrium was calculated using the following equation:

$$q_e = \frac{V(C_i - C_e)}{m} \tag{16}$$

where, q_e is the uptake of metal(mg/g); C_i and C_e are the initial and equilibrium metal concentrations in the water (mg/l), respectively; V is the volume of solution (l); and m is the mass of adsorbent (g).

Column experiments: Figure 2 shows a schematic diagram of fluidized bed system. The column is 7.5 cm in diameter and 1 m length.

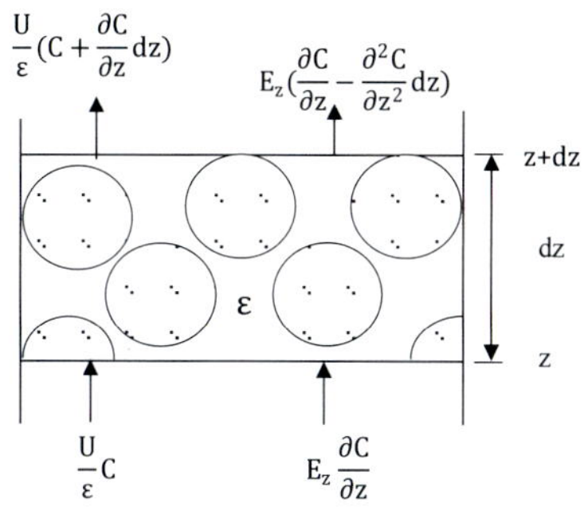

Figure 1: Model of a fluidized bed ion exchange reactor.

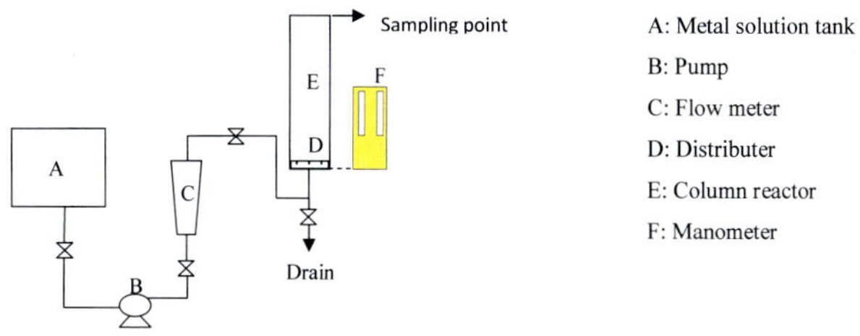

Figure 2: Fluidized bed setup.

A stainless steel distributor of 1 mm thickness and 0.2 mm holes dia. was installed at the bottom of the reactor to distribute the influent flow smoothly. The column was partially packed with a known weight of biomass. A U-tube manometer was connected to the reactor to record the pressure drop across the bed, the manometer has an inside diameter of 5 mm and length is 50 cm. Carbon Tetrachloride (CCl_4) with density equal 1590 kg/m^3 was used as the manometer liquid. A feed solution containing 50 mg/l for each metal ion was pumped upward through the column at selected velocities for each particle size. The pH of each metal solution was adjusted to the best value obtained from the batch experiments. Samples were taken from the effluent from the reactor at time intervals 1, 2, 5, 10, 20, 30, 40, 50, 60, 70, 80, 90, 120, 150, 180 min. During the experiment, the pressure drop, and fluidized bed height were measured at the corresponding time intervals.

Results and Discussion

Characterization of algal biomass

The physical properties of the algal biomass were measured and the results were tabulated in Table 1. These values are very important in the modeling of breakthrough curves. Figure 3 shows the results of the FTIR analyses of each algal biomass before biosorption (black line), and Pb^{2+} (gray line), Cd^{2+} (green line), Cu^{2+} (blue line) and As^{3+} (red line) loaded algae. The results of the FTIR spectrum showed that there were different functional groups detected on the surface of the algal biomass. Broad peaks detected in spectra were identified at 3414 cm^{-1}(hydroxyl and amine groups), 2966 and 2943 cm^{-1} (alkyl chains), 1797cm^{-1}(C=0 of the carboxylic groups or ester groups), 1647cm^{-1}(C=O of amide groups), 1427 cm^{-1}(COO^- of the carboxylate groups), and1300 cm^{-1}(S=O of the sulfonates groups and COO^- groups of the fatty acids). The peak value at 1033 cm^{-1} is attributed to the P-O-C links of the organic phosphate groups. Some bands in the fingerprint regions could be attributed to the phosphate groups. From Figure 3, it can be seen that

Parameter	Value	
Particle diameter (mm)	0.4-0.6	0.6-1
Bulk density(kg/m³)	474	400
Real density (kg/m³)	1120	1120
Surface area (m²/g)	1.88	1.65
Particle porosity	0.713	0.77
Bed void fraction	0.577	0.642

Table 1: Properties of the algal biomass used in fluidized bed experiments.

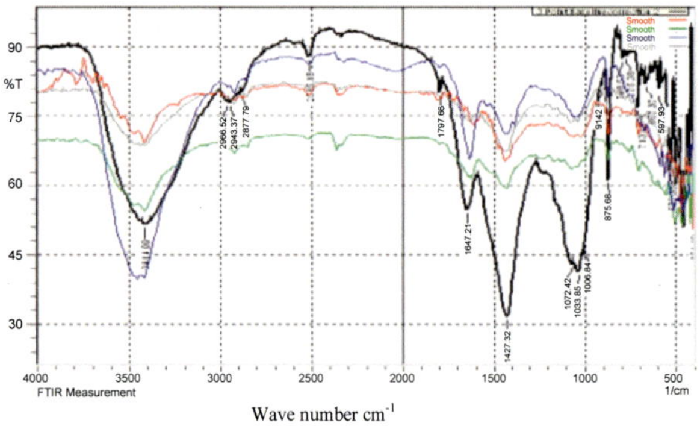

Figure 3: FTIR spectra obtained for algae before biosorption (black), and algae following adsorption of Pb^{2+}(grey), Cd^{2+}(green), Cu^{2+}(blue), As^{3+}(red).

the % transmittance (%T) values were shifted mainly at hydroxyl and carboxyl groups. These groups are very effective for capturing the heavy metals ions due to their negative charges [21].

Effect of pH

The pH of a solution affects metal ion solubility and biosorbent total charge [22]. Normally metal biosorption is directly dependent on pH value [12]. Figure 4b shows the biosorption efficiency of each metal at different pH values from 2-6. The highest removal of lead, cadmium and arsenic was achieved at pH of 4, while at pH of 5 the highest removal of copper was obtained. The maximum uptake (q_e) obtained at the optimum pH value and 1 g doses of metals were 44.5, 41, 39.5, and 35 mg/g for Pb^{2+}, Cu^{2+}, Cd^{2+}, and As^{3+}, respectively. At pH<3, the active surface sites of the adsorbet were either protonated [23], or dissociated [24], so that lowest biosorption was recorded at these pH values. Experiments beyond pH 5 were hampered due to precipitation of metals as hydroxides during the biosorption process.

Effect of contact time

Before the starting of the batch experiments, it was very important to know the required time to reach equilibrium biosorption. In other words, for all batch experiments, the contact time should be sufficient to ensure reaching an equilibrium concentration. Figure 4a shows the results of biosorption efficiency versus contact time for the four metal solutions. It can be seen that a 3 hour contact time is sufficient to reach equilibrium for all four heavy metals.

Effect of Temperature

The effect of temperature on the equilibrium sorption capacity has been investigated within a temperature range of 10-50°C. The pH of each metal solution was fixed at the optimum value determined from the pH study. Figure 4c shows the variation of percentage removal with temperature. It was observed that the maximum percentage removal for each of the four metals occurred at 20°C. Several authors showed that further increases in temperature (above 30°C) lead to a decrease the percentage removal. This decrease in biosorption efficiency may be attributed to: an increase in the relative desorption of the heavy metals from the solid phase to the liquid phase, deactivation of the biosorbent surface, destruction of active sites on the biosorbent surface due to bond disruption [25,26], or due to the weakness of the sorbent active site binding forces and the sorbate species and also between the adjacent molecules of the sorbed phase [27]. From Figure 4c, it can be seen that the variation of temperature from 20-30°C has minimal effect on the biosorption process, so that experiments can be carried out at room temperature without adjustment.

Effect of initial concentration

Figure 4d shows the variation of percentage removal at different initial metal concentrations. The results indicated that the percentage removal was not altered greatly from concentrations between 10-50 mg/l, this may be due to 1 g of algae containing enough bisorption surfaces and sites for this concentration range [28], thereafter the percentage removal rapidly decreased with the further increase in concentration to 100 mg/l.

Biosorption isotherms

Biosorption onto algae was assessed and compared to the ion exchange isotherm model. The ion exchange model was fitted with experimental isotherm data at the optimum pH value for each metal. The ion exchange model parameter (affinity constant, K) were measured and optimized by non-linear regression using the STATISTICA program. The isotherm curves obtained for Pb^{2+}, Cd^{2+}, Cu^{2+} and As^{3+} are depicted in Figure 5. This figure shows a good fit between the experimental and theoretical data as indicated by the coefficient of determination values (R^2) obtained. The value of the biomass affinity constant gives a good indication of the biosorption capacity of metals measured [17]. The calculated affinity constants were 16.55, 15.97, 10.52, and 7.45 for Pb^{2+}, Cu^{2+}, Cd^{2+}, and As^{3+} respectively. Yun and Volesky [29] concluded that for divalent ions the major factor affecting the exchange biosorption is the strength of the charge on the ions (electronegativity). The electronegativity (in Pauling scale) of Pb^{2+}, Cd^{2+}, Cu^{2+} and As^{3+} ions are 2.33, 1.69, 1.9, and 2.18, respectively. Pb^{2+} demonstrated the

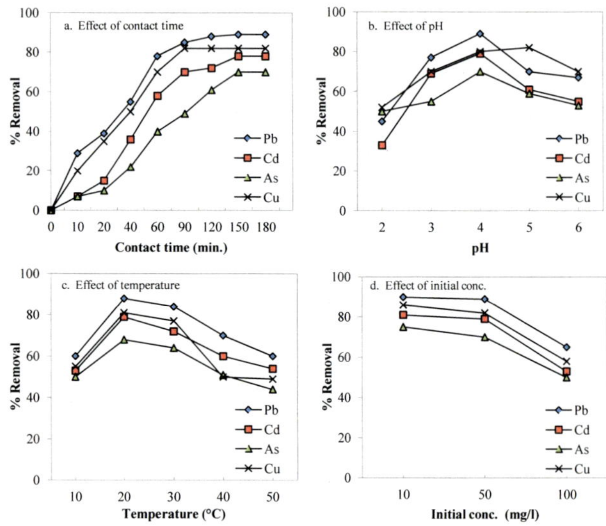

Figure 4: Effects of physical variables on heavy metal biosorption using algal biomass.

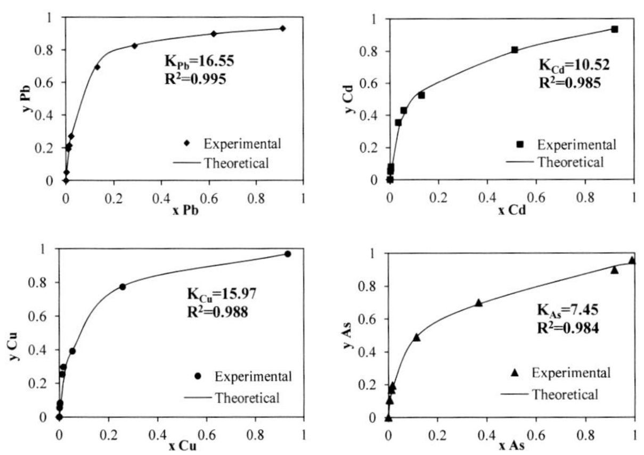

Figure 5: Ion exchange equilibrium isotherm for Pb^{2+}, Cd^{2+}, Cu^{2+}, and As^{3+}.

highest affinity compared to Cu^{2+}, Cd^{2+} and As^{3+}. This can be attributed to the high electronegativity of this metal ion. Despite the fact that As^{3+} has an electronegativity value greater than Cd^{2+}, and Cu^{2+} this metal demonstrated lowest affinity value which may be due to it's trivalent ion state needing more exchangeable sites on the biosorbant. Therefore, the sequence of metal biosorption from high to low was calculated as $Pb^{2+}>Cu^{2+}>Cd^{2+}>As^{3+}$. These findings are in good agreement with the results of Figueria MM et al. [10].

Minimum fluidization velocity

The minimum fluidization velocity (U_{mf}) was determined experimentally by measuring the pressure drop through the bed of algal particles. The column was partially filled with particles of known mass and then vigorously agitated with water in order to disperse the particles and break down any internal structure. After that the bed was left to settle, and the flow rate increased incrementally from 0 to 100 l/h. At each flow rate increment, the pressure drop was recorded using the manometer. Figure 6 shows the pressure drop across the bed against the superficial fluid velocity on a logarithmic scale. This figure shows that the pressure drop rises linearly below the minimum fluidization velocity in the packed bed region and then plateaus. The U_{mf} can be read from the sharp change in the pressure drop. In addition, it is important in the modeling of continuous systems to be able to establish the relationship between the superficial liquid velocity and the bed void fraction [30]. The bed void fraction (V_ε) can be found experimentally by subtracting the volume of the particles (V_p) from the total volume of the fluidized bed (V_b) using the following equation:

$$\varepsilon = \frac{V_\varepsilon}{V_b} = \frac{V_b - V_p}{V_b} = 1 - \frac{V_p}{V_b} = 1 - \frac{m_p}{\rho_p . V_b} \qquad (17)$$

where m_p represents the bed mass (kg), ρ_p is the density of particles (kg/m^3), A is the cross sectional area of the bed (m^2), H is the expanded bed height in meter sand depends on the liquid velocity.

Table 2 shows the values of minimum fluidization velocity, plateau pressure drop (ΔP), bed void fraction, and expanded bed height at minimum fluidization velocity (h_{mf}) for two different size particles that used in column biosorption processes.

Breakthrough curves

Several authors have proposed generalized correlations to predict the mass transfer coefficient. Park et al. [19] presented the following correlation obtained from the experimental data of a liquid-solid fluidized bed reactor:

$$\frac{Sh.\varepsilon^{0.5}}{Sc^{1/3}} = 0.62\,\mathrm{Re}_p + 0.6 \quad (18)$$

Where: Sh is the Sherwood number ($K_L.d/D_m$), K_L is the mass transfer coefficient (m/s), d is the particle diameter (m), D_m is the diffusivity (m^2/s), S_c is the Schmidt number ($\mu/\rho.D_m$), and Re_p is the Reynolds number for particle.

The diffusivity of each metal can be calculated from the following equation [31]:

$$D_m = 2.74 * 10^{-9} \left(M_w\right)^{-1/3} \qquad (19)$$

where: M_w is the molecular weight of each metal.

In this work, the previous two equations were used in the modeling of the breakthrough curves. The breakthrough curves for each metal were obtained by plotting C/C_i versus time. The experimental and predicted breakthrough curves are presented in Figure 7 (a,b,c,d). These figures show a good fit between the experimental and theoretical data according as indicated by the R^2 value. The operating time corresponding to 90% removal (or C/C_i=0.1) was chosen as a point for comparison of the removal efficiency of each metal. Also, from Figure 7 (a,b,c,d) it can be seen that Pb^{2+} has the greatest operating time compared to the other metals. The operating time corresponding to 90% removal at different operating conditions are listed in Table 3. The metal biosorption from highest to lowest can be arranged as $Pb^{2+}>Cu^{2+}>Cd^{2+}>As^{3+}$. These findings are in a good agreement with the results of the batch system.

The effect of bed height on the sorption process was investigated for U=1.1U_{mf}, C_i=50 mg/l, d_p=0.4-0.6 mm and bed weight 50, 100 and 150 g (corresponding to static bed heights of 2.5, 5 and 7.5 cm) and the results for breakpoint time were listed in Table 3. These results showed that with increases in the bed height of the algal biomass the time at which an effluent concentration reached equilibrium increased, this is due to the large contact time between the metal solution and particles

Particle size (mm)	Mass (g)	Static height(cm)	U_{mf}(mm/s)	ε	ΔP (pa)	h_{mf}(cm)
0.4-0.6	50	2.5	2.27	0.79	56.3	5
	100	5	2.27	0.79	80.1	10
	150	7.5	2.27	0.79	112	15
0.6-1	50	3	3.64	0.83	66.1	6
	100	6	3.64	0.83	103.3	12
	150	9	3.64	0.83	124.8	18

Table 2: Minimum Fluidization Velocity (U_{mf}), Plateau Pressure Drop (ΔP) and Expanded Bed Height at U_{mf} (h_{mf}) for two different size particles.

Parameter	Value	Operating time (min.) of 90% removal for each metal			
		Pb^{2+}	Cd^{2+}	Cu^{2+}	As^{3+}
Bed height	2.5 cm	30	17	27	15
	5 cm	38	22	36	19
	7.5 cm	44	37	41	33
Superficial velocity	1.1U_{mf}=2.5 mm/s	38	22	36	19
	1.5U_{mf}=3.5 mm/s	22	14	19	10
Particle diameter	0.4-0.6 mm	38	22	36	19
	0.6-1 mm	35	19	33	17

Table 3: Breakpoint time for each metal at different operating conditions.

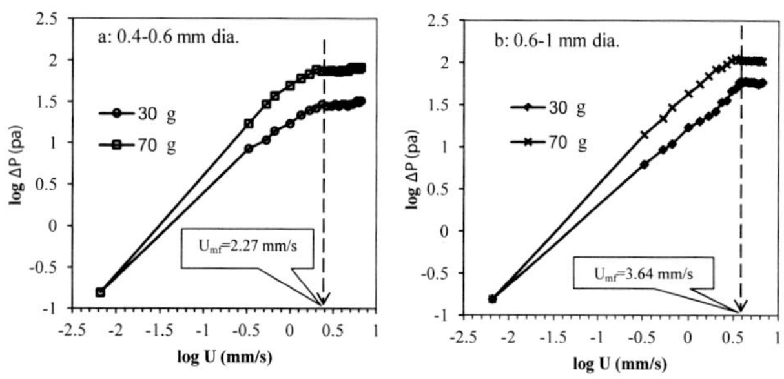

Figure 6: Pressure drop vs. superficial fluid velocity of two ranges of algal biomass particles.

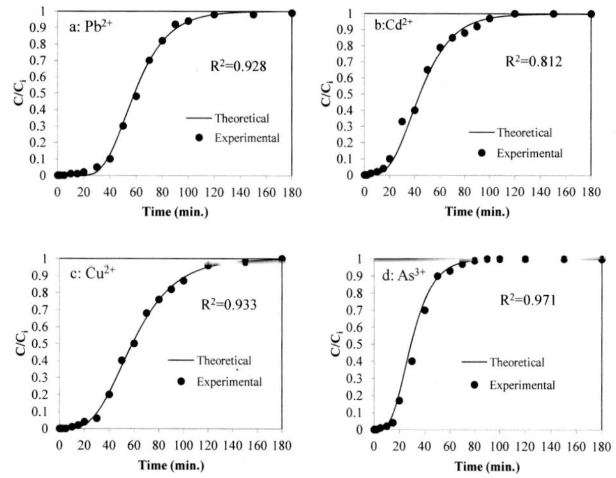

Figure 7: Experimental and theoretical breakthrough curves at 100 g algal biomass, pH=4, 25°C, 0.4-0.6 mm particle diameter and $U=1.1U_{mf}$.

at a high bed height. Smaller bed heights will be saturated in less time. Also, an increase in the bed depth will increase the surface area for the adsorption which will improve the adsorption process.

The fluid velocity is a major parameter in the design of fluidized column due to its effect on the contact time between the particles and the metal solution. Table 3 shows the breakpoint time for each metal at superficial liquid velocity of $U=1.1U_{mf}$ and $U=1.5U_{mf}$; bed height=2.5 cm, and C_i=50 mg/l. It can be seen that for all metals the breakthrough point appeared earlier with increasing fluid velocity due to the reduction in contact time for the metal ions to occupy the spaces within the particles.

Finally, the effect of particle diameter was investigated at 0.4-0.6 and 0.6-1 mm particle diameter, 100 g algal biomass weight, and $1.1U_{mf}$. The results were listed in Table 3 and it can be seen that an increase in particle size causes a decrease of the breakthrough time, which would be anticipated with the decrease in the surface to volume ratio and the subsequent decrease in the surface locations of the particles.

Conclusion

The present study evaluated the removal of Pb^{2+}, Cd^{2+}, Cu^{2+} and As^{3+} from wastewater using algal biomass as adsorbent material in fluidized bed reactor. Batch experiments showed that algal biomass can be successfully used for the removal of these metal ions from wastewater. The maximum biosorption capacities obtained were 44.5, 39.5, 41, and 35 mg/g for Pb^{2+}, Cd^{2+}, Cu^{2+} and As^{3+}, respectively. Biosorption was found to depend significantly on the pH of the solution and is optimal at pH values of 4 and 5. The equilibrium data showed a good fit to the ion exchange model with high correlation coefficients for each metal biosorption. The sequence of calculated affinity constants was $K_{Pb}>K_{Cu}>K_{Cd}>K_{As}$. FTIR analyses showed that hydroxyl and carboxyl groups could be significant for capturing these metals.

An ideal plug flow model has been adopted to characterize the fluidized bed reactor; this model fit well to the experimental data. In the fluidized bed system, an increase in the bed depth of algal biomass increased the breakthrough time. Increasing the solution flow rate decreased the breakthrough time due to the decrease in the contact time between the adsorbate and the adsorbent, as well as, at low flow rates the metal ions had a sufficient contact time to occupy the spaces within the particles. Also, it was found that an increase in particle size caused a decrease of the operating time due to the decrease in surface area of large particles relative to small particles.

References

1. Gin KY, Tang YZ, Aziz MA (2002) Derivation and application of a new model for heavy metal biosorption by algae. Water Res 36: 1313-1323.
2. Wang J, Chen C (2009) Biosorbents for heavy metals removal and their future. Biotechnol Adv 27: 195-226.
3. Kim H, Baek K, Kim B-K, Shin H-J, Yang J-W (2008) Removal characteristics of metal cations and their mixtures using micellar-enhanced ultrafiltration. Korean J of Chemical Engineering 25: 253-258.
4. McKay G, Ho YS, Ng JCY (1999) Biosorption of Copper from Waste Waters: A Review. Separation & Purification Reviews 28: 87-125.
5. Romera E, González F, Ballester A, Blázquez ML, Muñoz JA (2007) Comparative study of biosorption of heavy metals using different types of algae. Bioresour Technol 98: 3344-3353.
6. Sulaymon AH, Ebrahim SE, Abdullah SM, Al-Musawi TJ (2012) Removal of lead, cadmium, and mercury ions using biosorption. J Desalination and Water Treatment 24: 344-352.
7. Kratochvil D (1997) A study of the metal biosorption process utilizing Sargassum seaweed biomass, Ph.D Thesis, McGill University, Department of Chemical Engineering.
8. Brady JM, Tobin JM, Roux J-C (1999) Continuous fixed bed biosorption of Cu^{2+} ions: application of a simple two parameter mathematical model. J Chem Technol Biot 74: 71-77.
9. Romera E, González F, Ballester A, Blázquez ML, Muñoz JA (2006) Biosorption with algae: a statistical review. Crit Rev Biotechnol 26: 223-235.
10. Figueria MM, Volesky B, Ciminelli VST, Roddlick FA (2000) Biosorption of metals in brown seaweed biomass. Water Res 34: 196-204.
11. Davis TA, Volesky B, Mucci A (2003) A review of the biochemistry of heavy metal biosorption by brown algae. Water Res 37: 4311-4330.
12. Hossain MA, Ngo HH, Guo WS, Setiadi T (2012) Adsorption and desorption of copper(II) ions onto garden grass. Bioresour Technol 121: 386-395.
13. Fu Y, Liu D (2007) Novel experimental phenomena of fine-particle fluidized beds. Exp Therm Fluid Sci 32: 341-344.
14. Schiewer S, Volesky B (1996) Modeling Multi-Metal Ion Exchange in Biosorption. Environ Sci Technol 30: 2921-2927.
15. Diniz V, Volesky B (2005) Biosorption of La, Eu and Yb using Sargassum biomass. Water Res 39: 239-247.
16. Kratochvil D, Volesky B (1998) Advances in the biosorption of heavy metals. Trends Biotechnol 16: 291-300.
17. Diniz V, Weber ME, Volesky B, Naja G (2008) Column biosorption of lanthanum and europium by Sargassum. Water Res 42: 363-371.
18. Ho YS, Huang CT, Huang HW (2002) Equilibrium sorption isotherm for metal ions on tree fern. Process Biochem 37: 1421-1430.

19. Park YG, Cho SY, Kim SJ, Lee GB, Kim BH, et al. (1999) Mass Transfer in Semi-Fluidized and Fluidized Ion-Exchange Beds. Environ Eng Res 4: 71-80.

20. Ijzerman MM, Falkinham JO, Reneau RB, Hagedorn C (1994) Field evaluation of two colorimetric coliphage detection methods. Appl Environ Microbiol 60: 826-830.

21. Sheng PX, Ting YP, Chen JP, Hong L (2004) Sorption of lead, copper, cadmium, zinc, and nickel by marine algal biomass: characterization of biosorptive capacity and investigation of mechanisms. J Colloid Interface Sci 275: 131-141.

22. Romera E, González F, Ballester A, Blázquez ML, Muñoz JA (2007) Comparative study of biosorption of heavy metals using different types of algae. Bioresour Technol 98: 3344-3353.

23. Lu D, Cao Q, Li X, Cao X, Luo F, et al. (2009) Kinetics and equilibrium of Cu(II) adsorption onto chemically modified orange peel cellulose biosorbents. Hydrometallurgy 95: 145-152.

24. Lodeiro P, Cordero B, Grille Z, Herrero R, Sastre de Vicente ME (2004) Physicochemical studies of cadmium(II) biosorption by the invasive alga in Europe, Sargassum muticum. Biotechnol Bioeng 88: 237-247.

25. Saleem M, Pirzada T, Qadeer R (2007) Sorption of acid violet 17 and direct red 80 dyes on cotton fiber from aqueous solutions. Colloid Surface A 292: 246-250.

26. Meena AK, Mishra GK, Rai PK, Rajagopal C, Nagar PN (2005) Removal of heavy metal ions from aqueous solutions using carbon aerogel as an adsorbent. J Hazard Mater 122: 161-170.

27. Sari A, Tuzen M (2008) Biosorption of cadmium(II) from aqueous solution by red algae (Ceramiumvirgatum): equilibrium, kinetic and thermodynamic studies. J Hazard Mater 157: 448-454.

28. Anwar J, Shafique U, Waheed-uz-Zaman, Salman M, Dar A, et al. (2010) Removal of Pb(II) and Cd(II) from water by adsorption on peels of banana. Bioresour Technol 101: 1752-1755.

29. Yun YS, Volesky B (2003) Modeling of lithium interference in cadmium biosorption. Environ Sci Technol 37: 3601-3608.

30. Ngian KF, Martin WRB (1980) Bed expansion characteristics of liquid fluidized particles with attached microbial growth. Biotechnol Bioeng 22: 1843-1856.

31. Volesky B, Weber J, Park JM (2003) Continuous-flow metal biosorption in a regenerable Sargassum column. Water Res 37: 297-306.

Hydrodynamics of Liquid-Solid Semi-fluidized Bed with Regular Homogenous Ternary Mixture

SamalDK*, MohantyYK and Roy GK

Department of Chemical Engineering, Gandhi Institute of Engineering and Technology (GIET), Gunupur, Rayagada, Odisha-765022, India

Abstract

Hydrodynamics studies in Semi-fluidized bed relating to bed pressure drop, height of top packed bed, minimum and maximum semi-fluidization velocities with regular (spherical glass beads) homogeneous ternary mixtures, have been made in 0.01, 0.025, 0.037, 0.05, and 0.065 m internal diameter Perspex columns, with water as the fluidizing medium. Ternary mixtures of glass beads of different sizes (-8+10, -10+12 and -12+14 BSS) have been used as the semi-fluidized solids. Experimental studies relate to establish the effect of system parameters viz. column diameter, average particle size of semi-fluidized solid and initial static bed height and operating parameters viz. superficial liquid velocity and bed expansion ratio on the above mentioned system variables. Empirical and semi-empirical models have been developed using dimensional as well as statistical analyses. The calculated values from the predicted models have been compared with their experimental counterparts and fairly good agreement has been obtained. The results have also been compared with those available in the literature for irregular homogenous ternary mixtures.

Keywords: Semi-fluidization; Liquid-solid system; Homogeneous ternary mixtures; Regular particles; Hydrodynamics; Mathematical modeling

Nomenclature

BSS	British Standard Sieve
D, d	Diameter, *m*
G	Mass velocity, kg/m^2s
H, h	Height, *m*
P	Pressure, N/m^2
R	Bed expansion ratio
U	Velocity, *m/s*
Greek letters	
μ	Viscosity, *kg/m.s*
Δ	Difference
Subscripts	
av	average
c	column
f	fluid
l	liquid
mf	minimum fluidization
msf	maximum semi-fluidization
osf	onset of semi-fluidization
p	particle
pa	top packed bed
s	static/superficial
sf	semi-fluidization
w	water

Introduction

Semi-fluidization is a novel technique of fluid-solid contacting which is a stage between fluidization and pneumatic conveyance. Semi-fluidization begins when the expanded bed of fluidization is restricted at certain height by means of a restraint which only allows the fluid to pass. A semi-fluidized bed is thus obtained by increasing the fluid velocity beyond minimum fluidization velocity and thereafter arresting the movement of the solid particles with a restraint fixed at a suitable height towards the top of the conduit. This results in the formation of packed bed below the top restraint. The velocity at which the first particle of solid touches the top restraint is called minimum semi-fluidization velocity (U_{osf}). Similarly when all the bed materials form a packed bed below the top restraint, the corresponding fluid velocity is called the maximum semi-fluidization velocity (U_{msf}) [1]. The pioneering work as regards the concept and application of a semi-fluidized bed started by Fan et al. [2]. Hydrodynamic study on single size particle in solid-liquid system was highlighted by Fan and Wen [3]. Investigations relating to various aspects of semi-fluidized bed behavior involving different systems are available in literature. Considerable work on hydrodynamic studies in semi-fluidized bed has been reported by Roy et al. [4-6] and Ho et al. [7]. Murthy and Roy reviewed the various studies on semi-fluidization [8]. Of late, hydrodynamic investigations relating to mixtures (binary/ternary) of homogenous or heterogeneous nature has become essential in view of their potential applications in slurry and gas-solid reactions. A few articles on hydrodynamic characteristics of liquid-solid binary homogenous and heterogeneous mixtures in semi-fluidization have been reported [1,5,9,10]. The use of mixtures have the

***Corresponding author:** Deepak Kumar Samal, Department of Chemical Engineering, Gandhi Institute of Engineering and Technology(GIET), Gunupur, Rayagada, Odisha-765022, India

advantages in semi-fluidization as with less pressure drop both fluidized and the packed bed condition can be achieved compared to a particle [1]. The maiden investigations relating to hydrodynamics of liquid-solid semi-fluidized bed with irregular homogenous ternary mixtures have been reported by present author [1]. The correlations developed from dimensional analysis approach are as under:

For the bed pressure drop,

$$\Delta P_{sf}/\Delta P_{mf}=2\times10^{-7}(\rho_{sav}/\rho_f)^{1.675}(D_c/d_{pav})^{3.338}(G_{sf}/G_{mf})^{3.094}(H_s/D_c)^{-0.424}R^{-0.938} \quad (1)$$

For the top packed bed formation,

$$H_{pa}/H_s=1\times10^{-5}(\rho_{sav}/\rho_f)^{3.155}(D_c/d_{pav})_{1.510}(G_{sf}/G_{mf})^{2.426}(H_s/D_c)^{-0.524}R^{-1.676} \quad (2)$$

For minimum semi-fluidization velocity,

$$U_{osf}/U_{mf}=296.5(\rho_{sav}/\rho_f)^{-0.451}(D_c/d_{pav})^{-1.343}(H_s/D_c)^{0.228}R^{0.755} \quad (3)$$

For maximum semi-fluidization velocity,

$$U_{msf}/U_{mf}=0.001(\rho_{sav}/\rho_f)^{-0.680}(D_c/d_{pav})^{2.430}(H_s/D_c)^{0.613}R^{0.633} \quad (4)$$

Investigation relating to liquid-solid semi-fluidization regular homogenous ternary mixtures is almost absent. The objective of the present work is to study the hydrodynamics of liquid-solid semi-fluidization viz. the bed pressure drop, height of the top packed bed, minimum and maximum semi-fluidization velocities using homogenous ternary mixture of spherical glass beads. In addition the current study includes investigations carried out in five different column diameters with five different ternary mixtures of spherical glass beads of diameter 0.001854, 0.001504,0.001303 m. The experimental data have been correlated with the system parameters by two different approaches viz. the dimensional analysis and the statistical analyses.

Material and Methods

Schematic representation of the experimental setup is shown in Figure 1. The details of the experimental set up and its components are given elsewhere [1]. The scope of the experiment is presented in Table 1. Accurately weighed amount of material was fed into the column, fluidized and de-fluidized slowly and adjusted for a specific reproducible initial static bed height. Liquid was then pumped to the fluidizer through a rotameter and the temperature was maintained at 25 ± 5°C. At least two minutes were allowed to attain a steady state and then the readings of the manometers and the top packed bed height for each liquid flow rate were noted. The procedure was repeated for different values of initial static bed height, column diameter, the average particle size and the bed expansion ratio. In this work dimensional and statistical analyses approach have been used for predicting dimensionless responses like semi-fluidized bed pressure drop ($\Delta P_{sf}/\Delta P_{mf}$), height of top packed bed (H_{pa}/H_s), and minimum and maximum semi-fluidization velocities (U_{osf}/U_{mf} *and* U_{msf}/U_{mf}) as functions of the dimensionless parameters viz. H_s/D_c, D_c/H_s, d_{pav}/D_c, G_s/G_{mf}, and R. For statistical analysis, Central Composite Design (CCD) [11-13] has been used to develop correlations for responses for the four dependent variables in dimensionless form. The complete experimental range and the levels of the independent variables are given in Table 2. The design of the experiment is given in Table 3 (dimensional analysis) and Tables 4a and 4b (statistical analysis). For statistical analysis, a statistical software package Design-Expert-8.0.7.1, Stat-Ease, Inc., Minneapolis, USA, has been used for regression analysis of the semi-fluidized bed responses.

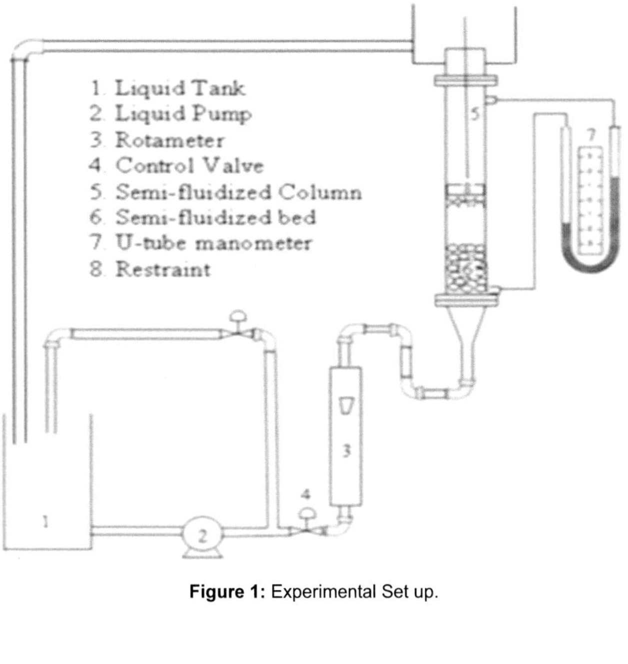

Figure 1: Experimental Set up.

a) System		Liquid-Solid
b) Liquid		Water
c) Solid		Regular
	i. Material	Glass beads (Spherical)
	ii. Density (ρ_{sav}), Kg/m³	2470
	iii. Size (d_p), m	d_{p1}=0.001854, d_{p2}=0.001504, d_{p3}=0.001303
	iv. Composition (wt. %)	i20:30:50, ii20:40:40, iii20:50:30, iv20:60:20, v20:90:10
d) Average particle diametr (d_{pav} × 10^3), m		i1.457, ii1.482, iii1.509, iv1.536, v1.565
e) Initial Static Bed Height (H_s), m		0.033, 0.06, 0.08, 0.10, 0.128
f) Bed Expansion Ratio, (R)		1.3, 2.0, 2.5, 3.0, 3.7
g) Column Diameter (Dc), m		0.01, 0.025, 0.037, 0.05, 0.065

Table 1: Scope of the experiment.

Variables	Symbol	-α	-1	0	+1	+α
H_s/D_c	X_1	0.917	1.673	2.222	2.771	3.527
D_c/H_s	X_2	0.125	0.324	0.469	0.613	0.813
d_{pav}/D_c	X_3	0.0405	0.0413	0.0420	0.0426	0.0435
G_s/G_{mf}	X_4	405	5.08	5.5	5.92	6.5
R	X_5	1.3	2.0	2.5	3.0	3.7

Table 2: Level of independent variables.

Results and Discussions

The minimum and maximum semi-fluidization velocities are determined by the extrapolation of H_{pa}/H_{sto} 1 on H_{pa}/H_{svs} U_{sf} plot which also used in present investigation. During the present investigation it has been found that maximum contribution for the semi-fluidized bed pressure drop is due to the top packed bed. Also it has been observed that due to the presence of fines in mixture though bulk density of bed increases, but the fines move faster to form top packed bed. The experimental investigations has established that the semi- fluidized bed responses viz. ΔP_{sf}, H_{pa}, U_{osf}, and U_{msf} not only depend on fluid velocity and the material properties, but also on other system parameters like initial static bed height, column diameter and bed expansion ratio. The detailed results on individual parameters and the correlations developed are presented below.

Sl.No	H_s/D_c	D_c/H_s	d_{pav}/D_c	G_{sf}/G_{mf}	R	$\Delta P_{sf}/\Delta P_{mf}$		H_{pa}/H_s				U_{mf}/U_{mc}			U_{mg}/U_{mf}		
						Cal. (eq. 4)	Expt.	Cal. (eq. 5)	Cal. (eq. 1a)	Cal. (eq. 1b)	Expt.	Cal. (eq. 6)	Cal. (eq. 2)	Expt.	Cal. (eq. 7)	Cal. (eq. 3)	Expt.
1	0.916	0.45	0.0419	5.5	2.5	18.828	18	0.731	3.643	0.699	0.75	3.081	3.381	3	6.502	8.130	6.6
2	1.666	0.45	0.0419	5.5	2.5	17.874	17.5	0.674	2.653	0.652	0.69	3.213	3.381	3.125	7.074	10.08	7
3	2.222	0.45	0.0419	5.5	2.5	17.432	16.8	0.648	2.217	0.628	0.67	3.278	3.381	3.2	7.367	11.18	7.3
4	2.777	0.45	0.0419	5.5	2.5	17.097	16	0.629	1.966	0.612	0.65	3.330	3.381	3.25	7.602	12.11	7.63
5	3.527	0.45	0.0419	5.5	2.5	16.745	25	0.609	1.744	0.596	0.61	3.386	3.381	3.3	7.863	13.20	8
6	2.222	0.125	0.0419	5.5	2.5	29.233	17.5	1.179	24.620	0.895	1	1.939	3.381	2.222	4.669	11.18	5.1
7	2.222	0.3125	0.0419	5.5	2.5	20.196	15	0.769	3.394	0.669	0.72	2.823	3.381	3	6.470	11.18	6.6
8	2.222	0.625	0.0419	5.5	2.5	15.297	13.5	0.556	0.508	0.506	0.55	3.751	3.381	4	8.281	11.18	9
9	2.222	0.8125	0.0419	5.5	2.5	13.733	11.5	0.492	0.315	0.471	0.45	4.177	3.381	4.3	9.092	11.18	9.8
10	2.222	0.45	0.040	5.5	2.5	19.166	18.5	764	2.438	0.639	0.75	3.223	3.425	3.157	6.120	11.31	6.315
11	2.222	0.45	0.041	5.5	2.5	18.307	17	0.706	2.353	0.635	0.7	3.250	3.403	3.173	6.694	11.24	6.938
12	2.222	0.45	0.042	5.5	2.5	16.613	15.8	0.596	1.994	0.617	0.6	3.307	3.358	3.25	8.094	11.11	8.1
13	2.222	0.45	0.043	5.5	2.5	15.787	15	0.546	1.795	0.607	0.55	3.337	3.335	3.3	8.943	11.04	9
14	2.222	0.45	0.0419	5.5	2.5	9.557	9	0.485			0.45						
15	2.222	0.45	0.0419	5.5	2.5	13.103	13	0.565			0.55						
16	2.222	0.45	0.0419	5.5	2.5	22.622	21	0.735			0.71						
17	2.222	0.45	0.0419	5.5	2.5	28.750	26.5	0.826			0.8						
18	2.222	0.45	0.0419	5.5	2.5	26.648	23	0.975	349.05	1.505	1	1.663	1.852	1.55	4.892	7.12	4.6
19	2.222	0.45	0.0419	5.5	2.5	20.148	18.5	0.745	9.586	0.814	0.75	2.600	2.753	2.6	6.406	9.58	6.1
20	2.222	0.45	0.0419	5.5	2.5	15.487	13.7	0.579	0.660	0.507	0.6	3.961	3.998	3.8	8.258	12.68	8
21	2.222	0.45	0.0419	5.5	2.5	13.516	11.5	0.508	0.129	0.383	0.48	4.925	4.849	4.8	9.416	14.65	9

Table 3: Experimental design matrix and responses (Dimensional analysis).

Sl.No	H_s/D_c	D_c/H_s	d_{pav}/D_c	G_{sf}/G_{mf}	R	$\Delta P_{sf}/\Delta P_{mf}$	H_{pa}/H_s
1	1.673	0.3242	0.04133	5.08	2	19.32	0.86
2	2.77	0.3242	0.04133	5.08	2	18.5	0.8
3	1.673	0.6132	0.04133	5.08	2	15	0.64
4	2.77	0.6132	0.04133	5.08	2	14.3	0.6
5	1.673	0.3242	0.04260	5.08	2	17.8	0.75
6	2.77	0.3242	0.04260	5.08	2	17	0.7
7	1.673	0.6132	0.04260	5.08	2	13.76	0.55
8	2.77	0.6132	0.04260	5.08	2	13.2	0.5
9	1.673	0.3242	0.04133	5.92	2	30.6	1
10	2.77	0.3242	0.04133	5.92	2	29.2	1
11	1.673	0.6132	0.04133	5.92	2	23.6	0.8
12	2.77	0.6132	0.04133	5.92	2	22.6	0.74
13	1.673	0.3242	0.04260	5.92	2	28.2	0.93
14	2.77	0.3242	0.04260	5.92	2	27	0.87
15	1.673	0.6132	0.04260	5.92	2	21.8	0.7
16	2.77	0.6132	0.04260	5.92	2	21	0.65
17	1.673	0.3242	0.04133	5.08	3	15	0.66
18	2.77	0.3242	0.04133	5.08	3	14.2	0.62
19	1.673	0.6132	0.04133	5.08	3	11.45	0.5
20	2.77	0.6132	0.04133	5.08	3	11	0.46
21	1.673	0.3242	0.04260	5.08	3	13.6	0.58
22	2.77	0.3242	0.04260	5.08	3	13	0.54
23	1.673	0.6132	0.04260	5.08	3	10.5	0.43
24	2.77	0.6132	0.04260	5.08	3	10.1	0.4
25	1.673	0.3242	0.04133	5.92	3	23.4	0.83
26	2.77	0.3242	0.04133	5.92	3	22.42	0.77
27	1.673	0.6132	0.04133	5.92	3	18.1	0.61
28	2.77	0.6132	0.04133	5.92	3	17.3	0.58
29	1.673	0.3242	0.04260	5.92	3	21.6	0.72
30	2.77	0.3242	0.04260	5.92	3	20.6	0.67
31	1.673	0.6132	0.04260	5.92	3	16.7	0.54
32	2.77	0.6132	0.04260	5.92	3	16	0.5
33	0.916	0.4687	0.04197	5.5	2.5	18.45	0.71
34	3.527	0.4687	0.04197	5.5	2.5	16.4	0.6
35	2.221	0.125	0.04197	5.5	2.5	29.1	1
36	2.221	0.8125	0.04197	5.5	2.5	13.7	0.49
37	2.221	0.4687	0.04047	5.5	2.5	18.8	0.75
38	2.221	0.4687	0.04347	5.5	2.5	15.5	0.53
39	2.221	0.4687	0.04197	4.5	2.5	9.4	0.47
40	2.221	0.4687	0.04197	6.5	2.5	28.2	0.8
41	2.221	0.4687	0.04197	5.5	1.3	26.1	0.95
42	2.221	0.4687	0.04197	5.5	3.7	13.2	0.5
43	2.221	0.4687	0..197	5.5	2.5	17	0.63
44	2.221	0.4687	0.04197	5.5	2.5	17.1	0.63
45	2.221	0.4687	0.04197	5.5	2.5	17.1	0.61
46	2.221	0.4687	0.04197	5.5	2.5	17	0.62
47	2.221	0.4687	0.04197	5.5	2.5	16.96	0.62
48	2.221	0.4687	0.04197	5.5	2.5	17	0.63
49	2.221	0.4687	0.04197	5.5	2.5	17	0.63

Table 4a: Experimental design matrix and responses for $\Delta P_{sf}/\Delta P_{mf}$ and H_{pa}/H_s (Statistical analysis).

Semi-fluidized bed Pressure drop (ΔP_{sf})

Figure 2 shows the variation of semi-fluidized bed pressure drop with respect to the superficial liquid velocity for different values of initial static bed height for average particle diameter of 0.001509 m at constant bed expansion ratio ($\underline{R}$=2) in the 0.05 m internal diameter column. From Figure 2, it has been observed that with increase in initial static bed height, the bed pressure drop increases. This is due to the increase in the bed weight and therefore an increase in the top packed bed formation with all the other variables remaining constant. Effect of fines in the mixture is shown in Figure 3 shows the variation of semi-fluidized bed pressure drop with superficial liquid velocity for different average particle diameter with constant bed expansion ratio (R=2), initial static bed height (H_s=0.08m) in 0.05 m internal diameter column. Increase of fines results in a relatively fast formation of top packed bed leading to higher values of semi-fluidized bed pressure drop, which is evident from Figure 3. Figure 4 shows the variation of pressure drop

Sl.No	H_s/D_c	D_c/H_s	d_{pav}/D_c	G_s/G_{mf}	R	$\Delta P_{sf} / \Delta P_{mf}$	H_{pa} / H_s
1	1.5695	0.297	0.04125	1.9	2	4.7	1.5695
2	2.8745	0.297	0.04125	1.9	2.1	5.1	2.8745
3	1.5695	0.641	0.04125	1.9	2.76	6.16	1.5695
4	2.8745	0.641	0.04125	1.9	2.88	6.7	2.8745
5	1.5695	0.297	0.04275	1.9	2	5.66	1.5695
6	2.8745	0.297	0.04275	1.9	2.15	6.2	2.8745
7	1.5695	0.641	0.04275	1.9	2.8	7.44	1.5695
8	2.8745	0.641	0.04275	1.9	2.94	8.1	2.8745
9	1.5695	0.297	0.04125	3.1	3.35	6.4	1.5695
10	2.8745	0.297	0.04125	3.1	3.5	7	2.8745
11	1.5695	0.641	0.04125	3.1	4.6	8.4	1.5695
12	2.8745	0.641	0.04125	3.1	4.8	9.11	2.8745
13	1.5695	0.297	0.04275	3.1	3.4	7.7	1.5695
14	2.8745	0.297	0.04275	3.1	3.55	8.4	2.8745
15	1.5695	0.641	0.04275	3.1	4.67	10.11	1.5695
16	2.8745	0.641	0.04275	3.1	4.9	11	2.8745
17	0.917	0.469	0.042	2.5	3.14	6.7	0.917
18	3.527	0.469	0.042	2.5	3.45	8	3.527
19	2.222	0.125	0.042	2.5	1.95	4.72	2.222
20	2.222	0.813	0.042	2.5	4.2	9.2	2.222
21	2.222	0.469	0.0405	2.5	3.28	6.24	2.222
22	2.222	0.469	0.0435	2.5	3.4	9.1	2.222
23	2.222	0.469	0.042	1.3	1.7	5	2.222
24	2.222	0.469	0.042	3.7	5	9.66	2.222
25	2.222	0.469	0.042	2.5	3.33	7.56	2.222
26	2.222	0.469	0.042	2.5	3.34	7.55	2.222
27	2.222	0.469	0.042	2.5	3.31	7.56	2.222
28	2.222	0.469	0.042	2.5	3.34	7.54	2.222
29	2.222	0.469	0.042	2.5	3.32	7.56	2.222
30	2.222	0.469	0.042	2.5	3.34	7.56	2.222

Table 4b: Experimental design matrix and responses for U_{osf}/U_{mf} and U_{msf}/U_{mf} (Statistical analysis).

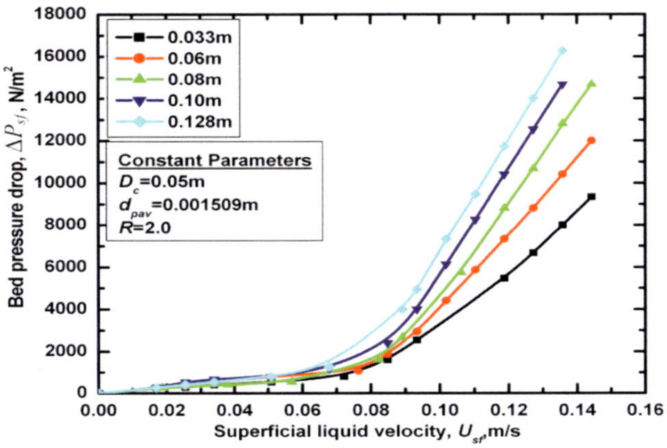

Figure 2: Variation of bed pressure drop with respect to superficial liquid velocity for different initial static bed height.

with superficial liquid velocity for different bed expansion ratio for an initial static bed height of 0.08 m, with a mixture of average particle diameter of 0.001457 m in 0.05 m internal diameter column. From Figure 4, it is clearly observed that with increase in bed expansion ratio the pressure drop decreases. This is due to the fact that, relatively less amount of particles reach the top restraint to form the top packed bed, which has also been observed in other systems. Figure 5 shows the variation of bed pressure drop with superficial liquid flow rate for different column diameters for average particle diameter of 0.01509 m with 0.08 m initial static bed height and a bed expansion ratio of 2.5. The superficial liquid velocity decreases as the column diameter increases for a constant liquid flow rate. Decrease in velocity results in

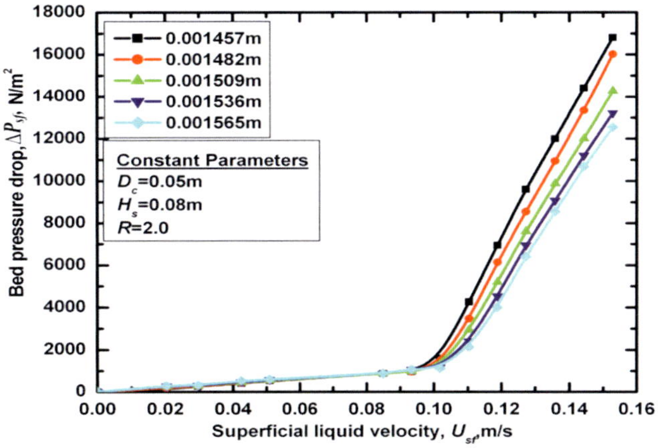

Figure 3: Variation of bed pressure drop with respect to superficial liquid velocity for different average particle diameter.

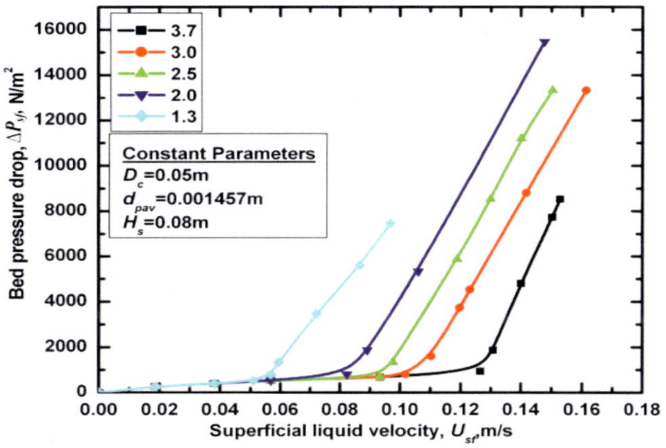

Figure 4: Variation of bed pressure drop with respect to superficial liquid velocity for different bed expansion ratio.

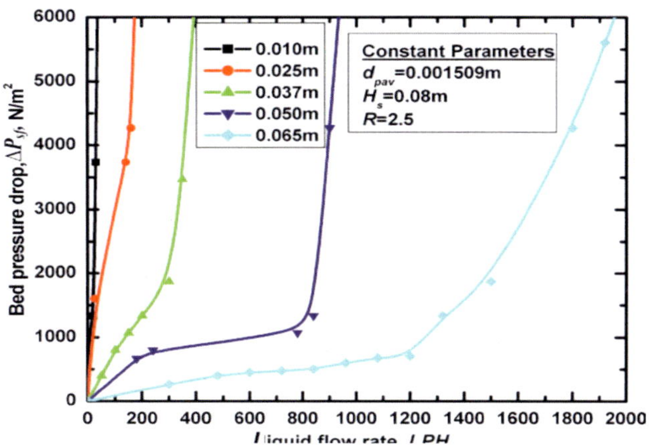

Figure 5: Variation of bed pressure drop with respect to liquid flow rate for different column diameters.

lesser transport of bed solids to the top with slower formation of top packed bed and thereby lower values of bed pressure drop.

Formation of Top packed bed (H_{pa})

Formation of the top packed bed in a semi-fluidized bed is controlled not only by the superficial liquid velocity but also by other variables like bed expansion ratio, initial static bed height, column diameter and average particle diameter. Figure 6 shows the variation of the height of top packed bed in semi-fluidized bed with respect to the superficial liquid velocity for different values of initial static bed height for average particle diameter of 0.001509 m at constant bed expansion ratio (R=2) in 0.05 m internal diameter column. From Figure 6, it has been observed that with increase in initial static bed height, the height of top packed bed increases. This is because of the increase in fines in the increased bed which help rapid formation of the top packed bed when the other variables remaining constant. Figure 7 shows the variation of height of top packed bed in semi-fluidized bed with superficial liquid velocity for different average particle size at constant bed expansion ratio (R=2) in 0.05 m internal diameter column. Increase of fines result in a relatively fast formation of top packed bed, which is evident from Figure 7. Figure 8 depicts the variation of height of top packed bed with superficial liquid velocity for different bed expansion ratio for an initial static bed height of 0.08 m, with a mixture of average particle diameter of 0.001457 m in 0.05 m internal diameter column. From Figure 8, it is clearly observed that with increase in bed expansion ratio the height of top packed bed decreases. This is due to the fact that the buoyancy force is insufficient to lift the particles when the restraint is placed a higher distance in the column with superficial liquid velocity remaining unchanged. The effect of column diameter is shown in Figure 9 where the variation of dimensionless height of top packed bed with superficial liquid flow rate for different diameter columns for an average particle diameter of 0.01509 m, 0.08 m initial static bed height and a bed expansion ratio of 2.5. Increase in column diameter decreases the superficial liquid velocity and thereby relatively fewer amounts of solids are transported to the top.

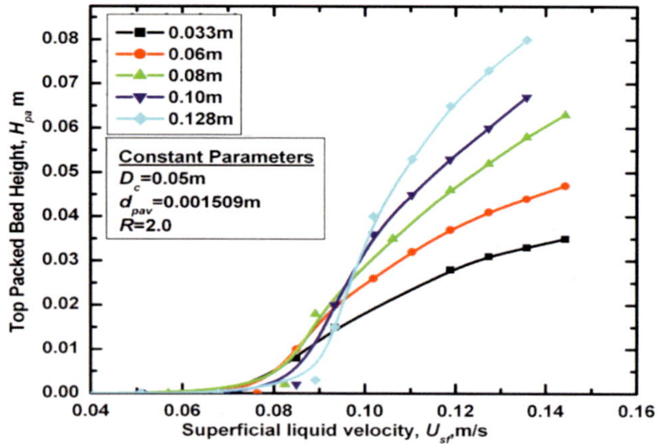

Figure 6: Variation of height of top packed bed with respect to superficial liquid velocity for different initial static bed height.

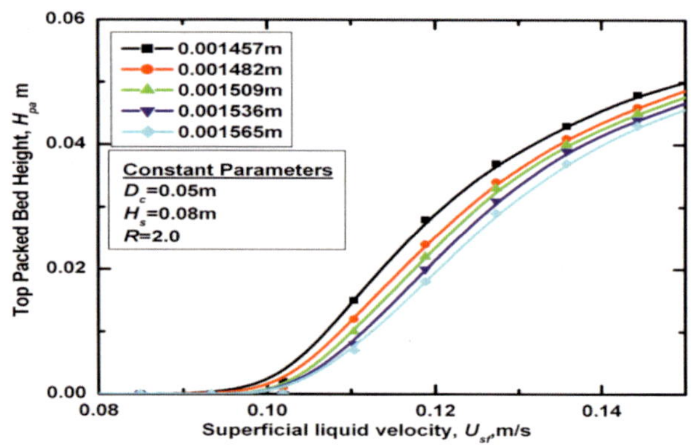

Figure 7: Variation of height of top packed bed with respect to superficial liquid velocity for different average particle diameter.

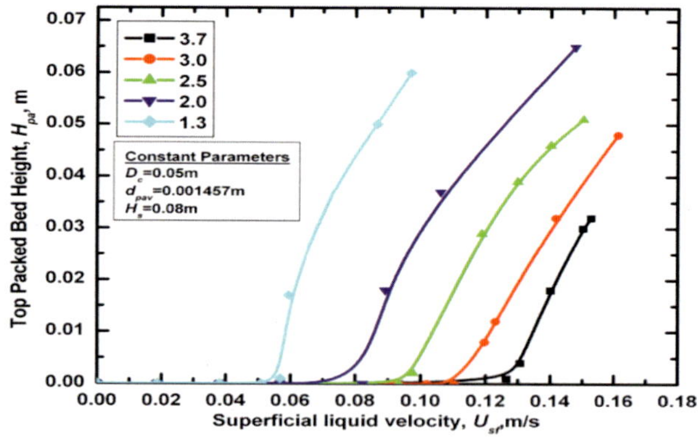

Figure 8: Variation of height of top packed bed with respect to superficial liquid velocity for different bed expansion ratio.

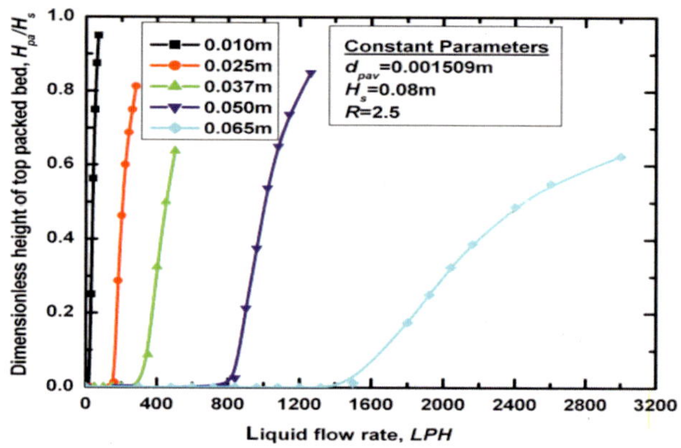

Figure 9: Variation of dimensionless height of top packed bed with respect to liquid flow rate for different column diameters.

Development of correlations by dimensional analysis

The dimensionless semi-fluidized bed pressure drop and top packed bed height are found to be dependent on five different variables viz. H_s, D_c, d_{pav}, G_{sf} and R while dimensionless minimum and maximum semi-fluidization velocities are functions of H_s, D_c, d_{pav}, G_{sf} and R. The values of the parameters and responses for the developing correlations are given in Table 3 and the developed correlations are represented as Eqs. (5) - (8).

For dimensionless semi-fluidized bed pressure drop,

$$\Delta P_{sf}/\Delta P_{mf}=2.73\times10^{-5}(H_s/D_c)^{-0.087}(D_c/H_s)^{-0.404}(d_{pav}/D_c)^{-2.712}(G_{sf}/G_{mf})^{2.995}R^{-0.649} \quad (5)$$

For dimensionless top packed bed height,

$$H_{pa}/H_s=2.6\times10^{-8}(H_s/D_c)^{-0.135}(D_c/H_s)^{-0.467}(d_{pav}/D_c)^{-4.688}(G_{sf}/G_{mf})^{1.447}R^{-0.623} \quad (6)$$

For dimensionless minimum semi-fluidization velocity,

$$U_{osf}/U_{mf}=7.74(H_s/D_c)^{0.070}(D_c/H_s)^{0.410}(d_{pav}/D_c)^{0.485}R^{1.038} \quad (7)$$

For dimensionless maximum semi-fluidization velocity,

$$U_{msf}/U_{mf}=1\times10^{8}(H_s/D_c)^{0.141}(D_c/H_s)^{0.356}(d_{pav}/D_c)^{5.304}R^{0.626} \quad (8)$$

Figure 10 shows the comparison between the experimental and calculated values (Eq. (1) and Eq.(5)) of $\Delta P_{sf}/\Delta P_{mf}$. The standard deviations and coefficients of correlation are 0.39 and 0.662 for Eq. (1) and 0.94 and 0.982 for Eq. (5) respectively. Figure 11 shows the comparison between the experimental and calculated values (Eq. (2) and Eq. (6)) of H_{pa}/H_s. The standard deviations and coefficients of correlation are 0.06 and 0.703 for Eq. (2), 0.043 and 0.967 for Eq.(6) respectively. Figure 12 shows the comparison between the experimental and calculated values (Eq.(3) and Eq. (7)) of U_{osf}/U_{mf}, which shows standard deviations and coefficients of correlation of0.64 and 0.792 for Eq.(3) and 0.13 and 0.984 for Eq. (7) respectively. Figure 13 shows the comparison between the experimental and calculated values (Eq. (4) and Eq. (8)) of U_{msf}/U_{mf}, which shows standard deviations and coefficients of correlation of 0.60 and 0.460 for Eq. (4) and0.32 and 0.973 for Eq. (8) respectively.

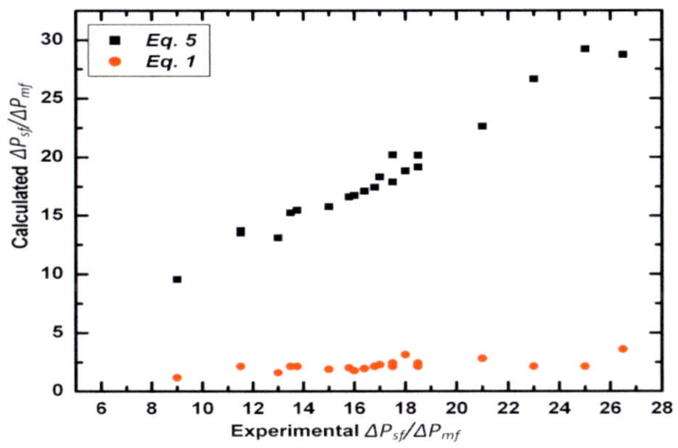

Figure 10: Comparison between the experimental and calculated (Eq. (1) and (5)) values of dimensionless semi-fluidized bed pressure drop.

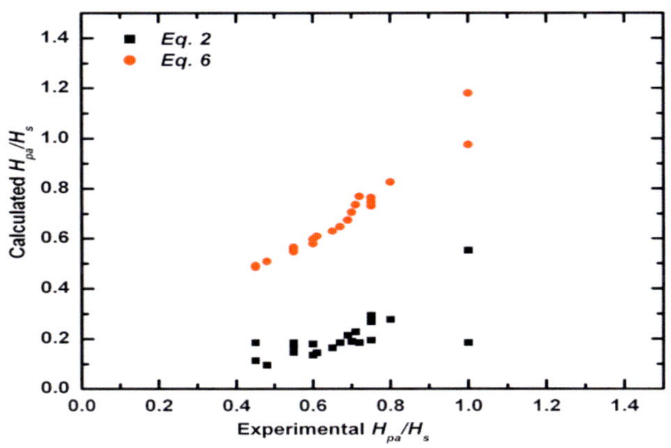

Figure 11: Comparison between the experimental and calculated (Eqs. (2) and (6)) values of dimensionless height of top packed bed.

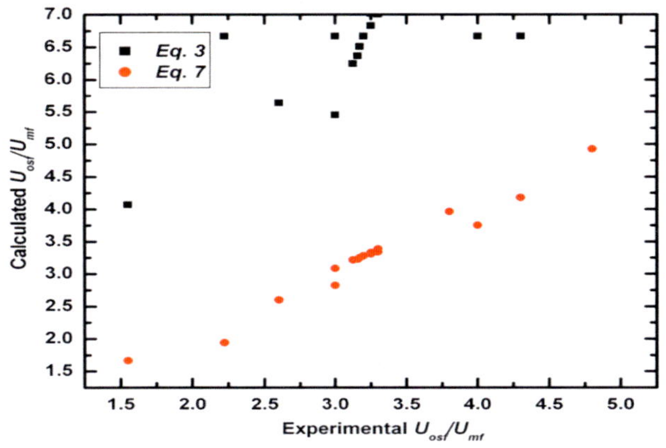

Figure 12: Comparison between the experimental and calculated (Eqs. (3) and (7)) values of dimensionless minimum semi-fluidization.

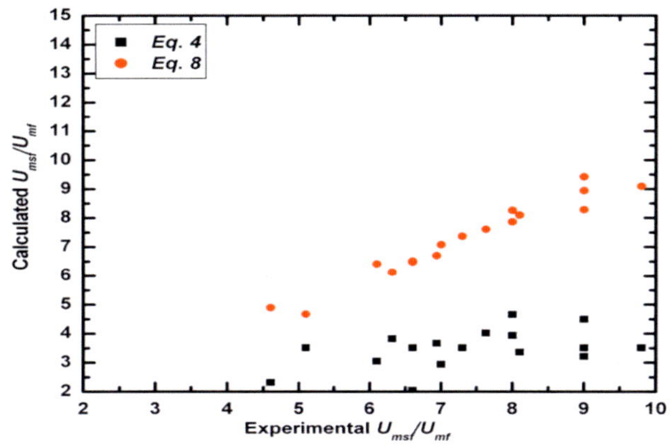

Figure 13: Comparison between the experimental and calculated (Eqs. (4) and (8)) values of dimensionless maximum semi-fluidization velocity.

Development of correlations by Statistical analysis

The method of experimentation is based on statistical design of experiments (Factorial Design Analysis) in order to bring out the interaction effects of variables, which would not otherwise be found by conventional experimentation and to explicitly find out the effect of each of the variables quantitatively on the response. In addition, the number of experiments required is far less compared to the conventional experiments [1].The equations developed by the statistical analysis approach are:

For dimensionless semi-fluidized bed pressure drop:

$$\Delta P_{sf}/\Delta P_{mf}=16.99-0.41\times X_1-2.58\times X_2-0.74\times X_3+4.09\times X_4-2.51\times X_5+0.068\times X_1\times X_2+0.028\times X_1\times X_3-0.086\times X_1\times X_4+0.048\times X_1\times X_5+0.11\times X_2\times X_3-0.53\times X_2\times X_4+0.301.50\times X_2\times X_5-0.14\times X_3\times X_4+0.081\times X_3\times X_5-0.56\times X_4\times X_5+0.034\times X_1^2+0.74\times X_2^2-0.014\times X_3^2+0.28\times X_4^2+0.43\times X_5^2 \quad (9)$$

In Figure 14, the experimental values of $\Delta P_{sf}/\Delta P_{mf}$ have been compared with the calculated values obtained from Eq. (9). The values of standard deviation and correlation coefficient are 0.55 and 0.993.

For dimensionless top packed bed height:

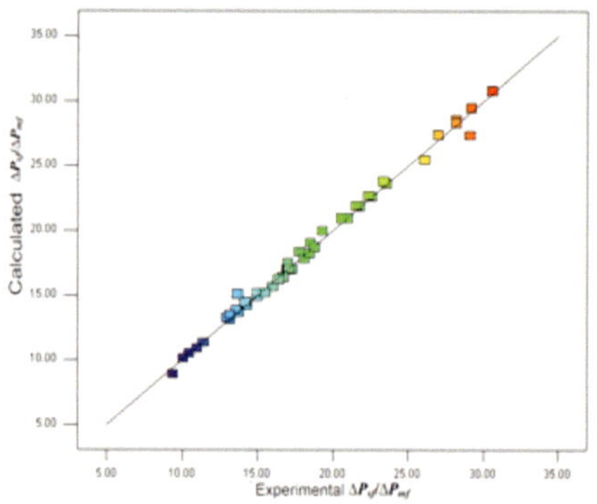

Figure 14: Comparison between the experimental and calculated (Eq. (9)) values of dimensionless semi-fluidized bed pressure drop.

$$H_{pa}/H_s=0.62-0.022\times X_1-0.100\times X_2-0.045\times X_3+0.072\times X_4-0.087\times X_5+6.25\times 10^{-4}\times X_1\times X_2-1.250\times 10^{-3}\times X_1\times X_3+0.000\times X_1\times X_4+1.250\times 10^{-3}\times X_1\times X_5+3.75\times 10^{-3}\times X_2\times X_3-7.500\times 10^{-3}\times X_2\times X_4+0.011\times X_2\times X_5-1.875\times 10^{-3}\times X_3\times X_4+4.375\times 10^{-3}\times X_3\times X_5-8.125\times 10^{-3}\times X_4\times X_5+5.510\times 10^{-3}\times X_1^2+0.021\times X_2^2+3.413\times 10^{-3}\times X_3^2+1.974\times 10^{-3}\times X_4^2+0.018\times X_5^2 \quad (10)$$

In Figure 15, the experimental values of H_{pa}/H_s have been compared with the calculated values obtained from Eq. (10). The values of standard deviation and correlation coefficient are 0.013and 0.995.

For dimensionless minimum semi-fluidization velocity:

$$U_{osf}/U_{mf}=3.33+0.077\times X_1+0.53\times X_2+0.027\times X_3+0.82\times X_5+8.75\times 10^{-3}\times X_1\times X_2+6.25\times 10^{-3}\times X_1\times X_3+0.014\times X_1\times X_5+7.50\times 10^{-3}\times X_2\times X_3+0.13\times X_2\times X_5+7.5\times 10^{-3}\times X_3\times X_5-7.083\times 10^{-3}\times X_1^2-0-.062\times X_2^2+4.167\times 10^{-3}\times X_3^2+6.667\times 10^{-3}\times X_5^2 \quad (11)$$

In Figure 16, the experimental values of Uosf/Umf have been compared with the calculated values obtained from Eq. (11). The values of standard deviation and correlation coefficient are 0.023and 0.999.

For dimensionless maximum semi-fluidization velocity:

$$U_{msf}/U_{mf}=7.55+0.32\times X_1+1.03\times X_2+0.7\times X_3+1.14\times X_5+0.035\times X_1\times X_2+0.034\times X_1\times X_3+0.047\times X_1\times X_5+0.095\times X_2\times X_3+0.15\times X_2\times X_5+0.097\times X_3\times X_5-0.041\times X_1^2-0.14\times X_2^2+0.039\times X_3^2-0.046\times X_5^2 \quad (12)$$

In Figure 17, the experimental values of U_{msf}/U_{mf} have been compared with the calculated values obtained from Eq. (12). The values of standard deviation and correlation coefficient are 0.091and 0.998.

The coefficient of correlations for statistical analysis is more than that of dimensional analysis.

Conclusion

Hydrodynamic parameters for the liquid-solid semi-fluidization using homogenous regular ternary mixture viz. the minimum and maximum semi-fluidization velocities, semi-fluidized bed pressure drop, height of top packed bed have the importance for knowing the limits of operational parameters when the semi-fluidized beds are in use. In the current study, investigations have been carried out to study the behavior of homogenous ternary mixtures of regular glass beads with the superficial liquid velocity in a liquid-solid semi-fluidized bed. To get the advantages of a semi-fluidized bed, the fines in the mixture have a great role as those help the fast formation of top packed bed, which is desirable. For small and large scale operations, conduits of different sizes are required. The effects of column diameter and the fines have been studied along with other process variables. Correlations for the calculation of semi-fluidized bed pressure drop, height of top packed bed, minimum and maximum semi-fluidization velocities have been proposed. The values calculated from the developed correlations have been compared with the experimental ones. The values of coefficients of correlation are found to be greater than0.96, thus emphasizing the validity of the developed correlations over the range of the operating parameters investigated. The statistical analysis approach can suitably be used for the development of model equations as it expresses the individual, interaction and cubic effects. Apart from this

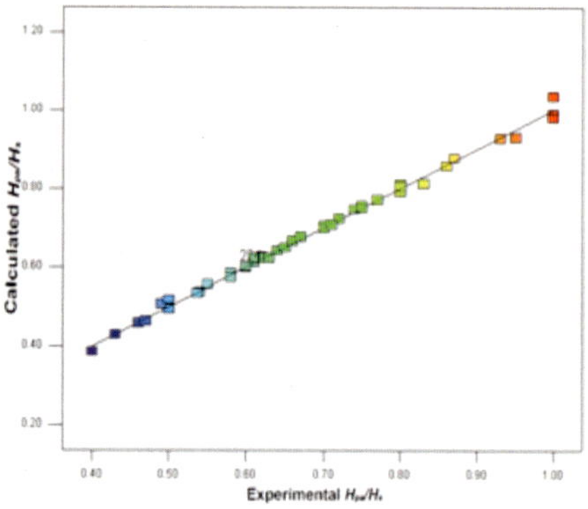

Figure 15: Comparison between the experimental and calculated (Eq. (10)) values of dimensionless height of top packed bed.

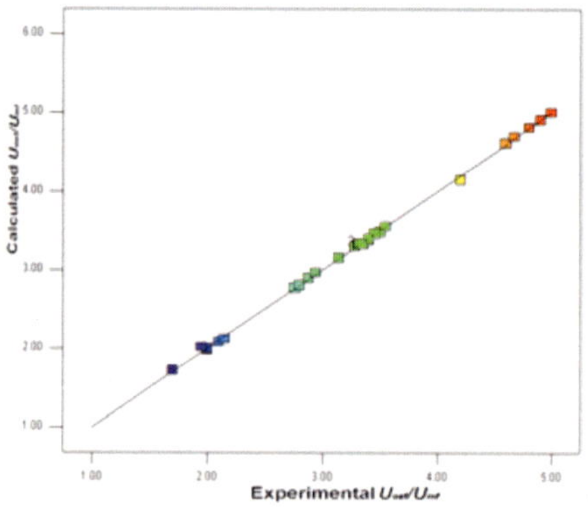

Figure 16: Comparison between the experimental and calculated (Eq. (11)) values of dimensionless minimum semi-fluidization velocity.

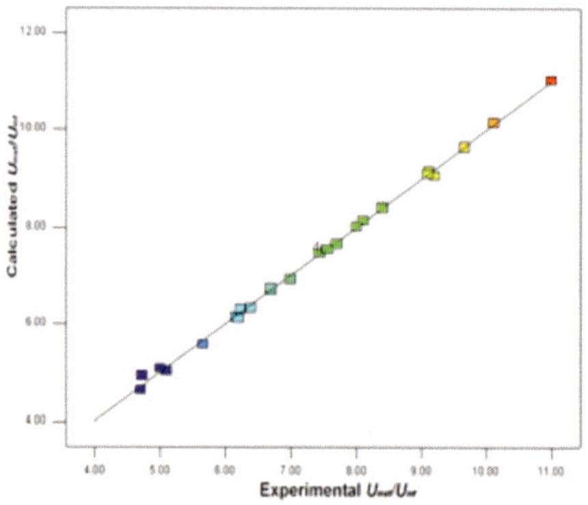

Figure 17: Comparison between the experimental and calculated (Eq. (12)) values of dimensionless maximum semi-fluidization velocity.

the requirement of number of experimental data is less as compared to the conventional method.

The hydrodynamics study conducted and the correlations developed thereof can find potential applications in physical and chemical processing. In the present investigation, the wide range in experimental parameters involved especially with respect to the mixture compositions and the column diameter makes it amenable to logical scale-up while considering design of liquid-solid semi-fluidization systems, involving ternary mixtures of homogenous particles.

Acknowledgement

The authors gratefully acknowledge the support and encouragement received from GIET, Gunupur, Odisha, India-765022.

References

1. Samal DK, Mohanty YK, Roy GK (2013) Hydrodynamics of liquid–solid semi-fluidized bed with irregular homogenous ternary mixture. Powder Technol 235: 921-930.
2. Fan LT, Yang YC, Wen CY (1959) Semi-fluidization: Mass transfer in semi-fluidized beds. AIChE J 5: 407-409.
3. Fan LT, Wen CY (1961) Mechanics of semi-fluidization of single size particles in solid liquid systems. AIChE J 7: 606-610.
4. Roy GK, Sarma KJR (1974) Relation between maximum semi-fluidization and minimum fluidization velocity in liquid-solid systems. J Inst Engrs (India) 54: 34-35.
5. Roy GK, Saratchandra HN (1976) Liquid-Solid Semi-fluidization of Heterogeneous mixtures: II: Prediction of Minimum Fluidization Velocity. Chem Eng J 12: 77-80.
6. Sarat Chandra HN, Roy GK (1977) Prediction of Maximum Semi-fluidization Velocity for Heterogeneous Mixtures in Liquid-Solid Systems. Indian Chem Engineer 19: 33-35.
7. Ho TC, Yau SJ, Hopper JR (1987) Hydrodynamics of semi fluidization in gas-solid systems. Powder Technology 50: 25-34.
8. Murthy JSN, Roy GK (1986) Semi-fluidization: a Review. Indian Chem Engineer 29: 9-22.
9. Mydlarz J (1987) Prediction of the packed bed height in liquid-solid semi-fluidization of homogeneous mixtures. Chem Eng J 34: 155-158.
10. Biswal KC, Sahoo SN, Verma P, Murthy JSN, Roy GK (1990) Dynamics of Gas-Solid Semi-Fluidization of Binary Homogeneous and Heterogeneous Mixtures. Chem Eng J 70: 61-64.
11. Box GEP, Hunter JS (1957) Multi-factor experimental design for exploring response surfaces. Ann Math Stat 28: 1-279.
12. Box GEP, Hunter WG (1961) The 2k_pfractional designs, parts I and II. Technometrics 3: 311-458.
13. Napier-Munn TJ (2000) The Central Composite Rotatable Design. JKMRC, The University of Queensland, Brisbane, Australia.

The Application of Transparent Glass Sponges for Improvement of Light Distribution in Photobioreactors

Anna Jacobi[1]*, Ethel Claudia Bucharsky[2], Karl Guenter Schell[2], Peter Habisreuther[3], Rainer Oberacker[2], Michael J. Hoffmann[2], Nikolaos Zarzalis[3] and Clemens Posten[1]*

[1]Karlsruhe Institute of Technology KIT, Institute of Engineering in Life Sciences, Department of Bioprocess Engineering, Fritz-Haber-Weg 2, Building 30.44, 76131 Karlsruhe, Germany
[2]Karlsruhe Institute of Technology KIT, Institute for Applied Materials – Ceramics in Mechanical Engineering, Haid-und-Neu-Str. 7, 76131 Karlsruhe, Germany
[3]Karlsruhe Institute of Technology KIT, Engler-Bunte-Institute, Division of Combustion Technology, Engler-Bunte-Ring 1, Building 40.11, 76128 Karlsruhe, Germany

Abstract

A new concept for improving light dilution and light distribution in photobioreactors by applying transparent, open-pored sponges was realized during cultivation experiments. A manufacturing process based on the polymer replica technique was established. A polyurethane template is impregnated with a nanoscaled SiO_2 powder suspension (solids loading of 60 wt% at around pH 10) and dried. The green body is prepared by burning-out the polymer at 800°C. Subsequently the sintering of the remaining SiO_2 structure to transparent cellular bodies is carried out at 1330°C. Many different factors influence the process. The slurry stabilization (viscosity, zeta-potential, pH) is important to generate a stable and self-supporting SiO_2 shell around the template and achieve appropriate green bodies. The sintering temperature determines the transition of amorphous (transparent) into crystalline silica. The 8 ppi sponge has a porosity of 0.9, the specific surface area of 568 m^{-1} means an increase of the (inner) surface to volume ratio of the reactor. Porous glass has a very high surface to volume ratio (about 0.26 m^2/g of glass at 10 μm pore size). Because of multiple and complex reflection of light into the porous glass, the algae can obtain light energy at any location of the total volume. First cultivation experiments in special designed Miniplate reactors show an increase in growth rate (about +25%) at low cell densities. The photo conversion efficiency was enlarged from 4.9% for the empty reactor to 5.6% and 5.9% for the 8 and 15 ppi sponge filled reactor. The effect and the improved efficiency of biomass build-up per applied light will be more apparent at high cell densities. Therefore concentration of the standard Tris-Phosphate (TP) medium (2.5 fold) and feeding are necessary.

Keywords: Microalgae; Photo-bioreactor; Bioprocess design; Growth kinetics; Photo conversion efficiency; Light distribution; Light dilution; Transparent open-pored sponge

Abbreviations: CFD: Computational fluid dynamics; LM: Light microscopy; MRI: Magnetic resonance imaging; OD: Optical density; P_A: Areal productivity; P_V: Volumetric productivity; PAR: Photosynthetic active radiation; PCE: Photo conversion efficiency; ppi: Pores per inch; RMS: Root mean square; SEM: Scanning electron microscopy; SFR: Surface to footprint ratio; SVR: Surface to volume ratio; TAP: Tris Acetate Phosphate medium; TP: Tris Phosphate medium; XRD: X-ray diffraction; μ: Growth rate

Introduction

Decreasing oil resources force the interest in renewables for biofuel production as well as systems for production of raw materials in chemical industry and high-value products [1-4]. For this purpose phototrophic processes are advantageous using light as energy and carbon dioxide as carbon source for biomass formation with only little additional mineral nutrient necessary. Besides higher land plants like sugarcane, maize, canola and corn, algae show high potential because of higher growth rates and productivities as well as higher photosynthetic efficiencies (efficiency of biomass build-up from sunlight energy) [5-7].

Microalgae are aquatic unicellular and simple multicellular, microbial organisms. They perform no cell differentiation and the lack of additional supporting structures made of structural biopolymers like roots, stems and leaves in land-based plants enhances the photosynthetic efficiency [8]. When grown in photobioreactors cultivation is also possible in arid regions (possible land coverage) and not in competition with food production. Faster growth rates and more generation cycles during the year are advantageous. Furthermore the possible recycle of water and unconsumed nutrients and the tolerance to a high CO_2 content in the gas phase which additionally increases growth kinetics make algae cultivations attractive compared to conventional agriculture.

Cultivation of microalgae in photobioreactors is state of the art and compared to open ponds highly controlled conditions can be applied (temperature, pH, aeration rate and CO_2 content, mixing, nutrients and additional feeding). However, not much large scale applications are in operation now, the few that exist produce high value products and no bulk materials e.g. for energetic use. Reasons for that are the high costs and the high demand of auxiliary energy e.g. for mixing. Although closed reactors are already better in light distribution compared to open ponds, there is still a need to reach the goal of reaching a homogeneous light distribution in the medium without the aforementioned disadvantages. Many different geometric modifications exist (e.g. flat plate or tubular reactors, airlift systems,...) all trying to overcome the problem of insufficient light supply for photosynthesis, but no optimal configuration has already been developed. The main problem of

***Corresponding author:** Clemens Posten, Karlsruhe Institute of Technology KIT, Institute of Engineering in Life Sciences, Department Bioprocess Engineering, Fritz-Haber-Weg 2, Building 30.44, 76131 Karlsruhe, Germany

phototrophic processes is the demand of light as energy source which is not miscible as other substrates. Only low biomass concentrations are achievable due to the inadequate light supply of the whole culture. The light attenuation shows a very fast exponential decay above the layer thickness because of light absorption and scattering by the algae. Large volume elements offer only light limiting conditions resulting in no/less photosynthetic growth whereas on the illuminated surface light saturation or also photo inhibition occurs. So light dilution on the illuminated surface is one issue and diminishing the gradient in the bulk phase is another.

To date many different technical solutions have been described in literature all having the aim to improve light supply, e.g. horizontal and vertical tubular reactors in many different configurations or flat plate reactors with a small layer thickness also in special arrangements. Moreover specific configurations like the Flat Panel Airlift reactor (Subitec) or cheap low-ceiling designs made of plastic foils (Soilx and Proviron) [9-11] are proposed. All this configurations try to improve the surface to volume ratio (SVR) of the photobioreactor as well as the surface to footprint ratio (SFR). Both aspects intend to distribute the light over a larger reactor surface and avoid dark volume elements and to use the incident sunlight as efficient as possible (volumetric and areal productivity P_V and P_A). An optimization of the incident angle of sunlight hitting the reactor surface is a further option for possible developments [10-12].

The concept of this new approach is the enlargement of the inner surface area of the reactor which can contribute to illumination, light distribution and light supply. For this purpose transparent sponges made of glass are employed. In the investigation presented here they are placed in special designed Miniplate reactors for testing the effect. Maximum growth rates μ and the photo conversion efficiency (PCE) which means the efficiency of buildup of biomass from captured light are determined. Transparent glass sponges are a new class of materials. They are prepared by the polymer replica technique [13,14] employing polyurethane sponges with cell sizes of 8 and 15 pores per inch as templates and commercial available nanoscaled SiO_2 powders.

Necessary conditions for obtaining transparent open-celled glass sponges were presented in [13] where topics such as slurry stabilization, temperature for burning-out the polymer and subsequent sintering of the remaining SiO_2 structure to transparent cellular bodies are discussed. It was found, that concentrated suspensions at around pH 10 offer suitable properties for both, bringing enough particles onto the support for the formation of a stable and self-supporting SiO_2 shell and for successful sintering of the particulate framework to transparent bodies. Therefore, an adjusted burning-out process and an adequate sintering regime is necessary.

In order to obtain information about the developed phases as a function of sintering temperature-in particular the transition of amorphous into crystalline silica-X-Ray diffractometry was performed. Beside the stability and transparency of the manufactured open-celled sponges, for the successful use of the cellular material, the knowledge of its geometrical properties is necessary. In this study, these parameters are derived from calculations based on datasets of magnetic resonance imaging. Volume Image Analysis on the basis of magnetic resonance imaging (MRI) of porous media has been used now for several decades beginning with the aim of characterizing soil material, e. g. sandstones [15]. Increased resolution has enabled even a tomographic insight into such materials together with the possibility to analyze the structure with respect to geometrically characteristic parameters [16]. Its application to reticulated structures can therefore give the most extended view into its details and has even been used recently to generate high resolution geometrical representations for computational fluid dynamics (CFD) calculations [17-19]. Additionally, scanning electron microscopy (SEM) and light microscopy (LM) were used for the characterization of the resulting transparent sponge structure.

Materials and Methods

Manufacturing and characterization of transparent sponges

Nanoscaled amorphous SiO_2 particles (AEROSIL OX50, SA of 50 m^2/g, Evonik Degussa GmbH, Essen, Germany) were dispersed in an aqueous media with a solids loading of 60 wt%. The pH of the slurries was adjusted at pH 10 employing 1 M tetramethylammonium hydroxide (TMAH, AlfaAesar, Karlsruhe, Germany). Details and characterization of suitable suspensions for the preparation of transparent sponges are described in [13].

As templates for the replica technique sponges (having 8 or 15 pores per inch (ppi), Foam Partner Fritz Nauer AG, Leverkusen, Germany) with dimensions 64 x 122 x 25 mm^3, necessary to cover half of the reactor volume, were used. The polymeric sponges were dip-coated employing well-dispersed slurries and compressed while submerged in order to fill all the pores. The impregnated polyester support was then removed from the slurry and the excess material was squeezed from the foams by means of a rolling mill. The coated polymers were dried for 24 h at 40°C and 80% humidity. For burning-out the polymer, the coated preforms were placed in an alumina crucible and heated in a furnace (VMK 135, Linn High Therm GmbH, Eschenfelden, Germany) under an air flow rate of approximately 150 – 200 L/h. To allow the gradual decomposition and diffusion of the polymeric material, heating was performed carefully up to 800°C. Subsequently, the samples were sintered at air atmosphere in a chamber furnace (Nabertherm GmbH, Lilienthal, Germany) at 1330°C with heating rates of 30 K/min and a dwell time of 10 min.

Characterization: For the determination of phase composition after thermal treatments, X ray diffraction (XRD) spectra were recorded using a Siemens D500 diffractometer, operating with CuKa radiation at 40 kV and 25 mA. Macroscopic analysis was performed by image analysis using a digital camera integrated on an optical microscope (Olympus AX70, Pittsburgh, USA). In addition, volume image analysis was performed using magnetic resonance imaging (MRI). In the present work the ^{1}H-MRI measurements were taken on a BrukerAvance 200 SWB tomograph (magnetic flux density, 4.7 T; micro-2.5 gradient system generating up to 1 T/m). Therefore, the sintered transparent sponge samples were cut into pieces of 8 x 16 x 20 mm^3 and filled with an aqueous solution of copper sulfate ($4x10^{-3}$ M) in order to accelerate the spin relaxation.

Volume imaging analysis of transparent foams: Starting point for the image analysis are the 3-d images of the sponges, consisting of gray values for the 256^3 voxels of the scan domain. Using this 3-d volume data of transparent sponge, a model of the solid surface could be extracted with digital filter routines for noise reduction and an individual threshold value to separate solid from void. The selected threshold directly influences the surface area and, thus, simply selecting a value is somewhat vague as this threshold additionally depends on the operational parameters when acquiring the raw data. In the present work an automatic selection procedure was used that is described in detail in [20]. The whole procedure was programmed using the freely available Visualisation Toolkit (VTK) [21]. The results of the procedure are discretized surfaces comprising typically 0.5 - 1 million connected triangles with the exact number depending on the surface area and resolution. After discretization, calculating the surface areas can be done by a simple addition of these triangles areas.

In order to determine the porosity the void space on one side of the interface area can be filled with tetrahedral volume elements and again summed up to result in the volumes of the void phase. The tetrahedral meshes for this purpose were generated in this work using the commercially available meshing tool ANSYS ICEMCFD as this meshes can easily be imported by many simulation toolkits. In order to give an overview, in Figure 1 an image of the resulting geometry is shown.

Based on the detailed geometry of the samples an estimation of the extents of single pores has been derived. The proposed procedure starts with an estimation of points in the void phase that are assumed to be the pore centers. These points are determined by first calculating the Euclidean distance of every point in the void phase to the nearest surface point by application of the algorithm according to Saito and Toriwaki [22] and then searching local maxima in the 3-dimensional distance field. Twice of the local distance at these local maxima then gives an estimation of a minimal pore diameter. Using these values, a statistical analysis of the pore sizes was made. As a result, in Table 1 the average and the RMS values of the pore diameters as well as porosity and specific surface area for both 8 and 15 ppi glass sponges are given. Up to now, the influence of cell size of the foams on light distribution is not clear, but the structural parameters are intended to form the basic data for further analysis. These further studies are necessary, e.g. employing Monte-Carlo-Simulation of multiple light scattering, in order correlate the parameters to the light distribution.

Cultivation in miniplate reactors

Strains and medium: *Chlamydomonas reinhardtii* wildtype wt13 [23] derived from the original Ebersold/Levine line 137c (CC-125 mt+) [24,25] was used as a model organism. For maintenance and precultur-ing the algae was continuously grown mixotrophic in liquid cultures of TAP media (Tris Acetate Phosphate) [26] under low light intensities (50 $\mu E\ m^{-2} \cdot s^{-1}$).

Miniplate reactors: Special Miniplate reactors were developed for testing the transparent glass sponges (Figure 2). These consist of stainless steel frameworks (bottom and top plate, two side plates) with the internal measurements of 140 x 100 x 20 mm and two glass panes bonded between the side plates using neutral crosslinking silicon rubber (SCRINTEC 600, RALICKS Industrie- und Umwelttechnik, Rees-Haldern, Germany). The bottom and top plate are sealed with a silicon gasket (1.5 mm VMQ-Elastomer, Angst + Pisterer, Zürich, Switzerland) and fixed with screws. For aeration and mixing a PTFE membrane (1 µm THOMAPOR-Poröse Platte, RTC ReicheltChemietechnik, Heidelberg, Germany) is fixed in the bottom plate. The inlet and exhaust gas passes 0.2 µm sterilefilters, the connection to the plate is done via LUER-LOCK. For sampling a syringe is used connected to a safe flow valve (B.Braun, Melsungen, Germany) and a needle. Light is supplied using 60 LEDs (MS6L083AT, Nichia, Japan) with 6° collimating lenses (IMM Photonics GmbH, Unterschleißheim, Germany) in 100 mm distance from the illuminated side of the reactor.

Photoautotrophic cell growth was investigated in a working volume of 0.2 L 2.5xTP medium (TAP without Acetat), pH 8.8 at varying light intensities. Inoculation was done after saturation of the medium with CO_2 with a four days old preculture grown in TAP; the starting biomass was adjusted to approximately 0.05 g/L. The cell suspension was purged continuously with a sterile mixture of air enriched with 5% CO_2 at an aeration rate of 50 mL/min controlled via two mass flow controllers (MassFlo, MKS Instruments, München, Germany). Temperature was kept constant at 25°C via an incubator (Flohr Instruments, The Netherlands). Light intensity was controlled using the control system BioProCon written as a routine in Lab View (National Instruments).

The installation of the glass sponges into the Miniplate reactor is shown schematically in Figure 6. The microalgae cells are grown in suspension within the pores of the sponges. The separation of algae after cultivation is easily done. The liquid suspension is just draining off from the solid sponges and thereby the algae.

Analytics and calculations:

- **Biomass determination:** Biomass concentrations were determined by offline optical density (OD) measurements at 750 nm using a spectrometer (T60 UV-Visible Spectrophotometer, PG instruments, Leicester, Great Britain).

  A correlation between OD and biomass was established by dry weight measurements.

- **Light measurements:** Measurements of light intensity were done using a planar quantum sensor (MQS-C, Walz connected with a LI-250 Light Meter, Licor, Walz Mess- und Regelungstechnik, Effeltrich, Germany). Light intensities up to 1500 $\mu E \cdot m^{-2} \cdot s^{-1}$ were applied.

- **Calculation of specific growth rate:** The specific growth rate µ was determined in the exponential growth phase under non-light limiting conditions (dry biomass <0.2 g/L) according to the following formula.

$$\frac{dx}{dt} = \mu_{max} \cdot x \tag{1}$$

with x biomass, t time and μ_{max} maximum specific growth rate.

This can be converted to

$$x = x_0 \cdot e^{\mu_{max} \cdot (t - t_0)} \tag{2}$$

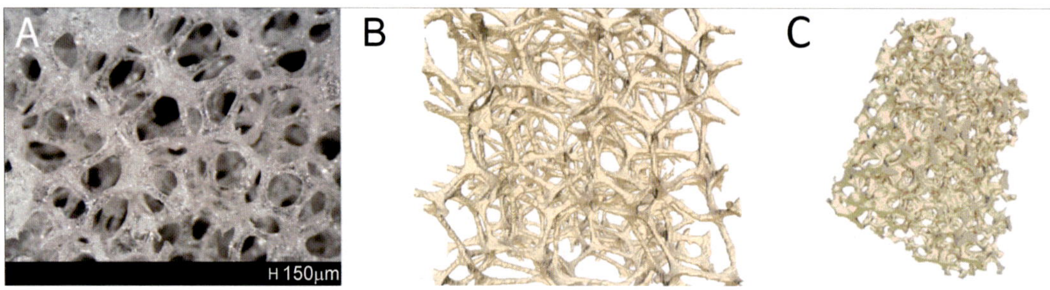

Figure 1: Real (A) and reconstructed (B, C) structure from 1HNMR image of the polymer template transparent sponge and sintered at 1330°C.

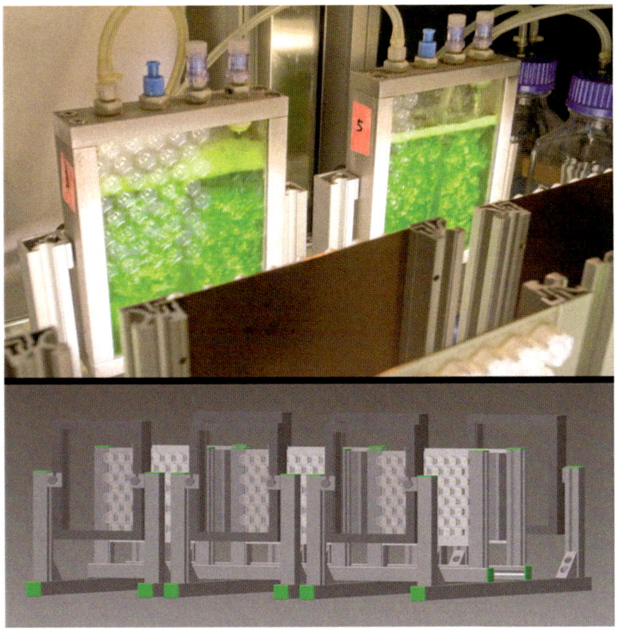

Figure 2: Mini-Plate Reactors as model system for testing transparent glass sponges.

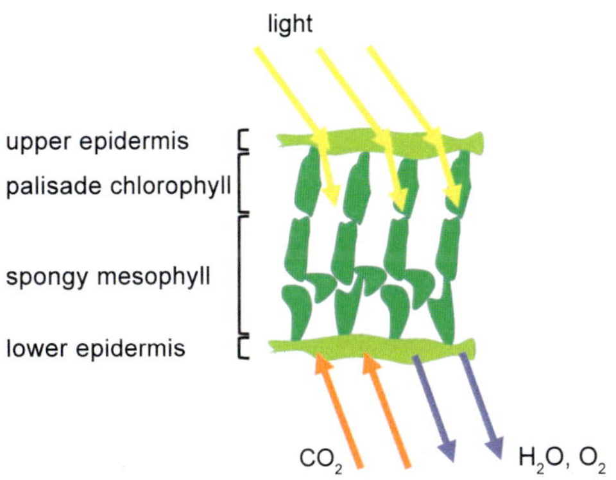

Figure 3: Schematic principle of leaf structure (leaf parenchyma) [10].

The values are averaged over at least three independent experiments.

- **Calculation of photo conversion efficiency (PCE):** The photo conversion efficiency (PCE) [J/J] describes the efficiency of biomass buildup form incident light energy. The following assumptions are used for calculation:
  - biomass calorific value 20 MJ/kg [12,27]
  - light energy conversion factor 1 W/m² = 4.75*µE·m⁻²·s⁻¹ (calculated from spectral distribution of used LEDs, for PAR photosynthetic active radiation 400-700 nm)

Finally a PCE based on PAR was calculated during the linear, light limited growth phase at higher cell densities (dry biomass <1 g/L) for defined light intensities. The values are also averaged over at least three independent experiments.

Results and Discussion

Anticipated effects of sponges for light dilution

The enlargement of the illuminated surface of photobioreactors is the main intention today to improve the efficiency of phototrophic cultivation systems. Data for SVR and SFR ratios of photobioreactors and open ponds are given in Table 2. "Natural photosynthetic systems" like trees could reach ten times higher values (Table 2). This can be explained by the "lime tree model" communicated by Pulz: the photosynthetic active leaf surface area accounts for 2500 m^2 and the footprint area for only 100 m^2. But comparing the illuminated surface area of photobioreactors with the leaf area of trees is only the first step for improvements.

The second step is based on an idea of nature inspired design published by Posten [10]. An additional effect on efficiency could be reached by enlargement of the inner surface of the reactor for light dilution and supply. Similar inner structures can be found in leafs of trees called the parenchyma (Figure 3). For this purpose transparent, open pored glass sponges are used as installations in photobioreactors. By inserting transparent 8 ppi sponges as light diluting materials into the reactor the SVR can be increased to a value of 568 m^{-1} (see Table 1). An additional effect of the sponges would be the reduction of light gradients inside the whole reactor volume. Based on this concept photobioreactors still have quite promising potential for further developments.

The necessity of light dilution and the reduction of light gradients in photobioreactors can be explained with the growth kinetics of microalgae which show the growth rate depending on light intensity. Ex-

		8 ppi(glass)	15ppi(glass)
Porosity	-	0.90	0.86
Specific surface area	m^{-1}	568	883
RMS pore diameter	mm	0.24	0.2073
average pore diameter	mm	3.56	2.79

Table 1: Foam parameters, obtained from volume analysis.

	Lime tree	PBR	Open Pond
SVR /m²/m³	<1000	<100	<4
SFR /m²/m²	<25	<10	1

Table 2: Values for surface to volume ratios (SVR) and surface to footprint ratios (SFR) for the lime tree model, photobioreactors and open ponds.

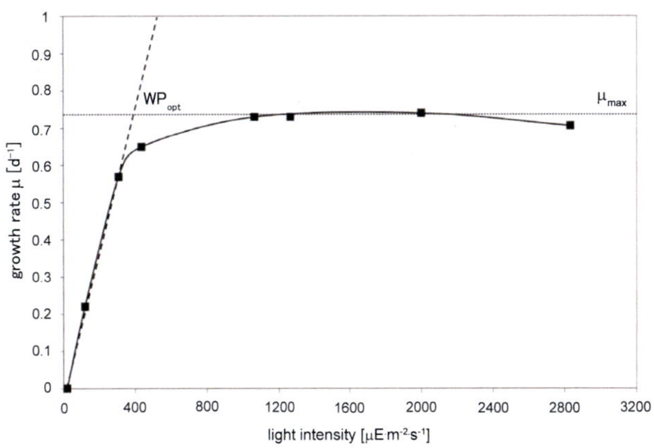

Figure 4: Growth kinetic of *Porphyridium purpureum* (modified based on [10]), with μ_{max} maximum growth rate and WP_{opt} optimal and most efficient working point of biomass production with highest increase of growth with rising light intensity.

emplarily the growth kinetic of the red algae *Porphyridium purpureum* is displayed in Figure 4. The different effects of light intensity on microalgal growth can be observed from this data. Growth is light limited at low light intensities up to about 400 $\mu E \cdot m^{-2} \cdot s^{-1}$. Then the optimal and most efficient working point of biomass production is reached with highest increase of growth with rising light intensity. At light intensities >400 $\mu E \cdot m^{-2} \cdot s^{-1}$ photosynthesis becomes light saturated and later even inhibited. That means the more light energy supplied to the cultures cannot be stored in biomass or products.

The low efficiencies of production systems today is the combined effect of different light conditions outside and inside the reactors on growth increased even due to the change of optical densities of the algae suspension during the cultivation process. Figure 5 depicts schematically conditions in plate reactors illuminated from one side at changing cell densities. Only at low cell concentrations when all algae obtain sufficient light for growth the maximum growth rate μ_{max} will be reached. With increasing optical density of the suspension volume elements with light limited conditions (μ_{lim}) or even no cell growth (μ=0) will exist. The overall growth rate measured under this condition will be a combination of all these effects.

For optimization of light dilution in photobioreactors transparent open pored sponges could be used as installations (for a schema see Figure 6). The algae grow in suspension within the pores of the sponge. The sponge structure dilutes light into the deeper, dark parts of the reactor by light conduction in the struts, as well as reflection and refraction at the surfaces due to the refractive index of glass which is higher than those of water. Therefore an amorphous SiO_2 glass structure has to be produced. Ideally sun light should be diluted to the optimal light intensity in the whole cultivation volume. Thereby the efficiency of production scale rectors could be improved by maximizing overall growth rates.

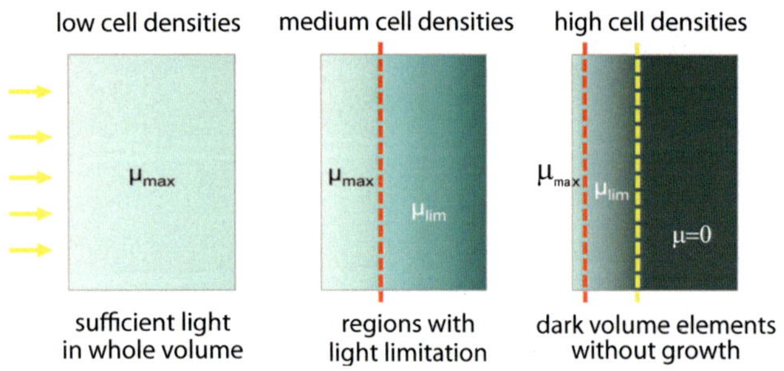

Figure 5: Schema of light conditions in a plate photobioreactor illuminated from one side at different cell densities, with μ_{max} maximum growth rate, μ_{lim} light limited growth rate, μ=0 no cell growth.

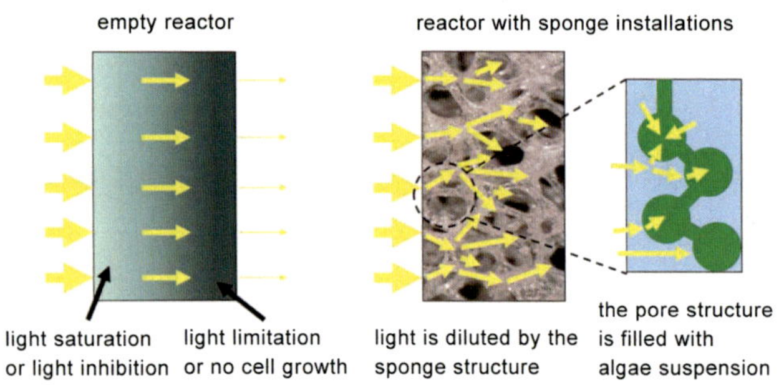

Figure 6: Installation of glass sponges into Mini-Plate Reactors, Schema of light dilution.

Preparation and characterization of transparent open-pored structures

Suspension: stability and viscosity: In order to identify a suitable pH-range for electrostatic stabilization of the silica particles ζ-potential measurements were performed [13]. Low-viscosity slurries are necessary for coating the polymer. In the system SiO_2 - water a compromise between a maximum of ζ - potential and suitable viscosity was obtained at a pH level higher than pH 10.

Using well stabilized slurries, an increase of SiO_2 leads to a moderate increase of viscosity and therefore a maximal concentration with acceptable viscosity values was obtained for slurries with solids loadings up to $\Phi < 60$ wt%.

Coated, dryer and burning out the polymer template: The aim of the impregnation process is to achieve a coating layer of defined thickness on the polymer template, consisting of dense and homogenously packed SiO_2 nanoparticles. Both, structure and thickness of the layer depend strongly on the suspension used. The dominating parameters which control these factors are the solids loading, the interparticle forces and as consequence the viscosity. With respect to a sufficient coating layer thickness, a high solids loading is desirable. A dense and homogeneous particle packing requires repulsive interparticle forces. Both can be achieved by adjusting the pH-value as discussed previously.

Burning out the polymer is a very important step for obtaining manageable green bodies. A special temperature regimen has to be applied and a sufficient concentration of SiO_2 on the struts is necessary to prevent the framework from collapsing. Therefore, the solids loading of the slurry should be high.

Regardless the used slurry, the recommended amount of SiO_2 should be higher than approximately 0.1 grams per cubic centimeter of template. Below this value for the polymer employed in this study [8 and 15 ppi] it was not possible to get the desired structure through the burning-out process [28]. But if the thermal treatment is successful, there is no significant change in structure visible to the naked eye and the color of the samples remain white.

Figure 7 depicts a coated sample before (A) and after burning out the polymeric framework (B). It is worth to note, that the geometrical density of the shown greenbody was >0.1 g/cm³. In the case of a density of 0.08 g/cm³, see Figure 8, the framework was pulverized during the burning-out step.

Sintering and characterization: After the burning-out step, the obtained green bodies can be densified by viscous-flow sintering, leading to shrinkage of about 20% in all directions. By carefully selecting the sintering regime, i.e. an appropriate heating rate, temperature and dwell time, the initial white green bodies can be transferred into transparent open-celled foams. Transparency indicates that the amorphous character of the initial powder is still present. To monitor the phase composition XRD spectra were performed after sintering at temperatures varying from 1150 to 1400°C. In Figure 9, typical XRD-diffraction patterns of amorphous SiO_2 after sintering at 1350°C are shown. Above this temperature, SiO_2 foams commence to crystallize, as indicated in the XRD- spectra (Figure 9). Hence, after sintering at 1400°C the XRD pattern corresponds to crystalline cristobalite as the only detectable phase.

The broad peak near 22° is attributed to the short-range order in amorphous silica, whereas the sharp peaks in Figure 9 (right side) show the formed crystalline structure of cristobalite.

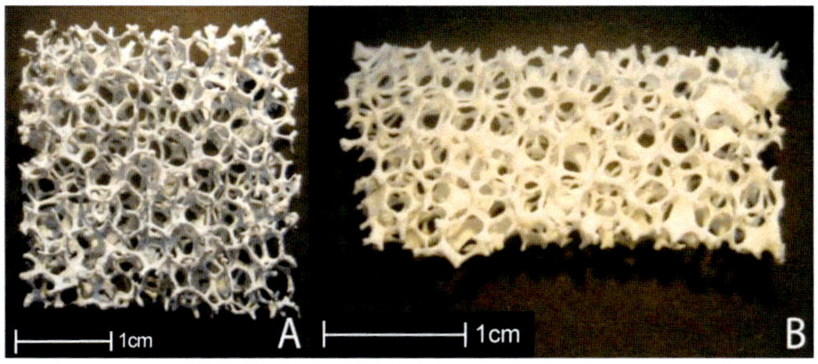

Figure 7: Coated sample before (A) and after burning out the polymeric framework (B), geometrical density of the greenbody >0.1 g/cm³.

Figure 8: Pulverization of the framework during the burning-out step, geometrical density of the greenbody 0.08 g/cm³.

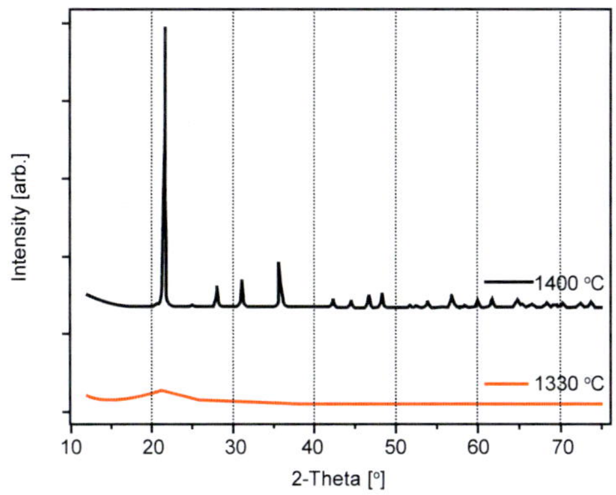

Figure 9: Typical XRD-diffraction patterns of amorphous SiO_2 after sintering at 1350°C.

Additionally to the XRD spectra, changes in structure and consequently on the transparency can be explained considering the apparent density and SEM analysis. These topics were extensively discussed in our previous work [28].

Cultivation experiments/ Determination of growth rates and photo conversion efficiencies (PCE)

The aim of experiments performed in Miniplate reactors is to test the transparent glass sponges as light diluting and light distributing installations in photobioreactors. Relevant parameters to estimate in this context are the growth rate μ in the exponential growth phase at low cell densities and non-limiting conditions and more important the photo conversion efficiency (PCE) at high cell densities. The latter is a measure for the efficiency of biomass buildup from the captured light energy.

The reference system is the "empty" reactor applied for growth experiments without any light distributing installations inside. To increase the final cell densities reached in the reactor already concentrated medium (2.5xTP) is used at the beginning of the cultivation with CO_2 supply by aeration (5%). Typical growth curves of batch experiments at a selected light intensity (500 $\mu E \cdot m^{-2} \cdot s^{-1}$) and the corresponding light transmission behind the reactor for 8 and 15 ppi glass sponges and the empty reactor are shown in Figure 10.

A characteristic growth curve of micro algal batch cultures is observed with a very short exponential phase until day one at low cell densities <1 g/L. Already at day one light becomes limiting but in contrast to other substrates it constantly further supplied which results in a transition into a linear growth phase. When Ammonium is limiting in the medium (ionchromatography measurements not shown) the maximum biomass concentration of about 3-3.5 g/L is reached during the third day.

The light transmission on the non-illuminated side of the reactor decreases in correlation with the increasing biomass concentrations resulting in light limiting conditions with transmission equals zero at day one. The light transmission at the beginning of the cultivation is lower with sponge filled reactors compared to the empty one. This is possibly due to a better light distribution inside the reactor caused by refraction and scattering. It could be a first indication for a better light dilution in the reactor by the sponge structure. It is also possible that by scattering at the sponge surface some amount of light is lost at the top, bottom and the small dark sides of the reactor. This amount is potentially bigger than in the empty reactor but could not be measured. The portion of scattering due to cells would be equal in all three systems. The measured reflection at the illuminated surface is less than <10% of the incident light intensity and no difference between the different configurations (empty or sponge filled reactor) can be observed. The influence of reflection at the reactor surface (typically 4% for air-glass) on incident light energy is eliminated by calibrating inside the reactor

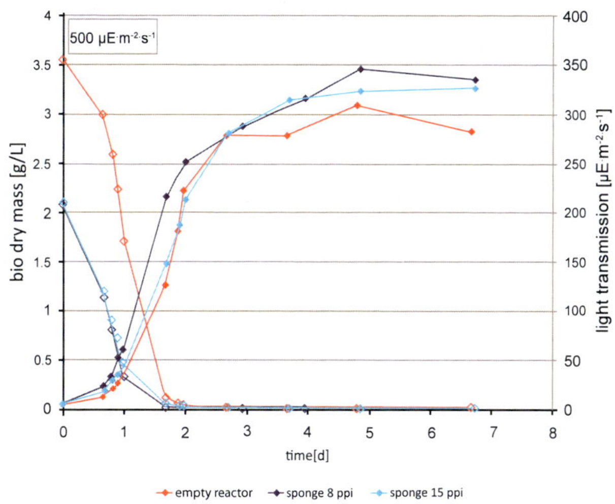

Figure 10: Bio dry mass concentration of wt13 and light transmission over cultivation time for 8 and 15 ppi glass sponges and the empty reactor 500 $\mu E \cdot m^{-2} \cdot s^{-1}$, 2.5xTP, 5% CO_2 supply by aeration.

behind the first glass plate. The absorption of light by glass is only in the rage of <0.1% per m light path.

Growth rates μ are determined in the exponential growth phase (until day one) and the resulting growth kinetic of *Chlamydomonas* wildtype depending on the applied light intensity in reactors with sponges as light diluting structures (pore size 8 and 15 ppi) compared to the empty reactor is depicted in Figure 11.

The growth kinetic of *Chlamydomonas* wt13 (empty reactor) illustrates the three possible light regimes in an algal culture. Light intensity is limiting until intensities up to 200-250 $\mu E \cdot m^{-2} \cdot s^{-1}$ and the growth rate μ increases straight linear with rising light energy. At light intensities >400 $\mu E \cdot m^{-2} \cdot s^{-1}$ light becomes saturating and the additional energy cannot be used for photosynthesis. At even higher intensities ≥1000 $\mu E \cdot m^{-2} \cdot s^{-1}$ the inhibiting effect of too much light can already be observed resulting in a decrease of growth rate μ.

The effect of the glass sponges was tested at 200, 500, 800 and 1000 $\mu E \cdot m^{-2} \cdot s^{-1}$ and a positive influence on the growth rates μ (about +25%) at low cell densities can be seen.

But more interesting would be the effect at higher cell concentrations. The photo conversion efficiency (PCE) can be used as a measure to assess the use of transparent glass sponges as installations in photobioreactors. The calculation can be done during the linear growth phase under light limiting conditions from the increase of biomass and the applied light intensity. However one major limiting factor is the medium and the short linear growth phase because of ammonium limitation at day two. Therefore optimization of medium composition and testing of different feeding strategies are in progress to increase final biomass concentrations and prolong the linear growth phase.

First calculations of PCEs at 500 $\mu E \cdot m^{-2} \cdot s^{-1}$ were done using a biomass calorific value of 20 MJ/kg [12,27] and a light energy conversion factor of 1 W/m^2 = 4.75*$\mu E \cdot m^{-2} \cdot s^{-1}$ (calculated from spectral distribution of used LEDs, for PAR photosynthetic active radiation 400-700 nm). The results indicate an increase from 4.9% for the empty reactor to 5.6% and 5.9% for the 8 and 15 ppi sponge filled reactor. These are promising results with potential for further improvement.

Concluding remarks

It has been shown that the new concept for improving light dilution and light distribution in photobioreactors by applying transparent, open-pored sponges was realized and tested during cultivation experiments.

A manufacturing process based on the polymer replica technique was established and important factors influencing the process to get stable and transparent probes like slurry stabilization (viscosity, zeta-potential, pH), sintering temperature and shrinkage were investigated. The 8 ppi sponge shows the following characteristics: porosity 0.9, specific surface area 568 m^{-1} which means an increase of the (inner) surface to volume ratio of the reactor.

Cultivation experiments showed an increase in growth rate (about +25%) at low cell densities and the PCE was increased from 4.9% for the empty reactor to 5.6% and 5.9% for the 8 and 15 ppi sponge filled reactor. The results show a first positive effect, but further progress is necessary. The effect will be more apparent at higher light intensities and higher cell densities. Therefore experiments to increase biomass concentrations by concentrating the TP medium (2.5 fold) and appropriate feeding are in progress. Furthermore a scale-up to larger volumes would be necessary to mimic conditions in production reactors which is limited to the producible sponge size at the moment.

The final target would be thin reactor systems with internal light diluting structures lying horizontally on the ground and supplying CO_2 by surface or membrane aeration. This would diminish cost for materials and supporting installations due to less water demand and decreased resulting hydrostatic pressure. Furthermore the need of auxiliary energy can be reduced when aeration and gas hold up are minimized.

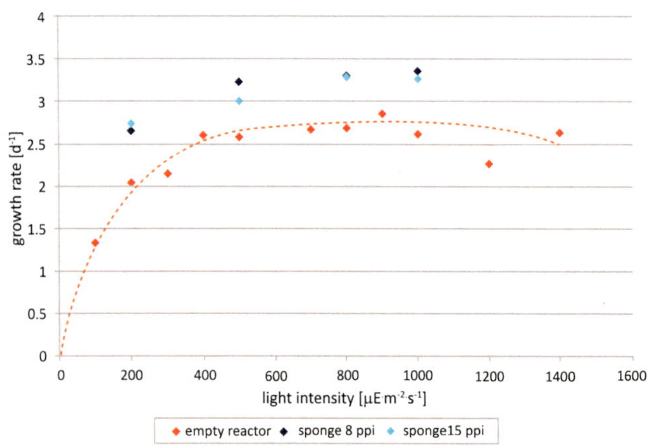

Figure 11: Growth kinetics of *Chlamydomonas reinhardtii* wt13 depending on light intensity determined in the Miniplate reactor with 8 and 15 ppi glass sponges as light conducting installations and the empty reactor for comparison in the exponential growth phase at low cell densities <1 g/L.

Acknowledgements

The authors would like to thank the German Research Foundation (DFG) for funding the Research Group FOR 583 "Solid Sponges—Application of monolithic network structures in process engineering."

We acknowledge support by Deutsche Forschungsgemeinschaft and Open Access Publishing Fund of Karlsruhe Institute of Technology.

References

1. Pulz O, Gross W (2004) Valuable products from biotechnology of microalgae. Appl Microbiol Biotechnol 65: 635-648.
2. Rosello Sastre R, Posten C (2010) Die vielfältige Anwendung von Mikroalgen als nachwachsende Rohstoffe. The Variety of Microalgae Applications as a Renewable Resource. Chem Ing Tech 82: 1925-1939.
3. Wijffels RH, Barbosa MJ (2010) An Outlook on Microalgal Biofuels. Science 329: 796-799.
4. Rosenberg JN, Oyler GA, Wilkinson L, Betenbaugh MJ (2008) A green light for engineered algae: redirecting metabolism to fuel a biotechnology revolution. Curr Opin Biotechnol 19: 430-436.
5. Dismukes GC, Carrieri D, Bennette N, Ananyev GM, Posewitz MC (2008) Aquatic phototrophs: efficient alternatives to land-based crops for biofuels. Curr Opin Biotechnol 19: 235-240.
6. Blankenship RE, Tiede DM, Barber J, Brudvig GW, Fleming G, et al. (2011) Comparing Photosynthetic and Photovoltaic Efficiencies and Recognizing the Potential for Improvement. Science 332: 805-809.
7. Carvalho AP, Silva SO, Baptista JM, Malcata FX (2011) Light requirements in microalgal photobioreactors: an overview of biophotonic aspects. Appl Microbiol Biotechnol 89: 1275-1288.
8. Tredici MR (2010) Photobiology of microalgae mass cultures: understanding the tools for the next green revolution. Biofuels 1: 143-162.
9. Xu L, Weathers PJ, Xiong XR, Liu CZ (2009) Microalgal bioreactors: Challenges and opportunities. Eng Life Sci 9: 178-189.

10. Posten C (2009) Design principles of photo-bioreactors for cultivation of microalgae. Eng Life Sci 9: 165-177.

11. Morweiser M, Kruse O, Hankamer B, Posten C (2010) Developments and perspectives of photobioreactors for biofuel production. Appl Microbiol Biotechnol 87: 1291-1301.

12. Lehr F, Posten C (2009) Closed photo-bioreactors as tools for biofuel production. Curr Opin Biotechnol 20: 280-285.

13. Bucharsky EC, Schell KG, Oberacker R, Hoffmann MJ (2010) Preparation of transparent glass sponges via replica method using high-purity silica. J Am Ceram Soc 93: 111-114.

14. Schwartzwaelder K, Sommer AV, Mich F (1963) Method of making porous ceramic particles. United States, Patent No. 3090094.

15. Timur A (1969) Pulsed nuclear magnetic resonance studies of porosity, movable fluid, and permeability of sandstones. J Petrol Technol 21: 775-786.

16. Fredrich JT, Menendez B, Wong TF (1995) Imaging the pore structure of geomaterials. Science 268: 276-279.

17. Petrasch J, Wyss P, Steinfeld A (2007) Tomography-based Monte Carlo determination of radiative properties of reticulate porous ceramics. J Quant Spectrosc Radiat Transf 105: 180-197.

18. Coles ME, Hazlett RD, Spanne P, Soll WE, Muegge EL, et al. (1998) Pore level imaging of fluid transport using synchroton X-ray microtomography. J Petrol Sci Eng 19: 55-63.

19. Habisreuther P, Djordjevic N, Zarzalis N (2009) Statistical distribution of residence time and tortuosity of flow through open-cell foams. Chem Eng Sci 64: 4943-4954.

20. Grosse J, Dietrich B, Garrido GI, Habisreuther P, Zarzalis N, et al. (2009) Morphological Characterization of Ceramic Sponges for Applications in Chemical Engineering. Ind Eng Chem Res 48: 10395-10401.

21. Kitware I (2006) The VTK users guide. Kitware Inc., Colombia, USA.

22. Saito T, Toriwaki JI (1994) New algorithms for euclidean distance transformation of an n-dimensional digitized picture with applications. Pattern Recogn 27: 1551-1565.

23. Schönfeld C, Wobbe L, Borgstädt R, Kienast A, Nixon PJ, et al. (2004) The Nucleus-encoded Protein MOC1 Is Essential for Mitochondrial Light Acclimation in Chlamydomonas reinhardtii. J Biol Chem 279: 50366-50374.

24. Harris EH (2001) Chlamydomonas as a model organism. Annu Review Plant Physiol Plant Mol Biol 52: 363-406.

25. Pröschold T, Harris EH, Coleman AW (2005) Portrait of a Species: Chlamydomonas reinhardtii. Genetics 170: 1601-1610.

26. Harris EH (2009) The Chlamydomonas sourcebook: Introduction to chlamydomonas and its laboratory use. (2nd edn.), Academic Press, Amsterdam.

27. Schaub G, Vetter A (2007) Biofuels - a review. Chem Ing Tech 79: 569-578.

28. Bucharsky EC, Schell KG, Habisreuther P, Oberacker R, Zarzalis N, et al. (2011) Preparation of Optically Transparent Open-Celled Foams and its Morphological Characterization Employing Volume Image Analysis. Adv Eng Mater 13: 1060-1065.

Multilineage Differentiation Potential of CNS Cell Progenitors in a Recent Developed Gilthead Seabream (*Sparus aurata* L.) Nervous Model

Maria Pia Santacroce[1]*, Antonella Tinelli[2], Anna Selene Pastore[1], Michele Colamonaco[3] and Giuseppe Crescenzo[3]

[1]*Unit of Aquaculture and Zooculture, Department of Veterinary Medicine, University of Bari "Aldo Moro", Italy*
[2]*Unit of Pathology, Deptartment of Veterinary Medicine, University of Bari "Aldo Moro", Italy*
[3]*Unit of Pharmacology and Toxicology, Department of Veterinary Medicine, University of Bari "Aldo Moro", Italy*

Abstract

Neural Progenitor Cells (NPCs) have gathered more and more attention in the field of Neural Stem Cells (NSCs). However, the multilineage differentiating behavior of these cells and their contribution to tissue regeneration, almost in lower vertebrate taxa, remain unknown. Since the early 1970s, many comparative studies have been performed using immunocytochemical screening on the brains of several vertebrate taxa, including teleosts, in order to identify these cells, even if the data are sometimes contrasting. This study aims: (1) to investigate *in vitro* the potential proliferative role of NPCs and Radial Glia Progenitors (RGP) in seabream neurogenesis; (2) to reveal the strict ability of fish NSCs to undertake the multilineage development and differentiation in neurons, astrocytes and oligodendrocytes.

By the use of double Immunofluorescence (IF) analysis and phase contrast microscopy, we identified the multilineage differentiation and the exact cell morphology. We demonstrated that NSC can self-renew and differentiate into different types of neurons or glial cells during extended culturing. Mature neurons expressed specific neuronal markers; they could differentiate during long term culturing, generating an extensive neurite growth. Glia was found highly mitotic and could developed mature astrocytes and oligodendrocytes. Glial cells were assessed by Glial Fibrillary Acidic Protein (GFAP) reactivity; neurons and myelinating oligodendrocytes were immunostained with cell-specific markers.

This work provide that the multilineage differentiation potential of seabream neural cell progenitors might be a useful tool for neurodegenerative diseases, being a promising approach for repairing the CNS injuries, also in other animals, as a new coming strategy for function recovery of damaged nerves.

Keywords: Teleost cell culture; *In vitro* fish neurogenesis; *Sparus aurata;* CNS cell progenitors

Introduction

The generation of new neurons in the adult Central Nervous System (CNS), so called adult neurogenesis, is a fundamental feature of the vertebrate brain during the post-embryonic neural growth [1]. It is well known that in mammals and birds, constitutive turnover of neurons occurs not only during embryogenesis but also during adult stages [2-5]. Adult neurogenesis has been observed in rodents and songbirds, where newborn neurons are added throughout life to highly restricted spatial regions in the telencephalon [5-9]. *In vivo* in mammals, endogenous Neural Progenitor Cells (NPCs), almost generated along the Sub Ventricular Zone (SVZ) in the adult brain, have been proposed as a potential source of newborn neurons for the neural tissue repair after various brain insults, such as ischemic stroke or traumatic brain injury [10,11].

Dissimilarly to birds and mammals, low non-mammalian vertebrates, as fish, amphibians and reptiles, display a considerable amount of adult neurogenesis in more numerous brain regions, even though such differences are poorly understood and investigated [11]. The adult CNS of teleost fish exhibits a high capacity for neuronal regeneration after injury [4]. Comparative studies in zebra fish and mammals show that the telencephalic Ventricular Zone (VZ) in the adult zebra fish brain generates NPCs that share characteristics with the NPCs in the mammalian SVZ: they migrate tangentially and differentiate into mature neurons into the Olfactory Bulb (OB) [5,11-15]. Recently, Kishimoto et al. [11] developed an *in vivo* fish model of telencephalic injury to demonstrate the strong ability of adult zebra fish to undergo neuronal regeneration. Prodigious regenerative processes were found in the injured zebra fish telencephalon, with the proliferation of endogenous NPCs, their lateral migration and neuronal differentiation at the injury site [11].

Studies on Neural Stem Cells (NSCs) defined these cells as self-renewing CNS cells that can differentiate into any of the three major neural cell lineages, specifically neurons, astrocytes and oligodendrocytes [16,17].

Even though the maintenance of vertebrate NSCs in the neural epithelium, as their controlled differentiation, is regulated by a conserved genetic network, many pro-differentiation factors have been explored for being involved in promoting the neural induction, as Fibroblast Growth Factor (FGF), Retinoic Acid (RA) and Epidermal Growth Factor (EGF) [18,19]. The mitogenic action of these growth factors can induce the *in vitro* proliferation and differentiation of NPCs that began to divide within the first week of culture *in vitro*. The first signal produced in the cluster of differentiating progenitors is the expression of nestin, an intermediate filament present in neuroepithelial stem cells, as revealed in cultures of albino mouse embryos [18].

***Corresponding author:** Maria Pia Santacroce, Unit of Aquaculture and Zooculture, Department of Veterinary Medicine, University of Bari "Aldo Moro", Italy

In zebra fish, as in rodents, neuroblasts become quiescent after larval development; while in the adult brain, often after brain injury, a Notch signaling pathway prompted by growth factors regulate the transition of NSCs cells to a proliferative state, into neuroblasts toward neurogenesis. However, the extrapolation on the NSCs self-renewing capabilities arose mainly from *in vivo* studies of rodents and mammals, and we are still lacking evidences on the truly proliferative ability of multipotent NSCs and whether these cells can undergo multilineage differentiation to develop mature and self-renewing neurons or glial cells in brain.

In mammals, NPCs have been isolated and characterized *in vitro* from various regions of the fetal, adult and post-mortem brain tissues of various species [20-25]. Accumulating evidence suggests that in teleost fish brain, CNS has the capability to generate new neurons in the adult CNS continuously and in large numbers, probably more than in any other vertebrate taxon [4,26-34].

Nevertheless, none of these studies on fish has addressed in a definite manner the involvement of radial glia (RDG) in the generation of new neurons, by interplaying with the full development of the fish nervous tissue in an *in vitro* microenvironment. Due to several discrepancies, arose the need for improving differentiation procedures able to generate highly pure populations of mature neurons and glial cells in *in vitro* culture systems of lower vertebrates.

For this reason, we recently developed a seabream nervous model to study neuronal differentiation and regenerative processes *in vitro*. Using this model, we aimed: (1) to investigate *in vitro* the potential proliferative role of NPCs and Radial Glia Progenitors (RGP) in fish neurogenesis; (2) to reveal underlying the strict ability of fish NSCs to undertake the multilineage development and differentiation in neurons, astrocytes and oligodendrocytes.

This study provides a novel seabream nervous model for studying a series of regenerative processes involved in fish brain neuronal regeneration, giving new insights for the treatment of injury-impaired neurogenesis, even in mammals.

Materials and Methods

SaGliPs isolation and differentiation

Spaurus aurata juveniles (35 ± 5 g mean body weight, *n*=50) were used to obtain the primary neuroglial culture (SaGliPs), according to a recently developed method [35].

On arrival, fish were anaesthetized by immersion in seawater plus 3-aminobenzoic acid ethyl ester (MS222, 0.02%; Sigma-Aldrich) and rapidly sacrificed by spinal transection with a scalpel blade for decapitation. Whole brain tissue samples (weighing approximately 0.1-0.2 g/brain) were removed aseptically, pooled and collected in a sterile pre-cooled Hank's Balanced Salt Solution (HBSS, Sigma-Aldrich) (without Ca^{2+} and Mg^{2+}) supplemented with 10 mM HEPES, 0.5 mM EDTA, 25 mM $NaHCO_3$, 200 IU/ml penicillin, 200 µg/ml streptomycin, 200 µg/ml amphotericin B and 100 µg/ml gentamicin (Washing Solution, WS) [36,37]. Unless specified all chemicals were from Sigma-Aldrich-Aldrich Ltd., Milan, Italy.

Primary neuroglial cultures were rapidly isolated from freshly-enucleated brains [35]. After three washings with three volumes of HBSS (WS), brains were crushed by pressing a glass pestle into a stainless-steel sieve with 380 µm mesh for mechanical disruption. Cell suspension was filtered and collected with Digestion Medium (DM) lacking enzymes (7 mM $CaCl_2$, 200 IU/ml penicillin, 200 µg/ml streptomycin, 200 µg/ml amphotericin B, 100 µg/ml gentamicin, 10 mM HEPES, 25 mM $NaHCO_3$ in Leibovitz's L-15). The homogenate was then extracted by adding a cocktail of four enzymes including 0.05% Collagenase Type IV, 0.03% Hyaluronidase Type IV-S, 0.3% Dispase Type II, 0.03% DNase Type I for 20 min dissociations at 20°C to the homogenate in DM. The enzyme digestion was blocked adding Leibovitz's L-15 medium supplemented with 10% Foetal Bovine Serum (FBS; Bio Whittaker, Lonza Walkersville Inc., Italy). The digestion mixture was filtered through 104 µm and 60 µm stainless-steel filters, supernatant was discarded, and the pellet was resuspended in cold 1X Phosphate-Buffered Saline (PBS) 5% Fetal Calf Serum (FCS) and centrifuged twice at 150 g for 10 min at 4°C. The white pellet was recovered, washed and dissociated by adding 0.45 µg ml^{-1} DNase I in cold PBS to dissociate cell clumps by Pasteur pipette, then centrifuged at 70 g for 5 min at 4°C. The final collected pellets were resuspended in 10% FBS/L-15 medium and cell number was counted. Viability of cells was estimated by trypan blue exclusion. Brain cell suspension yielded 5.2×10^7 viable cells/g per tissue weight with a viability of 92.8 ± 0.06%.

The suspension of purified neuroglia was adjusted to a density of 1×10^6 cells/ml in basal culture medium (BM) consisting of Leibovitz's L-15 with 2 mM L-glutamine, 10% FBS, 100 IU/ml penicillin, 100 µg/ml streptomycin, 100 µg/ml amphotericine, 50 µg/ml gentamycin, 1 mM Na Pyruvate, 5 mM D-Glu, 10 mM HEPES, 12 mM $NaHCO_3$, supplemented with 0.05% ITS *plus* (insulin/transferrin/sodium selenite plus oleic acid/linoleic acid/BSA), 0.01 mM MEM non-essential amino acid (MEM-NEAA; Bio Whittaker, Lonza Walkersville Inc., Italy), 0.01 mM MEM-vitamin mix (Bio Whittaker, Lonza Walkersville Inc., Italy) and 0.1 mM ascorbic acid. The BM osmolality was adjusted to seabream serum osmolarity (369 mOsm/kg) by adding 20 mMNaCl as reported by other Authors [38,39].

Cells were cultured in complete Leibovitz's L-15 (BM) and seeded in 0.01% collagen type I-coated plates and flasks at a density of 3×10^4 cells/cm², respectively, into 25 cm² Falcon Primaria™ culture flasks for morphological characterization of primary cell types by phase-contrast, and on 12-well plates Falcon BD Biosciences (BD Biosciences, Sismed s.r.l., Italy) for immunophenotypization by immunofluorescence. Firstly cells were allowed to attach for 12 hours, then fresh BM was added, thereafter replaced every 48 hours. Cells were cultured in a refrigerate incubator in humidified atmosphere at 97% air/3% CO_2 at 18°C (CO_2-170 Innova, New Brunswick Scientific, PBI, Milan, Italy). Cultures were used after approximately 3-4 weeks when the cells had become confluent monolayers. To obtain a glial cell-enriched culture, after the first week SaGliPs were cultured in Enriched Differentiation Medium (EDM) consisting in L-15 BM supplemented with 0.01 µg/ml Epidermal Growth Factor (EGF), 0.2 µg/ml hydrocortisone, 10 mM glucose, 10 µg/ml insulin and 4 mM glutamine. For assessment of cytostructural features, phase-contrast microscopy pictures were taken daily on adherent cultured SaGliPs with a CCD camera on a Motic AE31 inverted Epi-Fluorescence microscope.

SaGliPs characterization by immunofluorescence

Microscopy analysis was performed, for bright field and fluorescence, by a Motic AE31 Epi-Fluorescent Inverted Microscope, equipped with DAPI/TRITC/FITC fluorescence filter cube set (DAPI/Hoechst set: Exciter D350/50x /Emitter D460/50m, FITC/RSGFP/Fluo 3/DioAcridine set: Ex. D480/30x/Em. D535/40m, TRITC/DiI/Cy3 set: Ex. D540/25x/Em. D605/55m). Digital image capture was performed by Moticam 3000C Cooled CCD digital color camera (3.3 Megapixel, 1/2" CCD), capture system in origin Live Cam 1.0 (32-32) and Motic Images Advanced (V. 3.2) acquisition software (Motic, Seneco, Milan,

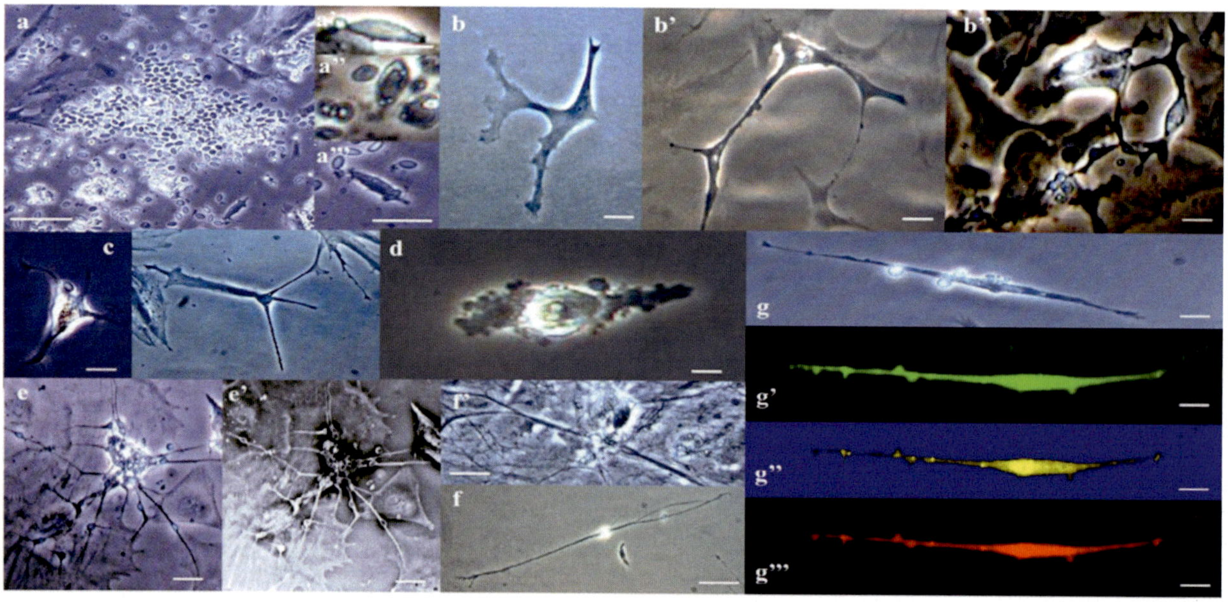

Figure 1: Morphological characterization of *S. aurata* primary neuronal cells. (a) Morphology of freshly isolated staminal neural cells (CNS) with the characteristic ovoid shape (a', a''), 1 day after seeding (×200). (b, b', b'') Early elongated attached glioblast progenitor after 2 days (×200). (b'') Negative inverted phase contrast image showing newly formed glioblasts from neuron precursor cells (NPC) in 6 day old culture (×200). (c) Glioblast (c') Neuroblast (d) Morphology of Neuron Precursor Cells (NPC) which generate neurons but not glia: here is visualized a colony showing new synthetized neurons (×200). (e, e') Morphology of neuronal Progenitor showing a cluster of newly synthetized neural cells, 4 day old culture (×200). (f) Morphology of a bipolar neuron after 2 day (f) and pseudo-unipolar neuron after 2 weeks of culture (f') showing the characteristic soma ovoid shape. (g) Bipolar shaped neuroblasts immunolabeled with GFAP-FITC (green, g'), with beta III tubulin-TRICT (red, g'''), merged image of g' and g''' (g''). Scale bar = 50 μm (a); 20 μm (f); 10 μm all.

Italy). Image analysis and assemblage was performed with Motic Images Advanced (V. 3.2) software (Motic, Seneco, Milan, Italy) and by using the Adobe Photoshop 8.0 (Adobe, Inc.) software for reunion from different filters.

For *indirect immunolabeling*, the following polyclonal (pAb) or monoclonal (mAb) primary antibodies were used at the dilutions indicated: mAb mouse IgG Anti-Glial Fibrillary Acidic Protein (GFAP) (G3893, clone No. G-A-5, Sigma-Aldrich, 1:500); pAb rabbit anti-UCHL1 (HPA005993, Sigma-Aldrich, 1:1000); mAb mouse IgM Anti-Vimentin, (V5255, clone VIM-13.2, Sigma-Aldrich, 1:200); pAb rabbit IgG Anti-Neurofilament 200 (NF), (N4142, Sigma-Aldrich, 1:200); mAb mouse Anti-Nestin [10C2] (ab22035, 1:250); pAb rabbit Anti-MBP (M3821, Sigma-Aldrich, 1:200); rabbit Anti-S100 beta antibody (S2644, Sigma); rabbit anti-b-III-tubulin antibody (Sigma). The secondary antibodies used were as follows: Goat Anti-Mouse IgG TRITC conjugate (T7028, Sigma); goat anti-rabbit IgG1 fluorescein isothiocyanate (FITC)-conjugate (F6005, Sigma); goat anti-rabbit conjugated with Alexa 594, Invitrogen (A11037); goat anti-rabbit conjugated with Alexa 594, Invitrogen (A11037); Goat Anti-Rabbit IgG-FITC (Product No. F0382); goat Anti-Mouse IgG FITC conjugate (F0257, Sigma); goat anti-rabbit conjugated with Alexa 594, Invitrogen; goat Anti-Rabbit IgG-FITC (F9887) at a 1:50 dilution for 45 minutes at RT. All primary and secondary antibodies were diluted in PBS/3% BSA. Cells were rinsed three times with PBS, fixed with 4% paraformaldehyde/PBS for 15 min at RT, washed twice with PBS and permeabilized with 0.2% Triton X-100/PBS for 10 min at RT. After three washes with PBS each of 5 min, cells were blocked in 3% BSA/PBS for 30 min, and then incubated in primary antibodies diluted in blocking buffer for 60 min at RT. For indirect labelling, after three PBS washes, cells were incubated with the appropriate secondary antibody for 60 min at RT. After other three PBS washes, nuclei were counterstained with 4',6-diamidino-2-phenylindole (DAPI; 1.5 μg/ml) for 10 min at RT. To avoid photo bleaching of fluorescent dyes, stained cells were embed with antifade mounting medium PBS/glycerine/1,4-diazabicyclo[2.2.2]octane (90% Glycerol/100 mg/ml DABCO/PBS), adding the antifade directly onto the wells before microscope visualization. As a negative control, cells were stained with the same procedure without the primary antibody labelling.

For *direct immunolabeling*, the following monoclonal (mAb) antibody was used at the dilutions indicated: mAb mouse anti Proliferating Cell Nuclear Antigen (PCNA) FITC-conjugate (F0167, clone PC10, Sigma-Aldrich, 1:100).

Results

Neuronal and radial glial cells

Studies on NSCs defined these cells as self-renewing CNS cells that can differentiate into any of the three major neural cell lineages, specifically neurons, astrocytes and oligodendrocytes [16,17]. To study the potential role of NPCs and Radial Glia Progenitors (RGP) in fish neurogenesis, the neuroglial primary culture of the fish *S. aurata* has been used by our laboratory as a teleost CNS model system.

Multilineage development and differentiation of neurons, astrocytes and oligodendrocytes from *S. aurata* neuroglial cells in primary culture (SaGliPs), previously isolated in our laboratory, were morphologically characterized by inverted phase-contrast microscopy, and analyzed by Motic Images Advanced (V. 3.2) acquisition software (Motic, Seneco, Milan, Italy) [35]. In the panel of Figure 1 are shown microscopy images indicating the appearance, the morphologic features and the immunophenotyping of the neuronal lineage, with NSCs (Figures 1a-1d) and immature neurons (Figures 1e-1g''').

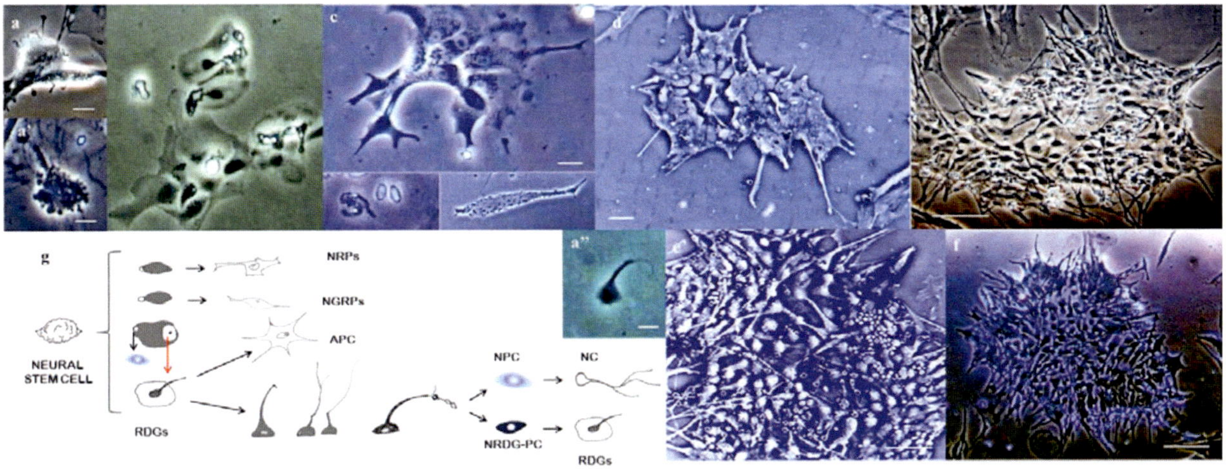

Figure 2: Morphological characterization of *S. aurata* primary neuroglial cultures. Phase-contrast microscopy pictures showing development and differentiation steps of Neuroglial Restricted Precursor cells (NGRPs). (a) NGRP cluster of stem cells at the early stage of neurosphere. (a') Highly branched neurosphere in adhesion morphology. (a'') Early stage of radial fiber generation: basal fiber cell of radial glia forming a short stub protrusive from the soma. (b, c, d) Early radial glia cluster producing basal glial cells; the NGRPs extend their basal process in a radial direction, thus gaining the typical form of radial glial rosette-like colony. (e) Phase-contrast image showing a rosette like colony of radial glia with early elongated protrusions after 2 weeks of culturing (×200). (e') An enlargement of cells in (e) showing a particular of the radial colony with areas of active neuronal proliferation and first dividing cells generating neurons (×400). (f) Phase-contrast image showing a rosette colony of radial glia with a blue filter enlightening newly formed neurons after 2 weeks of culturing. (g) Scheme represents development and differentiation of neuro-glial restricted precursor cells (NGRPs) astrocytes, oligodendrocytes, and radial glial cells. Scale bar = 10 μm (a-d); 50 μm (e-f).

In the panel of Figure 2 are presented microscopy images representative of the Neuro-Glial Restricted Precursors (NGRPs) in their appearance and morphologic features, with cluster of NSCs (Figure 2a) and glioblasts (Figures 2b-2f). Immunofluorescence images showed differentiated neuronal progenitor cells attached with pro-oligodendrocyte precursor cell expressing positive GFAP immunoreactivity (Figure 2b). In SaGliPs, after the first day of cell adhesion, the characteristic shape of neural stem cells appeared to the bottom of the flask as a round neurosphere rich of multi-extruding processes (Figure 2a); on the second day the shape of NGRPs changed both in morphology and in colour becoming flatten as in Figure 2a', while the first radial glial (RDG) rosette like colonies were observable (Figures 2e, 2f).

Within the first week after plating, the NGRPs extended their basal process in a radial direction, thus gaining the typical form of RDG rosette like colony (Figures 2b-2d). The contrast phase analysis has been shown that the new cell production occurred in specific mitotically active proliferation zones. Inside the RDG, a typical cell population of neuronal perikaryal appearance started to proliferate directly from the foot of cone-shaped radial cell, thus forming a cluster of highly proliferative cells (Figures 2e, 2e', 2f). From these proliferation zones, the newly formed young neuronal cells are guided, in their migration to the final target areas, by RDG fibers (Figures 3a,3b).

The neurogenic niches with astroglial morphology riches of highly replicating cells were observed during the 2nd week of differentiation (Figures 2 and 3). Fish neurogenic islands (generated by RGCs) showed a greater strong ability to regenerate neural cells, succeeding in a continuous generation of young neurons and RDGc renewal (Figures 2 and 3).

To determine whether RDGc were mitotically active, we tested their mitotic activity by fluorescence DNA markers (Figures 3e, 4d, 4e, 5e, 5f', 5h): (1) the UV-light excitable DAPI used as a DNA specific fluorochrome, as it appears to associate with AT clusters in the dsDNA; and (2) the Proliferating Cell Nuclear Antigen (PCNA), the most widely used expression marker of proliferation-associated proteins. PCNA is an antigen detected by antibody to cyclins D, E, A, and B which is specific for G1/2 and S, M phases of the mitotic cycle [40,41].

The morphology and distribution of the radial fiber indicates the type of dividing cells, as a consequence of different cleavage plane during cell cycle division of RDGc (Figures 3a-3f). During the early stage of fiber generation, the basal fiber of RDG remained elongated; it appeared like tired out forming a short radial stub projecting from the soma, that thins distally (Figure 2a'', 2b, 2g). Our results suggest that the vertical cleavage plane result only in neuronal production of daughter cells similar in shape (Figure 3a), while the horizontal cleavages produce two types of asymmetric divisions by generating (1) one new radial glial cell and one neuron progenitor (cell/cell horizontal cleavage) (Figure 3b); (2) one radial fiber and one neuron progenitor (fiber/cell horizontal cleavage) (Figure 3d). Furthermore, the contrast phase and immunofluorescence analysis revealed that young RDGc, during their cell division, generate two different types of daughter cells, one destined to become a long fiber process and the other destined to develop a neuron (Figures 3c, 3c', 3c''). In particular, the end of the fiber grew a thick process that would break, generating a neuron that started to create a line of newborn neurons by symmetric divisions (Figure 3). Such kind of neurogenesis was a peculiar characteristic of the seabream RDG model (Figure 3).

Two populations of glial fibers were apparent, one immunopositive for GFAP, the other for vimentin (Figure 4d). Immunocytochemistry analyses for double-or triple-immunofluorescent labelling were performed to define the status of differentiation of cultured cells. After 2 weeks of culturing, immunofluorescence revealed elongated DAPI-labelled nuclei, in close apposition to GFAP-labelled radial glial fibers (Figure 4d), thus signifying that NPCs grew closely opposed to the RDG fibers (Figure 4e). The GFAP-positive RDG fibers increased near the proliferating site of new neurons. It seems conceivable that these

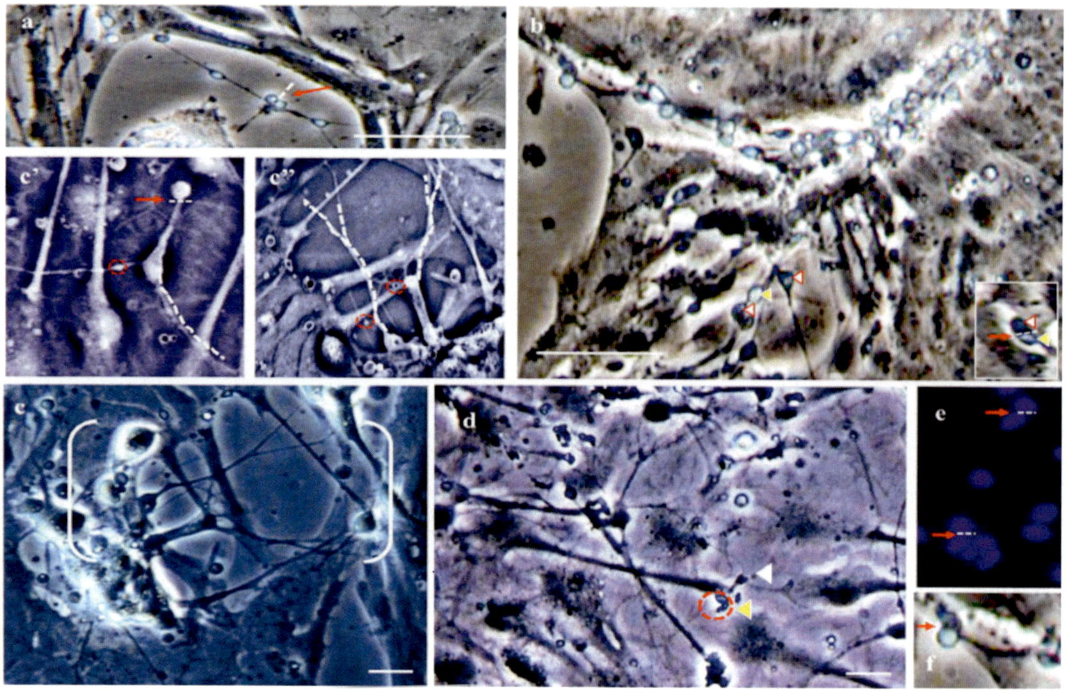

Figure 3: Symmetric and asymmetric division of *S. aurata* radial-glial cells (RDGs). (a) Phase contrast image showing symmetric divisions with vertical cleavages resulting only in neuronal production of daughter cells similar in shape. (b) Phase contrast image showing asymmetric divisions with horizontal cleavages by generating one new radial glial cell and one neuron progenitor (cell/cell). The bracketed area in (c) is shown in separate picture in (c') and (c"), as inverted phase contrast image displaying asymmetric division. Arrowheads in red indicate the horizontal cleavage plane generating the asymmetric RG cell division, white dotted lines indicate RG fibers, red circles indicate the neuron progenitor. (d) Phase contrast image showing asymmetric divisions on horizontal cleavage resulting in one radial fiber and one neuron progenitor (fiber/cell). Particularly (f) at the fiber extremity grows a thick process which collapses, generating a neuron that began to develop a line of newborn neurons by symmetric divisions. (e) DAPI counterstained nuclei show asymmetric divisions with horizontal cleavages

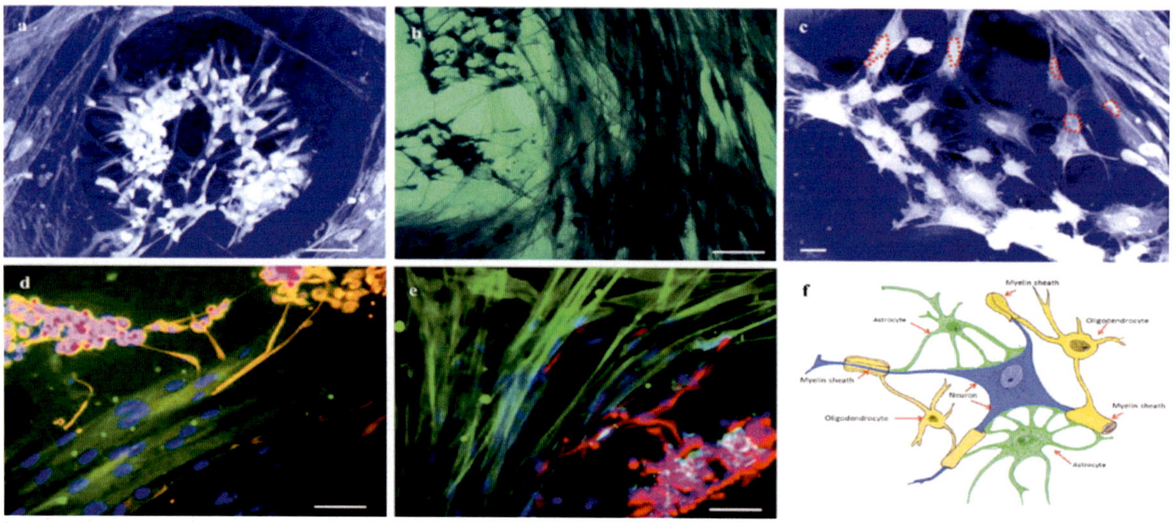

Figure 4: Mature radial glia from primary neuroglial cultures of *S. aurata* brain. Phase contrast microscopy images showing radial glia progeny from neural stem cells. (a, b) Rosette-like colonies of radial glia after 2 weeks of culture (200x). (c) Edge colony enlargement of (a); detail of the radial glial fibers, making an interlaced scaffold to which NPCs were closely apposed, showing the nuclei migration of new neurons (red dashed circles) within the radial migrating fiber towards the boundary (400x). (d) Immunofluorescence staining focusing GFAP positive astrocytes (red) differentiated into large cells with many filopodia while contacting multiple axons (green filament) in order to sustain oligodendrocytes during the myelination process (green spheres) (GFAP, red; vimentin, green filament; MBP, green spheres; DAPI/PCNA, blue/cyan). Merged picture (e) displaying colocalisation of MBP to label OPCs (green spheres) and GFAP to label astrocytes (red) near NF positive axons (green), with DAPI/PCNA (blue/cyan) to label cell nuclei. (d, e) RGCs can produce new generation of young neurons even in the old culture as revealed by PCNA within the inner cluster. (f) This diagram shows a schematic arrangement of nervous cells in culture. Astrocytes contact neurons being involved in metabolic exchange between neurons and blood; oligodendrocytes wrap themselves around axons forming the myelin sheath. Scale bar = 50 μm (a, b, d, e); 10 μm (c).

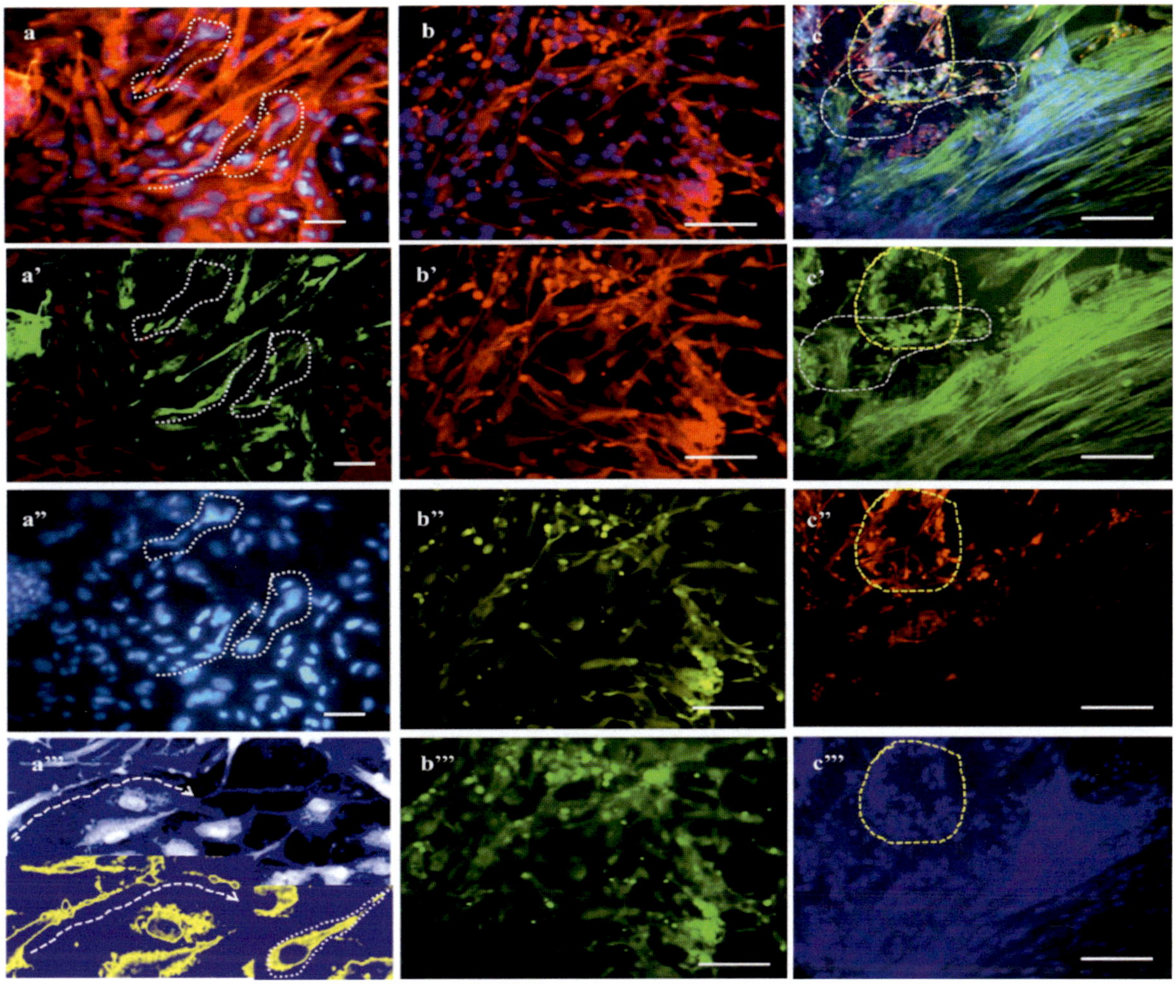

Figure 5: Indirect and direct immunofluorescence staining focusing development and differentiation in *S. aurata* brain cultured cells. (a) GFAP immunoreactivity at the tips of the astrocytes processes after 3 week of culture (×200); (a') Vimentin labeled fiber; Radial glia stained immunopositive for both glial fibrillary acidic protein (GFAP) and vimentin at 3 weeks old cultures; (a'') DAPI counterstained nuclei corresponding to the same picture of (a) and (a') (400x). The proliferative capacity of cells was shown by mitotic nuclei (white/cyan) double stained with Ab anti-PCNA-FITC and counterstained with DAPI (blue); the white dashed lines showed the actively dividing nuclei within the fiber extension. (a''') Phase contrast image showing basal RDG cells generating neurons (white dashed line showed the nuclear migration and fiber extension). (b-b') Intense fluorescence for Nestin here appeared to be highly expressed in the processes by most IPCs, 3 weeks old cultures. (b'') βIII-tubulin label immature neurons. (b''') NeuN was expressed nearly by mitotic figures. The mature post-mitotic neurons expressed in high density NeuN positive processes. (c-c''') Double immunofluorescence demonstrating colocalisation of neurons and astrocytes within the radial glial colony. (c) Several clusters of proliferative cells were observed (white and yellow circles); in particular, the central zone showed an intense granular production of neurons (UCHL1+/DAPI+/PCNA+) (white) inside the center of the colony of radial glia (GFAP+) (yellow circle). (c') 3 weeks old cultures, highly immunostained with UCHL1 (green), showed newborn neurons migrate along neuronal fiber. (c'') The outer boundary of the radial glia colony showed differentiated astrocytes with long GFAP positive filopodia (red) and neuronal fibers (c and c'), whereas in the inner zone there are clones of mitotic radial glia (c''', PCNA+/DAPI+). Scale bar = 50 µm.

neurons moved from end to the tip of the radial process by nuclear translocation, as visualized with the DNA marker (DAPI), thus reaching their final target site predominantly by the nucleus migration within the fiber (Figure 5).

Between the end of the second week and the beginning of the third of culturing, the seabream RDG colony assumed a specialized aspect, with RDG fibers arranged in an interlaced scaffold to which NPCs were closely apposed, possibly for their radial migration towards the final target site. The scaffold was organized with GFAP positive differentiated astrocytes with many filopodia contacting multiple axons (vimentin positive) in order to sustain oligodendrocytes during the myelination process (Figures 4d and 4e). Neurons were apparent when fish CNS cultures were at 2 weeks of culturing, emerging from the tip of RDG fibers (Figure 5). The morphology and the immunophenotyping of neural cells were displayed by the immunostaining with antineurofilament (NeuN), anti-βIII-tubulin, anti-nestin, anti-vimentin antibodies and the neuron-specific markers for ubiquitin carboxy-terminal hydrolase 1 (UCHL1) (Figure 5).

We tested neural differentiation of NPCs by examining the expression of maturation protein markers. To characterize the first early events of differentiation, we studied markers of NPCs during the immature status. After 1 week of culture, most immature precursors cells (IPCs) neural structures contained neuroepithelial cells immunoreactive for nestin. Nestin is an intermediate filament protein commonly expressed by many progenitor cells, and here appeared to be highly expressed in high density into neurospheres and IPCs processes (Figure 5). Most

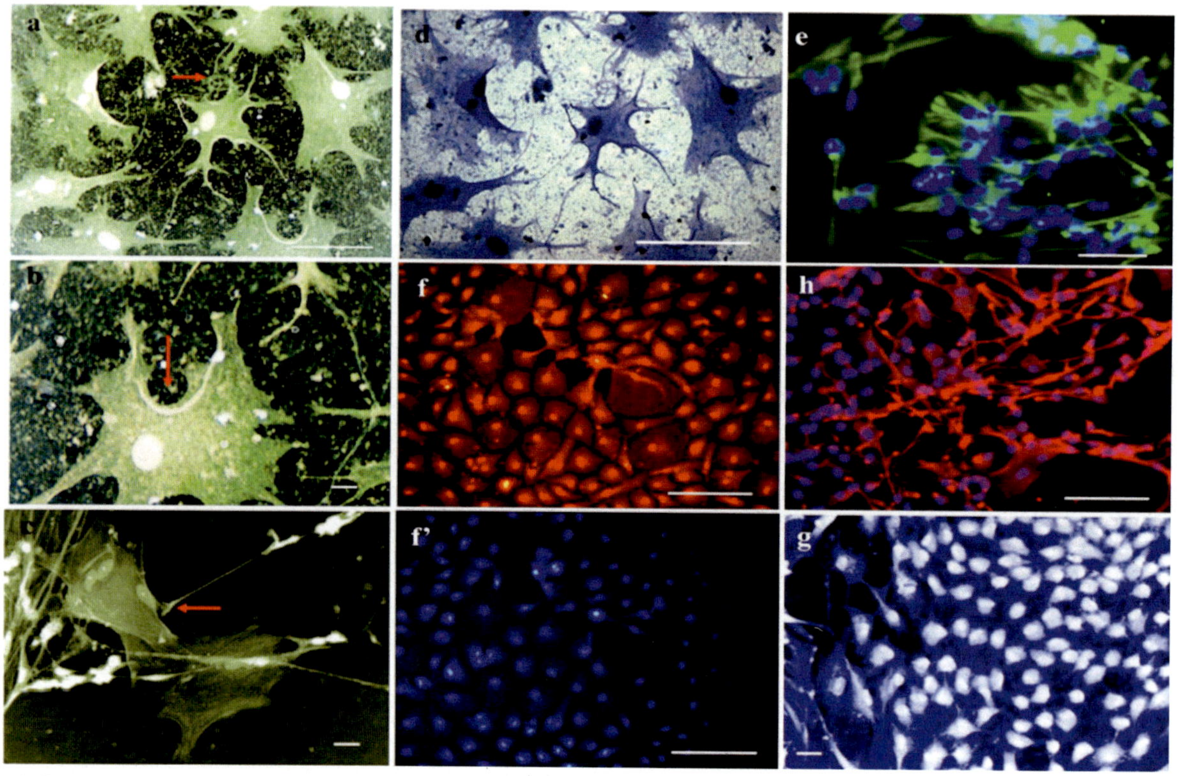

Figure 6: Characterization of *S. aurata* astrocyte monolayer. Phase contrast microscopy images showing the monolayer growth and differentiation in primary neuroglial cell cultures. (a) Astrocytes intercommunicated forming a monolayer after 2 weeks (×400), and (b) young neuron attached on the astrocyte border. (c, d) Phase contrast image showing fully differentiated communicating astrocytes connected with elongated neuritis overwhelming the cells after 1 month of culture (×400). (e) Mature astrocytes positive for UCHL1 labeling at the tips of the astrocytes processes (green) and DAPI (blue nuclei) after 1 week of culturing. (×400). (f, f') GFAP-immunoreactive mature astrocytes after 1 month of culture with the typical stellate shape and nuclei DAPI counterstained (×400). (g) Negative inverted phase contrast image showing 2 weeks old cultures of microglia (×400). (h) Mature astrocytes positive for GFAP labeling at the tips of the astrocytes processes (red) and DAPI (blue nuclei) after 1 week of culturing. (×400). Scale bar = 50 μm

nestin positive cells contained PCNA expression which colocalized with the DAPI counterstaining, thus counting for mitotic figures (Figure 5). RDGc can produce new generation of young neurons even in the old culture as revealed by PCNA within the inner cluster (Figures 4d and 4e). PCNA is a cell cycle dependent intranuclear protein, which assists the delta-DNA polymerase during DNA replication, whereby its expression is considered a marker of DNA synthesis, being detectable in G1/S phase [40]. By producing new neurons and guiding them to the growing edge of colony, radial glial cells confirmed their relatively strong ability as in the adult fish brain to regenerate neural tissues.

In a first early event, the basal fibers of RDGc supported the production and migration of new born neurons (Figure 5a-5a'''), around the edge of the RDG rosettes (Figure 5c and 5c'). Whereas during the phase of neuronal differentiation a new circuit formation of horizontal fibers occurred later, displaying an interlaced network, as a mesh woven of axonal fibres closely connected (Figure 4). Next, to evaluate the transition from IPCs to mature neurons, we tested two neuronal differentiation markers, βIII-tubulin to label immature neurons and NeuN the definite marker of mature neurons. After 14-20 days of differentiation, the cultures were highly enriched in neurons expressing the βIII-tubulin marker (Figure 5). The colocalization of βIII-tubulin/NeuN was expressed almost by differentiating neurons, while cells immunoreactive for NeuN were cells which completed the transition from immature to mature neurons. All differentiated neurons presented the multilineage neural phenotype, defined as coexpression of the following markers gradient oriented according the developmental status: GFAP, Nestin, Beta III-tubulin, NeuN, UCHL1 (Figure 5). During long term culturing the density of fibers markedly increased within 3 weeks, and differentiation up to 4 weeks achieved a complete coverage of the astrocyte bed with a massive arborisation of neurites (Figure 6). The density of vimentin-positive RDG fibers increased over the monolayer as long as the neuritis full-grown (Figure 6). By analyzing the sequence of events taken in the formation of astro-neuronal tissue *in vitro*, it appears apparent that the formation of the fiber network of RDG is a plastic prerequisite needed to enable the support of growth and migration of the newly generated neuronal cells (Figure 6), as confirmed by the presence of actively dividing nuclei, doublestained with Ab anti-PCNA-FITC and counterstained with DAPI (blue), within the fiber extension (Figure 5a"). Finally, even in the 3rd week old cultures were still observed mitosis both for newborn neurons migrating along neuronal fibers (UCHL1+/PCNA+/DAPI+) and within the inner rosette-like colony, suggesting the coexistence of differentiating astrocytes and neurons with newly formed cells.

Astroglial cells

Two specific progenitor cells, Glial-Restricted Precursor (GRP) cells and oligodendrocyte/type-2 astrocyte progenitor cells (O-2A/OPCs) have been studied as distinct glial precursor subtypes restricted to astrocyte development in mammals; they only differed in the differentiation potential and in the timing of their developmental

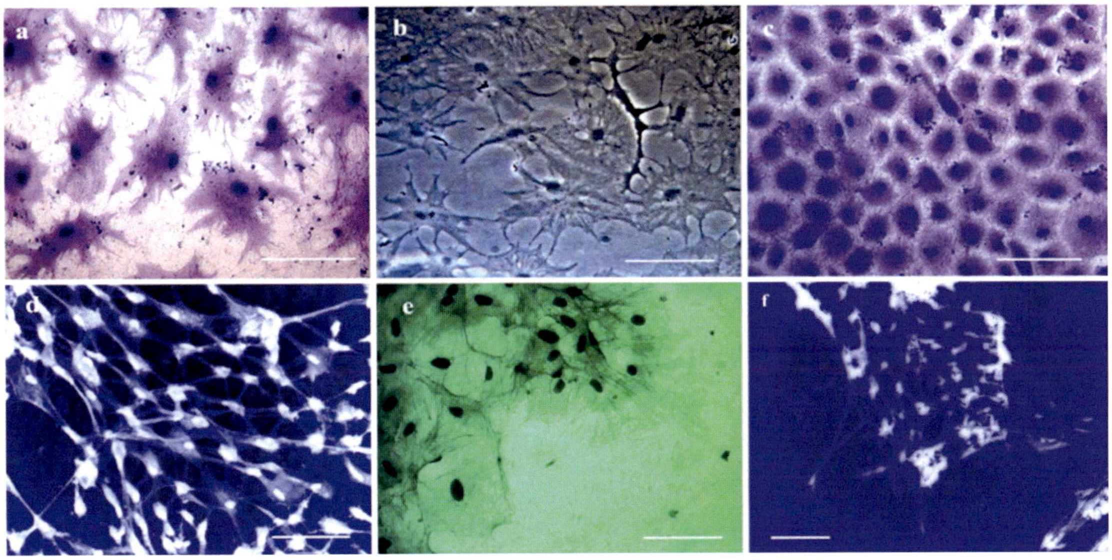

Figure 7: Characterization of *S. aurata* astrocyte monolayer. Phase-contrast microscopy pictures showing the type-I and type II mature astrocytes within the monolayer in SaGliPs. (a) Primary neuroglia cultures in enriched medium differentiated in type-II astrocytes during monolayer formation after 2 weeks of culturing. (b) Neural elongated stem cells (NS) while generating neurons and astrocytes (stellate cells) in 1 week old cultures. (c) Primary neuroglia cultures in enriched medium differentiated in type-I mature astrocytes forming a monolayer after 3 weeks of culturing. (d) Monolayer of astrocyte cells with dense dendrite and neurite branching, and neuron bodies after 4 weeks of culturing (×200). (e) Monolayer of mature type-II astrocytes. (f) Morphology of a radial glia growing islet after 5 day. Scale bar = 50 µm.

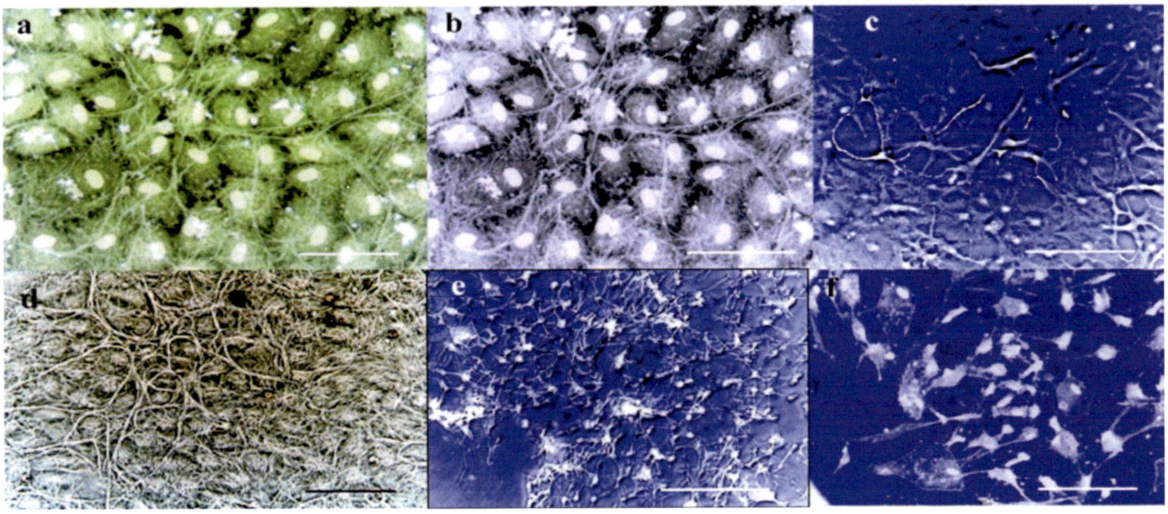

Figure 8: Morphological characterization of *S. aurata* primary neuronal cells. (a, b) Phase contrast image showing fully differentiated communicating neurons with elongated neuritis overwhelming the astrocytes bed after 1 month of culture (×400). (c) Inverted contrast phase image showing early elongated neuritis growing over the astrocytes bed after 1 week of culture (×200). (d) Phase contrast image showing fully differentiated neurons with elongated neuritis overwhelming the astrocytes bed, 1 month of culture (×200). (e) Negative inverted phase contrast image showing intercellular communications (vesicles) between neurons 2 weeks (×400). (f) Enlarged area of (e) picture. Scale bar= 50 µm.

appearance for the astrocyte subtypes produced, type 1 and type 2 astrocytes [42,43].

Neural seabream mixed cultures grew best in SaGliPs-enriched growth factors medium supplemented with serum, and could be maintained for at least 2 months. The astrocytes displayed a heterogeneous phenotype, including a highly branched fibrous astrocytes (type-II astrocytes) and polygonal stellate cells (type-II astrocytes) (Figures 7a, 7b, 7e and Figures 7c and 7d respectively). After 2 weeks, SaGliPs cultured in enriched medium started acquiring features of differentiation, such as the typical appearance of mature astrocyte forming a tight monolayer and generating neurons (Figure 7 and 8). According the lineage derivation the NGRPs-derived the type-I astrocytes, which firstly formed a confluent layer, and secondly the other cells grew on top; neurons grew in small clusters of RDGc with large bundles of axons. Next, after 3-4 weeks of culturing, neurons formed on the astrocyte bed a dense dendrite and neurite branching radiating outward (Figure 9).

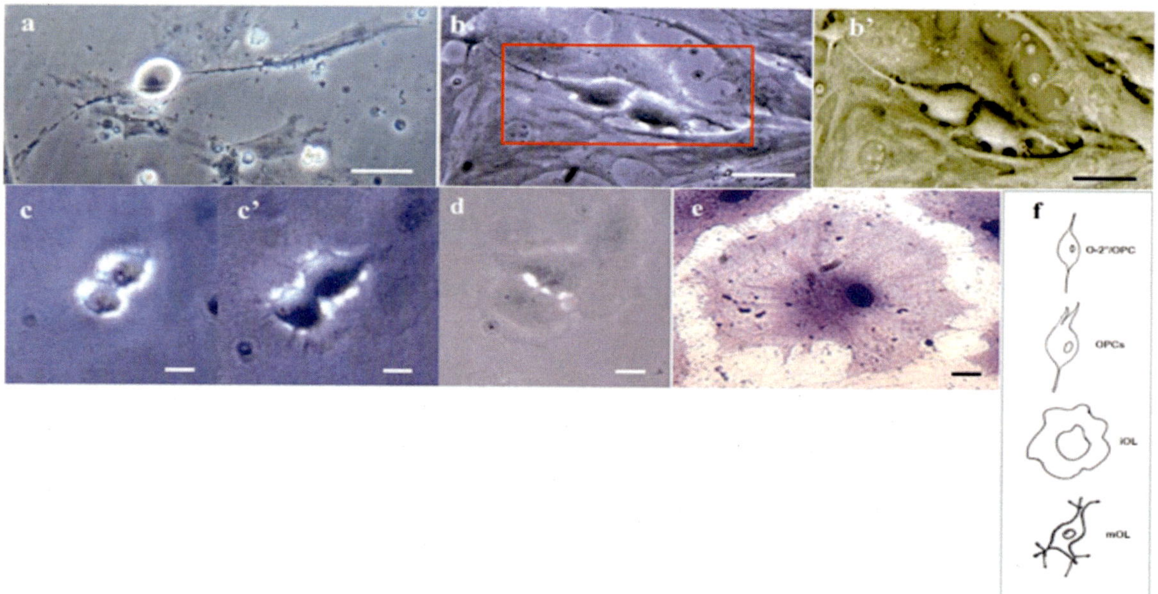

Figure 9: Oligodendrocyte lineage cell development and differentiation timing in neuroglial primary cultures from the brain of *S. aurata*. Phase-contrast microscopy images showing the lineage development of oligodendrocytes. (a) Glial-restricted precursor cells (GRPs); 2-day old cultures. (b, b') Oligodendrocyte precursors (OPCs); 1-week old cultures. (c, c') Enlargement of red area (d, e) Immature (iOL) and Mature Oligodendrocyte (mOL), 3 week old cultures.mOL showed a round cell body enriched of cytoplasmic granules and a dense filamentous meshwork in the peripheral areas often enwrapping other neuroglial processes. (f) Scheme of oligodendrocyte development. Scale bar = 50 µm (a, b, b'); 20 µm (c, c', d, e)e.

In seabream neural cell cultures, after 2 weeks of culturing, the growth velocity of fish astrocytes seemed to be higher, seabream cells developed quickly, reaching confluent culture within 2 weeks (Figure 7), fish cells were flat and stellate (Figure 7). The GFAP astroglial lineage marker was searched to highlight the presence of cultured astrocytes using anti-GFAP antibodies. In the fish model a typical astrocyte bed was depicted by the GFAP labelling after 3 weeks of culture.

Oligodendroglial cells

Oligodendrocytes, like other neural cells in the CNS, develop from the neuroepithelial precursors and are one of the last cell types to differentiate in the CNS.

Recently, some studies identified oligodendrocyte-type 2 astrocyte (O-2A) progenitor cells only generate type 2 astrocytes and oligodendrocytes [42,43].

In 1 week old cultures we observed cells with a simple bipolar form identified as O-2A progenitor cells undergoing a morphological transition during the 2nd week of culturing (Figure 9) to pro-oligodendrocytes (pro OLIGO), with a more complex and flattened form enriched of multiple processes, accompanied by a change in cell surface antigens through the S100B expression acquisition which is essentially restricted to the oligodendroglial cell lineage (Figure 10b).

We observed, by CF and microscopic time lapse imaging, that proliferative oligodendrocyte progenitor cells converted into mature oligodendrocytes and after 2 weeks oligodendrocytes were able to recognize neurons and wrap their target axons (Figures 9 and 10).

After 3 weeks of culturing, fish oligodendrocytes stopped to divide, completed their differentiation transforming in mature Oligodendrocyte (mOL), and enveloped axons. In panel of Figure 10 we showed all stages of lineage development including proliferation, differentiation and long-term survival of oligodendrocytes up to the ultimate progression into fully-fledged oligodendrocytes expressing myelin products released in membrane sheats (Figures 9 and 10). In Figure 10, mOL showed a round cell body enriched of cytoplasmic granules, with a dense filamentous meshwork, initially enwrapping neuroglial processes in the peripheral areas.

Discussion and Conclusion

Since a long time, studies on NSCs defined as self-renewing CNS cells that can differentiate into any of the three major neural cell lineages, specifically neurons, astrocytes and oligodendrocytes, have been attempt almost in mammals [16,17]. Because of the lack of data on the culture of fish neural cells, with the exception on studies from the adult zebra fish injured brain, we carried out the characterization and lineage differentiation of NPCs from seabream CNS, primarily isolated by our research group and known as SaGliPs [35]. This is the first well characterized *in vitro* CNS teleost model, getting the chance of producing newly neural cells for long time in culture. The differentiation of neuronal lineages in seabream has never been described before. Previous studies supported the notion that, also in fish, radial glia may be a potential source of new neurons. For instance, while lower vertebrates display the capacity to generate new retinal neurons in response to injury, in teleost new neurons arise mainly from RDG following neural induction, and their subsequent development and differentiation [44].

We detailed in seabream the early stage of RDG generation; firstly the basal cell producing the fiber of Radial Glia (RG) remained elongated, appearing like tired out from the base and forming a short radial stub projecting from the soma. Studies from Miyata et al. [45] demonstrated that, during RDGc division, each neural connected fiber becomes thin but is neither lost nor divided, and it is inherited with an asymmetrical cleavage by one daughter cell. In these divisions were produced a neuron and a progenitor cell; the neuron inherits the pial

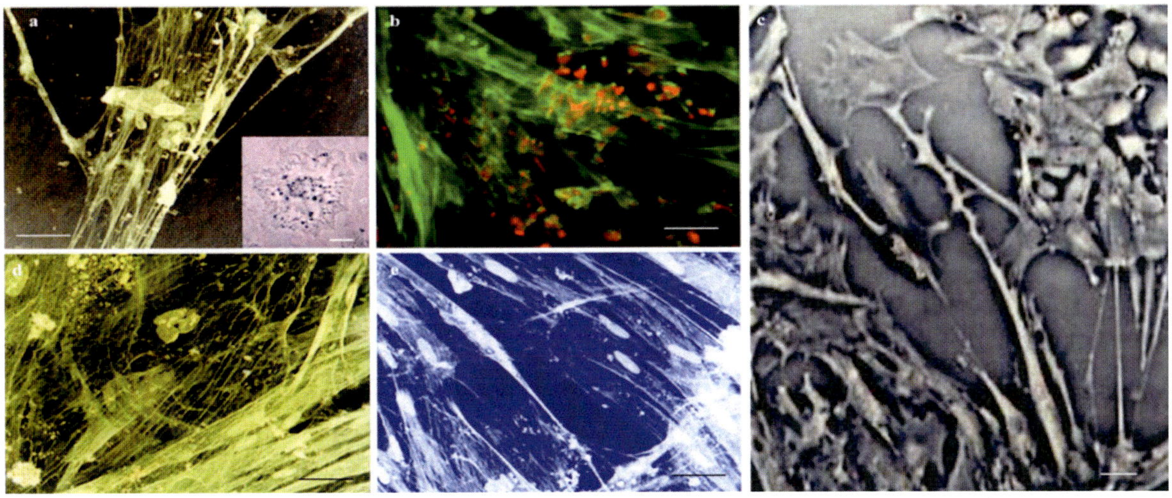

Figure 10: Differentiation of oligodendrocyte lineage from *S. aurata* brain cultured cells. (a) Early stages of fiber engulfment by oligodendrocytes by contrast phase inverted image. Insert within (a) showing the CF aspect of a mature Oligodendrocyte (mOL): mOL showed a flattened round cell body enriched with cytoplasmic granules while forming a dense filamentous meshwork in the peripheral areas; 3-week old cultures. (b) Double immunolabeling with S100B antibody (red) and β-III tubulin (green); mOLs (red) were closely apposed to neuron fiber while enwrapping the neuronal axons (green). The S100B expression is essentially restricted to oligodendroglial cell lineage after 5 days of differentiation. (c) mOL showed enwrapping the long axon of a neuronal cell. (d, e) particulars of wrapped layers and prominent inner/outer tongue processes of mOLs; CF inverted images. Scale bar= 20 μm (a, b, d, e); 10 μm (a', c).

fiber, and is therefore indistinguishable from the progenitor RDG. Our work confirmed results obtained from such recent *in vitro* and *in vivo* studies, examining the neurogenic potential of RDGc by characterizing the morphology and modes of neurogenic division planes as cited by Noctor et al. [33] Chenn and McConnel [47]. Moreover, our findings on the fate of the radial fibers, during RDGs development, revealed that in seabream the young RDGc, throughout their cell division, generate two different types of daughter cells, one destined to become a long fiber process and the other destined to develop a neuron. We found also that the cleavage plane will result only in neuronal production of two daughter cells similar in shape, while the horizontal cleavages will produce two types of asymmetric divisions by generating (1) one new RDGc and one NPC (horizontal cleavage versus cell/cell), and (2) one radial fiber and one NPC(horizontal cleavage versus fiber/cell). Recently, also in *in vivo* studies from the zebrafish, RDGc were found to be able to function as multipotent retinal stem cells that responded to the loss of photoreceptors by specifically regenerating the missing neurons [46]. The work of Chenn and McConnell [47] showed that as neurogenesis proceeded, vertical cleavages decreased, while horizontal cleavages increased. This appear very consistent with our finding, in that the vertical cleavages were symmetric divisions producing two similar precursor cells; whereas horizontal cleavages were asymmetric divisions, resulting in the production of one neuron and one RDG precursor cell [48].

In regards to the formation of fiber scaffold, after 2 weeks of culturing, we observed that NPCs grew closely apposed to the RDG fibers, thus providing strong evidence for their radial migration through the scaffold of RDG fibers, until cells reached the regenerating site and final confluence. We observed a characteristic fiber behavior: fibers initiate to grow from the tips of the protrusion of a basal cell, later propagate out of the tip origin, by splitting at the end into two lateral turns [49]. Afterward, they continue to run in two opposite lateral directions to the left and right side, generating a row of new neurons that resemble to a pearl necklace. This behavior is in accordance with the findings of Zupanc et al. [50,51] that observed, in the adult fish, RDGc actively divided to produce new neurons in the injured brain and guided them to the lesion site to regenerate the neural tissue [50,51]. Such evidences were supportive for the role exerted by RDGc in adult teleost brain to be responsible for the strong ability of neural tissue regeneration [52]. The possibility that RDGc might exist as radially oriented cells is absolutely dissimilar in mammals, where RDGc were transformed into multipolar astrocytes in the adult brain, no longer regenerating [52,53].

In seabream brain most dividing NPCs come from RDG progenitors. We demonstrated that RDGc maintained a vimentin-positive radial fiber throughout each stage of cell division, and an asymmetric inheritance of this fiber. Out of directing neuronal migration, radial fibers supported another important feature in the role of being neuronal progenitors; they were found highly mitotically active, and underwent nuclear migration. Radial glial divisions produced both additional radial glia and newly young neurons, while a separate population of neuronal precursors generated almost exclusively neurons.

Our findings overlap with the current understandings of rodent neurogenesis: two distinct types of NPCs were identified. The RG was the first type, being conceivably recognized as neuronal and glial progenitors also by other authors [30,33,45,54]. The island of RDGc behavior as a neuroepithelial niche, giving rise to radial clones of newborn neurons, where neuronal progenitor cells generated rounds of asymmetric division. The second type, known as (IPCs), was recently characterized as neuronal progenitors, generated from RG; it proliferate for one or two mitotic cycles, and produce neurons only [55-57]. In fish we found that RDGc generate IPCs, which divide to produce pairs of neurons. This is an intriguing result for the potential use of fish neural stem cells in regenerating nervous tissue [58]. In recent times, GRPs have been so well studied for the development of astrocyte transplantation therapies, highlighting new exciting potentials for regenerative therapies in the injured adult nervous tissue [48,49].

The self-renewing ability of NSCs which can differentiate neuroepithelial precursors into neural cell lineage of oligodendrocytes it has been recently demonstrated [16,17]. In mammals, oligodendrocytes develop from the neuroepithelial precursors that line the so-called VZ. Soon after the migration, they can differentiate into either oligodendrocytes type or "type-2" astrocytes *in vitro*, depending on the composition of the culture medium. This is the reason for these cells are referred to as O-2A progenitor cells [59,60]. Dissimilarly to other neuronal progenitors, O-2A glial progenitors continue to divide after they leave the VZ, then they undergo a transition to pro-oligodendrocytes. Peculiar modifications characterized their morphology: they changed from a simple bipolar form to a more complex form with multiple processes, accompanied by a change in cell surface antigens [61]. Others Authors found that, in proximity of the site of terminal differentiation, the pro-oligodendrocytes stop to divide and enwrapped axons; here, they ultimate the lineage development by expressing myelin gene products and mature into fully-fledged oligodendrocytes [62]. All the previous discoveries overlapped with our findings; as a result we demonstrated all stages of lineage development including proliferation, differentiation and long-term survival. In the CNS, multiple precursor cells were found to develop both the neuronal precursor cells and glial precursors; among the latter, Glial-Restricted Precursor (GRP) cells and oligodendrocyte/type-2 astrocyte progenitor cells (termed also oligodendrocyte precursor cells, O-2A/OPCs) represent the two well-studied distinct glial precursor subtypes in mammals [42]. Although GRP cells and O-2A/OPCs were restricted to astrocyte development, they both differed in the differentiation potential, in the timing of developmental appearance and in the astrocyte subtypes produced (type 1 and type 2 astrocytes); they were also able to generate oligodendrocytes [42,43].

In seabream, multiple myelinating areas of single axons were easily identified by using fluorescent antibodies directed against components of myelin (i.e. S100B, MBP). Also, phase contrast analysis gave evidence of axonal myelin shapes, with some loosely wrapped layers and protruding processes. The excellent remyelinating fish ability of mOLs might be a promising approach for repairing the CNS injuries, also in other animals, as a new coming strategy for function recovery of damaged nerves.

References

1. Kaslin J, Ganz J, Brand M (2008) Proliferation, neurogenesis and regeneration in the non-mammalian vertebrate brain. Philos Trans R Soc Lond B Biol Sci 363: 101-122.
2. Alvarez-Buylla A, García-Verdugo JM, Mateo AS, Merchant-Larios H (1998) Primary neural precursors and intermitotic nuclear migration in the ventricular zone of adult canaries. J Neurosci 18: 1020-1037.
3. Doetsch F, Scharff C (2001) Challenges for brain repair: insights from adult neurogenesis in birds and mammals. Brain Behav Evol 58: 306-322.
4. Zupanc GK (2001) Adult neurogenesis and neuronal regeneration in the central nervous system of teleost fish. Brain Behav Evol 58: 250-275.
5. Grandel H, Kaslin J, Ganz J, Wenzel I, Brand M (2006) Neural stem cells and neurogenesis in the adult zebrafish brain: origin, proliferation dynamics, migration and cell fate. Dev Biol 295: 263-277.
6. Lois C, Alvarez-Buylla A (1994) Long-distance neuronal migration in the adult mammalian brain. Science 264: 1145-1148.
7. Hack MA, Saghatelyan A, de Chevigny A, Pfeifer A, Ashery-Padan R, et al. (2005) Neuronal fate determinants of adult olfactory bulb neurogenesis. Nat Neurosci 8: 865-872.
8. Kirn JR, Alvarez-Buylla A, Nottebohm F (1991) Production and survival of projection neurons in a forebrain vocal center of adult male canaries. J Neurosci 11: 1756-1762.
9. Nottebohm F, O'Loughlin B, Gould K, Yohay K, Alvarez-Buylla A (1994) The life span of new neurons in a song control nucleus of the adult canary brain depends on time of year when these cells are born. Proc Natl Acad Sci U S A 91: 7849-7853.
10. Kernie SG, Parent JM (2010) Forebrain neurogenesis after focal Ischemic and traumatic brain injury. Neurobiol Dis 37: 267-274.
11. Kishimoto N, Shimizu K, Sawamoto K (2012) Neuronal regeneration in a zebrafish model of adult brain injury. Dis Model Mech 5: 200-209.
12. Adolf B, Chapouton P, Lam CS, Topp S, Tannhäuser B, et al. (2006) Conserved and acquired features of adult neurogenesis in the zebrafish telencephalon. Dev Biol 295: 278-293.
13. Lam CS, März M, Strähle U (2009) gfap and nestin reporter lines reveal characteristics of neural progenitors in the adult zebrafish brain. Dev Dyn 238: 475-486.
14. März M, Chapouton P, Diotel N, Vaillant C, Hesl B, et al. (2010) Heterogeneity in progenitor cell subtypes in the ventricular zone of the zebrafish adult telencephalon. Glia 58: 870-888.
15. Ganz J, Kaslin J, Hochmann S, Freudenreich D, Brand M (2010) Heterogeneity and Fgf dependence of adult neural progenitors in the zebrafish telencephalon. Glia 58: 1345-1363.
16. Temple S, Alvarez-Buylla A (1999) Stem cells in the adult mammalian central nervous system. Curr Opin Neurobiol 9: 135-141.
17. Henrique D, Bally-Cuif L (2010) A cross-disciplinary approach to understanding neural stem cells in development and disease. Development 137: 1933-1938.
18. Reynolds BA, Tetzlaff W, Weiss S (1992) A multipotent EGF-responsive striatal embryonic progenitor cell produces neurons and astrocytes. J Neurosci 12: 4565-4574.
19. Stavridis MP, Collins BJ, Storey KG (2010) Retinoic acid orchestrates fibroblast growth factor signalling to drive embryonic stem cell differentiation. Development 137: 881-890.
20. Ryder EF, Snyder EY, Cepko CL (1990) Establishment and characterization of multipotent neural cell lines using retrovirus vector-mediated oncogene transfer. J Neurobiol 21: 356-375.
21. Reynolds BA, Weiss S (1992) Generation of neurons and astrocytes from isolated cells of the adult mammalian central nervous system. Science 255: 1707-1710.
22. Gage FH, Coates PW, Palmer TD, Kuhn HG, Fisher LJ, et al. (1995) Survival and differentiation of adult neuronal progenitor cells transplanted to the adult brain. Proc Natl Acad Sci U S A 92: 11879-11883.
23. Roy NS, Wang S, Jiang L, Kang J, Benraiss A, et al. (2000) *In vitro* neurogenesis by progenitor cells isolated from the adult human hippocampus. Nat Med 6: 271-277.
24. Palmer TD, Schwartz PH, Taupin P, Kaspar B, Stein SA, et al. (2001) Cell culture. Progenitor cells from human brain after death. Nature 411: 42-43.
25. Taupin P (2008) Potential of neural stem cells for the treatment of brain tumors. Clin Med Oncol 2: 451-454.
26. Rakic P (1972) Mode of cell migration to the superficial layers of fetal monkey neocortex. J Comp Neurol 145: 61-83.
27. Alvarez-Buylla A, Nottebohm F (1988) Migration of young neurons in adult avian brain. Nature 335: 353-354.
28. Alvarez-Buylla A, Theelen M, Nottebohm F (1990) Proliferation "hot spots" in adult avian ventricular zone reveal radial cell division. Neuron 5: 101-109.
29. Doetsch F, Caillé I, Lim DA, García-Verdugo JM, Alvarez-Buylla A (1999) Subventricular zone astrocytes are neural stem cells in the adult mammalian brain. Cell 97: 703-716.
30. Malatesta P, Hartfuss E, Götz M (2000) Isolation of radial glial cells by fluorescent-activated cell sorting reveals a neuronal lineage. Development 127: 5253-5263.
31. Malatesta P, Hack MA, Hartfuss E, Kettenmann H, Klinkert W, et al. (2003) Neuronal or glial progeny: regional differences in radial glia fate. Neuron 37: 751-764.
32. Hartfuss E, Galli R, Heins N, Götz M (2001) Characterization of CNS precursor subtypes and radial glia. Dev Biol 229: 15-30.

33. Noctor SC, Flint AC, Weissman TA, Dammerman RS, Kriegstein AR (2001) Neurons derived from radial glial cells establish radial units in neocortex. Nature 409: 714-720.

34. Tamamaki N, Nakamura K, Okamoto K, Kaneko T (2001) Radial glia is a progenitor of neocortical neurons in the developing cerebral cortex. Neurosci Res 41: 51-60.

35. Centoducati G, Zacchino V, Narracci M, Iaffaldano N, Santacroce MP (2013) Neuroglial cells in long-term primary cultures from the gilthead seabream (*Sparus aurata* L.): new functional *in vitro* model from bony fish brain. Ital J Anim Sci 12: 28-38.

36. Centoducati G, Santacroce MP, Conversano MC, Crescenzo G (2009) Biotechnological process for isolation of hepatocytes from marine organisms. EP2103686 (A1).

37. Santacroce MP, Zacchino V, Casalino E, Merra E, Tateo A, et al. (2011) Expression of a highly differentiated phenotype and hepatic functionality markers in gilthead sea bream (*Sparus aurata* L.) long-cultured hepatocytes: first morphological and functional *in vitro* characterization. Rev Fish Biol Fisheries 21: 571-590.

38. Pagés T, Gomez E, Suner O, Viscor G, Tort L (1995) Effects of daily management stress on haematology and blood rheology of the gilthead seabream. J Fish Biol 46: 775-786.

39. López-Castejón G, Sepulcre MP, Mulero I, Pelegrín P, Meseguer J, et al. (2008) Molecular and functional characterization of gilthead seabream *Sparus aurata* caspase-1: the first identification of an inflammatory caspase in fish. Mol Immunol 45: 49-57.

40. Boldogh I, Milligan D, Lee MS, Bassett H, Lloyd RS, et al. (2001) hMYH cell cycle-dependent expression, subcellular localization and association with replication foci: evidence suggesting replication-coupled repair of adenine:8-oxoguanine mispairs. Nucleic Acids Res 29: 2802-2809.

41. Pozarowski P, Darzynkiewicz Z (2004) Analysis of cell cycle by flow cytometry. Methods Mol Biol 281: 301-311.

42. Herrera J, Yang H, Zhang SC, Proschel C, Tresco P, et al. (2001) Embryonic-derived glial-restricted precursor cells (GRP cells) can differentiate into astrocytes and oligodendrocytes *in vivo*. Exp Neurol 171: 11-21.

43. Wang J, Bushman J, Wang X, Mayer-Proschel M, Johnson M, et al. (2013) Oligodendrocyte/type-2 astrocyte progenitor cells and glial-restricted precursor cells generate different tumor phenotypes in response to the identical oncogenes. J Neurosci 33: 16805-16817.

44. Reh TA, Levine EM (1998) Multipotential stem cells and progenitors in the vertebrate retina. J Neurobiol 36: 206-220.

45. Miyata T, Kawaguchi A, Okano H, Ogawa M (2001) Asymmetric inheritance of radial glial fibers by cortical neurons. Neuron 31: 727-741.

46. Bernardos RL, Barthel LK, Meyers JR, Raymond PA (2007) Late-stage neuronal progenitors in the retina are radial Müller glia that function as retinal stem cells. J Neurosci 27: 7028-7040.

47. Chenn A, McConnell SK (1995) Cleavage orientation and the asymmetric inheritance of Notch1 immunoreactivity in mammalian neurogenesis. Cell 82: 631-641.

48. Davies SJ, Shih CH, Noble M, Mayer-Proschel M, Davies JE, et al. (2011) Transplantation of specific human astrocytes promotes functional recovery after spinal cord injury. PLoS One 6: e17328.

49. Ketschek AR, Haas C, Gallo G, Fischer I (2012) The roles of neuronal and glial precursors in overcoming chondroitin sulfate proteoglycan inhibition. Exp Neurol 235: 627-637.

50. Zupanc GK, Ott R (1999) Cell proliferation after lesions in the cerebellum of adult teleost fish: time course, origin, and type of new cells produced. Exp Neurol 160: 78-87.

51. Zupanc GK, Zupanc MM (2006) New neurons for the injured brain: mechanisms of neuronal regeneration in adult teleost fish. Regen Med 1: 207-216.

52. Kishimoto N, Alfaro-Cervello C, Shimizu K, Asakawa K, Urasaki A, et al. (2011) Migration of neuronal precursors from the telencephalic ventricular zone into the olfactory bulb in adult zebrafish. J Comp Neurol 519: 3549-3565.

53. Voigt T (1989) Development of glial cells in the cerebral wall of ferrets: direct tracing of their transformation from radial glia into astrocytes. J Comp Neurol 289: 74-88.

54. Heins N, Malatesta P, Cecconi F, Nakafuku M, Tucker KL, et al. (2002) Glial cells generate neurons: the role of the transcription factor Pax6. Nat Neurosci 5: 308-315.

55. Haubensak W, Attardo A, Denk W, Huttner WB (2004) Neurons arise in the basal neuroepithelium of the early mammalian telencephalon: a major site of neurogenesis. Proc Natl Acad Sci U S A 101: 3196-3201.

56. Miyata T, Kawaguchi A, Saito K, Kawano M, Muto T, et al. (2004) Asymmetric production of surface-dividing and non-surface-dividing cortical progenitor cells. Development 131: 3133-3145.

57. Noctor SC, Martínez-Cerdeño V, Ivic L, Kriegstein AR (2004) Cortical neurons arise in symmetric and asymmetric division zones and migrate through specific phases. Nat Neurosci 7: 136-144.

58. Reynolds R, Wilkin GP (1988) Development of macroglial cells in rat cerebellum. II. An in situ immunohistochemical study of oligodendroglial lineage from precursor to mature myelinating cell. Development 102: 409-425.

59. LeVine SM, Goldman JE (1988) Spatial and temporal patterns of oligodendrocyte differentiation in rat cerebrum and cerebellum. J Comp Neurol 277: 441-455.

60. Raff MC, Miller RH, Noble M (1983) A glial progenitor cell that develops *in vitro* into an astrocyte or an oligodendrocyte depending on culture medium. Nature 303: 390-396.

61. Bansal R, Pfeiffer SE (1992) Novel stage in the oligodendrocyte lineage defined by reactivity of progenitors with R-mAb prior to O1 anti-galactocerebroside. J Neurosci Res 32: 309-316.

62. Pfeiffer SE, Warrington AE, Bansal R (1993) The oligodendrocyte and its many cellular processes. Trends Cell Biol 3: 191-197.

Permissions

All chapters in this book were first published in JBB, by Omics International; hereby published with permission under the Creative Commons Attribution License or equivalent. Every chapter published in this book has been scrutinized by our experts. Their significance has been extensively debated. The topics covered herein carry significant findings which will fuel the growth of the discipline. They may even be implemented as practical applications or may be referred to as a beginning point for another development.

The contributors of this book come from diverse backgrounds, making this book a truly international effort. This book will bring forth new frontiers with its revolutionizing research information and detailed analysis of the nascent developments around the world.

We would like to thank all the contributing authors for lending their expertise to make the book truly unique. They have played a crucial role in the development of this book. Without their invaluable contributions this book wouldn't have been possible. They have made vital efforts to compile up to date information on the varied aspects of this subject to make this book a valuable addition to the collection of many professionals and students.

This book was conceptualized with the vision of imparting up-to-date information and advanced data in this field. To ensure the same, a matchless editorial board was set up. Every individual on the board went through rigorous rounds of assessment to prove their worth. After which they invested a large part of their time researching and compiling the most relevant data for our readers.

The editorial board has been involved in producing this book since its inception. They have spent rigorous hours researching and exploring the diverse topics which have resulted in the successful publishing of this book. They have passed on their knowledge of decades through this book. To expedite this challenging task, the publisher supported the team at every step. A small team of assistant editors was also appointed to further simplify the editing procedure and attain best results for the readers.

Apart from the editorial board, the designing team has also invested a significant amount of their time in understanding the subject and creating the most relevant covers. They scrutinized every image to scout for the most suitable representation of the subject and create an appropriate cover for the book.

The publishing team has been an ardent support to the editorial, designing and production team. Their endless efforts to recruit the best for this project, has resulted in the accomplishment of this book. They are a veteran in the field of academics and their pool of knowledge is as vast as their experience in printing. Their expertise and guidance has proved useful at every step. Their uncompromising quality standards have made this book an exceptional effort. Their encouragement from time to time has been an inspiration for everyone.

The publisher and the editorial board hope that this book will prove to be a valuable piece of knowledge for researchers, students, practitioners and scholars across the globe.

List of Contributors

Luis Beltrán Ramos-Sánchez
Department of Chemical Engineering, University of Camagüey "Ignacio Agramonte Loynaz" Circunvalación norte, Km 5.5, 74650 Camagüey, Cuba

Mario César Cujilema-Quitio
Department of Chemical Engineering, University of Camagüey "Ignacio Agramonte Loynaz" Circunvalación norte, Km 5.5, 74650 Camagüey, Cuba

Maria Caridad Julian-Ricardo
Department of Chemical Engineering, University of Camagüey "Ignacio Agramonte Loynaz" Circunvalación norte, Km 5.5, 74650 Camagüey, Cuba

Jesus Cordova
Department of Chemistry, University of Guadalajara, Blvd. M. García Barragán 1421, CP 44430, Guadalajara, Jalisco, Mexico

Patrick Fickers
Biotechnology and Bioprocess Unit, Université libre de Bruxelles, Av. F.-D Roosevelt, 50, CP165 / 61, 1050 Brussels, Belgium
Microbial Processes and Interactions, Gembloux Agrobiotech, Univeristy of Liège, Passage des déportés, 2, 5030 Gembloux, Belgium

Cunxun Wang
State Key Laboratory of Bioreactor Engineering, New World Institute of Biotechnology, East China University of Science and Technology, Shanghai 200237, China

Kefeng Ni
State Key Laboratory of Bioreactor Engineering, New World Institute of Biotechnology, East China University of Science and Technology, Shanghai 200237, China

Xu Zhou
State Key Laboratory of Bioreactor Engineering, New World Institute of Biotechnology, East China University of Science and Technology, Shanghai 200237, China

Dongzhi Wei
State Key Laboratory of Bioreactor Engineering, New World Institute of Biotechnology, East China University of Science and Technology, Shanghai 200237, China

Yuhong Ren
State Key Laboratory of Bioreactor Engineering, New World Institute of Biotechnology, East China University of Science and Technology, Shanghai 200237, China

Ze-Jian Wang
State Key Laboratory of Bioreactor Engineering, East China University of Science and Technology, Shanghai 200237, China

Hui-Yuan Wang
State Key Laboratory of Bioreactor Engineering, East China University of Science and Technology, Shanghai 200237, China

Ping Wang
State Key Laboratory of Bioreactor Engineering, East China University of Science and Technology, Shanghai 200237, China

Yi-ming Zhang
Department of Chemical and Biological Engineering, Chalmers University of Technology, Kemivägen 10, SE-412 96 Göteborg, Sweden

Ju Chu
State Key Laboratory of Bioreactor Engineering, East China University of Science and Technology, Shanghai 200237, China

Ying-Ping Zhuang
State Key Laboratory of Bioreactor Engineering, East China University of Science and Technology, Shanghai 200237, China

Si-Liang Zhang
State Key Laboratory of Bioreactor Engineering, East China University of Science and Technology, Shanghai 200237, China

R Ilamathi
Department of Biotechnology, SNIST, Yamnampet, Hyderabad, Andhra Pradesh, India

GS Nirmala
Chemical Engineering Division, School of Mechanical and Building Sciences, Vellore Institute of Technology, Vellore, India

L Muruganandam
Chemical Engineering Division, School of Mechanical and Building Sciences, Vellore Institute of Technology, Vellore, India

Anjuvan Singh
Department of Biotechnology, Lovely School of Sciences, Lovely Professional University, Phagwara, India

K. M. Purohit
Department of Chemistry, National Institute of Technology, Rourkela -769008, Orissa, India

H. Laribi-Habchi
National Polytechnic school of Algiers, B.P. 182-16200, El Harrach, Algiers, Algeria

M. Biche
Institut National Agronomique d'Alger, El Harrach, Algeria

N. Drouiche
National Polytechnic school of Algiers, B.P. 182-16200, El Harrach, Algiers, Algeria
Centre de Recherche en Technologie des Semi-conducteurs pour l'énergétique (CRTSE). 2, Bd Frantz Fanon BP140, Alger - 7 Merveilles, 16000, Algeria

W. Khalfi
Institut National Agronomique d'Alger, El Harrach, Algeria

N. Abdi
National Polytechnic school of Algiers, B.P. 182-16200, El Harrach, Algiers, Algeria

N. Mameri
National Polytechnic school of Algiers, B.P. 182-16200, El Harrach, Algiers, Algeria

E. Bigdelian
Department of Food Science and Technology, Shahre kord Branch, Islamic Azad University, Shahre kord, Iran

SH Razavi
Department of Food Science and Technology, Faculty of Biosystem Engineering, Campus of Agriculture and Natural Resources, The University of Tehran, Tehran, Iran

Carl Parson
Department of Biology and Pharmaceutical Sciences, Elizabeth City State University-University of North Carolina, Elizabeth City, NC 27909, USA

Valerie Smith
Department of Biology and Pharmaceutical Sciences, Elizabeth City State University-University of North Carolina, Elizabeth City, NC 27909, USA

Christopher Krauss
Department of Biology and Pharmaceutical Sciences, Elizabeth City State University-University of North Carolina, Elizabeth City, NC 27909, USA

Hirendra N Banerjee
Department of Biology and Pharmaceutical Sciences, Elizabeth City State University-University of North Carolina, Elizabeth City, NC 27909, USA

Christopher Reilly
Department of Biomedical Sciences and Pathobiology, Virginia-Maryland Regional College of Veterinary Medicine, Virginia Polytechnic Institute and State University, Blacksburg, Virginia 24061, USA

Jeanette A Krause
Department of Chemistry, University of Cincinnati, Cincinnati, OH 45221, USA

James M Wachira
Department of Biology, Morgan State University, Baltimore, MD 21251, USA

Dipak Giri
Department of Chemistry, Morgan State University, Baltimore, MD 21251, USA

Angela Winstead
Department of Chemistry, Morgan State University, Baltimore, MD 21251, USA

Santosh K Mandal
Department of Chemistry, Morgan State University, Baltimore, MD 21251, USA

BMA Carvalho
Department of Chemistry, Biotechnology and Bioprocess Engineering, Federal University of São João Del Rei, Ouro Branco, MG 36420-000, Brazil

LM Carvalho
Department of Veterinary, Federal University of Viçosa, Viçosa, MG 36570-000, Brazil

Silva Jr WF
Department of Chemistry, Biotechnology and Bioprocess Engineering, Federal University of São João Del Rei, Ouro Branco, MG 36420-000, Brazil

LA Minim
Department of Food Technology, Federal University of Viçosa, Viçosa, MG 36570-000, Brazil

GGP Carvalho
Department of Animal Science, Federal University of Bahia, Salvador, BA 40110-909, Brazil

Günter Müller
Department I, Genetics, Biocenter, Ludwig-Maximilians-University Munich, 82152 Martinsried near Munich, Germany

Abraham R. Martin
Dept. of Chemical Engineering, Michigan Technological University, Houghton, MI 49931, USA

David Shonnard
Dept. of Chemical Engineering, Michigan Technological University, Houghton, MI 49931, USA

Sachin Pannuri
Dept. of Chemical Engineering, Michigan Technological University, Houghton, MI 49931, USA
Cambrex North Brunswick, 661 Highway One, North Brunswick, NJ 08902, USA

Sanjay Kamat
Cambrex North Brunswick, 661 Highway One, North Brunswick, NJ 08902, USA

Deepika Dave
Centre for Aquaculture and Seafood Development, Marine Institute of Memorial University of Newfoundland, St. John's, Newfoundland, Canada

Vegneshwaran V Ramakrishnan
Centre for Aquaculture and Seafood Development, Marine Institute of Memorial University of Newfoundland, St. John's, Newfoundland, Canada

Sheila Trenholm
Centre for Aquaculture and Seafood Development, Marine Institute of Memorial University of Newfoundland, St. John's, Newfoundland, Canada

Heather Manuel
Centre for Aquaculture and Seafood Development, Marine Institute of Memorial University of Newfoundland, St. John's, Newfoundland, Canada

Julia Pohling
Centre for Aquaculture and Seafood Development, Marine Institute of Memorial University of Newfoundland, St. John's, Newfoundland, Canada

Wade Murphy
Centre for Aquaculture and Seafood Development, Marine Institute of Memorial University of Newfoundland, St. John's, Newfoundland, Canada

Junaid Muneer Raja
Department of Chemical and Process Engineering, Faculty of Engineering and Built Environment, University Kebangsaan Malaysia, 43600 UKM, Bangi, Selangor, Malaysia
InnoBiologics Sdn. Bhd., Lot 1, Persiaran Negeri, 71800 Nilai, Negeri Sembilan, Malaysia

Nurina Anuar
Department of Chemical and Process Engineering, Faculty of Engineering and Built Environment, University Kebangsaan Malaysia, 43600 UKM, Bangi, Selangor, Malaysia

Badarulhisam Abdul Rahman
InnoBiologics Sdn. Bhd., Lot 1, Persiaran Negeri, 71800 Nilai, Negeri Sembilan, Malaysia

Jamaliah Md Jahim
Department of Chemical and Process Engineering, Faculty of Engineering and Built Environment, University Kebangsaan Malaysia, 43600 UKM, Bangi, Selangor, Malaysia

Jairam Choudhary
Division of Microbiology, Indian Agricultural Research Institute, New Delhi 110 012, India

Saritha M, Lata Nain
Division of Microbiology, Indian Agricultural Research Institute, New Delhi 110 012, India

Anju Arora
Division of Microbiology, Indian Agricultural Research Institute, New Delhi 110 012, India

Sherif MAS Keshk
King Khalid University, Faculty of Science, Chemistry Department, P.O. Box 9004, Abha 61413- Saudi Arabia
Ain-Shams University, Faculty of Environmental Studies and Research, Basic Science Department, Abbassia, Cairo 11566, Egypt

Pooja Kohli
Department of Biotechnology, Himachal Pradesh University, Summer Hill, Shimla-171005, INDIA

Manmohit Kalia
Department of Biotechnology, Himachal Pradesh University, Summer Hill, Shimla-171005, INDIA

Reena Gupta
Department of Biotechnology, Himachal Pradesh University, Summer Hill, Shimla-171005, INDIA

El-Sayed G Khater
Agricultural Engineering Department, Faculty of Agriculture, Benha University 13736, Egypt

Adel H Bahnasawy
Agricultural Engineering Department, Faculty of Agriculture, Benha University 13736, Egypt

Arunachalam Thirunavukkarasu
Lecturer, Dept. of Biotechnology, PSG College of Technology, Coimbatore, India

Rajarathinam Nithya
Asst.Professor, Faculty of Biotechnology, PRIST University, Thanjavur, India

Moses Jeyakumar Rajesh
Deparment of Biomedical Engineering, Jimma Institute of Technology, Jimma University, Jimma, Ethiopia

Leelavathy Rajesh
Research lab, GIET School of Pharmacy, NH-5, Chaitanya nagar, Rajahmundry-533294, India

Vanapalli V Satya veni
Research lab, GIET School of Pharmacy, NH-5, Chaitanya nagar, Rajahmundry-533294, India

G.Thirumurugan
Research lab, GIET School of Pharmacy, NH-5, Chaitanya nagar, Rajahmundry-533294, India

Rajaram mohan Sivasubramanian
School of Chemical and Biotechnology, SASTRA University, Tanjore, Tamilnadu, India

Omar A Hewedy
Genetics Department, Faculty of Agriculture, Minoufiya University, Shibin El-Kom, Egypt

Ragaa A Eissa
Genetics Department, Faculty of Agriculture, Minoufiya University, Shibin El-Kom, Egypt

AM Elzanaty
Genetics Department, Faculty of Agriculture, Minoufiya University, Shibin El-Kom, Egypt

HH Nagaty
Genetics Department, Faculty of Agriculture, Minoufiya University, Shibin El-Kom, Egypt

MI Abd Elbary
Genetics Department, Faculty of Agriculture, Minoufiya University, Shibin El-Kom, Egypt

Michael Mueller
Department of Biosystems an Agricultural Engineering, Oklahoma State University, 111 Agricultural Hall, Stillwater, Oklahoma, USA 74078

Mark R. Wilkins
Department of Biosystems an Agricultural Engineering, Oklahoma State University, 111 Agricultural Hall, Stillwater, Oklahoma, USA 74078

Ibrahim M. Banat
School of Biomedical Sciences, University of Ulster, Coleraine, BT52 1SA, Northern Ireland, UK

Shujing Li
Institute of Agro-Products Processing Science & Technology, Chinese Academy of Agricultural Sciences, Key Laboratory of Agricultural Product Processing, Ministry of Agriculture, P.O. Box 5109, Beijing 100193, People's Republic of China

Yingquan Zhang
Institute of Agro-Products Processing Science & Technology, Chinese Academy of Agricultural Sciences, Key Laboratory of Agricultural Product Processing, Ministry of Agriculture, P.O. Box 5109, Beijing 100193, People's Republic of China

Yimin Wei
Institute of Agro-Products Processing Science & Technology, Chinese Academy of Agricultural Sciences, Key Laboratory of Agricultural Product Processing, Ministry of Agriculture, P.O. Box 5109, Beijing 100193, People's Republic of China

Wei Zhang
Institute of Agro-Products Processing Science & Technology, Chinese Academy of Agricultural Sciences, Key Laboratory of Agricultural Product Processing, Ministry of Agriculture, P.O. Box 5109, Beijing 100193, People's Republic of China

Bo Zhang
Institute of Agro-Products Processing Science & Technology, Chinese Academy of Agricultural Sciences, Key Laboratory of Agricultural Product Processing, Ministry of Agriculture, P.O. Box 5109, Beijing 100193, People's Republic of China

S. Sandhya
CSIR-National Environmental Engineering Research Institute, CSIR-Complex, Chennai, India

Sadhana Rayalu
CSIR-National Environmental Engineering Research Institute, Nagpur, India

Benedict Bruno
CSIR-National Environmental Engineering Research Institute, CSIR-Complex, Chennai, India

Esra Aydemir
Department of Genetics and Bioengineering, Faculty of Engineering and Architecture, Yeditepe University, Turkey

Selami Demirci
Department of Genetics and Bioengineering, Faculty of Engineering and Architecture, Yeditepe University, Turkey

Ayşegül Docan
Department of Genetics and Bioengineering, Faculty of Engineering and Architecture, Yeditepe University, Turkey

Ali Özhan Aytekin
Department of Genetics and Bioengineering, Faculty of Engineering and Architecture, Yeditepe University, Turkey

Fikrettin Sahin
Department of Genetics and Bioengineering, Faculty of Engineering and Architecture, Yeditepe University, Turkey

Fengqiang Wang
Sterile Product and Analytical Development, Merck Research Laboratories, Kenilworth, NJ, USA

Hong Li
Process Development and Engineering, Merck Research Laboratories, Kenilworth, NJ, USA

Zhi Chen
Sterile Product and Analytical Development, Merck Research Laboratories, Kenilworth, NJ, USA

John P Welsh
Process Development and Engineering, Merck Research Laboratories, Kenilworth, NJ, USA

Douglas Richardson
Bioprocess Technology and Expression, Merck Research Laboratories, Kenilworth, NJ, USA

Jennifer M Pollard
Process Development and Engineering, Merck Research Laboratories, Kenilworth, NJ, USA

Daisy Richardson
Sterile Product and Analytical Development, Merck Research Laboratories, Kenilworth, NJ, USA

Mohammed Shameem
Sterile Product and Analytical Development, Merck Research Laboratories, Kenilworth, NJ, USA

Chinnathambi Velu
Research Associate, ARMATS BIOTEK Research Institute, Kottur, Chennai, India-600085

Ezhilarasan Veeramani
Research Associate, ARMATS BIOTEK Research Institute, Kottur, Chennai, India-600085

Sridhar Suntharam
Assistant Professor, Jeppiaar Engineering College, Jeppiaar nagar, Chennai, India-600119

Kasinathan Kalimuthu
Research Associate, ARMATS BIOTEK Research Institute, Kottur, Chennai, India-600085

Żaneta Bargańska
Gdansk University of Technology, Faculty of Chemistry, Department of Analytical Chemistry, Narutowicza 11/12, 80-233 Gdańsk, Poland

Ewa Olkowska
Gdansk University of Technology, Faculty of Chemistry, Department of Analytical Chemistry, Narutowicza 11/12, 80-233 Gdańsk, Poland

Tomasz Dymerski
Gdansk University of Technology, Faculty of Chemistry, Department of Analytical Chemistry, Narutowicza 11/12, 80-233 Gdańsk, Poland

Jacek Namieśnik
Gdansk University of Technology, Faculty of Chemistry, Department of Analytical Chemistry, Narutowicza 11/12, 80-233 Gdańsk, Poland

Nihat Alpagu Sayar
Department of Bioengineering, Marmara University, Kadikoy 34722, Istanbul, Turkey

Selcen Durmaz-Sam
Department of Bioengineering, Marmara University, Kadikoy 34722, Istanbul, Turkey

Gizay Odabasi
Department of Bioengineering, Marmara University, Kadikoy 34722, Istanbul, Turkey

Ahmet Alp Sayar
Department of Bioengineering, Marmara University, Kadikoy 34722, Istanbul, Turkey

Abbas H. Sulaymon
Professor and Head of Energy Engineering Department, Baghdad University, Iraq

Ahmed A. Mohammed
Assistant Professor, Environmental Engineering Department, Baghdad University, Iraq

Tariq J. Al-Musawi
Environmental Engineering Department, Baghdad University, Iraq

DK Samal
Department of Chemical Engineering, Gandhi Institute of Engineering and Technology (GIET), Gunupur, Rayagada, Odisha-765022, India

YK Mohanty
Department of Chemical Engineering, Gandhi Institute of Engineering and Technology (GIET), Gunupur, Rayagada, Odisha-765022, India

GK Roy
Department of Chemical Engineering, Gandhi Institute of Engineering and Technology (GIET), Gunupur, Rayagada, Odisha-765022, India

Anna Jacobi
Karlsruhe Institute of Technology KIT, Institute of Engineering in Life Sciences, Department of Bioprocess Engineering, Fritz-Haber-Weg 2, Building 30.44, 76131 Karlsruhe, Germany

Ethel Claudia Bucharsky
Karlsruhe Institute of Technology KIT, Institute for Applied Materials - Ceramics in Mechanical Engineering, Haid-und-Neu-Str. 7, 76131 Karlsruhe, Germany

Karl Guenter Schell
Karlsruhe Institute of Technology KIT, Institute for Applied Materials - Ceramics in Mechanical Engineering, Haid-und-Neu Str. 7, 76131 Karlsruhe, Germany

Peter Habisreuther
Karlsruhe Institute of Technology KIT, Engler-Bunte-Institute, Division of Combustion Technology, Engler-Bunte-Ring 1, Building 40.11, 76128 Karlsruhe, Germany

Rainer Oberacker
Karlsruhe Institute of Technology KIT, Institute for Applied Materials - Ceramics in Mechanical Engineering, Haid-und-Neu-Str. 7, 76131 Karlsruhe, Germany

Michael J. Hoffmann
Karlsruhe Institute of Technology KIT, Institute for Applied Materials - Ceramics in Mechanical Engineering, Haid-und-Neu-Str. 7, 76131 Karlsruhe, Germany

Nikolaos Zarzalis
Karlsruhe Institute of Technology KIT, Engler-Bunte-Institute, Division of Combustion Technology, Engler-Bunte-Ring 1, Building 40.11, 76128 Karlsruhe, Germany

Clemens Posten
Karlsruhe Institute of Technology KIT, Institute of Engineering in Life Sciences, Department of Bioprocess Engineering, Fritz-Haber-Weg 2, Building 30.44, 76131 Karlsruhe, Germany

Maria Pia Santacroce
Unit of Aquaculture and Zooculture, Department of Veterinary Medicine, University of Bari "Aldo Moro", Italy

Antonella Tinelli
Unit of Pathology, Deptartment of Veterinary Medicine, University of Bari "Aldo Moro", Italy

Anna Selene Pastore
Unit of Aquaculture and Zooculture, Department of Veterinary Medicine, University of Bari "Aldo Moro", Italy

Michele Colamonaco
Unit of Pharmacology and Toxicology, Department of Veterinary Medicine, University of Bari "Aldo Moro", Italy

Giuseppe Crescenzo
Unit of Pharmacology and Toxicology, Department of Veterinary Medicine, University of Bari "Aldo Moro", Italy